D1327689

THE TREASURY SOLICITOR
LIBRARY AND INFORMATION SERVICE
Queen Anne's Chambers
28 Broadway
London, SW1H 9JS
Tel: 020 7210 3044/3102
E-mail: library@treasury-solicitor.gsi.gov.uk
Please return this publication on or before the latest date shown below

0000004973

CONTAMINATED LAND: THE NEW REGIME

PART IIA OF THE ENVIRONMENTAL PROTECTION ACT 1990

AUSTRALIA
LBC Information Services
Sydney

CANADA and U.S.A.
Carswell
Toronto Ontario

NEW ZEALAND
Brooker's
Auckland

SINGAPORE and MALAYSIA
Sweet & Maxwell Asia
Singapore and Kuala Lumpur

CONTAMINATED LAND: THE NEW REGIME

PART IIA OF THE ENVIRONMENTAL PROTECTION ACT 1990

by

Stephen Tromans

and

Robert Turrall-Clarke

London
Sweet & Maxwell
2000

Published in 2000 by
Sweet & Maxwell Limited of
100 Avenue Road
London NW3 3PF
Phototypeset by LBJ Typesetting Ltd. of Kingsclere
Printed in Great Britain by MPG Books Ltd, Bodmin, Cornwall

No natural forests were destroyed to make this product;
only farmed timber was used and replanted

A CIP catalogue record for this book is
available from the British Library

ISBN 0-421-66120-8

All rights reserved.
UK statutory material in this publication
is acknowledged as Crown copyright.

No part of this publication may be reproduced or transmitted in any form or by any means, or stored in any retrieval system of any nature without prior written permission, except for permitted fair dealing under the Copyright, Designs and Patents Act 1988, or in accordance with the terms of a licence issued by the Copyright Licensing Agency in respect of photocopying and/or reprographic reproduction. Application for permission for other use of copyright material including permission to reproduce extracts in other published works shall be made to the publishers. Full acknowledgement of author, publisher and source must be given.

© Stephen Tromans and Robert Turrall-Clarke 2000

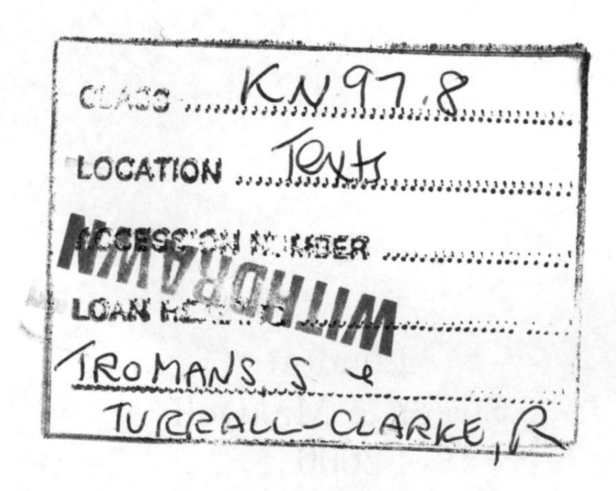

Preface

Contaminated land is perhaps one of the thorniest of environmental issues faced during the past decade by, first, the Conservative administration which enacted Part IIA of the Environmental Protection Act 1990 and, second, by the Labour government which then had to implement it. If volume of legislative material is any indicator of complexity, then this is indeed a subject of substantial difficulty. The primary legislation runs through the alphabet from section 78A to 78YC, and requires to be read together not only with Regulations, but also a Circular of over one hundred and fifty pages, large portions of which have the status of prescriptive guidance.

Despite this complexity, the questions which are inevitably posed are simple: is land sufficiently contaminated so as to require clean up, and if so, what is to be done to it and who is to pay for the requisite works? The purpose of this book is to provide those affected by the new regime, including those charged with enforcing it, with practical answers to these questions and other less fundamental ones arising from the implementation of the new regime. For convenience of use, we have included the legislation and guidance as Appendices, and have also attempted to provide some precedents for the many forms that the legislation will require. At the same time, it must be remembered that Part IIA is only one part of the legal mosaic which has contaminated land as its subject; this book should therefore be used in conjunction with our other publication, *Contaminated Land*, and its Supplement.

The law is stated as a April 1, 2000, the perhaps unfortunate date on which the new regime comes into force in England. Its fate in Scotland and Wales depends on the devolved administrations there, but the indications are that the guidance is likely to follow that for England in terms of main content; the primary legislation is of course the same.

We have been greatly assisted by Mr Mark Poustie, of the School of Law at the University of Strathclyde, who provided material and analysis on the issue of the appeals process in Scotland. Responsibility for all the material of course rests with us as authors.

Stephen Tromans
Robert Turrall-Clarke

April 1, 2000

TERMINOLOGY

Part IIA has created its own terminology, with words such as "remediation". To this, the Guidance contained in the Circular has added another layer of terms, such as "pollutant linkage", "common actions", "collective actions", and so forth. Annex 6 to the Circular provides a full glossary of terms, running to seven pages, to which reference should be made. We list below the key abbreviations which we use in the text:

"the 1990 Act": the Environmental Protection Act 1990
"the Circular": DETR Circular 02/2000
"the Regulations": the Contaminated Land (England) Regulations 2000 No. 227
"the Guidance": the statutory guidance contained in Annex 3 of the Circular

Table of Contents

Table of Cases

Table of Statutes

Table of Statutory Instruments

TABLE OF EUROPEAN LEGISLATION

Treaties and Conventions

Directives

Chapter 1

Background and Introduction

Summary

Part IIA of the Environmental Protection Act 1990 formulates a new regime for dealing with the problems arising from contaminated land. It imposes new duties on local authorities to inspect their areas so as to identify contaminated land, and creates a complex scheme of strict and retrospective liability for remediation of land identified as contaminated. Part IIA came into force in England on April 1, 2000 by S.I. 2000 No. 340.

This chapter describes the background to Part IIA and introduces its scheme, concluding with a guide as to the key questions for those involved in the process.

Contaminated land: the problem

1.01 The main purpose of this book is to describe in some detail the new regime for contaminated land introduced by Part IIA of the Environmental Protection Act 1990 (inserted by section 57 of the Environment Act 1995) rather than to provide an explanation of the broader range of legal and technical problems associated with contaminated land.[1] Such explanation can be found in *Contaminated Land*[2] and in its *1998 Supplement*,[3] to which readers are referred for further detail. Whilst this book is comprehensive in its treatment of the new regime, it is dangerous to address contaminated land problems without reference to the wider legal context, particularly civil liability. Practitioners should, therefore, continue to make use of *Contaminated Land* and the *Supplement* to obtain an overall perspective of the subject.

---

[1] Such as civil liability, access to information, investigation and appraisal, development, and commercial transactions.

[2] *Contaminated Land* by Stephen Tromans and Robert Turrall-Clarke, London, Sweet & Maxwell, 1994.

[3] *Contaminated Land—First Supplement* by Stephen Tromans and Robert Turrall Clarke, London, Sweet & Maxwell, 1998.

1.02 The regime introduced in Part IIA is essentially a public law one, aimed at the cleaning up of contaminated land and the protection of controlled waters in the public interest; it is regulated by public bodies acting under a statutory code. The contamination of land, however, may well also give rise to civil claims in contract as between successive landowners, and in tort as between neighbours. The inter-relationship between the regime under Part IIA and civil suits is thus a fundamental consideration for the practitioner. Further, many active industrial and waste disposal sites which may be classed as contaminated both by the experts and in the public mind may be subject to controls under Parts I and II of the 1990 Act or the Pollution Prevention and Control Act 1999, to the exclusion of Part IIA.[4]

1.03 It is clear that contaminated land presents problems to legislators and policy makers that are somewhat different to other types of environmental issues. This is by virtue of the fact that land is permanent, may be put to many different uses, and may be subject to numerous successive ownerships or current interests. The issues that need to be considered, and which are addressed by Part IIA, are:

(a) When is land to be regarded as "contaminated" so as to require remedial action to be taken?
(b) If it is regarded as contaminated, then to what standard does it need to be cleaned up, or what other actions are required?
(c) Over what timescale should those measures be undertaken?
(d) Who should bear the legal and financial responsibility for undertaking those measures?

1.04 Whilst capable of being simply stated, these are complex and controversial questions, which other countries such as The Netherlands and the USA have found difficult to answer. There are in fact no easy answers, and an important function of Part IIA is to introduce a framework of public law to tackle these problems and answer these questions. However, before outlining the scheme of Part IIA, it may be helpful to describe briefly the policy background which led to its enactment.[5]

Pressure for action and the section 143 registers

1.05 At the end of the 1980s, concern was focussed on contaminated land in two reports of the House of Commons Select Committee on the Environment dealing with toxic waste and contaminated land: *Toxic Waste*[6] and *Contaminated Land*.[7] In the second of these Reports, the Select Committee

---

[4] See Chapter 2 below.
[5] The description that follows is taken largely from *The Environment Acts 1990–95* by Stephen Tromans (assisted by Mike Nash and Mark Poustie), London, Sweet & Maxwell, 1996, pp. 196–202.
[6] Session 1988–89, Second Report, H.C. 22 I–II.
[7] Session 1989–90, First Report, H.C. 170 I–III.

recognised the difficult policy issues raised by any statutory scheme of liability, but suggested that nonetheless action was needed:

> "... urgent attention [must] be given to the question of creating statutory liability for damage caused by contamination to land—particularly where this causes damage to neighbouring property or to the environment. We recognise that this will raise complex questions as to retrospection, insurance cover and limitation periods in particular, but we believe that the present lack of clarity in relation to civil liability hampers the development of appropriate policies on the issue of contaminated land."

1.06 The initial response of the Government was not to address the issue of statutory liability directly, but to introduce a provision (section 143 of the Environmental Protection Act 1990) for public registers of land which had been subject to a "contaminative use", *i.e.* a use which might cause land to be contaminated with noxious substances. This approach was intended, with "minimal demands" on local authority resources[8] to provide a means of alerting interested parties to the possible existence of contamination without "extending planning blight in those areas of the country with a legacy of industrial land use".[9]

1.07 This proved to be a somewhat optimistic assumption. Section 143 was intended to be implemented late in 1991. But, following an initial consultation exercise as to the range of uses to be prescribed as "contaminative", the Government announced on March 10, 1992 that it was postponing the introduction of the registers in view of the concerns that had been expressed over them, relating to the likelihood of serious blight in an already depressed property market.

1.08 A second consultation process ensued, with a markedly reduced list of contaminative uses, and with a proposed timetable which would have led to the opening of the registers to public inspection in about April 1994. This exercise enabled the Government to identify three serious grounds for criticism of the proposed registers. First, because they were to be based on current or former uses of land rather than actual contamination, they would include a number of sites which were not actually contaminated, whilst excluding others which were. Furthermore, the logical consequence of a register based on historical fact as to land use would have been the inability to have the entry removed from the register following completion of clean-up. Finally, the system would have left it unclear what action should be taken and by whom, *i.e.* whether the land should be cleaned up and, if so, who should pay and how much. The registration system was therefore essentially designed to alert the parties interested in a given site to a potential problem, leaving them to address it within a contractual relationship; the wider public interest in ensuring the land was actually cleaned up was not necessarily recognised.

---

[8] *Hansard*, H.L. Vol. 520, col. 2269.
[9] DoE News Release No. 279, April 30, 1990.

Withdrawal of the registers and further consultation

1.09 These misgivings over section 143 led the Government to withdraw the proposal for registers, a decision announced by then Secretary of State for the Environment, Michael Howard, on March 24, 1993. Further, he announced the institution of a "wide-ranging review" of the legal powers of public bodies to control and tackle contaminated land, to be conducted by an inter-departmental group under the chairmanship of the Department of Environment. Following similar consultation exercises in Scotland, the Secretary of State for Scotland announced an equivalent review on the same date.

1.10 As part of this review process, the Government's first step was to issue the consultation paper, *Paying for our Past: the Arrangements for Controlling Contaminated Land and Meeting the Cost of Remedying Damage to the Environment* (March 1994). In Scotland, a similar though not identical consultation paper entitled *Contaminated Land Clean-up and Control* (March 1994) was issued by the Scottish Office Environment Department. Its conclusions did not differ in any material respect from those of the consultation paper issued in England and Wales. Neither paper presented any firm proposals but rather set out a number of "preliminary conclusions" and posed a series of questions for respondents, grouped under seven issues:

(a) What should the policy objectives be?
(b) How should the statutory framework meet the objectives?
(c) What relationship should the statutory framework have with the common law and civil liability?
(d) Should there be any extension of strict liability?
(e) Who should pay for putting right any damage?
(f) How should markets be provided with information?
(g) What roles should public sector bodies have?

"Framework for Contaminated" Land and new legislation

1.11 The March 1994 Consultation Papers, inconclusive even by consultation standards, instituted another phase of the debate, which culminated on November 24, 1994 with the publication of the outcome of the review, *Framework for Contaminated Land* (and in Scotland, *Contaminated Land Clean-up and Control: Outcome of Review*) and the subsequent passage of the Environment Act 1995, s.57, which introduced into the 1990 Act a new Part IIA dealing with contaminated land.

1.12 The central planks of the *Framework*, which were mirrored by those of its Scottish counterpart, were said to be:

(1) The maintenance of the "suitable for use" approach, requiring remedial action only where the contamination poses unacceptable

actual or potential risks to health or the environment, and where there are "appropriate and cost-effective means to do so, taking into account the actual or intended use of the site".

(2) To deal with "urgent and real problems, but in an orderly and controlled fashion with which the economy and large and individual business and land-owners can cope".

(3) The creation of greater clarity and certainty than the law currently provides, so as to assist in the development of an efficient market in land which is contaminated and in land which has been subject to remedial works.

(4) Replacing the existing statutory nuisance powers, which "have provided an essentially sound basis for dealing with contaminated land", with a modern, specific contaminated land power. Here the position in Scotland differs, and the *Outcome of Review* indicated that the introduction of the contaminated land regime was to coincide with the extension to Scotland of the statutory nuisance provisions already existing in England and Wales, replacing existing provisions. This extension is achieved by section 107 of and Schedule 17 to the Environment Act 1995.

1.13 Time will no doubt tell to what extent the actual drafting of the relevant provisions achieves those objectives. Substantial amendments were made to the provisions during the Parliamentary process, in response to concerns expressed by the industrial, financial, land-owning and legal communities. Taken together, these amendments have had the effect of mitigating somewhat the harshness of the proposed liability regime and perhaps making its provisions more palatable. Nonetheless, the fact remains that the new sections constitute a regime of strict and retroactive liability for historic contamination, and that liability can extend in certain circumstances not only to the original polluter, but also to an "innocent" landowner or occupier.

1.14 As mentioned throughout this work, much reliance is placed on Government guidance as a means of mitigating the potentially harsher elements in operation of the legislation, for example what degree of contamination or risk is sufficient to justify action, and how liability is to be apportioned between a number of potentially liable parties. At the time of the passage of the 1995 Act, none of this guidance was in final form, and much had not emerged even in draft. The shape of the guidance was only to become clear over the next five years.

Uncertainties after enactment

1.15 The heavy dependence of the new regime on statutory guidance for such fundamental questions as what constitutes contaminated land and how liability for clean-up is to be allocated and apportioned, meant that implementation was unlikely to be rapid or straightforward. It certainly proved to be neither. Various informal consultations took place on draft statutory guidance, which was ultimately issued for formal consultation in September 1996. In the meantime, the Royal Commission on Environmental Pollution published its Nineteenth Report, *Sustainable Use of Soil*[10] which included consideration of contaminated land issues, and the House of Commons Environment Select Committee published a short report on the new legislation.[11] A research project was commissioned by the Department of Environment to assess how the draft Guidance would work in practice[12] and various seminars with interested parties took place.

1.16 Further progress was halted temporarily by the 1997 General Election, followed by a period during which the new Administration considered whether the legislation enacted by its predecessor did indeed provide "broadly the right framework" for the future,[13] concluding that it did.[14] However, it was announced in December 1997, by way of letter from the Minister for the Environment to the Chairman of the House of Commons Select Committee, that there were a number of "points of detail" (some quite substantial) where changes to the approach in the draft guidance could be expected.[15]

1.17 The question of resources also became involved, with the cost of implementing the new duties on enforcing authorities being addressed as part of the 1997–1998 Comprehensive Spending Review, the results of which were announced in July 1998.[16] The outcome was provision of £14 million for implementation of Part IIA by local authorities in 1999–2000, rising to £18 million for each of the following two years. Sums of £14 million (1999–2000), £15 million (2000–2001) and £15 million (2001–2002) were allocated for capital funding on contaminated land. The Environment Agency was allocated £13 million additional funding over the three years, at least in part to support its role in dealing with "special sites" and providing technical guidance to local authorities.

1.18 A selective consultation exercise on the revised draft guidance followed in October 1998, on substantive and presentational changes to the 1996 version. It was however not until October 1999 that the Government proceeded to formal consultation on the full package of statutory guidance, regulations and associated material, with the intention of bringing the new regime into force from April 1, 2000.

---

[10] Cm. 3165, February 1996.

[11] Session 1996–97, Second Report, *Contaminated Land*, H.C. 22 I, December 11, 1996.

[12] *A Study of the Effectiveness of the Liability Regime in the Draft Statutory Guidance on Contaminated Land*, Stanger Science and Simmons & Simmons, March 1997.

[13] DETR News Release 539, December 22, 1997.

[14] See *H.C. Written Answer*, December 22, 1997, col. 439.

[15] The letter is summarised in *Environmental Law Bulletin* No. 42, Sweet & Maxwell, February 1998, p. 22.

[16] Cm. 4011, Chancellor of the Exchequer, July 1998.

The scheme of the legislation

1.19 The provisions of Part IIA follow a sequence from the initial identification of contaminated land to securing its remediation. The relevant sections are as follows:

Section 78A—Definitions.
Section 78B—Duty of local authorities to inspect their area to identify contaminated land, and notification to relevant persons.
Section 78C—Identification and designation of "special sites".
Section 78D—Referral of disputed decisions as to whether land is a "special site" to the Secretary of State.
Section 78E—Duty to require the remediation of contaminated land by service of a "remediation notice".
Section 78F—Determination of who is the appropriate person to bear responsibility for remediation.
Section 78G—Grant of rights of entry or other rights for the purpose of complying with a remediation notice, and compensation for the grant of such rights.
Section 78H—Restrictions and prohibitions on service of a remediation notice.
Section 78J—Restrictions on liability in relation to pollution of controlled waters and water from abandoned mines.
Section 78K—Provisions with respect to contaminating substances which escape to other land.
Section 78L—Appeals against remediation notices.
Section 78M—Offences of non-compliance with a remediation notice.
Section 78N—Powers of enforcing authorities themselves to carry out remediation.
Section 78P—Powers for enforcing authorities to recover and secure their costs incurred in carrying out remediation under s.78N.
Section 78Q—Procedures relating to special sites.
Section 78R—Public registers.
Section 78S—Exclusion from registers of information affecting national security.
Section 78T—Exclusion from registers of commercially confidential information.
Section 78U—Publication by the Environment Agencies of reports on the state of contaminated land.
Section 78V—Provision of site-specific guidance by the Environment Agencies.
Section 78W—Provision of guidance to the Agencies by the Secretary of State.
Section 78X—Miscellaneous supplementary provisions dealing with sites having a combined effect, land outside the area of a local authority which affects its own area, and the protection of insolvency practitioners.

Section 78Y—Application of provisions to the Scilly Isles.
Section 78YA—Procedures for the issue of guidance by the Secretary of State.
Section 78YB—Interaction of contaminated land provisions and other provisions relevant to contaminated land.
Section 78YC—Non-application of provisions to harm caused by radioactivity.

1.20 However, these statutory provisions are in many senses only the starting point. The Contaminated Land (England) Regulations 2000 (S.I. 2000 No. 227) deal with various important procedural matters, including the content and service of remediation notices and the appeals process. The vast DETR Circular 02/2000, *Contaminated Land*, is also vital to the working of the new regime. It contains some six Annexes, the most important of which is Annex 3. This contains the statutory guidance in accordance with which enforcing authorities are required to act. Annex 1 comprises a general statement of Government policy, Annex 2 a description of the new regime, and Annex 4 a guide to the Regulations. Annex 5 is a brief guide to the commencement order, and Annex 6 is a glossary of the somewhat arcane terminology of Part IIA and the Guidance.

Relationship of Part IIA and previous law

1.21 During the passage of the 1995 Act, the Government was anxious to stress on a number of occasions that the provisions on contaminated land were not intended to create new categories of liability, and simply reflected the pattern of powers and duties under previous law, most notably statutory nuisance.[17] Certainly the scheme of liability bears many striking similarities to that for statutory nuisance, with the primary responsibility resting with the originator of the contamination (the person "responsible for the nuisance") and residual liability with the current owner or occupier. Whether it was wise to frame legislation intended to deal with the complex issues presented by contaminated land on the basis of a statutory nuisance code remains open to doubt.

1.22 Specifically, steps have been taken to restrict liability where the harm or risk presented by the contaminated land takes the form of water pollution.[18] Here liability does not extend to the owner or occupier purely in their capacity as owner or occupier; this accords broadly with the position under section 161 of the Water Resources Act 1991 (or, in Scotland, under section 46 of the Control of Pollution Act 1974) which refers to those causing or knowingly permitting pollution but not to the owner or occupier.[19] Similarly, restrictions apply in relation to remediation in respect

---

[17] See, for example, H.L. Vol. 560, col. 1461, and H.L. Vol. 562, col. 1054.
[18] Section 78J.
[19] See H.C. Standing Committee B, May 23, 1995, col. 354.

of water from abandoned mines, so as to reflect the position under the Water Resources Act 1991 (or, in Scotland, under the Control of Pollution Act), as modified by the Environment Act 1995.

1.23 Indeed, the then Conservative Government could with some justice point out that there were safeguards such as the requirements for consultation before service of a notice, the role of Government guidance and the provisions on financial hardship, which made Part IIA preferable to statutory nuisance from the perspective of anyone facing potential liability.[20] There is also no provision in Part IIA corresponding to that in section 82 of the 1990 Act, which permits any person aggrieved by the existence of a statutory nuisance to take proceedings by way of complaint direct to a magistrates court.

1.24 Similarly, the line taken in Annex 1 of the Circular is that although Part IIA is new, it largely replaces existing regulatory powers and duties, with their origins in the mid-nineteenth century legislation which created the concept of statutory nuisance.[21]

1.25 Despite these similarities and safeguards, Part IIA cannot with entire credibility be presented as nothing more than a mere extension to the ambit of previous law. The new provisions on liability exist within an overarching framework of statutory duties to seek out, identify, prioritise and remediate contaminated land as such. If those duties are not performed, or are perceived by local residents or an environmental group as being performed inadequately, they may well provide the basis for actual or threatened judicial review proceedings. Overall, it seems unlikely simply to be "business as usual" so far as contaminated land is concerned.[22] This was certainly the view of Baroness Hilton of Eggardon in debate on the Bill:

> "For the first time local authorities will have an explicit duty to inspect their areas in order to identify contaminated land. The existing provisions, which are much more vague and tenuous, require only that they identify nuisances".[23]

1.26 One measure of the significance of this new explicit duty is to consider the extent to which local authorities had in place active programmes to identify contaminated land prior to the introduction of the regime. A survey of local authorities in England and Wales by the Chartered Institute of Environmental Health (CIEH), to which 303 out of a total of 405 responded, indicated that only 64 out of the 303 respondents had such a programme in 1993–1994.[24]

---

[20] See H.L. Vol. 56, col. 1055.

[21] Annex 1, para. 23.

[22] The duty to inspect has been described as "undoubtedly an innovation [which] . . . if duly observed, will lead to the realisation of liabilities which might otherwise have remained notional". Christopher Miller, *Environmental Rights: Critical Perspectives*, Routledge, London, 1998, p. 156.

[23] H.L. Vol. 565, col. 1499.

[24] CIEH, *Report on Environmental Health*, 1993/1994.

Likely cost implications

1.27 The final regulatory impact assessment of the new regime, signed by the Minister on January 27, 2000 accepts that until local authorities begin to inspect their areas for contaminated land, and the results can be compiled, there can be no reliable estimate of the amount of land falling within the new statutory definition; nor any easy way to assess how businesses with a current or past connection with that land might be affected. However, the definition of contaminated land is much more tightly drawn than previous definitions, so that earlier estimates of land which may be subject to contamination should not be relied upon.

1.28 There is no "typical" business which might be affected by the new regime: such businesses may range from major industrial and utilities companies through to small family businesses. Also the cost of remediation works will vary widely according to their nature or scale. The assessment provides three hypothetical examples at Annex D.

Example 1: A petrol filling station

1.29 The example assumes that leakage from underground storage tanks has led to hydrocarbon contamination in a nearby water supply well. No remediation notice is served because, after consultation, the operator agrees to carry out voluntary remediation. He agrees to undertake further assessment to determine the scale of the problem and what needs to be done: the cost for this is £12,000. Assuming that the contamination is restricted to an area around the tank, the action could be the removal of the tank and the contaminated soil, followed by further monitoring of boreholes for six months; the cost of this, including professional fees would be some £60,000. If more extensive migration of contaminants had occurred, requiring some form of *in-situ* treatment, the cost of remediation could be in the order of £0.5 to £1 million.

Example 2: Housing built on a closed landfill which is generating gas

1.30 The example here is of housing built on part of a closed landfill, of some six hectares, which has been used for the deposit of mixed domestic and building/commercial refuse over a 30-year period, prior to licensing requirements under the Control of Pollution Act 1974. Gas has been detected in some of the housing, but has not migrated off-site. No explosive concentrations of gas have been detected, but there is a significant risk requiring action. The remedial action consists of the installation of passive venting pipes under the floor slabs of the affected houses, at a cost of £2,000 per house.

Example 3: A large derelict chemical works

1.31 This example assumes a derelict 40 hectare site used from the turn of the century for the manufacture of dyes and chemicals. Production wastes were deposited on the site and the land and structures are contaminated by many different substances. The site is determined to be contaminated by reason of pollution of an adjoining canal and nearby river, exposure of those who may gain access the site to asbestos, lead paint and organics, and risks arising from windborne asbestos. Redevelopment is not in prospect. A remediation notice is served requiring further assessment at a cost of £40,000 to clarify the characteristics of the pollutant risks and enable sensible decisions to be made on what remediation is required. Remediation actions might involve excavation and disposal of contaminated land and structures, the installation of barriers to prevent migration of contaminants, and measures to prevent unauthorised access to the site. The range of costs would be very broad, "possibly between £1 million to £20 million or more, depending on exactly what action was necessary". The requirements, and hence the costs, would be phased over several years.

1.32 It will be appreciated that even in these relatively simple scenarios, it is difficult to give accurate estimates of likely cost; nor do those figures include what could be the very significant costs involved in legal advice, in obtaining any necessary access to neighbouring land, and possible costs relating to business disruption.

The sustainable development context

1.33 One of the problems of a regime which takes as long as Part IIA to come into force, is that in the meantime the overall policy context may have moved on. This is reflected in the statement of Government policy which forms Annex 1 to the Circular. This statement seeks to set the new provisions in the overall context of current Government thinking on sustainable development. As para. 6 points out, the existence of contamination presents its own risks to sustainable development:

(a) it impedes social progess, depriving local people of a clean and healthy environment;
(b) it threatens wider damage to the environment and to wildlife;
(c) it inhibits the prudent use of land, in particular by obstructing the recycling of previously developed land and increasing development pressures on greenfield areas; and
(d) the cost of remediation represents a high burden on companies, home and land owners, and the economy as a whole.

These comments are applicable to contaminated land in its broadest sense, not simply the narrow meaning of the term as used in Part IIA. The Environment Agency has estimated that there may be some 300,000 hectares of land in the U.K. affected to some degree by natural or industrial contamination.

1.34 The Government's stated objectives, set out at para. 7 of Annex 1, are threefold:

(a) to identify and remove unacceptable risks to human health and the environment;

(b) to seek to bring damaged land back into beneficial use; and
(c) to seek to ensure that the cost burdens faced by individuals, companies and society as a whole are proportionate, manageable and economically sustainable.

These three objectives feed into the "suitable for use" approach, which the Government considers the most appropriate way to achieve sustainable development in this field.

The suitable for use approach

1.35 Annex 1 to the Circular goes on to describe the "suitable for use" approach as based on a site-specific assessment of risk. Para. 10 ascribes to it three distinct elements:

(a) Ensuring that land is suitable for its current use, so that the acceptability or otherwise of the risks presented by the land are assessed on the basis of its current use and circumstances. The use of the word "circumstances" is important here, since land can present risks to vulnerable targets offsite, regardless of what use the land itself happens to be put.
(b) Ensuring the land is made suitable for any new use, as planning permission is given for such use. This is primarily the role of the planning and building control systems rather than Part IIA.
(c) Limiting requirements for remediation to the work necessary to prevent unacceptable risks to human health or the environment in relation to the current use or any future use for which planning permission is being sought. This approach, of relating risk to actual rather than hypothetical uses, is distinguished from the "multifunctionality" approach, of seeking to clean up land to a standard suitable for the most sensitive uses. That approach risks carrying out premature work (so distorting priorities) or unnecessary work (so wasting resources).

1.36 The thrust of the approach is therefore to avoid wasting resources by cleaning up land so as to make it fit for any purpose for which it might conceivably be needed in future. At one level, this is about risk assessment, but more fundamentally it is about the differing philosophical approaches of pragmatically doing enough for present purposes as against a complete and comprehensive solution here and now. The difficulty with the latter approach is partly one of uncertainty as to whether such complete solutions can ever be achieved at all, and partly one of sheer expense, as countries which have taken that approach have found.

1.37 Annex 1 indicates that there is an exception to the "suitable for use" approach, which applies where contamination has resulted from a specific breach of an environmental licence or permit. In such circumstances it will generally be appropriate for the polluter to remove the contamination completely (para. 15). Otherwise the regulatory regimes which are aimed at preventing new contamination would be undermined.

Objectives of the new regime

1.38 The Government's primary objectives for introducing the new regime, set out at para. 26 of Annex 1, are:

(a) improving the focus and transparency of the controls, and ensuring authorities take a strategic approach to the problem;
(b) enabling all problems resulting from contamination to be handled as part of the same process, rather than by separate regulatory action to protect human health and the water environment;
(c) increasing the consistency of approach taken by different authorities; and
(d) providing a more tailored regulatory mechanism, including liability rules, which is better able to reflect the complexity and range of circumstances found on individual sites.

To achieve most, if not all of these objectives, the Guidance is essential, as providing a detailed set of principles or rules to be applied to the individual circumstances of each site.

1.39 An important secondary objective of the creation of a detailed liability framework is to encourage voluntary remediation. The Government wishes to see 60 per cent of new housing built on previously developed land, and will be responding to the proposals for action by the Urban Task Force (*Towards an Urban Renaissance*) with an Urban White Paper. The hope is that landowners and investors, faced with a clearer liability regime, will be able to assess the likely requirements of the enforcement authorities and make their own plans to carry out remediation in advance of regulatory intervention. Such plans may well involve redevelopment in order to provide the funds to meet the costs of clean up.

1.40 These aims of the Labour Government in bringing the legislation into effect are, it will be appreciated, not widely different from the aims of the previous Conservative Government in enacting it, as described above. It is however important to keep these general policy statements in perspective. What matters ultimately are the words of the statute, and of the Guidance which has statutory force under it. How the present Government views the objectives of legislation which it did not enact cannot affect the meaning of that legislation as a matter of law; nor for that matter can the Guidance promulgated by the present Government widen the ambit or alter the effect of the primary legislation from which it derives its legitimacy and force. Such policy statements do however provide the context within which the legislation will operate and (subject to challenge on administrative law grounds) will no doubt have an influence on the way in which decisions by the relevant regulatory bodies are taken.[25]

Commencement in England

1.41 The Environment Act 1995 (Commencement No. 16 and Saving Provision) (England) Order 2000 (S.I. 2000 No. 340) brings the Part IIA regime fully into force in England as from April 1, 2000. It also consigns sections 61 and 143 of the 1990 Act to legal history, repealing them before they ever entered into force (Article 2(c)).

Scotland and Wales

1.42 A further political development between the enactment of Part IIA and its implementation has been devolution for Scotland and Wales. The ministerial functions of making the regulations and guidance necessary to bring the new regime into force and to determine its effect in practice rest with Scottish ministers under section 53 of the Scotland Act 1998 and with the Welsh Assembly under section 22 of the Government of Wales Act 1998. Draft Guidance for Scotland was published for consultation shortly after the English version in 1999, and was (with the exception of differences attributable to Scots law) in the same form. At the time of implementation in England, the Scottish Executive was still working to produce the final version, with the intention of bringing Part IIA into force around May 2000. Progress in Wales has been somewhat slower, with no consultation draft issued at the time the English and Scottish versions were promulgated. Accordingly, the text of this book is based on the English Regulations and Guidance. Whilst it seems likely that the Scottish and Welsh versions will not differ significantly, it cannot be guaranteed that they will be identical, and readers should bear that in mind.

Key questions and where to find the answers

1.43 The new regime will undoubtedly present some difficult questions to those charged with enforcing the legislation and those subject to such enforcement. The public, for whose protection the legislation is enacted, may also have questions. This Chapter concludes by setting out what seem to be the most likely questions, and indicating where in the text the relevant material is to be found.

---

[25] On the approach of the courts to determining the meaning and application of policy, see James Maurici [1998] J.R. 85.

Enforcing authorities (local authorities, the Environment Agency and SEPA)

1.44

1. How is the local authority's duty of inspecting for and identifying contaminated land to be fulfilled? See Chapter 4.
2. At what risk of potential liabilities are local authorities in performing those functions? See Chapter 4.
3. What is the procedure when contaminated land is identified? See Chapter 5.
4. Which sites are special sites and how are they dealt with? See Chapter 6.
5. What action should be taken before serving a remediation notice? See Chapter 9.
6. What are the procedures in case of urgency? See Chapter 9.
7. How should the appropriate remedial action be determined? See Chapter 12.
8. How should the remediation notice be drafted? See Chapter 12.
9. On whom should the remediation notice be served? See Chapter 13.
10. How should voluntary remediation be dealt with? See Chapter 11.
11. What are the procedures where the local authority's own land is involved? See Chapter 19.
12. How is compliance with a remediation notice to be enforced? See Chapter 22.
13. How do the powers under Part IIA relate to other statutory powers and other enforcing agencies? See Chapter 2.
14. What are the requirements for publicity in relation to procedures under Part IIA? See Chapter 25.

Persons involved with contaminated land (as original polluters, owners or occupiers)

1.45

1. How should I respond to site investigation by the enforcing authority? See Chapter 4.
2. What are my rights to be consulted before a remediation notice is served? See Chapter 9.
3. Can I avoid service of a remediation notice? See Chapter 11.
4. Can I argue the remediation notice should be served on someone other than me? See Chapter 13.
5. How do I appeal against a remediation notice? See Chapter 23.
6. What other means of challenging the notice may be open to me? See Chapter 24.
7. What should I do if I think someone other than me should be responsible rather than me or as well as me? See Chapter 23.
8. Can I avoid damaging publicity which may devalue my land or make it unsaleable? See Chapter 25.
9. What if I can't afford to comply with the remediation notice? See Chapter 22.

The general public and NGOs

1.46

1. How can I bring contaminated land to the attention of the enforcing authority? See Chapter 4.
2. How can I make the authority comply with its duties under Part IIA? See Chapter 4.
3. What information should be made available to the public on contaminated land and on action taken under Part IIA? See Chapter 25.
4. Can I take direct action to enforce Part IIA myself? See Chapter 1.

Chapter 2

The Ambit of the Regime: What is "Contaminated Land"?

Summary

The term "contaminated land" is used in a specific sense in Part IIA. The definition is given at section 78A(2) and is based on two main criteria: significant harm and the pollution of controlled waters. In applying those criteria, to determine whether land is in fact contaminated, local authorities must act in accordance with guidance issued by the Secretary of State. The approach of the Government is based on the twin pillars of the "suitable for use" test and the application of risk assessment; this approach is reflected in the Guidance. The Guidance also introduces the concept of a "pollutant linkage", that is the relationship between the potentially harmful substances, available pathways, and vulnerable receptors. This concept is a recurrent theme in much of the Guidance.

Certain situations are excluded from the Part IIA regime entirely and as such are "off limits" to local authorities acting under those powers. These are, specifically, situations relating to processes subject to integrated pollution control, waste management licensing, fly-tipped waste, consented discharges to controlled waters, and radioactive contamination.

Problems of definition and the "suitable for use" approach

2.01 "Contaminated land" as used outside Part IIA is an imprecise term, which can cover many different situations of varying degrees of seriousness. Discussion on the issue can be found in more detail elsewhere,[1] but unless reliance is placed on numerical criteria as to the concentrations of specified

[1] See Stephen Tromans and Robert Turrall-Clarke, *Contaminated Land* (Sweet & Maxwell, London, 1994) pp. 3–6.

substances present, the normal approach is to consider whether by virtue of the substances present, the land presents an unacceptable risk of harm. This immediately begs a number of questions, however:

(a) Harm to what?
(b) What constitutes harm?
(c) What level of risk is acceptable?
(d) Within what context is risk to be judged? The land in its current use, or likely future uses, or all possible uses?[2]

2.02 These are the issues which will need to be considered in relation to specific sites within the framework of legislation and guidance provided by Part IIA, and the answers to which may have very significant legal and financial effects. The Government's stated intention[3] is that the definition and supporting guidance are consistent with the "suitable for use" approach, which recognises that risk needs to be assessed on a site-by-site basis, and which limits remediation costs to what is necessary to avoid unacceptable risks.

2.03 On these principles, what matters is the current use of the land in question, or the actual presence of vulnerable receptors. Where a change of use is proposed, the planning and building control regimes will continue to play a vital role. The focus is on the range of uses to which the land is actually likely to be put.

What is "contaminated land"?

2.04 "Contaminated land" is defined by section 78A(2) as meaning:

> ". . . any land which appears to the local authority in whose area it is situated to be in such a condition, by reason of substances in, on or under the land, that:
>
> (a) significant harm is being caused or there is a significant possibility of such harm being caused; or
> (b) pollution of controlled waters is being, or is likely to be, caused;
>
> and, in determining whether any land appears to be such land, a local authority shall . . . act in accordance with guidance issued by the Secretary of State . . . with respect to the manner in which that determination is to be made".

2.05 It will therefore be appreciated that first the definition is a technical one, and may well exclude land which in a broad sense would be regarded

---

[2] See the discussion in Chapter 1. See also Stephen Tromans and Robert Turrall-Clarke, *Contaminated Land: First Supplement* (Sweet & Maxwell, London, 1998) pp. 121–2 for discussion of changing approaches to this issue in The Netherlands.

[3] DETR Circular, *Contaminated Land,* Annex 1.

as contaminated. Secondly, Government guidance has a critical role to play in the determination of the issue. Effectively, the legislation makes an implicit distinction between land which is contaminated, and land which is so contaminated that it requires clean-up in the public interest. The terminology of the statutory definition is considered in this Chapter, while the legal process of making the determination is covered in Chapter 5.

2.06 It is also important to read the definition in conjunction with the supplementary provisions of section 78X, which allows the effects of two or more sites to be considered together, and which deals with the situation where land adjoining or adjacent to the boundary of the relevant local authority is contaminated.

The "pollutant linkage" concept

2.07 The concept of risk assessment is fundamental to the determination of whether land is "contaminated land".[4] In particular, the Guidance applies existing principles of considering whether there is a contaminant, a relevant receptor (target), and a possible pathway between the two.[5] The contaminating substances in question are the potential source of harm, and the receptor is the living organisms, ecological systems or property which may be harmed, or the controlled waters which may be polluted. The pathway is the route or means by which the receptor is being exposed to or affected by the contaminants, or by which it could be so exposed or affected. The Guidance states that the pathway may be identified on the basis of general scientific knowledge without the need for direct observation.[6] The source and the receptors must be specific and actual, not hypothetical.

2.08 The relationship between the contaminant, the pathway and the receptor are termed by the Guidance "a pollutant linkage". Without identification of all three elements of a pollutant linkage, land should not be identified as contaminated.[7]

2.09 The Guidance makes it clear that there may be more than one pollutant linkage on any given piece of contaminated land.[8] This follows from the fact that a single polluting substance may potentially affect a number of different potential receptors, and may do so through various pathways. The situation can be made more or less complex by the specificity with which the source, pathway and target are described. The Guidance refers to this issue only briefly, stating that the local authority may treat two or more substances as being a single substance, in any case where they are compounds of the same element, or have similar molecular structures, and

---

[4] Guidance, Annex 3, para. A.9.
[5] *ibid.*, paras A.10–A.21. See also the explanation at Annex 2, para. 2.6.
[6] *ibid.*, para. A.15.
[7] *ibid.*, para. A.17.
[8] *ibid.*

where it is the presence of that element, or type of molecular structure, that determined the harmful effect on the relevant receptor.[9] This is a more precise test than in earlier drafts of the Guidance, which simply referred to the same "chemical species". In other words, there is no necessity in that case to find a separate pollutant linkage for each specific substance, where a common "active" element or molecular structure can be identified.

Substances

2.10 The presence of a substance or substances is the starting point of the statutory definition. "Substance" is defined by section 78A(9) as meaning: "any natural or artificial substance, whether in solid or liquid form or in the form of a gas or vapour". The definition is comprehensive, and could for example cover methane or other gases deriving from biological processes affecting degradable materials. In the context of establishing liability, section 78F(9) expressly contemplates the possibility that it may be the secondary substances arising from chemical reactions or biodegradation which in fact give rise to the harm.

2.11 It is also clear that no distinction is drawn between natural or artificial substances in the definition. This is of potential significance, since in some instances significant harm may be presented by naturally occurring substances such as arsenic.[10] There is nothing in the legislation to confine its scope to substances which are present as a result of human activity, although in section 78F(2) the draftsman clearly had it in mind that someone would have caused or knowingly permitted them to be present. The point is of potential significance since the legislation "bites" first on an "appropriate person" who caused or knowingly permitted the substance to be present; but if no such person can be found, then the current owner or occupier may be responsible.[11]

Land

2.12 "Land" is not a defined term in Part IIA, and so the definition in Schedule 1 to the Interpretation Act 1978 will be relevant: this provides that land ". . . includes buildings and other structures, land covered with

---

[9] *ibid.*, para. A.18.

[10] It is not a fanciful possibility that such substances can give rise to serious health effects. Arsenic-bearing sediments in the River Ganges have been linked with the poisoning of groundwater wells in Bangladesh—see *The Times*, October 21, 1998. The unknown factor is what circumstances are responsible for the arsenic becoming liberated.

[11] See Chapter 13.

water, and any estate, interest, easement, servitude or right in or over land."

2.13 For the purposes of the above definition of contaminated land, "land" would in principle include the sub-soil to any depth; disused deep mineshafts have sometimes been the recipients of contaminating material in the past.[12] The issue of deep contamination arose in the Parliamentary debates on the Bill, where the Government indicated that its intention was that in general, where contamination existed in sub-surface mine workings, it would be the owner of the mine or mineral rights who would be regarded as "owner" (and thus potentially an appropriate person) rather than the surface owner.[13]

2.14 River beds or ponds with contaminated silts would be included within the definition,[14] as would the seashore,[15] which for these purposes would fall within the local authority's area.[16] Accretions from the sea and into the sea (for example, reclaimed land) will also fall within the jurisdiction of the local authority in this respect.[17] Docks and harbours may be the subject of private Acts, which may make express provision to the same effect.[18]

2.15 This definition of land also includes buildings and other structures. "Other structures" are presumably man-made by either building or engineering works, but would not appear to include plant or equipment which is not of a structural nature, for example oil drums or parked vehicles. Clearly a building such as hospital, factory or research institute could be contaminated by hazardous materials which have been handled there, or by building materials such as asbestos. The drains of such premises may be particularly contaminated and are clearly either a structure or are part of the building. It would be artificial to exclude a building of this sort from the definition of "land"; if the building in question is declared surplus to requirements, as sometimes happens in the case of older Victorian hospitals at the present time, and demolition or dereliction follows, contamination in the upper floors of the building could, if proper care is not exercised, soon find itself on the surface of the land, or on adjacent land. The question is whether, in the current condition of the

---

[12] See for example *R. v. British Coal Corporation, ex p. Ibstock Building Products Ltd* [1995] Env. L.R. 277, a case on environmental information relating to the possible disposal of wartime munitions into a mineshaft.

[13] *Hansard*, H.L. Vol. 562, cols. 165–166.

[14] *Hansard*, H.L. Vol. 560, col. 1425.

[15] An example of contamination of the seashore is Orford Ness in Suffolk, where previous military activities have led to significant contamination of various types.

[16] The seaward boundaries of local authority areas will in general either be fixed by local acts, or by reference to the limit of median tides, or by statutory instrument under section 70 of the Local Government Act 1972.

[17] Section 72 of the Local Government 1972; see also *R. v. Easington District Council, ex p. Seaham Harbour Dock Company Limited* 1999) 1 L.G.L.R. 327.

[18] *R. v. Easington District Council*, above at note 17.

building or structure, there is a source of contamination, a pathway and a receptor.[19]

"In, on or under the land"

2.16 Contaminated land is often thought of as land where contaminants are directly present in the soil or groundwater. However, the use of the words "in, on or under", combined with the fact that the term "land" can include buildings or structures, means that the definition is in principle sufficiently wide to catch substances present in underground or surface structures or containers, provided the relevant conditions of harm or risk of harm are fulfilled. Examples might include exposed and friable asbestos within a building, harmful substances within drainage systems, chemicals stored in corroded drums, or oil stored in a leaking underground tank. The corroded drums would not themselves be "land", but they and their contents are present on the land. What has to be considered is the condition of the land by reason of the substances present on it: does that condition involve a significant risk?

"Appears to the local authority"

2.17 The definition of contaminated land, with its use of the words "appears to the local authority", seems on its face to be a subjective judgment for the local authority.[20] The appearance of subjectivity is deceptive, given the requirement to act in accordance with Government guidance as to the manner in which the determination is made, and the requirement to determine the key questions in accordance with such guidance.

Harm

2.18 "Harm" is defined in section 78A(4) to mean "harm to the health of living organisms or other interference with the ecological systems of which they form part and, in the case of man, includes harm to his property". The question of whether or not that harm is "significant" is to be determined in accordance with guidance. The potential for harm is the subject of rather

---

[19] Such an assessment should take account of the existence of other means of exercising proper control over demolition activities, for example, the Health and Safety at Work, etc. Act 1974, the duty of care under section 34 of the 1990 Act and the Special Waste Regulations 1996 (S.I. 1996 No. 972).

[20] The word "appears" indicates that something less than complete certainty may be sufficient—see *Ferris v. Secretary of State for the Environment* [1988] J.P.L. 777, considering the word in the context of planning enforcement proceedings.

more convoluted wording: "there is a significant possibility of such harm (*i.e.* significant harm) being caused". The phrase "significant possibility of such harm" requires a judgment as to the degree of potential for harm. It is a different question from that which may arise in the context of civil proceedings as to whether the harm in question is too remote in terms of foreseeability. In any event, the question whether the possibility of significant harm being caused is "significant" is a matter for guidance—see section 78A(5)(b) discussed below.

2.19 The definition of "harm" set out above is important. The word "health" and "living organisms" are not defined, but could clearly include microscopic organisms which can be said to be "living". The definition goes on to deal with interference with ecological systems of which living organisms form part. Thus in particular the ecological systems supporting mammals, fish, birds, insects, and other fauna and flora would be included. However, the property contemplated in the section is specifically man's property; for example unowned property, the general natural or built heritage, or an area of outstanding natural beauty or nature conservation interest do not as such fall within this definition of "property". They may of course still be protected by the legislation in the sense of being the habitat or ecological support systems of plant or animal life.

2.20 "Organism" is a term used in biological science to denote animals, fungi and micro-organisms. As such it is a very broad term and will cover such organisms at all stages of development—in that sense an embryo would be an organism.

2.21 There is also the question of the scope of "other interference with ecological systems of which they form part." Such interference is clearly something other than direct harm to living organisms. It presumably includes interference which diminishes their access to breeding or feeding grounds in terms of area or of available fodder, or which affects their corridors for migratory movements. Whether that type of harm would be "significant" will depend on criteria set out in the Guidance—see section 78A(6).

2.22 Government guidance does not deal with the definition of "harm". It is whether the harm is "significant" which the Guidance addresses, and in this respect it can provide for different degrees of importance to be assigned to different descriptions of living organisms or ecological systems or for certain such descriptions to be disregarded—see section 78A(6).

Significant harm: generally

2.23 We have noted that the word "harm" is defined in section 78A(4), and that whether certain types of harm are to be regarded as "significant" will be determined in accordance with guidance (section 78A(5)). However, without prejudice to this general guidance, there may be further and more detailed guidance making provision for different degrees of importance to be assigned to, or for the disregard of, (a) different descriptions of living organisms or ecological systems; (b) different descriptions of places; or (c) different descriptions of harm to health or property, or other interference. The word "descriptions" in section 78A(6) presumably means "categories" as described in the guidance. If the word "categories" is inserted in place of the word "descriptions" then this section becomes the more readily intelligible.

2.24 Much human activity has the effect of interference with one ecological system or another. What the Guidance seeks to do is identify and set priorities so that there comes a point at which the interference with the ecological system in question is so important that it becomes "significant harm", or alternatively it may require certain types of harm to be disregarded as being less than significant. The same applies to interference with living organisms. This is primarily an area for the environmental scientist and ecologist rather than the lawyer, and may turn for example, upon whether the living organism or ecological system would be capable, having suffered the harm, of renewing itself, or whether the damage would be irreparable and permanent; and if so, whether such loss is significant, taking into account other ecological systems of the same type which would not be affected. For example, if there are millions of beetles of a particular species within ecological systems throughout the country, damage to a few of their number may not be "significant" in the overall context. These are the types of issues which the Guidance is intended to address.

2.25 The importance of section 78A(6) lies in the discretion which it confers on the Government to distinguish between species, or places, or types of harm, on the grounds of perceived importance. The national perspective may be very different to the local one here. "Degrees of importance" clearly will not be left entirely to the views of individual local authorities, but will be influenced by the national public perception of importance as informed by experts. This distinction between the view of the public (as represented by elected members, magistrates or inspectors) and the views of experts may in itself give rise to conflict. It also leaves scope for divergences of national priorities between the policy-makers in England, Scotland and Wales.

What harm is to be regarded as "significant"

2.26 What harm is and is not to be regarded as "significant" is to be determined in accordance with statutory Guidance. The issue is dealt with in Annex 3, Chapter A, Part 3 of the DETR Circular, *Contaminated Land*, which is reproduced in the Appendices of this book, and to which reference should be made for the detailed wording. The relevant Guidance is given in the form of a table (Table A) which lists types of receptors and describes in relation to each those types of harm which are to be regarded as significant. The authority should regard as significant only harm which is both:

(a) to a receptor of a type listed in Table A; and
(b) within the description of harm specified for that type of receptor.[21]

2.27 The types of receptor listed are:

1. human beings;
2. ecological systems, or living organisms forming part of such systems, within certain listed types of location, such as SSSIs,[22] national nature reserves,[23] marine nature reserves,[24] European Sites within the Conservation (Natural Habitats, etc.) Regulations 1994[25] or Areas of Special Protection for Birds[26];
3. property in the form of crops; timber; produce grown domestically or on allotments for consumption; livestock; other owned or domesticated animals; or wild animals that are the subject of shooting or fishing rights; and
4. property in the form of buildings.[27]

2.28 The types of effects which are relevant in each case are listed in Table A and are described generically as:

(a) human health effects;
(b) ecological system effects;
(c) animal or crop effects; and
(d) building effects.

2.29 In each case, it is clear that the intention is to confine the scope of the legislation to what are clearly, and by any standards, serious effects. So for humans, the types of harm in question are death, disease, serious injury, genetic mutation, birth defects, or reproductive impairment. "Disease" is to be taken to mean an unhealthy condition of the body or part of it, and by way of example can include cancer, liver dysfunction, or extensive skin

---

[21] DETR Circular, *Contaminated Land,* Annex 1, para. A.23.
[22] Notified under section 28 of the Wildlife and Countryside Act 1981.
[23] Declared under Section 35 of the 1981 Act. Also covered are nature reserves established under section 21 of the National Parks and Access to the Countryside Act 1949.
[24] Designated under section 36 of the 1981 Act.
[25] S.I. 1994 No. 2716. This will cover European Special Areas of Conservation and Special Protection Areas. Candidate or potential areas which are protected as such under PPG9 on *Nature Conservation* are also covered, *i.e.* cSACs, pSPAs and listed Ramsar sites.
[26] Established under section 3 of the 1981 Act.
[27] Defined by reference to section 336(1) of the Town and Country Planning Act 1990, so as to include any structure or erection and any part of a building, including any part below ground level; but not plant or machinery comprised in a building.

ailments, but not mental illness, except insofar as it is attributable to the bodily effects of a pollutant on the person concerned.[28]

2.30 Ecological system effects means harm resulting in "an irreversible adverse change, or in some other substantial adverse change, in the functioning of the ecological system within any substantial part of that location". It also covers harm which affects any species of special interest within that location, and which endangers the long-term maintenance of the population of that species at that location (the "beetles" point referred to at paragraph 2.24 above). Additionally, in the case of a designated or candidate European Site, it covers any harm which is incompatible with the favourable conservation status of habitats at that location, or species typically found there. The enforcing authority should have regard to the advice of the relevant national nature conservation agency for this purpose, and to the requirements of the Conservation (Natural Habitats, etc.) Regulations 1994.

2.31 In relation to animal or crop-effects, the issue is whether there is a substantial loss in yield or substantial diminution in value resulting from death, disease or physical damage. A substantial loss in value should be regarded as occurring only when a substantial proportion of the animals or crops are dead or no longer fit for their purpose.[29] For domestic pets, death, serious disease or serious physical damage is enough; loss of value is not relevant. A benchmark of 20 per cent diminution in value or loss in yield is suggested as indicating what constitutes a substantial diminution or loss, though this is not phrased in absolute terms.

2.32 For building effects, reference is made in Table A to structural failure, substantial damage or substantial interference with any right of occupation. "Substantial damage" or "substantial interference" in this sense is said to refer to cases where any part of the building ceases to be capable of being used for the purpose for which it is or was intended.[30] This might cover the situation where, for example, a house becomes uninhabitable because of the risks presented by explosive vapours or gases.

2.33 It will be appreciated that contamination may have many adverse effects which will not fall within those categories of "significant harm". In particular, where property is concerned, actual physical harm or interference with enjoyment is necessary—the fact that property may be seriously and adversely affected by stigma or blight is not relevant. The Guidance makes it clear that "significant" harm is limited to the receptors

---

[28] The stress of living close to a known contaminated site has been shown to result in adverse impacts on stress levels and mental health in some cases, for example causing vulnerability to a range of disorders not specifically related to exposure to any chemical—see *Risk Based Contaminated Land Investigation and Assessment* by Judith Petts, Thomas Cairney, Mike Smith, (John Wiley & Son, Chichester 1997) p. 13.

[29] Food should be regarded as not fit for its purpose when it fails to comply with the provisions of the Food Safety Act 1990.

[30] In the case of scheduled ancient monuments the test is whether the damage significantly impairs its historic, archaeological or other cultural interest.

and to the types of harm mentioned in Table A: for example, harm to ecological systems other than those referred to as "ecological system effects" should be disregarded.[31]

2.34 The Guidance again stresses the "suitable for use" approach in applying Table A, in that the authority should disregard any receptors which are not likely to be present, given the "current use" of the land, or any other land that might be affected.[32] "Current use" means any use currently being made or likely to be made of the land, which is consistent with any planning permission or is otherwise lawful under planning legislation. This principle is qualified by the Guidance in four respects[33]:

(a) current use is taken to include any temporary use permitted under planning law to which the land is being put, or is likely to be put, from time to time;
(b) current use includes future uses which do not require a new or amended grant of planning permission;
(c) nonetheless, current use also includes any likely informal recreational use, whether authorised by the owners or not. A prime example would be children trespassing on waste land. However, in considering the likelihood of such use, due attention should be paid to measures taken to prevent or restrict access.[34] On the basis of this test, the authority would not have regard to the risks presented to gypsies or other travellers unlawfully occupying contaminated land, since this goes beyond "informal recreational use". This may be an area where the Guidance comes into conflict with the Human Rights Act 1998, since the right to life (Article 2) and the right to respect for home and family life (Article 8) are not dependent on the occupation of the land being lawful.[35]
(d) for agricultural land, the current agricultural use should not be taken to extend beyond the growing or rearing of the crops or animals "habitually grown or reared on the land". This principle could prove problematic. It is possible that particular types of crops could be especially sensitive to certain types of pollution. The hypothetical possibility of growing such crops ought not, it is submitted, to be regarded. But what if the farmer genuinely decides to start growing such crops, and they are actually affected? It is not clear why this harm should be disregarded simply on the basis that there has been a change in farming practice.

---

[31] Guidance, para. A.24.
[32] Guidance, para. A.25.
[33] Guidance, para. A.26.
[34] Though not expressly stated, the likelihood of such measures being breached or circumvented should also be taken into account.
[35] See *Buckley v. U.K.* (1996) 23 E.H.R.R. 101.

Significant possibility of significant harm: generally

2.35 The question of whether the possibility of significant harm being caused is itself significant is to be determined in accordance with statutory guidance. "Significant" is perhaps a curious word to use in this context when trying to assess the scale of risk, which makes the statutory guidance all the more important in establishing exactly what is meant.

2.36 The issue is effectively one of risk assessment,[36] on which there are generally a number of publications, in particular the reports in the DETR's research series[37] as well as the Department's general *Guide to Risk Assessment and Risk Management for Environmental Protection*[38] and the HSE's Report, *Use of Risk Assessment within Government Departments*.[39] The Department issued in May 1998 a number of draft model procedures for the management of contaminated land, including MP1 on *Risk Assessment Procedures*, covering the stages of hazard identification, hazard assessment, estimation of risk, and risk evaluation. Work on the Model Procedures has subsequently been taken forward by the Environment Agency. In addition, the CLEA model, developed at Nottingham University, seeks to derive human health guideline values based on nine different pathways. The issue is one of considerable subtlety and complexity, where many different views are possible.[40]

2.37 The Guidance requires the authority to take into account the following factors in deciding whether the possibility of significant harm being caused is significant[41]:

(a) the nature and degree of harm;
(b) the susceptibility of the receptors to which the harm might be caused; and
(c) the timescale within which the harm might occur.

2.38 The more detailed content of the Guidance on this issue is considered below, but it must be said at the outset that the Guidance is here entering what may transpire to be a minefield of scientific and legal technicality. To

---

[36] DETR Circular, *Contaminated Land*, Annex 3, para. A.27.

[37] This is the "CLR" series of technical reports. See *Contaminated Land*, Chapter 4 on "Investigation and Appraisal", and the Noter-up to that Chapter in the 1998 *First Supplement*.

[38] HMSO, 1995.

[39] Report prepared by the Interdepartmental Liaison Group on Risk Assessment 1996.

[40] Detailed discussion can be found in *Risk Based Contaminated Land Investigation and Assessment* by Judith Petts, Tomas Cairney and Mike Smith (John Wiley & Sons, Chichester, 1997) pp. 249, 301. See also more generally, the Twenty-first Report of the Royal Commission on Environmental Pollution, *Setting Environmental Standards* (Cm. 4053, October 1998) and *Pollution Risk Assessment and Management* (ed. Peter E.T. Douben, John Wiley & Sons Ltd, Chichester, 1998).

[41] DETR Circular, *Contaminated Land*, Annex 3, para. A.28.

be relevant at all, the possible harm must be "significant" in the sense used in the Guidance. However, for the purposes of the possibility test, it seems that some types of significant harm may be more significant than others. The general principle stated in the Guidance is that the more severe the harm would be, or the more immediate its effect, or the greater the vulnerability of the receptor, the lower is the degree of possibility to be regarded as significant. A relatively low possibility of harm might, therefore, be regarded as significant if it would affect a substantial number of people, or if it would be the result of a catastrophic single incident or exposure. In an attempt to give some precision to what is, of the essence, an imprecise exercise of judgment, the Guidance provides a Table which for various types of harm seeks to express conditions for there being a significant possibility of such harm. These in turn, refer to notions of acceptability or unacceptability of levels of risk, either from exposure to toxicological properties, or the risk of harm by, for example, fire or explosion. For other types of harm, the proposed test is "more likely than not". In all cases, reference is made to the need to make the assessment in relation to relevant, appropriate, authoritative and scientifically based information.

2.39 Given that this assessment of the local authority is likely to be subject to scrutiny and possible challenge, it is important that the authority is able to demonstrate how they have applied the Guidance in reaching their decision. The form of instructions to any consultants or other advisers will be vital in this respect.

Guidance on what possibility of harm is significant

2.40 As mentioned in the previous paragraph, the Guidance provides a table (Table B) which in relation to the various types of significant harm listed in Table A, sets out the conditions for there being a significant possibility of such harm. The authority should regard as significant any possibility which meets the relevant conditions.[42]

2.41 Table A subdivides human health effects into those arising from the intake of, or direct bodily contact with, a contaminant and, secondly, all other human health effects (particularly by way of explosion or fire). For the first category the question is whether the intake or exposure would be unacceptable when assessed in relation to relevant information on the toxicological properties of the contaminant. Effectively, this requires risk assessment of the likely exposure and its effects, based on what is known of the toxicological properties of the substance or substances.[43] Such an assessment should according to Table B take into account:

---

[42] DETR Circular, *Contaminated Land*, Annex 3, para. A.30.
[43] Discussed further below.

(a) the total likely intake or exposure to the polluting substance or substances from all sources, including the pollutant linkage in question;
(b) the relative contribution of the pollutant linkage in question to the likely aggregate intake or exposure; and
(c) the duration of the intake or exposure resulting from the pollutant linkage.

2.42 It will be appreciated therefore that the issue is not simply the extent to which the contaminated land may expose an individual to toxic substances—cumulative effects must also be considered. Toxicological properties are for this purpose stated to include carcinogenic, mutagenic, teratogenic, pathogenic, endocine-disrupting and other similar properties.

2.43 For all other human health effects, the issue is whether the probability or frequency of occurrence of the significant harm is unacceptable, assessed on the basis of relevant information, and taking into account the levels of risk which have been judged unacceptable in other similar contexts. Weight should be given to cases where the harm would be irreversible, would affect a substantial number of people, would result from a single incident, or would be likely to result from less than 24-hour exposure. An obvious example here would be the risk of death or injury caused by an explosion from a build-up of landfill gas. The issue therefore is what is the scientifically-based probability of that consequence, and whether that level of risk is deemed "acceptable".

2.44 What Guidance exists on this issue in other contexts? The Health and Safety Executive has suggested that a level of risk of death for the general public which can be regarded as "acceptable" (*i.e.* the cost of further risk reductions is not justified) is one in 1 million per annum.[44] For large industrial plant a risk of one in 10,000 per annum is generally regarded as "tolerable", *i.e.* the risk should be kept under review with a view to reducing it if possible.[45] Such figures can of their nature only provide broad assistance in what may be a sensitive exercise, given local concerns as to a specific contaminated site. In this respect the Guidance effectively passes the key decision back to the local authority.

2.45 For ecological system effects the test is whether the relevant harm is more likely than not to result, taking into account the relevant information for the type of pollutant linkage in operation, particularly in relation to ecotoxicological effects. This test then appears to be on the balance of probabilities, *i.e.* a chance of more than 50 per cent and as such a very different standard to the "acceptable risk" used for human effects. However, a second limb of the test is that there is a reasonable possibility

---

[44] See HSE, *The Tolerability of Risk from Nuclear Power Stations* (revised 1992) para. 175. By way of comparison a risk of 1:1,000,000 per annum is about the risk of being electrocuted in the home, compared with the annual risks of dying in a traffic accident (1:10,000) or of contracting fatal cancer (1:300).

[45] *ibid.*, paras. 10. 173.

of such harm, and if it occurred it would result in damage to the features of special interest such that they would be beyond any practicable possibility of restoration. The data available on ecotoxicological effects is likely to be far from complete, and to involve many uncertainties.

2.46 For animal and crop effects and for building effects, the test is whether the relevant harm is more likely than not to result. In the case of buildings, the relevant period for that assessment is over the maximum expected economic life of the building.[46] No such period is stated for animal or crop effects, (or for that matter, ecological effects). Obviously the longer the period in question the higher the probability of harm may such be (*e.g.* persistent exposure of an ecosystem to leaching chemicals).

2.47 As indicated above, in the case of human health effects, particular weight should be given to harm which is irreversible, or affects large numbers of people, or would result from a single incident such as fire or explosion, or from short-term (less than 24-hour) exposure. These criteria essentially go to the acceptability of risk, and are the type of circumstances which the public are likely to regard as particularly unacceptable in those terms. Para. A.32 of the Guidance requires a similar approach to be adopted to other types of harm, in that where the harm would result from a single incident or less than 24-hour exposure, the test of "more likely than not" becomes one of "reasonable possibility". Again, this accords with common perceptions of risk, in that most people would regard the reasonable possibility of their building being destroyed by an explosion as "significant", even if it were less than a 50 per cent risk. The cumulative effect of the terms of Table B and this aspect of the Guidance for non-human effects is a two-tier test: reasonable possibility for some types of harm, balance of probabilities for others.

"Relevant information"

2.48 Table B refers to taking into account relevant information for the type of pollution linkage in question. Para. A.31 defines such information as that which is:

(a) scientifically-based;
(b) authoritative;
(c) relevant to risk assessment; and
(d) appropriate to the determination under the Guidance, in that it is consistent with providing the level of protection from risk set out in the Table B criteria.

The issue of such information, in terms of literature and numeric guidelines, is discussed below.

---

[46] For ancient monuments, the relevant period is the forseeable future.

Possibility of harm and the relevance of the current use of land

2.49 When considering whether harm from contaminants is a possibility it is vital to be clear as to the circumstances and assumptions against which the determination is being made. In particular, is the possibility to be taken into account that the current use of the land (or of adjoining land) may change in such a way as to increase or decrease the risk of harm? The Guidance seeks to answer this question in a number of ways. First, receptors which are not present or likely on land in its current use are to be disregarded.[47] Secondly, in considering the timescale within which harm might occur, the authority is to take into account any evidence that the current use of the land will cease in the foreseeable future.[48] Thirdly, the possibility of significant harm being caused as a result of a use of any land which is not the current use is not to be regarded as significant.[49] Fourthly, in considering future uses, the assumption should be that any future use will be in accordance with any extant planning permission: accordingly it should be assumed that any remediation required by planning conditions will be carried out, and that the planning authority will exercise any powers of approval under the conditions to ensure adequate remediation.[50]

2.50 The intention behind these statements is to avoid arguments that land should be judged as contaminated (and accordingly cleaned up) if it fails to meet standards based on highly-sensitive hypothetical end uses. Consequently it becomes crucial to define clearly what is meant by "current use", as explained above, both for the contaminated land itself and for any adjoining land which may be affected. Uses which would require any new or amended grant of planning permission will in general fall outside the definition of "current use" in the Guidance. The reasoning behind this formulation is no doubt that if a use of land which is more sensitive than the current use of contamination is proposed, and requires planning permission, then under current policy such permission is not likely to be forthcoming unless the risks from contamination can be reduced to an acceptable level, and then only subject to conditions to ensure that the appropriate steps are taken. In some cases, development or a change of use may take place without planning permission (*e.g.* under permitted development rights) in which case such alternative use will be relevant.

2.51 The reference in the Guidance to the possibility that a current use of the land may cease in the foreseeable future sits somewhat uncomfortably with this reasoning: cessation of the use may or may not involve the extinguishment of existing use rights,[51] and if those rights are not extinguished there

---

[47] DETR Circular, *Contaminated Land*, Annex 3, para. A.26 (see above).
[48] *ibid.*, para. A.29.
[49] *ibid.*, para. A.33.
[50] *ibid.*, para. A.34.
[51] See S. Tromans and R. Turrall-Clarke, *Planning Law, Practice and Precedents* (London, Sweet & Maxwell, 1991) para. 8.46.

is no reason why the use should not be resumed at any time. Also, even where land has planning permission it may still be unlikely that the permission will be implemented because of access or infrastructure constraints, or restrictive covenants.

2.52 What the Guidance means in practice is that an adjoining landowner who is concerned that the development potential of his land is blighted by contamination on neighbouring land will find it difficult to secure the use of the Part IIA provisions as a means of alleviating that blight, where receptors on the land in its current use are not affected. His redress, if any, will lie at common law.[52]

The use of risk assessment and numeric guidance

2.53 Essentially, and at the risk of over simplification, there are two main approaches to assessing the risks of contaminated land: generic and site specific.[53] The generic approach involves the use of numerical values set by an authoritative body and intended to assist the exercise of appropriate professional judgment. Such values have the advantages of being convenient and of aiding consistency; however they will not be available for all contaminants and cannot always cater for unusual circumstances such as mixtures of contaminants or high local background levels.[54] If qualitative risk assessment involving such generic guidelines is used, then it is vital to understand the basis for those values and the assumptions within them which may be conservative or "worst case" in nature.[55] Such guidance has improved over the years[56] and is now often based on risk estimation models taking account of a number of exposure routes.[57] However, such an approach is unlikely to be acceptable as the sole means of making decisions under Part IIA, given the following comment on the generic approach[58]:

> "It must be stressed that it is not possible to make a direct statement about an estimated site risk to a defined target through the use of generic Guidelines, and the latter may not provide fully for the source-pathway-target scenarios relevant to a site."

---

[52] See S. Tromans and R. Turrall-Clarke, *Contaminated Land* (London Sweet & Maxwell, 1994), paras 2–03 *et seq.*

[53] *Risk Based Contaminated Land Investigation and Assessment*, by Judith Petts, Tom Cairney and Mike Smith (John Wiley & Sons, Chichester, 1997) p. 22.

[54] *ibid*, p. 23; see also *Pollution Risk Assessment and Management* (ed. Peter E.T. Douben, John & Wiley & Sons, Chichester 1998) pp. 91–201 and Chapter 9.

[55] *ibid.*, p. 206.

[56] *ibid.*, p. 210, referring to the replacement of ICRCL Guidance with DETR Guideline Values in CLR Report 10 (1997).

[57] *ibid.*, p. 211.

[58] *ibid.*, p. 212.

2.54 The Royal Commission on Environmental Pollution, in its 1998 Report, *Setting Environmental Standards*, identifies four possible sources of error in assessing pathways and exposures, all of which have potential relevance to contaminated land, and many of which are reflected in the Guidance[59]:

1. The form in which a substance is released is not necessarily the form in which it remains; transformation processes may occur.
2. The location in which a substance is released is not necessarily that in which it has its most important effect.
3. Behaviour of a substance in one location does not guarantee the same behaviour in another.
4. There may be simultaneous exposure to the same substance from different media and sources.

2.55 Similarly, the approach of risk ranking or semi-quantified risk assessment, which involves a scoring system to give an indication (rather than an assessment) of relative risk is unlikely to be an appropriate method for Part IIA, helpful though it may be as an aid to commercial decision making.[60]

2.56 Site specific or quantified risk assessment leans heavily on professional judgement as to what is an acceptable level of risk, involving as it does choices on what assumptions should be made on matters such as exposure periods, reference data on toxicity or cancer potency, risk estimation models, and site data.[61] Modelling techniques are often used, but have their own difficulties in terms of reliability.[62] As the Royal Commission Report puts it: "Confidence in the scientific principles on which [exposure models] are built sometimes gives those who construct such models an excessive faith in their predictions."[63] Other forms of exposure also have to be considered,[64] as does uncertainty analysis[65] and the quality of data inputs.[66] In short[67]:

> "Risk assessment of contaminated sites is primarily an "art" with abundant opportunities for personal interpretation and bias. The latter can be detrimental to the outcome, but also provide the underpinning professional judgments that are inherent to an effective process. Risk assessment cannot be reduced to standardised institutions and guidelines".

2.57 The requirements of the Guidance that information, in order to be relevant to the determination process, must be scientifically based, author-

---

[59] Cm. 4053, Box 2E.
[60] *ibid.*, p. 233.
[61] *ibid.*, p. 250.
[62] *ibid.*, p. 252.
[63] Cm. 4053, para. 2.56.
[64] Petts, Cairney and Smith, *op. cit.*, p. 276.
[65] *ibid.*, p. 278.
[66] *ibid.*, p. 295.
[67] *ibid.*, p. 301.

itative and relevant does not answer these questions in itself. The use of guideline values is addressed further in the context of the determination process itself. Paragraph B.48 of the Guidance requires attention to be paid to the assumptions underlying any numerical values, any conditions relevant to their use, and any adjustments that need to be made to reflect site-specific circumstances.

Some relevant literature

2.58 An enforcement authority considering the issues of harm and risk described above will undoubtedly need expert assistance and will also need to take into account the diverse range of available literature on the subject, given the requirement of the *Guidance* to make use of appropriate, authoritative and scientifically based information. The following list is not intended to be comprehensive, but is indicative of the types and sources of material available:

1. Environment Agency Model Procedures for the management of contaminated land, especially MP1 on *Risk Assessment Procedure*; MP2 on *Evaluation and Selection of Remedial Measures*; and MP3 on *Implementation of Risk Management Action*.
2. DETR Reports in the CLR Series, for example CLR 9 (Tox) *Collation of Toxicological Data and Intake Values for Humans*; CLR 10 *The Contaminated Land Exposure Assessment Model (CLEA)*; CLR 10 (GV) *Guideline Values for Various Substances in Soils*.
3. American Society for Testing and Materials (ASTM) *Risk Based Corrective Action at Petroleum Release Sites* (Std. Guide E1739–95).
4. Building Research Establishment (Construction Industry Research Association) various publications on effects of contaminants on building materials.
5. DETR Toxic Substances Division, TSD Reports on substances such as toluene, TCE and benzene.
6. Environment Agency R&D Technical Reports, *e.g.* clean up to protect water resources, leaching tests.
7. Welsh Development Agency, *The Remediation of Contaminated Land* (Manual, 1993).
8. HSE, *Occupational Exposure Limits* (EH40/96).

Pollution of controlled waters

2.59 Under section 78A(2)(b), if pollution of controlled waters is being caused or is likely to be caused by the condition of land by reason of substances present in, on or under it, that land is contaminated. The question then is whether the water pollution is being caused by the condition of land, which as pointed out above, is defined to include buildings or other structures.[68]

---

[68] See para. 2.05.

This raises some interesting issues as to the relationship with powers under water pollution control legislation. For example, if a spill occurs in the course of handling chemicals and water pollution results, this will not be within the definition. Equally clearly, substances leaching from land into controlled waters will be within the definition. But what of the situation where the pollution, or risk of it, stems from defective bunding, drainage systems or other fixed structures? It could be argued that such pollution is caused by a combination of the substances present and the defective condition of the land. Whether the draftsman had this situation in mind is perhaps another matter.

2.60 "Pollution" is defined simply as "the entry into controlled waters of any poisonous, noxious or polluting matter or any solid waste matter".[69] The pollution need not be "significant" in this context. "Controlled waters" are defined in section 78A(9) as having the same meaning as in Part III of the Water Resources Act 1991 or, in relation to Scotland, the Control of Pollution Act 1974 as amended.[70] The two definitions are similar in that they both cover:

(a) territorial waters;
(b) coastal waters;
(c) inland waters; and
(d) groundwaters.

2.61 There are some differences, however. The definition applicable to Scotland expressly includes within the definition of groundwaters water in wells, boreholes and other excavations. The definition applicable to England and Wales makes it clear that reference to waters of lakes, ponds and watercourses includes reference to the bottom, channel or bed which is for the time being dry.[71]

2.62 Under section 78A(8) controlled waters are "affected by contaminated land if (and only if) it appears to the enforcing authority that the contaminated land in question is, for the purposes of subsection 78A(2), in such a condition, by reasons of substances in, on or under it, that pollution of those waters is being or is likely to be caused". In respect of these subsections it is helpful to look at section 78J which deals with restrictions on liability relating to the pollution of controlled waters.

2.63 In respect of pollution of controlled waters under section 78A(2)(b) there is no need to show "harm" or "significant harm". The question arises, therefore, whether any pollution is enough. As stated above, pollution of controlled waters is defined simply by reference to the entry of certain things into the controlled waters. In one case it was held that the offence of causing polluting matter to enter controlled waters was not committed by

---

[69] See Environmental Protection Act 1990, s. 78A(9).
[70] *ibid*.
[71] See *R. v. Dovermoss Ltd* [1995] Env. L.R. 258; *Environment Agency v. Brock Plc* [1998] Env. L.R. 607.

stirring up silt that was already on the bed of a river, so as to cause the silt to enter the water.[72]

2.64 The question of whether pollution of controlled waters is being caused, or is likely to be caused, is to be determined in accordance with the statutory guidance—see subsection 78A(5)(c). The guidance is, however, extremely brief on this issue, simply stating that the authority should be satisfied that a substance is continuing to enter controlled waters or is likely to do so, and that "likely" in this context means more likely than not.[73] The explanation for this brevity is that there is in fact no power to issue guidance on what constitutes the pollution of controlled waters.[74] It may be that this position will be modified by future changes to the primary legislation.

2.65 The Guidance also deals with the situation where the entry of pollutants into ground water or surface water has already occurred. This is dealt with below under the heading, "Water pollution which has already occurred" (see paragraphs 2.69 *et seq.*).

2.66 The key issue may in many cases turn on what is meant by polluting matter. Current case-law suggests that the term "polluting matter" is referable to the potentially harmful nature of the matter in question, not whether it had any demonstrably harmful effect on the controlled waters. Thus in *National Rivers Authority v. Eggar U.K. Limited*[75] discolouration of water by a visible but transient brown stain was held sufficient to constitute the offence of causing or knowingly permitting poisonous, noxious, or polluting matter to enter controlled waters. Similarly, in *R. v. Dovermoss Ltd* the Court of Appeal rejected a submission that pollution must involve harmful effects and adopted the Oxford English Dictionary definition of "to make physically impure, foul or dirty; to dirty, stain, taint, befoul."[76]

2.67 The Government is aware of the possibility that relatively minor levels of water pollution could trigger the "contaminated land" definition, and the Circular suggests that the Government is proposing to review the wording of the legislation on this issue with a view to amendment.[77] But in any event, if the contamination is only slight, then the remediation requirements ought to be correspondingly low.[78]

---

[72] *National Rivers Authority v. Biffa Waste Services Ltd* [1996] Env. L.R. 227.

[73] DETR Circular, *Contaminated Land*, Annex 3, para. A.36. An alternative approach is to focus on the seriousness of the results that might occur if the risk materialised. If that risk is sufficiently serious, then "likely" might mean simply "a real risk, a risk that should not be ignored": see *In re M (Minors)* [1996] A.C. 563; *R. v Whitehouse*, *The Times*, December 10, 1999.

[74] See DETR Circular, *Contaminated Land*, Annex 2, para. 2.9.

[75] Newcastle-upon-Tyne Crown Court, June 15–17, 1992, unreported.

[76] [1995] Env. L.R. 258.

[77] Annex 2, para. 6.30.

[78] *ibid.*, para. 6.31.

Likelihood of pollution of controlled waters

2.68 The Guidance on the definition of "contaminated land" does not deal with what is meant by the likelihood of pollution of controlled waters being caused, beyond saying that it depends on the authority judging it more likely than not to occur. However, more detailed criteria are set out in Chapter B of the Guidance, dealing with the determination process, and are dealt with in Chapter 5 of this book on the identification of contaminated land. Essentially, however, the approach is one of source, pathway and target, involving the mobility of the pollutant, the geological or other pathways to controlled waters, and the existence of any suitable risk management arrangements.

Water pollution which has already occurred

2.69 One potential difficulty with the scheme of Part IIA is drawing a clear distinction between the subsurface of land and groundwater. Water in the saturated zone, or contained in underground strata, could well be regarded as part of the land and accordingly, contaminants present in such water would be present under the land and so could result in a determination that the land is contaminated.

2.70 However, the Act itself seems to pose a clear distinction between land and water, in that it is water which is affected by the presence of substances in the land—see section 78A(8). The statutory Guidance reinforces this distinction by stating that land should not be designated as contaminated where a substance is already present in controlled waters, entry into the water has ceased and it is not likely that further entry will take place.[79] For this purpose, substances are to be regarded as having entered controlled waters where they are dissolved or suspended in them, or if they are immiscible with water and have direct contact with it on or beneath the surface of the land.[80] "Continuing to enter" means any additional entry to that which has already taken place.[81] Accordingly, any determination should be based on the existence of substances in, on or under land which are continuing to enter controlled waters, or are likely to do so in future. The requirement to consider not only the present position but also what is likely in the future may serve to avoid the anomalous results which could occur with the status of land as "contaminated" varying according to whether the water table happens to be high or low at a particular time.

2.71 The practical consequence is therefore that the contaminated land provisions will not apply to land under which there is contaminated

---

79 DETR Circular, *Contaminated Land,* Annex 3, para. A.37.
80 *ibid.*, para. A.38.
81 *ibid.*, para. A.39.

groundwater, unless the condition of the land is making, or is likely to make, that contamination worse, or unless the presence of the substances in the groundwater means that the "significant harm" part of the definition is applicable. This does not mean that there is no way of securing clean-up of the groundwater: the provisions of sections 161 and 161A–D of the Water Resources Act 1991 may apply to cases where polluting matter is already present in controlled waters but where entry of pollution from the land has ceased or there is no likelihood that further entry will take place.[82]

Cumulative effect of sites

2.72 It is possible that a pollution problem may be the result of the cumulative impact of contamination from a number of sites, in circumstances where it could not be said that each individual site would itself have a sufficiently significant effect. Perhaps the most obvious example is where a number of sites together contribute to a critical load of heavy metals or organic pollutants to water.

2.73 Section 78X(1) deals with this situation by providing that where it appears to a local authority that two or more different sites, when considered together, are in such a condition as to satisfy the tests of significant harm or pollution of controlled waters, then Part IIA shall apply in relation to each of them whether or not they would be "contaminated land" if considered alone. It does not seem to follow from the section that they have to be dealt with as if they were one site; indeed they may not necessarily be adjacent. Rather the provisions of Part IIA apply "to each of them", implying they may be dealt with separately.

Matters excluded from the contaminated land regime

2.74 There are other types of problems which are excluded from the contaminated land regime, so that they are to be dealt with by the appropriate enforcing authority under other powers. These in particular relate to prescribed processes, licensed waste facilities, fly-tipped waste, consented discharges to controlled waters, and radioactive contamination. The local authority should be mindful of these exclusions in exercising its duties of inspection and enforcement. The authority will also need to bear in mind the issue of overlap with other powers, discussed in Chapter 7.

Contamination from prescribed processes

2.75 Processes falling within Part I of the Environmental Protection Act 1990 (integrated pollution control and local authority air pollution control) are dealt with by section 78YB(1), which may preclude a remediation notice

---

[82] As to the relationship between Part IIA and the Water Resources Act 1991 provisions, see Chapter 7.

under Part IIA being served, or may restrict its content. Processes authorised under Part I of the 1990 Act are not automatically excluded from the remediation notice regime; rather the restriction covers situations where it appears that the powers of the appropriate Agency may be exercised under section 27 of the 1990 Act in relation to the significant harm or pollution of controlled waters by reason of which the land in question is contaminated land.

2.76 Section 27 is applicable where the harm arose from the commission of an offence under sections 23(1)(a) or (c) of the 1990 Act, *i.e.* carrying on a prescribed process without authorisation, or otherwise than in accordance with applicable conditions, or failing to comply with or contravening a requirement or prohibition imposed by an enforcement or prohibition notice under sections 13 or 14 of the 1990 Act. The power of the appropriate Agency is to arrange for any reasonable step to be taken to remedy the harm and to recover the cost of taking those steps from any person convicted. The Agency needs the written approval of the Secretary of State to exercise such powers, together with the permission of any third party in occupation of land which will be affected.

2.77 Thus where contamination results from an offence committed by the operator of a prescribed process, it will need to be considered whether the Agency can act under section 27 before proceeding to enforcement under Part IIA of the 1990 Act. In that respect, the Part IIA powers may be regarded as something of a "long stop" where section 27 powers are not available (which of course they generally will not be in relation to historic contamination). As the wider regime of IPPC is introduced under the Pollution Prevention and Control Act 1999, with its greater emphasis on restoration of sites at the end of their operational life, so the role of Part IIA may come to assume less importance in relation to such sites. Essentially, the Government's intention is that the PPC regime will have the same relationship to Part IIA as IPC.[83]

2.78 The prohibition relates only to serving a remediation notice, not to the initial duty to cause land to be inspected. Operational IPC sites have no immunity in that respect and may be identified as contaminated.

Licensed waste sites

2.79 In contrast to the approach to prescribed processes, described above, Part IIA is entirely disapplied by section 78YB(2) in relation to land in respect of which there is for the time being in force a waste site licence under Part II of the 1990 Act. The expression "site licence" will cover not only licences granted under Part II, but also those originally granted under Part I of the Control of Pollution Act 1974 which were statutorily converted into such licences.

[83] DETR Circular, *Contaminated Land,* Annex 1, para. 54.

2.80 Not only is the service of a remediation notice prevented in such cases, but also the statutory duty of inspection does not apply. The rationale is no doubt that any problems from licensed waste activities are best dealt with under waste licensing inspection and enforcement powers so long as the licence remains extant. Section 39 of the 1990 Act will prevent surrender of the licence until the Agency is satisfied that environmental pollution or harm to human health are unlikely. However, the contaminated land regime will apply to the extent that the relevant harm or pollution are attributable to causes other than:

(a) breach of the licence conditions; or
(b) the carrying on of activities authorised by the licence in accordance with the conditions of the licence.

2.81 The effect of this rather complex wording is therefore that the contaminated land regime will apply to certain problems affecting licensed waste sites, in particular contamination arising from pre-existing activities, or from other sources than the licensed waste activities.[84] Thus the local authority as enforcing authority under Part IIA will not be obliged to inspect licensed waste sites to discover contamination caused by the licensed activities: it will, however, remain under a statutory duty to inspect for historic contamination or contamination from other causes. An example might be a licensed transfer station or waste treatment facility which is located on a previously contaminated site: such land will enjoy no protection from a remediation notice by reason of its current licensed use.

Unlawfully deposited (fly-tipped) waste

2.82 Where waste is deposited on land except under and in accordance with a waste management licence and in consequence the land becomes contaminated, it may be possible to require its removal or other remedial action under section 59 of the 1990 Act ("Powers to require removal of waste unlawfully deposited"). Section 78YB(3) provides that to the extent it appears to the enforcing authority that such powers may be exercised by the appropriate Agency, a remediation notice may not be served. The intention of the Government in giving precedence to the section 59 powers was to preserve for the benefit of innocent occupiers the defence provided by section 59(3), *i.e.* that they neither caused nor knowingly permitted the deposit of waste. However, one problem is that the applicability and efficacy of section 59 powers in relation to contaminated land is itself by no means clear.[85] In particular, the occupier of land who was not involved in

---

[84] *ibid.*, para. 56.

[85] See Stephen Tromans and Robert Turrall-Clarke, *Contaminated Land* (London, Sweet & Maxwell, 1994) para. 2–59.

the initial deposit may have a complete defence to any notice served under these powers, and the original "depositor" may no longer exist or may not be capable of being traced.

2.83 It is worth noting that the scope for action under section 59 is at least in one respect wider than Part IIA, in that by section 59(7) action need only be necessary to remove or prevent pollution of land, water or air, or harm to human health: the issue of "significant" harm does not figure in section 59.

2.84 All of this means that the wording of section 78YB(3) will be subject to careful scrutiny, given the substantial degree of protection it may afford to the occupier. In order to come within section 59, the relevant contamination must result from controlled waste having been deposited in breach of section 33 of the 1990 Act.[86] If this is the case, a remediation notice is precluded if and to the extent that it appears that section 59 powers may be exercised. The question is whether they may be exercised, not whether they are likely to be exercised successfully, nor which regime would provide the more expedient remedy.

2.85 The wording of section 78YB(3) is in that respect problematic. The enforcing authority under Part IIA must act *Wednesbury* reasonably in deciding whether it appears that the section 59 powers *may* be exercised, not whether they *should* be exercised. On whom is the burden of proof that the land was actually contaminated by fly-tipping? It will almost certainly be to the advantage of the owner or occupier to argue that section 78YB(3) is applicable: indeed, if the section is widely interpreted, it would open up a very large gap in the operation of the Part IIA regime, leaving the problem to be dealt with at the public expense either under Part IIA or under the powers of section 59(7) to remove the waste and remedy the consequences of its deposit.

2.86 For example, take the situation where the enforcing authority discovers asbestos buried in a field; it is not known who placed it there. Under Part IIA where the original pollutor cannot be found, the owner will be liable. The owner however suggests that the asbestos must have been dumped there before he acquired the land, either by the previous owner or by persons unknown—it is therefore controlled waste unlawfully deposited on land and as such within section 59. If the enforcing authority accepts that this is the case, then by section 78YB(3) they cannot serve a remediation notice. A notice requiring removal of the asbestos may be served on the occupier by section 59(1) but in the circumstances any notice is likely to be quashed under section 59(3)(a) on the basis that the occupier neither deposited it nor knowingly caused or permitted its deposit. The waste regulation or waste collection authority will then be thrown back on their power to deal with the situation themselves under section 59(7): the waste not only becomes their problem, it also becomes their property on removal by section 59(9).

---

[86] A key question may be whether this is to be read as also including waste deposited in breach of the previous legislation, the Control of Pollution Act 1974, and the Deposit of Poisonous Waste Act 1972.

2.87 Further, although section 59 is colloquially referred to as relating to "fly-tipping", it applies to all waste unlawfully deposited, and is not limited to fly-tipping as commonly understood. There is a possible argument that any spillage of material could constitute the deposit of waste, since once spilt the material is no longer usable.

Consented discharges to controlled waters

2.88 By Section 78YB(4), a restriction on the content of remediation notices applies so as to prevent any notice requiring anything which would impede or prevent the making of a discharge pursuant to a consent under the Water Resources Act 1991 or, in Scotland, under the Control of Pollution Act 1974. If the effect of a consented discharge from contaminated land is to cause significant harm, then the appropriate course will lie with the appropriate Agency to exercise its powers to revoke or vary that consent.

2.89 Consents which may be relevant to groundwater contamination can also be granted under the Groundwater Regulations 1998.[87] Under regulation 18, the Agency may grant authorisation for the disposal or tipping of List I or List II substances[88] in circumstances which might lead to an indirect discharge to groundwater. Similarly, under regulation 19, activities on or in the ground which may lead to such discharges to groundwater, may be the subject of a notice by the Agency authorising the activity subject to conditions. Such authorisations constitute a defence of authorised discharge to offences under the water pollution legislation.[89] The same principle as in section 78YB(4) should presumably therefore apply to such authorisations.

Harm or pollution attributable to radioactivity

2.90 By section 78YC, except as provided by Regulations, Part IIA does not apply to harm or to pollution of controlled waters so far as this is attributable to radioactivity possessed by any substance. Where a radioactive substance has other harmful properties, such as toxicity, Part IIA will be applicable in relation to those properties. The keeping, use and disposal of radioactive substances and wastes is regulated by the appropriate Agency under the Radioactive Substances Act 1993. Whilst the Government has indicated that it views Part IIA as providing a suitable basis for dealing with radioactive contamination on old industrial sites, it considered that changes to the detail might be required in view of the particular

---

[87] S.I. 1998 No. 2746.

[88] These are the substances listed in the Groundwater Directive 80/68/EEC and specified in the Schedule to the Regulations.

[89] Groundwater Regulations, reg. 14(2).

scientific problems involved. A consultation paper published in February 1998, *Control and Remediation of Radioactively Contaminated Land*, discusses the proposed extension of the regime to such land; the enforcing authority would be the Environment Agency/SEPA. Such extension will involve consideration of how to apply established concepts and principles of radiological protection to contaminated land: this issue was the subject of discussion by the National Radiological Protection Board in its 1998 publication, *Radiological Protection Objectives for Land Contaminated with Radionuclides* (Documents of NRPB, Vol. 9, No. 2, 1998).

Civil remedies

2.91 This Chapter has concerned itself with the new regime for the clean up of contaminated land in the public interest, in the context of statutory public law codes. The enforcing authority represents that public interest, supported by criminal sanctions and administrative powers. However, as with statutory nuisance, the common law remains in place to provide remedies as between the parties to a civil suit[90]; the public law liabilities under Part IIA must therefore be considered in that context. As was suggested in *Contaminated Land*,[91] the damage from contamination may only come about when stricter regulatory standards are applied: the new liabilities under Part IIA may therefore be a trigger for civil actions which would otherwise not have been brought. The Act is concerned with actual or prospective physical harm or damage, whereas often the concern of a landowner affected by nearby contaminated land will be property blight or diminution in value. The common law remains the appropriate redress (to the extent that it recognises such losses) for these concerns.[92]

Relationship to the identification process

2.92 The approach taken in the Guidance, which in turn follows the structure of Part IIA, is to give separate guidance on the meaning of "contaminated land" (Chapter A) and on the process of identifying land as contaminated (Chapter B). The distinction is not watertight, however, and some aspects of the substantive definition are amplified or glossed in the guidance on determination. The essential point, therefore, is that Chapters A and B of the Guidance need to be read together, as do the corresponding Chapters in this book (Chapters 2 and 5).

---

[90] S. Tromans and R. Turrall-Clarke, *Contaminated Land* (London, Sweet & Maxwell, 1994) Chapter 2.

[91] *ibid.*, para. 2–36.

[92] See, *e.g. Blue Circle Industries plc v. Ministry of Defence* [1998] 3 All E.R. 385.

Chapter 3

Enforcing Authorities

Summary

The contaminated land regime is enforced by local authorities at unitary and district level, except for "special sites", where the enforcing authority is the Environment Agency or SEPA. The environmental Agencies have powers of direction on a site specific basis in cases where they are not the enforcing authority. A strong level of central control is exerted through the ability of the Secretary of State to issue guidance which is mandatory in effect. Such Guidance goes to some of the most fundamental issues of the system, for example what constitutes "contaminated land", how inspection functions should be carried out, and who at the end of the day is liable for remediation.

Generally, therefore, remediation notices are served by the local enforcing authority, with a right of appeal to the magistrates' court. Where however a "special site" has been designated, the notice is served by the Environment Agency or SEPA, in which case appeal lies to the Secretary of State.

The Enforcing Authority

3.01 The general principle is that the task of identifying contaminated land and serving remediation notices falls to local authorities, defined to mean[1]:

1. In England and Wales:
   - (a) unitary authorities;
   - (b) district councils; and
   - (c) the Common Council of the City of London and the relevant officers of the Inner and Middle Temples.
2. In Scotland, a council for an area constituted under section 2 of the Local Government, etc., (Scotland) Act 1994.

---

[1] Section 78A(9).

3.02 However, a different approach applies to "special sites" in respect of which a special procedure for designation applies[2] and for which the appropriate Agency (the Environment Agency or the Scottish Environment Protection Agency) is the enforcing authority.[3]

Site specific guidance by the appropriate Agency

3.03 It is open to the appropriate Agency to issue guidance to any local authority[4] with respect to the exercise of performance of that authority's powers or duties under the contaminated land provisions in relation to any particular contaminated land; where such guidance is issued it must be considered by the local authority concerned in performing their duties or exercising their powers.[5] However, the guidance should be consistent with any guidance issued by the Secretary of State, and insofar as it is not, it must be disregarded.[6] The appropriate Agency also has power to require the local authority, by written request, to furnish such relevant information as the Agency may require for the purpose of enabling it to issue site-specific guidance.[7] The information covers not only that which the local authority may already have received in the exercise of its functions under Part IIA, but also such information as it may be reasonably expected to obtain under those powers.[8]

3.04 The use of these powers of site-specific guidance could prove controversial in practice, though there may well be cases where a local authority would welcome or indeed invite such guidance. It is presumably guidance of a technical nature that was primarily in the mind of the draughtsman of the section, but it is not specifically limited in that way, and could on its face extend to guidance as to how liability is to be allocated or apportioned. It seems somewhat unlikely in practice that the Agencies would wish to offer guidance of that type. However, the ability of the Agency to give such guidance does in principle present an additional way in which a landowner or other potentially liable party may seek to challenge demands of local authority which it regards as excessive.

3.05 Where an Agency has given guidance which is incorporated in a remediation notice and is then subject to appeal, the Agency will no doubt wish to participate in the appeal process in order to justify its position.[9]

---

[2] See Chapter 6.
[3] Section 78A(9).
[4] This presumably means any local authority in England and Wales for the Environment Agency, and any local authority in Scotland for SEPA: this follows from the definition of "appropriate Agency" in section 78A(9).
[5] Section 78V(1).
[6] Section 78V(2).
[7] Section 78V(3).
[8] Section 78V(4).
[9] Failure by the Agency to participate may present procedural problems that are considered in Chapter 23.

Relationship of central and local regulatory functions: Secretary of State's guidance

3.06 As already stated, with the exception of special sites, it is at the local level that the legislation on contaminated land is enforced. However, local enforcing authorities do not enjoy a measure of discretion comparable with many other local authority regulatory functions. An important, and novel, feature of Part IIA of the 1990 Act is the provision of statutory guidance by the Secretary of State; the local enforcing authority or appropriate Agency then is required to act in accordance with such guidance. In normal circumstances under other codes, Government guidance is one factor to which local authorities must have regard; there is accordingly an inherent discretion in applying such guidance. With the prescriptive guidance under Part IIA the position is different; the only discretion is any which is built into the Guidance itself. The procedure for issuing such Guidance reflects its unusual statutory status, in that the Guidance is not only required to be issued for consultation before issue, but must also be laid before Parliament in draft for 40 days, subject to a negative resolution by either House that it should not be issued.[10] The issues on which such Guidance is relevant are:

1. Whether land is contaminated within the statutory definition.[11]
2. Whether for that purpose harm or the possibility of harm are significant.[12]
3. The inspection function of local authorities.[13]
4. Which of the two of more persons who may be "appropriate persons" to bear the responsibility for remediation is to be treated as not being an "appropriate person", *i.e.* exclusion from liability.[14]
5. Where two or more persons are "appropriate persons" to bear responsibility, in what proportions they are to bear the cost of remediation, *i.e.* apportionment of liability.[15]
6. The inspection and review of special sites and termination of designation as special sites. [16]

3.07 It will be appreciated that these issues go to the fundamentals of the operation and impact of the new system. In addition to this prescriptive guidance, there is the more familiar type of guidance to which enforcing authorities must have regard: such guidance must be subject to consultation, but does not require to be laid before Parliament.[17] It covers the following issues:

---

[10] Section 78YA.
[11] Section 78A(2).
[12] Section 78A(5).
[13] Section 78B(2).
[14] Section 78F(6).
[15] Section 78F(7).
[16] Section 78Q(6).
[17] Sections 78YA(1) and (2). Guidance given to the Agencies under s.78Q(6) also seems to fall into this category, though it is prescriptive in effect.

1. What is to be done by way of remediation, the standard of remediation required, and what is or is not to be regarded as reasonable by way of remediation.[18]
2. Whether to recover all or part of remediation costs incurred by the enforcing authority from the appropriate person.[19]
3. General guidance to the Agency with respect to the exercise or performance of its powers of duties.[20] This provision does not apply to local enforcing authorities.

Progress on draft guidance

3.08 Draft statutory guidance by the Secretary of State was issued for consultation on September 18, 1996, with separate consultation in Scotland. The format adopted in that version took a novel approach in distinguishing between prescriptive and non-prescriptive guidance, by using different typefaces. Numerous comments were received, and in October 1998 a revised working draft was circulated to certain individuals and groups who had commented in detail in 1996. However, it was not until October 11, 1999 that a further full consultation exercise was launched. This dropped the distinctive typographical approach of its predecessor in favour of a more traditional circular format.

The final guidance and its structure

3.09 The final statutory guidance is contained in the DETR Circular, *Contaminated Land*. It is Annex 3 of that Circular which contains the statutory guidance, Annexes 1 and 2 and 4–6 containing explanatory material and a glossary of terms. Annex 3 is divided into five Chapters, each dealing with one aspect of the guidance:

Chapter A – the definition of contaminated land.
Chapter B – identification of contaminated land.
Chapter C – remediation of contaminated land.
Chapter D – liability exclusion and apportionment.
Chapter E – recovery of costs for remediation.

3.10 The regulatory impact assessment (RIA) provided with the Guidance gives some insight into the thinking of the Government as to the content of the Guidance. Annex C of the RIA lists four possible options in this respect:

---

[18] Section 78E(5).
[19] Section 78P(2).
[20] Section 78W(1).

Option A ("maximum requirements") This would involve regular and intensive site investigations, emphasis on contamination guideline values rather than actual risks presented by the site, wide interpretations of what constitutes "significant harm", remediation to a higher standard than current use would require, relatively little weight given to cost and reasonableness issues, and crude guidance on allocation of liability so as to make it easy for authorities to enforce.

Option B ("minimum requirements") At the opposite extreme, this would require inspection only "as and when" required by external events or known problems, narrower interpretation of the key definitional terms, greater weight on cost of remediation rather than effectiveness, and liability rules which placed greater demands on enforcing authorities, both in the range and complexity of factors to be considered, and in the broader range of situations where the enforcing authority would have to bear the costs of remediation.

Option C ("medium requirements") This option seeks a balance between the various interests involved, with guidance on inspection that is flexible, a risk-based approach, focusing on situations where there is real risk of harm, and avoiding excessive costs. It allows for differing local situations and circumstances, and for future changes in knowledge or standards.

Option D ("no guidance") This was not really an option at all, unless a political decision had been taken not to implement Part IIA. "Minimal guidance", giving the barest legal minimum, would have left enforcing authorities and appropriate persons "unsupported" in seeking to understand their duties and potential liabilities.

3.11 Given that Option D was not a realistic option, it perhaps comes as little surprise that the Government, following the approach of its Conservative predecessor, has chosen the best of all possible worlds, the third way of Option C.

Land falling outside a local authority's area

3.12 In general, a local enforcing authority's functions relate to contaminated land within its administrative area. However, contamination is no respecter of local government boundaries, and land situated outside a local authority's boundaries may quite conceivably represent a threat of harm within its area. An example would be the migration of landfill gas across a boundary so as to affect residential property. Another possible example would be a river which flows through the area of a number of authorities, which is affected by pollutants leaching from contaminated land upstream.

3.13 To deal with such problems, section 78X(2) states that where it appears to a local authority that land outside, but adjoining or adjacent to, its area is effectively within the definition of "contaminated land" and the significant harm or pollution of controlled waters is or will be caused within its area, then the authority may exercise its functions under Part IIA as if the land were situated within its area. The provision is stated to be without prejudice to the functions of the local authority in whose area the land is actually situated.

3.14 The relevant land must be "adjoining or adjacent to" the authority's area. If the use of the two terms together were not sufficient to indicate that the land can be adjacent to an authority's area without adjoining it, then case law confirms this. On the meaning of "adjacent" it was said in one case that it "is not a word to which a precise and uniform meaning is attached by ordinary usage. It is not confined to places adjoining, and it includes places close to or near. What degree of proximity would justify the application of the word is entirely a question of circumstances".[21] In construing the section it should be borne in mind that where pollution of surface or ground water is concerned, contaminants can be transported for considerable distances. Therefore if the authority can reasonably form the view that significant harm is being caused in its area, or there is a significant risk of such harm, there is a good argument that there is sufficient proximity to satisfy section 78X(2).

3.15 As stated above, the power of an affected authority to take action under this section is without prejudice to the functions of the local authority in whose area that land is in fact located. That authority may itself be required to take enforcement action because of harm or pollution within its own area. Indeed, there seems to be nothing in the legislation to limit the consideration by the local authority to harm or pollution within its own area. Accordingly there must be the risk of an appropriate person being faced with possibly conflicting remediation requirements by two local authorities. It is suggested that to avoid this situation, any authority considering exercising its powers under section 78X(2) should as a matter of good practice consult the local authority in whose area the relevant land is located.

3.16 As a matter of practice it may be wise for local authorities in setting up their procedures under Part IIA to arrange for the establishment of a cross-boundary joint liaison sub-committee with neighbouring authorities, in anticipation of this eventuality.[22] This may also present the possibility of the sharing of costs of investigation, thereby allowing financial savings.

---

[21] *Wellington v. Lower Hutt* [1904] A.C. 733, in which two New Zealand boroughs were held to be adjacent for the purpose of making statutory contributions to the cost of building a bridge, despite being nowhere closer to each other than six miles apart); see also *Stanward Corpn. v. Denison Mines Ltd* 67 D.L.R. (2d) 743, in which a similar approach was taken but two mining claims one and a quarter miles apart were held not to be adjacent. In Scotland, it has similarly been held that the word does not require actual contact: *Anderson v. Lochgelly Iron & Coal Ltd* (1904) 7F.187, where an accident which took place on a private railway connecting a mine to the public railway some 800 yards from the mine was held (with the Lord Justice-Clerk dissenting) to be "adjacent to the mine in terms of the statutory definition in the Coal Mines Regulation Act 1887; see also *Dunbeath Eastate Ltd v. Henderson* 1989 S.L.T. (Land Ct.) 99.

[22] See also the issue of delegation, explained in the next paragraph.

Delegation of functions and contamination straddling boundaries

3.17 Delegation of an enforcing authority's functions under Part IIA will be important in practice, and may arise in two main contexts:

1. Delegation to officers or committees;
2. Delegation to other authorities.

3.18 With regard to delegation to officers or committees, section 101 of the Local Government Act 1972 gives the power for a local authority to arrange for the discharge of its functions by a committee, sub-committee or officer, but not the chairman of a committee alone. Decisions such as whether to serve a remediation notice may therefore be delegated, as may functions such as inspecting land, serving a notice, or taking enforcement action. In relation to the powers of entry and investigation under section 108 of the Environment Act 1995, there must be specific authorisation in writing to a person who appears suitable, to exercise the relevant powers.[23]

3.19 In relation to administrative matters, such as the preparation and issue of a notice, it may be the case that formal delegation is not strictly necessary on the basis of *Provident Mutual Life Assurance Association v. Derby City Council*[24] and other cases.[25] However, it remains desirable for the avoidance of doubt that standing orders should address this issue.[26] In particular, because the Guidance within which the authority must act is so detailed and prescriptive, a court may well be persuaded that formal prior delegation is necessary.[27]

3.20 Given the legal significance and likely costs involved in identifying land as contaminated it seems most unlikely in practice that the relevant decision making functions will be delegated to an officer. However, arrangements may need to be made for sub-committees to meet to deal with urgent cases, for example, an urgent action sub-committee. The execution of a decision, once made, is a different matter entirely, and local authorities will probably in practice leave the execution of matters such as the detailed preparation and service of remediation notices to their offices and staff.

---

[23] Section 108(1) of the 1995 Act.

[24] (1981) 79 L.G.R. 297, HL.

[25] *Cheshire County Council v. Secretary of State for the Environment* [1988] J.P.L. 30; *R. v. Southwark London Borough Council, ex p. Bannerman* (1989) 22 H.L.R. 459; *Fitzpatrick v. Secretary of State for the Environment* (1990) 154 L.G. Rev.72, CA.

[26] For example, see S.H. Bailey, *Cross on Principles of Local Government Law* (2nd edition, London, Sweet & Maxwell, 1997) p. 263; Andrew Arden, Q.C. *et al.*, *Local Government Constitutional and Administrative Law* (London, Sweet & Maxwell, 1999), p. 229ff.

[27] See *R. v. Edmundsbury Borough Council, ex p. Walton* [1999] Env. L.R. 879.

3.21 The second issue involves delegation of the enforcing authority's functions to some other authority. By section 101 of the Local Government Act, arrangements may be made for another local authority to discharge the functions; effectively, though not necessarily accurately, as an "agency arrangement". This power may be useful, for example where a contaminated site straddles the boundary of two local authorities, and it is desired that one authority should take the lead in dealing with it. The alternative is either to constitute a formal joint committee for that purpose under section 102 of the Local Government Act 1972, or alternatively to set up a joint liaison committee, which has no power, usually, to pass binding resolutions. One authority may supply professional or technical services, or the use of plant, to another under the Local Authorities (Goods and Services) Act 1970.[28]

Use of external advisers

3.22 The technically difficult nature of some types of problems arising from contaminated land may mean that some local authorities, who have not developed internal expertise in this field, may need to rely on environmental consultants in order to fulfil their statutory duties under Part IIA.

3.23 Whilst there is no objection to a local authority seeking advice from external sources in order to enable it to make the necessary decisions, there are dangers in relying too heavily on such advice. This is particularly so where the functions involve making decisions of a quasi-judicial nature as to which party should be liable,[29] and to what extent, for the clean-up of contamination.[30] The incidental powers of section 111 of the Local Government Act 1972 will allow an authority to employ consultants as agents or contractors to provide advice or, indeed, to exercise functions on the authority's behalf.[31] The authority will not be allowed to delegate the ultimate decision-making power on whether land is contaminated, whether a notice should be served and on whom, to external consultants: the best course will be to use the consultant to obtain and present factual data, scientific conclusions, judgments and predictions, and make recommendations to the appropriate committee of the enforcing authority to allow the committee to reach its own decision.[32]

---

[28] See *R. v. Yorkshire Purchasing Organisation, ex p. British Educational Supplies Association* (1997) 95 L.G.R. 727, 732.

[29] Often the decision will be a mixed one of fact, scientific judgment, conclusions, prediction, and law.

[30] See *R. v. Chester City Council, ex p. Quietlynn Ltd* (1984) 83 L.G.R. 308, CA.

[31] *Credit Suisse v. Allerdale Borough Council* [1997] Q.B. 306 at 346 and 359. The authority will need to take account of its "best value" obligations under Part I of the Local Government Act 1999 in procuring such services: see DETR Circular 10/99.

[32] See S.H. Bailey, *Cross on Principles of Local Government Law* (London, Sweet & Maxwell, 1997) p. 262 and cases cited there.

3.24 One practical question is the extent to which consultants should conduct actual negotiations with the owner and other potentially liable persons: the authority will need to make the scope of the consultant's brief clear in this regard. There should, however, certainly be *consultation* between the local authority's consultants and the landowner[33]; exchange of reports and other material may be more problematic and should be a topic addressed in the contract appointing the consultant.

3.25 The authority may also turn to the Agency for advice. Section 37(3) of the Environment Act 1995 allows the Agency to provide advice or assistance as respects any matter in which it has skill or experience. Where a "special site" is involved, the Agency has of course a more formal role. The Agency has power to delegate its own functions to members, officers, employees, committees or sub-committees under paragraph 6 of Schedule 1 to the 1995 Act.

---

[33] See Chapter 9 on the issue of consultation.

Chapter 4

Inspection of Land

Summary

Section 78B places local authorities under a duty to cause their area to be inspected from time to time for the purpose of identifying contaminated land and deciding whether, if it is contaminated, it is a "special site". This function is to be performed in accordance with guidance issued by the Secretary of State.

Essentially, local authorities will need to set up a committee structure with delegation of functions decided at an initial stage and with a written strategy formulated in accordance with guidance.

They will need to consider how best to fulfil their inspection duties in the light of the Guidance and their strategy. Practical issues on the management of the relevant information will arise, as will the potential for liability of authorities if they are perceived to be failing in their duties or otherwise "getting it wrong".

Local authority inspection: the duty

4.01 Under section 78B(1) every local authority shall cause its area to be inspected from time to time for the purpose:

(a) of identifying contaminated land; and
(b) of enabling the authority to decide whether any such land is land which is required to be designated as a special site.

4.02 In performing its duty under this subsection by section 78B(2) it is mandatory that the authority acts in accordance with guidance issued for the purpose by the Secretary of State. The relevant Guidance is to be found at Chapter B of Annex 3 to the DETR Circular, *Contaminated Land*. Government policy is that the approach should be a flexible one, which is proportionate to the seriousness of any actual or potential threat. The difficulty is of course that the object of inspection is precisely that of establishing the significance of such harm or potential harm: thus inspection should be seen as an iterative process during which such significance will become clearer and which will constantly require re-appraisal at each stage of the process.

Guidance on the strategic approach

4.03 The essence of the approach required by the Guidance is that local authorities should take a strategic approach to inspection: this approach is to be set out in a written strategy, to be published within 15 months of the issue of the Guidance (*i.e.* by June 2001).[1] This approach is intended to enable authorities to identify, in a rational, ordered and efficient manner, the land which merits detailed inspection, identifying the most pressing and serious problems first and concentrating resources on the area where contaminated land is most likely to be found.[2]

4.04 The Guidance states that in carrying out its inspection duty, the local authority should take a strategic approach which should[3]:

(a) be rational, ordered and efficient;
(b) be proportionate to the seriousness of actual or potential risk;
(c) seek to ensure that the most pressing and serious problems are located first;
(d) ensure that resources are concentrated on investigating in areas where the authority is most likely to identify contaminated land; and
(e) ensure that the local authority efficiently identifies requirements for the detailed inspection of particular areas of land.

4.05 These matters are hardly controversial, and at one level may be said to represent little more than administrative "good sense". However, they are useful as a checklist, and local authorities must expect to have their approach scrutinised against those criteria. In that sense, the strategy is a fundamental issue, from which much else flows.

4.06 The Guidance also stresses the necessity to reflect local circumstances (para. 8), in particular any available evidence that significant harm or significant pollution of controlled waters is actually being caused; the extent to which any relevant receptor is likely to be found in different parts of the area and is likely to be exposed to a particular pollutant in view of geological or hydrogeological factors; the extent to which information is already available; past contaminative uses in the area and their nature and scale; the nature and timing of any past redevelopment; previous or likely future remedial action; and the extent to which other regulatory authorities are likely to be considering such harm).[4]

---

[1] DETR Circular, *Contaminated Land*, Annex 3, para. B.12.
[2] *ibid.*, Annex 2, para. 3.3.
[3] *ibid.*, Annex 3, para. B.9.
[4] *ibid.*, para. B.10.

4.07 Again, this may be thought to be at least to some extent stating the obvious, but there is no blueprint for a strategy that will fit every local authority. The approach of an authority in a traditional area of heavy industry which has been subject to extensive residential development is likely to be different to that of an authority whose area is a sparsely populated rural one. That is not to say that the latter case will not have examples of serious contamination problems, for example from old landfills, military bases or pesticide stores—simply that the approach will be different. The geology and groundwater vulnerability of the area will also be relevant and in this respect consultation with the Environment Agency or SEPA will be vital. The authority must also consult other appropriate public bodies, such as the county council (if any), statutory regeneration bodies, English Nature, English Heritage, and MAFF,[5] and of course their Scottish or Welsh equivalents.

Written strategy

4.08 The Guidance requires the formal adoption and publication of a written strategy setting out an ordered approach to the identification of contaminated land.[6] This is to be published within 15 months of the issue of the Guidance, and is to kept under periodic review. While strategies are likely to vary between authorities and between different parts of their areas, the strategy should include include a lengthy list of issues[7]:

(a) a description of the particular characteristics of the area and how that influences its approach;
(b) the authority's particular aims, objectives, and priorities;
(c) appropriate timescales for the inspection of different parts of its area (no doubt related to those priorities);
(d) arrangements and procedures for:
  (i) considering land for which the authority may itself have responsibilities by virtue of its current or former ownership or occupation;
  (ii) obtaining and evaluating information on actual harm or pollution controlled waters;
  (iii) identifying receptors and assessing the risk to them;
  (iv) obtaining and evaluating existing information on the possible presence of contaminants and their effects;
  (v) liaison with, and response to information from, other statutory bodies, including the appropriate Agency, the nature conservation authorities and the Ministry of Agriculture, Fisheries and Food;

---

[5] *ibid.*, para. B.11.
[6] DETR Circular, *Contaminated Land*, Annex 3, para. B.12.
[7] *ibid.*, para. B.15.

(vi) liaison with and response to information from the owners or occupiers of land and other relevant interested parties;
(vii) responding to information or complaints from members of the public, businesses and voluntary organisations;
(viii) planning and reviewing a programme for inspecting particular areas of land;
(ix) carrying out the detailed inspection of particular areas of land;
(x) reviewing and updating assumptions and information previously used to assess the need for detailed inspection of different areas, and managing new information; and
(xi) managing information obtained and held in the course of carrying out its inspection duties.[8]

4.09 Two further issues will be relevant in many cases. First, any strategy will need to take into account the relevant development plan policies of the authority. Many development plans will now contain a chapter on pollution and contamination, and in some cases specific reference to Part IIA. While Part IIA is not directly concerned with the clean-up of land for regeneration purposes, authorities will no doubt wish to consider how the strategy relates to the wider agenda for urban renewal. Secondly, the written strategy has cost implications with budgetary consequences; accordingly it may well have to be debated in full council.

Information from other statutory bodies

4.10 Other regulatory authorities may be able to provide information relevant to the identification of contaminated land. In particular, these may include the Environment Agency in its water protection functions. Local authorities should consider making specific arrangements with such bodies to avoid duplication of effort.[9] The role of the local authority in this respect is to receive and consider the information: the authority cannot abdicate its ultimate responsibility for determining whether land is contaminated.

Information from the public, businesses and voluntary organisations

4.11 Earlier drafts of the Guidance contained non-prescriptive material on dealing with information received from the public on possible contamination. While, as indicated above, the local authority's strategy should deal with that issue, the Guidance is silent on it.

---

[8] See Chapter 25.
[9] DETR Circular, *Contaminated Land*, Annex 3, para. B.16.

4.12 It takes little imagination to appreciate that this aspect of investigation may potentially give rise to acute controversy. National or local non-governmental organisations may well wish to see local authorities adopting a vigorous approach to inspection, either generally or on a site-specific basis. Such bodies might well see complaints or the provision of information as a means to attaining that end.[10] Less attractively, complaints might be made or information provided maliciously, or for ulterior commercial motives, such as impeding a rival development scheme. Whilst there was nothing to prevent such information being provided under previous law, the fact that the local authority is now under a duty to cause its area to be inspected means that some response will need to be made. The quality of information provided may also vary hugely, ranging from detailed material through to vague allegations that "some drums were buried somewhere in this area sometime in the past."

4.13 The early drafts of the Guidance stated that the appropriate response to such information would vary according to the nature of the information, its detail, and the credence which could be given to it. It was suggested that the authority "may wish to discourage if possible any anonymous representations, particularly where these are unfounded or potentially malicious." It was not stated, however, how this should be done, and in particular it is difficult to see how an authority can establish whether an allegation is unfounded without at least investigating it to some extent. Perhaps these difficulties led to the removal of these passages from later versions of the Guidance. It is certainly not the case that every complaint will necessarily have to result in intrusive investigation. It may be possible to rule out the risk of land being contaminated by considering whether there are any vulnerable receptors, or by reviewing the alleged adverse effects.

4.14 Earlier versions of the draft Guidance also acknowledged the possibility that a prospective purchaser of land, where there was some evidence of contamination, might ask the authority to inspect it. It is clearly not the objective of Part IIA or the Guidance to shift the cost of pre-acquisition site investigations from the private to the public purse: accordingly, the authority's approach should be determined by its strategy, not by such requests.

The process of inspection

4.15 It is important to note that the authority is not exempt from its inspection duty while its strategy is in the course of preparation. The authority should not await publication of its strategy before commencing

---

[10] Such information might however have to be disclosed under the Environmental Information Regulations 1992—see *R. v. British Coast Corporation, ex p. Ibstock Building Products Ltd* [1995] Env. L.R. 277.

detailed inspection work on particular areas of land, where this appears necessary.[11]

4.16 What is required by the Act is an inspection of the local authority's area rather than simply an inspection of registers, documents, plans or other purely "desktop" matters; the inspection should be a physical one. However, as mentioned above, such desktop matters will be relevant in the context of the local authority's strategy in informing and prioritising the process of inspection. The normal process of site investigation will involve a desktop study as the first stage. Secondly, it is clear that an inspection can be carried out other than by the authority itself, provided they cause it to be carried out. In this connection, the reputation and level of insurance cover of any consultants appointed to that task is obviously a material consideration.

4.17 The Guidance deals with the process of physical inspection at paragraphs B.18–B.25. The starting point is particular areas of land where a pollution linkage may exist. In other words, there may be present a pollutant which would be capable of affecting a recognised relevant receptor through an environmental pathway.[12] The authority should carry out detailed inspections of areas where it is possible that a pollutant linkage exists to obtain sufficient information to allow it to determine whether the land appears to be contaminated land and, if so, whether it is a special site.[13] The information which will be required to make determination includes evidence of the actual presence of the pollutant[14]: intrusive investigation is the most reliable way through which such evidence can be obtained. However, it is not the only way, and will rarely be admissible as the first step. As the Guidance states, detailed inspection may include all or any of the following[15]:

(a) the collation and assessment of documentary information or other information from other bodies;
(b) a visit to the site for visual inspection and limited sampling (for example, surface deposits); or
(c) intrusive investigation.

4.18 The Guidance also considers the use of statutory powers of entry[16] and stresses that any intrusive investigations should be carried out in accordance with appropriate technical procedures and ensuring that all reasonable precautions are taken to avoid harm, water pollution or damage to natural resources that might be caused.[17] Authorities will almost certainly wish to take specialist technical advice on these issues.

---

[11] DETR Circular, *Contaminated Land*, Annex 3, para. B.14.
[12] See Chapter 2 generally.
[13] DETR Circular, *Contaminated Land*, Annex 3, para. B.18.
[14] *ibid.*, para. B.19.
[15] *ibid.*, para. B.20.
[16] See below.
[17] DETR Circular, *Contaminated Land*, Annex 3, para. B.24.

4.19 It is of course possible that land may previously have been subject to investigation so that reports as to its condition already exist. If such information provides an appropriate basis for the authority to make its determination as to whether the land is contaminated, then it should not press for further intrusive investigation—the same is true where an owner or occupier offers to provide such information themselves within a reasonable timescale.[18] In those circumstances, the duty of the authority becomes rather that of checking, or getting checked, the information or report provided by the landowner, before deciding if further action is required.

4.20 Having once embarked on contaminated land investigation, it can be difficult to know when to stop. The aim of intrusive investigation is not to produce a complete characterisation of the nature and extent of pollutants, pathways and receptors, going beyond what is necessary to make the statutory determination. The question is whether the authority has enough information to come to a reasoned conclusion. If, at any stage, on the basis of information obtained, the authority considers there is no longer a reasonable possibility that a particular pollutant linkage exists, it should no longer carry out any further detailed inspection for that linkage.[19]

Special sites

4.21 Land which is designated as a "special site" becomes the responsibility of the Environment Agency as enforcing authority. The Guidance makes it clear that the Agency should have a formal role at the inspection stage for such land. Where it is clear from the information available before carrying out any detailed inspection that the land would be a special site, then the authority should in all cases seek to make arrangements with the Agency to carry out inspection on its behalf.[20] This may be the case, for example, where the land is known to have been put to a use which would bring it within the definition of a special site. The Guidance also requires the authority to seek to make arrangements for inspection by the Agency where there is a reasonable possibility that a particular type of pollutant linkage may be present, which would make the land a special site.[21] An example would be particular types of pollution of controlled waters. These circumstances could quite possibly only become apparent once the authority has already embarked on the investigation process. The guidance does not address this situation specifically, but it is submitted that the authority should involve the Agency as soon as possible, and should consider whether in the circumstances it is practicable for the Agency to take over the investigation process.

---

[18] *ibid.*, para. B.23.
[19] *ibid.*, para. B.25.
[20] DETR Circular, *Contaminated Land*, Annex 3, para. B.28.
[21] *ibid.*, para. B.29.

Powers of entry and investigation

4.22 The powers of local authorities to obtain access to land for the purpose of investigation are found in section 108 of the 1995 Act: a local authority acting in its capacity as enforcing authority for the purpose of Pt IIA is a "local enforcing authority" within that section.[22] Accordingly, the local authority may authorise in writing any person who appears suitable to exercise the powers conferred by section 108. That person need not be an employee or officer of the authority, so in the context of contaminated land investigations there is no reason why an authority should not authorise employees of a consultancy engaged by the authority. Similarly, in the case where the land in question is likely to be designated as a special site, the authorised person may be an officer of the appropriate Agency. The relevant powers include[23]:

(a) entry of premises;
(b) entry with other authorised persons and with equipment or materials;
(c) examination and investigation;
(d) direction that premises be left undisturbed;
(e) taking measurements, photographs and recordings;
(f) taking samples of air, water and land;
(g) subjecting articles or substances suspected of being polluting to tests;
(h) taking possession of and detaining such articles;
(i) requiring persons to answer questions;
(j) requiring production of records or the furnishing of extracts from computerised records;
(k) requiring necessary facilities or assistance to be afforded; and
(l) any other power conferred by Regulations.

4.23 These powers are broad, and in each case the detailed conditions as to their exercise will need to be considered.[24] An authority might, for example, require consultants who have been involved in the previous investigation of a site to answer questions. Failure, without reasonable excuse, to comply with requirements imposed under section 108, or obstructing the use of the powers, can be an offence.[25] The powers may not be used to compel the production of documents covered by legal professional privilege.[26]

---

[22] Section 108(15).
[23] Section 108(4).
[24] See the commentary to s.108 in *The Environment Acts 1990–95: Text and Commentary* (London, Sweet & Maxwell, 1996) for further detail.
[25] Section 110(1).
[26] Section 108(13).

4.24 The investigation of land by physical sampling may well require the use of heavy equipment. In such cases at least seven days' notice is required, and the entry can only take place with the consent of the occupier, or under the authority of a warrant.[27] These requirements apply in all cases to residential property, whether heavy equipment is involved or not.[28] Subsections 108(5)(a) and (b) provides express power to carry out "experimental borings" or other works and to install, keep or maintain monitoring or other apparatus on land. Both of these powers may be necessary in contaminated land investigations. However, the subsection states that they are exercisable for determining whether "pollution control enactments" have been complied with, whereas inspection by a local authority under Pt IIA is not to determine compliance with that legislation. The local authority will therefore need to rely on its more general powers under subsections 108(4)(b) and (f) as to the taking onto land of equipment and the taking of samples.

4.25 Before using these statutory powers, the authority will also need to consider carefully the statutory guidance on the subject, which may restrict the circumstances in which those powers should be exercised. Before using its powers of entry, the authority should be satisfied that there is a reasonable possibility that a pollutant linkage site exists.[29] For intrusive inspection the test is more onerous, in that the authority must be satisfied on the basis of information already obtained that it is likely that the contaminant is actually present and that the receptor is actually present or likely to be present.[30] In other words, the authority cannot use its investigative powers for a pure "fishing exercise". Nor, as explained above, can it insist on using those powers where adequate information has already been provided, or will be within a reasonable timescale.

Securing information

4.26 One of the most important practical problems in relation to the new legislation relates to the cost of investigation. Sampling and analysis costs in relation to a large site can easily exceed £50,000. Once a remediation notice is served, the recipient may be required to carry out works to assess the condition of the land in question, controlled waters or adjoining or adjacent land.[31] However, the authority will itself have to bear the cost of such investigations as are necessary to reach the view that the land is contaminated land: mere suspicion as to the presence of contaminants or the existence of harm will not be sufficient for this purpose.[32] It seems

---

[27] Section 108(6).
[28] Section 108(6).
[29] DETR Circular, *Contaminated Land*, Annex 3, para. B.22.
[30] *ibid.*
[31] This follows from the definition of "remediation" at s.78A(7), which includes these matters.
[32] See the statement by Ministers in *Hansard,* H.L. Vol. 562, col. 175; H.C. Standing Committee B, 11th sitting, col. 341. See also para. 4.25 above.

inherently unlikely at first sight that a landowner will wish to volunteer to bear the cost of investigations which may ultimately result in the imposition of statutory clean-up liabilities, unless there is some valid commercial incentive for so doing, such as the need to obtain planning permission for the development of land, or to achieve a sale or other disposal of the site. However, larger companies may well have in place environmental policies which mandate a proactive and co-operative approach to potential contamination; there may therefore be cases where the authority is able to secure voluntary co-operation.

4.27 In the absence of such co-operation, a local authority may be attracted to the idea of using its statutory powers to obtain existing reports or data which demonstrate the land to be contaminated, as a cheaper course than carrying out its own physical investigations. There may well be a number of reports in existence which have been commissioned by companies for internal management purposes or in the context of transactions or development proposals: these reports may indicate that the land is contaminated.

4.28 Assuming that the authority has at least some basis to believe the land may be contaminated the question then might be whether the authority could compel disclosure of such data or findings as "records" under section 108(4)(k) of the 1995 Act, or even ask the persons involved in such investigations (whether as company employees or officers or as external consultants) to answer questions under section 108(4)(j). Such a course might be regarded as somewhat heavy-handed; it would also raise the question of whether such reports could be regarded as "records" falling within the statutory power, and of whether the relevant documents are protected from disclosure by legal privilege. On the first issue there is authority to suggest that the term "records" refers only to primary or original sources of information, and that reports which summarise or opine on such material are not "records".[33] On that basis, consultant's reports might not be available, but the underlying data might be. Documents covered by legal privilege cannot be compelled to be produced under section 108.[34] With the heightened awareness of environmental liability risks in recent years, many reports have been produced with the involvement of lawyers with a view to seeking privilege. Whether this will be successful in actually securing privilege is another matter; this will depend on whether the information was produced and has since remained on a confidential basis, and whether its purpose was to provide legal advice. Mere involvement of lawyers in the process is unlikely to be enough of itself.

---

[33] See *R. v. Tirado* (1974) 59 Cr.App.R. 50; *R. v. Gwilliam* [1968] 1 W.L.R. 1839; *R. v. Schering Chemicals Ltd* [1983] 1 W.L.R. 143; *Savings and Investment Bank v. Gasco Investments (Netherlands) B.V.* [1984] 1. W.L.R. 271.

[34] Section 108(13).

Handling information

4.29 One of the components of the authority's written strategy must include procedures for the management of information obtained and held in the course of carrying out its inspection duties.[35] This is a somewhat different issue from the provision of information on the statutory registers required by section 78R. Those registers relate to the various notices and other procedural steps consequent upon determination that land is contaminated. However, the authority will also amass a potentially huge volume of information prior to, and in the process of making, that determination. Much of that information will be voluminous, indigestible and in a form which may be largely unintelligible to the general public, for example readings of borehole logs, and toxicological data. However, it may also have significant commercial implications in affecting the value of land, even if the ultimate determination is that the land is not "contaminated" in the statutory sense. Moreover, it may be useful information for a plaintiff in any civil proceedings involving the land in question. As part of the local search process on property transactions, local authorities may be asked to provide information about land which has not yet been identified as contaminated land; for example, whether the authority has at any stage investigated that land and, if so, the details of the outcome of those investigations.

4.30 Care will be needed in the practical processes necessary for collating and managing this information, whether in electronic or paper form. One legal issue is whether such information is within the Environmental Information Regulations 1992[36] and as such must be provided to the public on request. It is certainly information relating to the environment (concerning as it does the state of water or soil[37] and is held by a public body.[38] In particular, the authority will have to take a view on whether any of the information falls within the categories of confidentiality in regulation 4. It could be argued that it is confidential as relating to land which may be the subject of legal proceedings,[39] *i.e.* service of a remediation notice and subsequent appeal or enforcement. If so, then under the 1992 Regulations it may be treated as confidential but need not necessarily be. The authority will thus have to decide whether the information should be disclosed, bearing in mind the public interest in the state of land being known.

---

35 DETR Circular, *Contaminated Land*, Annex 3, para. B.15(d)(xi).

36 S.I. 1992 No. 3240.

37 1992 Regulations, reg. 2(1)(a) and (2)(a). A broad approach to interpretation of this issue was taken in *R. v. British Coal Corporation, ex p. Ibstock Building Products Ltd* [1995] Env. L.R. 277.

38 1992 Regulations, reg. 2(1)(b) and (3).

39 1992 Regulations, reg. 4(2)(b). But compare *R. v. British Coal Corporation (supra)* where the possibility of an appeal being made was held not to constitute prospective legal proceedings.

4.31 These issues were considered in the case of *Maile v. Wigan Metropolitan Borough Council*[40] where the authority had commissioned a baseline study of potentially contaminated sites in its area from Nottingham Trent University. The exercise had been put "on hold" and no report had been released. The applicant sought access to a database compiled for the study, using the 1992 Regulations as the basis of his request. The authority argued that the information fell within regulation 4(c) (information relating to the confidential deliberations of a public body) or 4(d) (information contained in a document or record still in the course of completion). It was accepted by the court that the information fell within these exempting provisions. The data was still incomplete and possibly inaccurate: disclosure of what was still a speculative database could cause unnecessary alarm to local citizens and landowners. Presumably however, in any such exercise there will come a point where information is complete and refusal to disclose it can no longer be justified on that basis.

4.32 Consideration must also be given to the provisions of Part VA of the Local Government Act 1972 on access to reports and background papers for meetings of committees and sub-committees of the authority. In *Maile v. Wigan B.C.* (above) an attempt to secure disclosure of the baseline study under these provisions failed. The database of potentially contaminated sites had not been used for preparing the report which had gone to the authority and, accordingly, was not "background papers" falling within section 100D. It should also be noted that the categories of "Exempt Information" under Schedule 12A to the Act include:

> "Information which, if disclosed to the public, would reveal that the authority proposes . . . to give under any enactment a notice . . . by virtue of which requirements are imposed on a person."

It is certainly arguable that information which shows that land is in such a condition that a remediation notice will be served falls within that category; however, at the stage of identifying land as contaminated, it may not be clear that a remediation notice will be served. Again, the authority will have to take a view on this in each case, in the light of individual circumstances.

4.33 Another issue is that of information provided voluntarily by a present or former owner or occupier—if the person providing the information was not under and would not have been put under a legal obligation to supply it, and has not consented to its disclosure, then it must be treated as confidential.[41] The key question may be whether the supplier of the information could have been made to provide it, which is not necessarily a clear-cut issue.[42] Also, there may be the suggestion that there is an express

---

[40] High Court, May 27, 1999; noted at ENDS Report No. 294, p55 and *Environmental Law Bulletin* 58, p. 21.

[41] 1992 Regulations, reg. 4.

[42] See para. 4.27.

or implied contractual duty of confidence; such duty may also be claimed under estoppel or in equity.[43] The prudent course would be for the authority to seek express consent to disclose it if requested.

4.34 The Government views and proposals on freedom of information generally will also need to be considered in this respect,[44] as will the brief non-prescriptive Guidance on the issue[45] As mentioned above, the key is that the authority's policies on information should be part of its written strategy, and as such published and consulted on in draft form.

Resource implications

4.35 There can be few local authorities who will not be concerned about the potentially major costs of instigating and maintaining a programme of inspection of its area for contaminated land. Although the language used echoes the long-standing provisions on statutory nuisance by referring to inspection "from time to time", a very different approach is likely to be necessary for contaminated land. Whereas many types of statutory nuisance will be all too self-evident, the risks and problems of contaminated land may only be yielded up on detailed examination. The words of no less a commentator than Lord Irvine of Lairg seem apposite here:

> "Public authorities in the 1990s take an ever-growing number of potentially reviewable decisions, often under considerable pressure, and in an environment where the time and resources to reach and reflect on them is reducing."[46]

4.36 There is no requirement for review of land at specified intervals; during the passage of the 1995 Act, the Government rejected an amendment proposing a requirement for quinquennial review, stating that local authorities should concentrate resources on areas where there were likely to be problems.[47]

4.37 It must be kept firmly in mind that the function of inspection is a duty, and as such a local authority which treats it as a discretion, to be dispensed with on resource, economic or other grounds, will be vulnerable to judicial review.[48] On the other hand, it is a duty which is couched in broad terms,

---

[43] See *R. v. Secretary of State for the Environment, Transport and the Regions, ex p. Alliance Against the Birmingham Northern Relief Road* [1999] Env. L.R. 447.

[44] See Cm. 3818 *Your Right to Know* 1997.

[45] DETR Circular, *Contaminated Land*, Annex 2, paras 3.36–3.40.

[46] *Judges and Decision-Makers: the Theory and Practice of Wednesbury Review* [1996] P.L. 59. See also the reference of Lord Browne-Wilkinson at *R. v. East Sussex County Council, ex p. Tandy* [1998] 2 W.L.R. 884, p. 886 to the "unenviable position" of local authorities.

[47] See *Hansard*, H.L. Vol. 562, col. 229.

[48] *R. v. Carrick District Council, ex p. Shelley* [1996] Env. L.R. 273.

so that apart from extreme cases, it may be difficult to categorise a particular course of conduct as being in breach of the duty. The reality is that local authorities must act within a framework of limited resources and competing calls on those resources, and even a matter which is stated in terms of a duty may not necessarily be treated as absolute in its effect where this would lead to a result which it would be difficult to attribute to Parliament's intention.[49] The key to this dilemma may lie in the adoption by the authority of an acceptable strategy which accords with the Guidance. That Guidance itself recognises that resources need to be prioritised.

4.38 On the other hand, where the legislation bases duties on objective criteria, the courts will often be reluctant to dilute the duty to a power by allowing the local authority to plead scarcity of resources: this would be to take into account irrelevant considerations, and the only proper course for the authority in such circumstances may be to divert resources from discretionary spending.[50]

4.39 This is where the Government's statutory guidance on inspection is important: provided a local authority can show that it has formulated a reasoned strategy on inspection which is consistent with that Guidance, the risk of a successful challenge may be relatively low. Different considerations may arise where the allegation is that the authority has failed to take action with respect to a specific site, and these are considered below. In any event, it is clear that more than a simply "paper" approach is needed; the authority must be able to show that it has put its "paper" strategy into effect and has kept it monitored, with regular reports to committee.

Enforcing local authority duties

4.40 The inspection, identification and ultimate remediation of contaminated land is an iterative process, beginning with general inspection of the area, and progressing through investigation of specific sites to the service of a remediation notice and, if necessary, the taking of enforcement action. Decisions or inaction of a local authority may be susceptible to challenge at each of these stages, but it is probably at the later ones, with specific sites in mind, that the duties will bite hardest.

4.41 As is demonstrated by *R. v. Carrick District Council, ex p. Shelley*,[51] once it is apparent that land may well be contaminated within the meaning of the Act, either on the basis of complaints by third parties or on the local authority's own investigations, judicial review for failure to serve a remediation notice may become a real possibility. The normal rules on standing, delay, and the exercise of judicial discretion will of course need to be carefully borne in mind by those considering such action.

---

[49] *R. v. Gloucestershire County Council, ex p. Barry* [1997] A.C. 584; *R. v. Norfolk County Council, ex p. Thorpe* [1998] C.O.D. 208.

[50] *R. v. East Sussex County Council, ex p. Tandy* [1998] 2 W.L.R. 884, 891–2.

[51] [1996] Env. L.R. 273.

4.42 There are also other courses which may be contemplated. Unlike statutory nuisance, the Part IIA provisions do not provide a direct complaint procedure for persons aggrieved by the existence of contaminated land. However, the possibility exists of making representation to the appropriate Agency with a view to site-specific guidance being given to the local authority.[52] Another possibility is of course a complaint to the Local Government Ombudsman, or to apply indirect pressure to the local authority through the DETR or local MPs.

Civil liability in relation to inspection functions

4.43 The existence of a statutory duty to cause land in its area to be inspected inevitably raises the issue of potential liability for failure to inspect or for alleged faulty inspection. Would the authority, for example, be liable to an adjoining landowner, or to an owner downstream of a plume of pollution, or to a subsequent owner of the contaminated land itself, for failure to identify land that should have been identified as contaminated? The statute makes no express provision as to liability in the event of breach of the statutory duty to cause the authority's area to be inspected from time to time. Whilst the duty of inspection is mandatory, rather than discretionary, it is not an absolute duty. The words "from time to time" imply a degree of discretion as to the timing and prioritisation of inspection, which clearly accords with the terms of the Guidance. Provided therefore that an authority adheres to that guidance in formulating its inspection strategy, it may, as explained in the previous section, be difficult to demonstrate any breach of duty which gives rise to liability.

4.44 Three main situations can be foreseen. The first is where a local authority has failed to inspect a particular site, and accordingly has not identified it as contaminated: essentially, what is complained of is an omission on the part of the authority. The second such situation is where an authority has inspected land (or has caused it to be inspected by a third party) but the inspection has allegedly been carried out negligently. The third situation is where the land has been identified as contaminated, but it is alleged that the authority has failed to take proper and effective enforcement action to remediate the relevant risks.

Failure to inspect

4.45 In the first type of case, that of failure to inspect or identify land as contaminated, the analysis of the majority of the House of Lords in *Stovin v. Wise and Norfolk County Council*[53] is relevant. The argument that the

---

[52] Section 78V.
[53] [1996] A.C. 923.

authority has a duty to act in any given situation depends on the public nature of its powers, duties and funding.[54] As mentioned above, the duty to inspect from time to time under section 78B does not appear to be absolute in nature. In order to challenge successfully a decision by a local authority not to inspect a particular site at a specific time, it would be necessary to show, on *Wednesbury* principles, that no authority, properly directed as to the facts and the law, and having before it all the information that this particular authority had, could possibly come to the decision they did. An example might be an authority which had failed to develop a strategy at all, or had failed to act in accordance with their own strategy without adequate reason, or had misconstrued the mandatory guidance or its legal duties.

4.46 Even in such a case, it would not follow that breach of the public law duty should necessarily give rise to a cause of action for damages at common law, either for breach of statutory duty or for negligence.[55] This will involve a question of statutory construction, bearing in mind the increased burden on public funds involved in giving a cause of action (which in the case of the contaminated land provisions could obviously be significant). As *Stovin v. Wise* itself graphically demonstrates, it is possible for distinguished judges to have different views on this issue,[56] and doubtless any decision on section 78B would be coloured to a degree by the specific factual background to the case before the court.

4.47 The concept of general reliance, as discussed by Lord Hoffmann in *Stovin v. Wise*, may help to provide some guidance, *i.e.* whether there could be said to be a general expectation as to the exercise of the duty by the authority, or a general reliance or dependence on its exercise.[57] If the strategy is well-drafted, and consulted on and published in draft, the "general expectation" in this respect cannot or should not exceed the terms of the strategy. The examples of general expectation given in *Stovin v. Wise* include the control of air traffic, the safety inspection of aircraft and the fighting of fire. All of these issues would seem to have a greater degree of predictability and uniformity involved than the inspection function under section 78B. On the analysis of Lord Hoffmann,[58] it would also be relevant to consider the extent to which members of the public might reasonably be expected to protect themselves against the relevant risks by other means—which in this context might involve commissioning their own site investigation before acquiring land,[59] or by taking action themselves under civil law against the owners of contaminated land which affects their own interests.

---

[54] *ibid.*, pages 935–936, 946.
[55] *ibid.*, page 952.
[56] The decision was a 3/2 majority, Lord Slynn and Lord Nicholls of Birkenhead dissenting.
[57] *ibid.*, page 953 *et seq.*
[58] *ibid.*, pages 954–955.
[59] They will of course have no statutory powers of access to do so.

Defective inspection

4.48 Somewhat different, but related, arguments would arise in a case where the authority had inspected land but had negligently failed to identify it as contaminated. The question here, applying the principles of *Stovin v. Wise and Norfolk County Council*[60] and *Barrett v. Enfield London Borough Council*,[61] is whether the authority owes a common law duty to care to those who may be affected by failure to observe the appropriate standard of care, or whether any such duty may be enlarged by the statutory context to create a relationship of "proximity" which would not otherwise exist. Inspection of land for contamination is a significantly more complex exercise than, say, inspecting the adequacy of foundations. There are many ways in which inspection for contamination may be carried out, and there is no guarantee that any investigation will provide total certainty. There is therefore a degree of discretion as to how the inspection duty is carried out, and a court applying the analysis of Lord Hoffmann in *Stovin v. Wise* might hesitate before imposing liability where a local authority for policy reasons had decided to carry out its duty in a certain way. Indeed, if the authority could be seen to have acted consistently with the relevant Guidance and its own strategy, it may be doubtful that there is negligence at all.

4.49 The issue of whether a duty of care exists will depend on the well-established but unpredictable principles of foreseeability of harm, analogy with existing common law duties, policy, fairness and justice.[62] The existence and ambit of any duty of care is likely to be influenced by the statutory framework within which the authority is operating. The distinction between policy and operational matters, enunciated in *Anns v. Merton London Borough Council*,[63] is unlikely to be conclusive in this respect. While the inspection of land might be said to be an operational matter, as mentioned above, there is inevitably a discretion as to how the inspection is performed. A dimension which may be present in this type of case is that of specific reliance by an individual on the outcome of an inspection, as opposed to the general reliance of all members of the public referred to above. However, even specific reliance will not necessarily imply a duty of care if policy considerations dictate otherwise.

4.50 The allegation may of course be not that the authority has been remiss in detecting contamination, but rather that it has been over-zealous in its enforcement functions to the detriment of the owner or occupier, perhaps resulting in disruption of business or the devaluation of the property. Again, there are serious obstacles in the way of a successful action in negligence in such cases. In particular, given that the purpose of the statutory functions under Part IIA is to protect the public's safety and the

---

[60] [1996] A.C. 923.
[61] [1999] 1 L.G.L.R. 829.
[62] *Barrett v. Enfield London Borough Council* (n. 61 above); *Caparo Industries plc v. Dickman* [1990] 2 A.C. 605.
[63] [1978] A.C. 728.

public interest in protecting the environment, a court is likely to be reluctant to find that a duty of care is owed to the owner or occupier of the land by the local authority in carrying out those functions.[64] To impose such a duty would be likely to result in excessive caution and reluctance to act on the part of the authorities.[65] The legislation by virtue of its inherent nature may have adverse effects on property values, and contains its own checks and safeguards against unreasonable action by enforcing authorities.[66]

Inadequate enforcement

4.51 A third possible liability situation is where it is alleged that the enforcing authority has failed to make proper use of its powers to secure the remediation of contaminated land, in consequence of which a third party has suffered loss or damage—perhaps because their health or their adjoining property has been affected by the contamination. A similar issue arose in relation to the use of planning and statutory nuisance powers in *R. v. Lam and Brennan (t/a Namesakes of Torbay) and Torbay Borough Council*.[67] In that case it was alleged that the local authority had (*inter alia*) failed to take enforcement action against a workshop which emitted fumes from chemicals and paint sprays. The owner of a nearby Chinese restaurant brought an action in negligence against the local authority for damage caused to his business and for health effects on his family. The alleged negligence lay in failure by officers to make proper recommendations to committee, and in failing to take enforcement proceedings.

4.52 The Court of Appeal, referring to the principles of *X and Others (Minors) v. Bedfordshire County Council*,[68] and *Stovin v. Wise*,[69] held that neither the planning legislation nor the provisions on statutory nuisance were such as to create a duty of care at common law with regard to the exercise or non-exercise of those powers. The provisions were regarded as being for the benefit of the public at large, and as not derogating from the right of affected individuals to take their own action in nuisance against the source of the problem.[70] Nor would it be said that this position was affected by arguments of "assumption of responsibility" going beyond the simple performance of the relevant statutory functions.[71]

---

[64] *Harris v. Evans* [1998] 3 All E.R. 522, CA.

[65] *X and others (minors) v. Bedfordshire County Council* [1995] 2 A.C. 633, HL.

[66] Compare the position, however, where the legislative powers are used in an unlawful fashion to impose inappropriate action: *Welton v. North Cornwall District Council* [1997] 1 W.L.R. 570 (distinguished in *Harris v. Evans*).

[67] [1998] P.L.C.R. 30.

[68] [1995] 1 A.C. 633.

[69] [1996] A.C. 923.

[70] [1998] P.L.C.R. 30, pages 48–49.

[71] *ibid.*, at page 50.

4.53 These arguments could no doubt also be applied to the functions of local authorities under Part IIA. The purpose of Part IIA in dealing with contamination for the public good seems not to be particularly different from the purpose of the planning and statutory nuisance regimes. Yet it is not difficult to conceive of circumstances where such a result might be thought unjust. For example, contaminated land may present significant and alarming risks to local residents, whose health may be jeopardised. Part IIA has provided a series of duties on the relevant authorities to deal with those risks; in reality the local residents will be relying on the proper performance of those duties to safeguard their health, in a way which the generality of the public will not. Also, whereas it may be a reasonable argument in the case of nuisance from fumes or noise that those affected can deal with the problem themselves by an action for private nuisance or by a complaint as a person aggrieved under section 82 of the 1990 Act, the same argument is not so strong in the case of a contaminated site. Securing effective redress under common law is likely to be much less straightforward, and Part IIA contains no provision for direct enforcement action by citizens.

The Human Rights Act 1998

4.54 In these circumstances, local residents may well feel rightly aggrieved if the law offers them no remedy where the enforcing authority, through inertia or ineptitude, fails to protect them. Should the common law not provide a remedy in these circumstances (as in the current state of the authorities it may not) then it may be possible to rely on the relevant provisions of the European Convention on Human Rights as requiring effective action to protect the health and safety of individuals, and their homes. Under the Human Rights Act 1998, primary and subordinate legislation must be read and given effect to, so far as possible, in a way which is consistent with the Convention rights set out in Schedule 1 of the Act.[72] It is unlawful for public authorities to act in a way which is incompatible with convention rights,[73] and the victim of such an unlawful act may bring proceedings against the authority.[74]

4.55 Case law of the European Court of Human Rights suggests that the rights to life under Article 2 and to the respect for home and private life under Article 8(2) may be capable of applying to situations where an authority fails to take effective action to protect individuals from harm caused by pollution, or to provide them with adequate information on the risks of such harm.[75] It remains to be seen whether these authorities can be

[72] Section 3(1).
[73] Section 6(1).
[74] Section 7(1).
[75] *Lopez Ostra v. Spain* (1994) 20 E.H.R.R. 277; *Guerra v. Italy* (1998) 26 E.H.R.R. 357.

used in relation to contaminated land situations, but the possibility must exist.[76] A local authority which is aware that it has contaminated land in its area, yet for whatever reason it fails to act effectively, could well find itself faced with proceedings under the Human Rights Act 1998 from a local resident who claims to be a "victim" of the contamination.

Position of local authority consultants

4.56 It seems likely in practice that many local authorities will use the services of environmental consultants to investigate, assess and advise on the remediation of contaminated land. An authority is perfectly entitled to do so, but the question may then arise as to the position of those consultants if negligence is alleged by a landowner, local resident, or other person who claims to have suffered damage as a result of that negligence.

4.57 The question will be whether the consultant and the person affected were in a relationship of sufficient proximity, in legal terms, to give rise to a duty of care. This seems unlikely, since the consultant's duty of care is owed to the authority, and there is no duty to the person whose past activities or whose land are the subject of such advice, or to the general public who may be affected by decisions made on the basis of that advice.[77] Such a duty would only arise if it was clear on the facts that the consultant, as well as performing his duty to the authority, had assumed personal responsibility for the relevant individual owner, occupier, etc.[78] Use of a consultant will not, of course, release the authority from its own legal responsibilities, such as they are; if the authority is sued, it may well join its consultant as a third party.

How should landowners react?

4.58 The practical question arises as to how a landowner who finds their land the subject of scrutiny by an enforcing authority under Part IIA should react. The risk that land will be subject to investigation may appear from the authority's draft or final strategy for inspection: either the strategy may identify particular areas of search, or it may prioritise types of contaminative use for early inspection.

---

[76] For a general discussion, see J. Thornton and S. Tromans, *Human Rights and Environmental Wrongs* [1999] J.E.L. 35.

[77] *Kapfunde v. Abbey National plc and Daniel* [1998] I.R.L.R. 583, CA. No duty owed by doctor examining prospective employees on behalf of company. See also the example given by Lord Browne-Wilkinson in *X (Minors) v. Bedfordshire County Council* [1995] 2 A.C. 633 at page 752 of a doctor employed by an insurance company to examine an applicant for life insurance.

[78] See *Phelps v. Hillingdon London Borough Council* [1999] 1 W.L.R. 500; *Jarvis v. Hampshire County Council* [2000] E.L.R. 36.

4.59 To what extent should the owner co-operate with the authority? This will depend on many factors, both environmental and commercial. The immediate response may be one of dismay or indignation that this particular site should be singled out for attention. A broader view is necessary however. There is little doubt that the authority has the legal powers—and probably the resources—to follow through the investigative process. Given that statutory duties are involved, this is a problem which is not likely simply to "go away". The question therefore is the extent to which the owner merely lets the authority get on with the investigation at their expense, or whether the owner seeks to pre-empt the process by commissioning their own investigations and actively co-operating at their own cost.

4.60 Two points should be borne in mind here. First, it is not a foregone conclusion that land which is subject to investigation will ultimately be identified as contaminated within the statutory definition. For the landowner who becomes actively involved, at least there may be opportunities to influence that process. The owner will be more in control of the situation, and may also benefit from reassurance that the investigation work is being carried out to a satisfactory standard. The current and projected future use of the land may also be relevant factors here.

4.61 Secondly, nor is it a foregone conclusion that the owner will ultimately be the person with financial responsibility for any clean-up. If there is an original polluter who is identifiable, the owner may wish to involve that person at this stage, since they will have an interest in whether the site is identified as contaminated, and may have greater resources to deploy in safeguarding their interests.

4.62 A number of other factors may also be relevant to the landowner. These will include the desire to minimise disruption to operations on site; whether site investigation is already in hand or is proposed for other reasons (*e.g.* a prospective sale); the desire to maintain good relations with the authority; the desire to avoid bad publicity or to appear responsible; existing corporate environmental policies and management systems; and any contractual restrictions or freedom of action (for example, as to confidentiality, or the terms of any indemnities that may have been entered into).

4.63 The situation may also be complicated by transactions which are in progress at the time the issue arises. Here careful attention will need to be paid to the respective interests and any contractual rights of the parties concerned. These are tactical issues which will need to be thought through carefully and discussed in each specific case. If the site is to be redeveloped, then as part of the development process the local planning authority will probably require full investigation at the landowner's expense, together with appropriate remediation measures imposed by conditions or by a section 106 obligation.

Chapter 5

Identifying Land as Contaminated

Summary

The process of inspection may lead to the local authority identifying land as contaminated land, in which case it must give notice of that fact to the appropriate Agency and to the owner, occupier and any other person who it appears may be liable for remediation.

The Guidance addresses the criteria for making that determination, and the process by which the determination is made. The determination must be made in accordance with that Guidance.

Guidance on identification: generally

5.01 The determination that land is contaminated is to be made in accordance with the statutory Guidance.

The Guidance at Annex 3, Chapter B, Part 4 deals separately with the four possible situations where that determination may be made, namely:

(a) significant harm is being caused;
(b) there is a significant possibility of significant harm being caused;
(c) pollution of controlled waters is being caused; and
(d) pollution of controlled waters is likely to be caused.

5.02 The *Guidance* also deals with the issues of the physical extent of land to be covered by the determination, consistency with the approach of other statutory bodies, the actual process of making the determination and the preparation of a written record of determination.

5.03 In dealing with these issues, the Guidance at the outset reminds the authority that whilst it may receive and rely on information or advice from other persons, such as the Environment Agency or an appointed consultant, the actual determination remains the sole responsibility of the authority and cannot be delegated other than in accordance with statutory powers of delegation.[1] This applies even where it appears that the land, if contaminated, would be a special site, so that inspection has been undertaken by the Agency.

Determining that significant harm is being caused

5.04 According to the Guidance, the authority should determine that land is contaminated land on the basis that significant harm is being caused where[2]:

(a) it has carried out an appropriate scientific and technical assessment of all the relevant and available evidence; and

(b) on the basis of that assessment, it is satisfied on the balance of probabilities that significant harm is being caused.

5.05 The key points of this guidance are the nature of the scientific and technical assessment which is necessary (discussed below), and the fact that the test is based on the balance of probabilities. This is of course the standard of proof generally required in civil cases and involves the authority being able to say they think it more probable than not that the relevant circumstances exist; if the probabilities are equal, or it is not possible to say which is more likely, then this standard is not met.[3] Annex 2 of the Circular deals with the situation where there is uncertainty.[4] There may be situations either where the investigation yields insufficient information to make a determination that the land is or is not contaminated, or where on the "balance of probabilities" test it is not contaminated, but there remains the possibility that it might be.[5] In such cases, the authority will need to consider if further investigation or analysis might assist: if not, then the authority will have to determine that the land is not contaminated.

Determining that there is a significant possibility of significant harm being caused

5.06 The Guidance on this issue states that the authority should determine that land is contaminated land on the basis that there is a significant possibility of significant harm being caused where[6]:

---

[1] DETR Circular, *Contaminated Land,* Annex 3, para. B.31. Delegation powers arise under section 101 of the Local Government Act 1972 or Section 56 of the Local Government (Scotland) Act 1973.

[2] *ibid.*, para. B.44.

[3] *Miller v. Minister of Pensions* [1947] 2 All E.R. 372 at pp. 373–4; *Hornal v. Neuberger Products Ltd* [1957] 1 Q.B. 247.

[4] Annex 2, paras 3.23 and 3.24.

[5] For example, if the mean concentration of a containment lies just below an appropriate guideline value.

[6] DETR Circular, Annex 3, para. B.45.

(a) it has carried out a scientific and technical assessment of the risks arising from the pollutant linkage or linkages, according to appropriate, authoritative and scientifically based guidance on such risk assessments;
(b) the assessment shows that there is a significant possibility of significant harm being caused; and
(c) there are no suitable and sufficient risk management arrangements in place to prevent such harm.

5.07 Point (b) above will involve the authority referring to that part of the Guidance which defines what is meant by "significant possibility" for the various types of harm.[7] Point (c) could be important in practice, in that there may be cases where a risk of harm is present, but the risk is controlled to acceptable levels. An example might be the presence of explosive landfill gases within range of occupied premises, but where those premises are fitted with appropriate gas detection and venting equipment. There is a source, a pathway and a target, and certainly the harm in question if it occurred would be "significant", but so long as the measures are properly maintained the risk should not be regarded as "significant". The authority will have to form a view as to whether the measures are suitable and sufficient, which may in itself involve technical assessment. The measures must be "in place", *i.e.* present and operational. A promise to instal them at some point in the future will not be sufficient. Point (a) raises the most potentially difficult issues and is discussed in the next paragraph.

Assessment of risk and use of guidelines

5.08 The assessment of the significance of risk, following the words of the Guidance, must be "scientific and technical", according to "relevant, appropriate, authoritative and scientifically based guidance." The authority will therefore have to be prepared to defend its decision-making process against those criteria, and will require appropriate technical support in this. In particular, if using any external guidance on risk, the authority must satisfy itself that such guidance is relevant to the actual circumstances in hand, and that appropriate allowances have been made for particular circumstances.[8]

5.09 The Guidance countenances the possibility that, in order to simplify the assessment process, the authority may use "authoritative and scientifically based" guideline values for assessing the acceptability of certain concentrations of substances; the test remains essentially the same where such guidelines are used.[9] The potential problems with this generic approach of

[7] See Chapter 2.
[8] DETR Circular, Annex 3, para. B.46.
[9] *ibid.*, para. B.47.

quantitative assessment have been described above.[10] The Guidance accepts that care is needed in using such values. In particular, the authority must satisfy itself of a number of matters when using guideline values.[11] These are:

(a) the relevance of the values to the circumstances in question and, in particular, to the judgment of whether the pollutant linkage in question consitutes a significant possibility of harm;
(b) the relevance to those circumstances of any assumption underlying the values[12];
(c) the observance of any conditions relevant to the use of the values[13]; and
(d) the making of any appropriate adjustments to allow for such differences in factual circumstances and underlying assumptions.

5.10 It is also stated that the authority should be prepared to reconsider any determination based on the use of such guideline values if it is demonstrated to the authority's satisfaction that under some other more appropriate method of assessing the risks the determination would not have been made.[14]

5.11 The way in which the Guidance is written clearly places the onus of exploring the validity and applicability of any guideline values on the authority. It is not the case that the authority can simply rely on such values, leaving others to show why they are inappropriate. The overall impact is that an authority will not be able to rely exclusively or simplistically on such generic values, which will often reflect political choices as much as hard science.[15] As already indicated, it may however appoint consultants to use their own expertise to evaluate the risk within the framework of such generic values, and to report upon it.

Determining that pollution of controlled waters is being caused

5.12 The Guidance states that the authority should determine that land is contaminated land on the basis that pollution of controlled waters is being caused where[16]:

---

[10] See Chapter 2.
[11] DETR Circular, Annex 3, para. B.48.
[12] For example, assumptions on soil conditions or land-use patterns, behaviour of contaminants or presence of pathways.
[13] For example, number of samples or methods of preparation and analysis.
[14] DETR Circular, Annex 3, para. B.49.
[15] See *Risk Based Contaminated Land Investigation and Assessment* by Judith Petts, Tom Cairney and Mike Smith (John Wiley & Sons, Chichester, 1997).
[16] DETR Circular, Annex 3, para. B.50.

(a) it has carried out an appropriate scientific and technical assessment of all the relevant and available evidence, having regard to any advice provided by the appropriate Agency; and

(b) on the basis of that assessment, it is satisfied on the balance of probabilities that:

(i) a potential pollutant is present in, on or under the land in question, which constitutes poisonous, noxious or polluting matter, or which is solid waste matter; and

(ii) that potential pollutant is entering controlled waters by the pathway identified in the pollutant linkage.

5.13 As explained above, the entry of the pollutant into controlled waters must be a present, not a past, phenomenon for this determination to be made.[17]

Determining that pollution of controlled waters is likely to be caused

5.14 The Guidance provides that the authority should determine that land is contaminated on the basis that pollution of controlled waters is likely to be caused where[18]:

(a) it has carried out an appropriate scientific and technical assessment of all the relevant and available evidence, having regard to any advice provided by the appropriate Agency;

(b) on the basis of that assessment it is satisfied that, on the balance of probabilities, all of the following circumstances apply:

(i) a potential pollutant is present in, on or under the land in question, which constitutes poisonous, noxious or polluting matter, or which is solid waste matter;

(ii) the potential pollutant in question is in such a condition that it is capable of entering controlled waters;

(iii) taking into account the geology and other circumstances of the land in question, there is a pathway by which the potential pollutant can enter controlled waters;

(iv) the potential pollutant is more likely than not to enter the controlled waters and when it does so will be in a form that is poisonous, noxious or polluting, or solid waste matter; and

(v) there are no suitable and sufficient risk management arrangements relevant to the pollution linkage in place to prevent such pollution.

---

[17] See Chapter 2.

[18] DETR Circular, Annex 3, para. B.50.

5.15 These requirements follow closely the criteria for what constitutes likelihood of pollution of controlled waters earlier in the Guidance and reference should be made to that discussion. The Guidance on this issue creates a rather curious double standard of proof, namely that the authority must be satisfied on the balance of probabilities that entry of the pollutant into the waters is more likely than not. One of those elements might arguably be thought to be redundant. The concept of "suitable and sufficient risk management arrangements" is again potentially important, but should also be seen in the wider context of whether there is a pathway at all. For example, a cut-off ditch which prevents pollutants reaching a river might be said not only to constitute such arrangements but also, more fundamentally, to prevent there being any pathway at all.

The physical extent of the contaminated land

5.16 The determination that land is contaminated land will have to be made in relation to a specific area. This may have important consequences in terms of the subsequent service of any remediation notice, so the decision is critical and may in practical terms be one of the most difficult decisions for the authority.

5.17 The Guidance states that the primary consideration should be the extent of land which is contaminated.[19] This is at one level simply stating the obvious, but implies that the precise extent of the contaminants present will be known with some certainty. However, the degree of certainty will in fact depend on the conditions encountered and on the level of investigation. Clearly, it will not be possible to sample every square metre of an extensive site, for reasons both of cost and practicability. Some areas (for example buildings in use) may be inaccessible for intrusive survey purposes. A measure of pragmatism, informed by expert judgment, will be necessary here. The Guidance suggests that the authority may need to begin by reviewing a wider area, subsequently refining it down to the "precise" areas which meet the statutory tests, and using these as the basis for its determination.[20]

5.18 A somewhat different, but related, issue is the case where a large area may properly be identified as contaminated, but separate designations of different parts of it may lead to simplicity in the actions which will follow. The Guidance states that three factors must be taken into account by the authority in seeking to achieve simplicity of administration, namely[21]:

(a) the location of the pollutants;
(b) the nature of the remediation that might be required; and

---

[19] DETR Circular, Annex 3, para. B.32.
[20] *ibid.*, para. B.33.
[21] *ibid.*, para. B.32.

(c) the likely identity of those who may be appropriate persons to bear responsibility for the remediation (where this is reasonably clear at this stage: which of course it may well not be).

5.19 The Guidance suggests that in practice this is likely to mean that the land to be covered by a single determination is likely to be the smallest area covered by a single remediation action, which cannot sensibly be broken down into smaller actions. Subject to this the land will be the smaller of[22]:

(a) the plots which are separately recorded in the Land Register or are in separate ownership or occupation; and
(b) the area of land in which the presence of significant pollutants has been established.

5.20 This part of the Guidance is perhaps not as clearly worded as might be desirable, but essentially the premise is that subdivision into smaller areas is a good thing, so far as that is consistent with sensible remediation solutions.

5.21 This assumption is in fact questionable. Determinations made separately in relation to a number of small plots may be helpful in avoiding the need to serve notices apportioning liability on a number of different owners or occupiers. But, equally, it may not be desirable to serve multiple notices on a single appropriate person in relation to small plots of land. Ownership may in fact not be relevant at all if the original polluter can be found. The problem is that the authority at the time of making the determination may well not have the information available to make sensible decisions on the matter.

5.22 The potential difficulties may be seen by considering the following simple case. Contamination by organic solvents is found to exist on part of a disused industrial site. It has spread to varying extents into the back gardens of adjoining residential properties. Should a single determination be made in relation to the entire area of contamination, or should separate determinations be made for the relevant parts of the industrial area and of each garden? Clearly the determination should not extend to a wider area than that where the contamination is physically present. However, whether the land can sensibly be subdivided into separate ownerships will depend to some extent on the nature of the proposed remediation actions. If what is required is (for example) simply to dig up and remove contaminated material, that could practicably be carried out separately by each individual owner, though economies of scale might well make cooperation sensible. On the other hand, if the remediation consists of pumping and treating contaminated groundwater, or inserting some form of artificial barrier, it will be entirely impracticable or indeed impossible to treat each

[22] *ibid.*, para. B.34.

individual garden separately. The authority will therefore need to have a reasonably clear idea of what the appropriate remediation will be in order to take sensible decisions on this issue.

5.23 Another factor which may have an important bearing on the issue is the likely identity of the appropriate person or persons to undertake the remediation. As we shall see later, a key issue is whether the original polluter can still be found. The statutory rules in section 78K on the escape or migration of contaminants will also be relevant, since the gardens were contaminated by substances escaping from elsewhere. If the application of those rules leads to the conclusion that a single former polluter is the appropriate person to clean up the entire area of contamination, it makes little sense to make separate determinations for smaller plots, based on current ownership. On the other hand, such a determination may be more appropriate where the original polluter cannot be found and where each separate owner or occupier will be potentially responsible. The authority may well not have sufficient information at the time of determining whether the land is contaminated land to make sensible decisions on this.

5.24 One possible solution to the problem is that there may be an element of flexibility to subdivide the relevant area at the time of serving remediation notices. This issue is considered later.[23] On the basis that there is such flexibility it might be more sensible for the authority to concern itself with the practicalities of remediation rather than with land ownership at the time of making the determination, while retaining the flexibility to subdivide on the basis of liability or tenure at a later stage.

Consistency with other bodies

5.25 The Guidance requires the authority to adopt a consistent approach to other relevant regulatory authorities. In practice, this means that in making any determination which relates to an ecological system effect the authority should consult English Nature, the Countryside Council for Wales or Scottish National Heritage and have regard to its comments.[24] Similarly, where the determination relates to the pollution of controlled waters the Environment Agency or SEPA should be consulted and regard given to its comments.[25]

General points on making the determination

5.26 The Guidance also lays down some general requirements on the determination process. In particular:

---

[23] See Chapter 12.
[24] DETR Circular, Annex 3, para. B.42.
[25] *ibid.*, para. B.43.

(a) A particular pollutant linkage or linkages must form the basis of the determination. All three elements of the linkage (pollutant, pathway and receptor) must be present.[26]

(b) The authority should consider any evidence that additive or synergistic effects between potential pollutants (whether the same substance on different areas of land or between different substances) may result in a significant pollutant linkage.[27] This, if done thoroughly, will not necessarily be a straightforward or easy exercise.

(c) The authority should also consider whether a significant pollutant linkage may result from a combination of several different pathways (inhalation, plus dermal contract, plus oral ingestion, for example.)[28]

(d) the authority should address the issue of whether there is more than one significant pollutant linkage on any land; if so, then each linkage must be considered separately, since different people may be responsible for their remediation.[29]

5.27 This last point is particularly important. It stresses the foundational nature of the pollutant linkage concept, which though it appears nowhere in the legislation derives its statutory force from the prescriptive Guidance.

The legal process of determination

5.28 The identification of land as "contaminated land" is an important step which has significant legal consequences, triggering the duty under section 78E(1) to serve a remediation notice. As such it is likely that in practice the matter will go to a committee of the authority for decision, rather than being delegated to an officer to determine. The report to committee is likely to be a crucial document in this respect. The question which the decision-making body should ask itself is whether the statutory criteria for contaminated land are satisfied, not whether it is appropriate or expedient to take enforcement action.[30] The officer's report should make clear what the legal tests are, and how the facts relate to those tests and to the statutory guidance.[31] Failure in this respect to report the legal tests accurately may lead to the undermining of the authority's action at a later date.

5.29 If the authority forms the view that the land is not contaminated land, then the resolution should state the basis on which that decision is reached; whilst there may be other ways of addressing the problem than

---

[26] *ibid.*, para. B.40.
[27] *ibid.*, para. B.41(a).
[28] *ibid.*, para. B.41(b).
[29] *ibid.*, para. B.41(c).
[30] *R. v. Carrick District Council, ex p. Shelley* [1996] Env. L.R. 266.
[31] *ibid.* pp. 280–281.

service of a remediation notice, that is not a correct basis for determining to take no action.[32] In short, the issue for the authority is not whether they feel it is or is not appropriate to serve a remediation notice (which may present all sorts of local, legal or financial problems) but simply whether or not the land in question is contaminated within the statute. There may be other factors that will preclude service of a remediation notice (for example, hardship, or the fact that remediation will be undertaken in any event), but consideration of these issues comes at a later stage.

Notice to appropriate Agency and others

5.30 Under section 78B(3) the local authority must give notice of the identification of any contaminated land in its area to:

(a) the appropriate Agency;
(b) the owner of the land;
(c) any person who appears to the authority to be in occupation of the whole or any part of the land; and
(d) each person who appears to the authority to be an appropriate person.

5.31 The system of notification in section 78B(3) makes early dialogue between the local authority and the present owners a prerequisite. Provisions for locating the owner and occupier might presumably have followed the procedure laid down in the Town and Country Planning Act prior to the service of enforcement notices, namely section 330, where a requisition for information as to interests in land may be served on the occupier and on any person in receipt of rent. There are, however, two non-specific powers which may assist. The first is the general power to require persons to answer questions under section 108(4)(j) of the Environment Act 1995. The second is the general power under section 16 of the Local Government Act (Miscellaneous Provisions) Act 1976 to serve notice on the occupier of the land, on any person having an interest in it or who receives rent for it, and on any manager or letting agent, a notice requiring information on his own interest and that of others interested in the land. This seems likely to be a widely used power in relation to Part IIA.

5.32 The authority will also need to consider making use of the information on current freehold ownership and on certain leasehold interests which is available from the Land Registry. Further information on past ownership and interests may also be available at the discretion of the Chief Land Registrar.

5.33 As to identifying an "appropriate person", the authority must for the purposes of the notice determine the appropriate person in accordance

---

[32] *ibid*. p. 284.

with the statutory provisions, under section 78F. It follows that the process of identifying the "appropriate persons" should be carried out at an early stage. However, the authority may not at the initial stage of identifying contaminated land, be able to establish who falls into these categories.[33] The authority must proceed on an iterative basis on whatever information it has at the relevant time. Special provision is accordingly made to deal with "latecomer" appropriate persons. Under section 78B(4) if the enforcing authority later discover that another person is an "appropriate person", then they must also serve notice on that person. Whereas the initial notice under section 78B(3) will in all cases be served by the local authority, the term "enforcing authority" in subsection 78B(4) will embrace not only the local authority but also the appropriate Agency in relation to special sites.

5.34 No provision is made for the situation where, having served a notice, the authority concludes that the recipient is not in fact an appropriate person, though presumably it is open to the authority to withdraw the notice, without prejudice to the service of a second or subsequent one. Such a power to withdraw a notice in appropriate cases seems likely to be implied on the analogy with the power in Part III of the 1990 Act, established by the Court of Appeal in *R. v. Bristol City Council, ex p. Everett.*[34]

Record of determination

5.35 The authority is required by paragraph B.52 of the Guidance to prepare a written record of any determination that land is contaminated land. The record should include (if necessary by means of reference to other documents):

(a) a description of the particular significant pollutant linkage, identifying each of the three components of pollutant, pathway and receptor;
(b) a summary of the evidence on which the determination is based;
(c) a summary of the relevant assessment of this evidence; and
(d) a summary of the way in which the authority considers that the relevant requirements of Chapters A and B of the Guidance have been satisfied.

5.36 Because of the prescriptive nature of the Guidance, the record of determination has effectively a statutory status. In any challenge to the determination of the authority, whether on appeal or by judicial review, it is likely to be a key document. It must therefore be drafted carefully and precisely, with adequate and accurate cross reference to any supporting reports or other documents. If properly prepared, the officer's report to

[33] DETR Circular, Annex 2, para. 4.2.
[34] [1999] Env. L.R. 587, 600.

committee will provide the basis for much of the material in the record of determination.

5.37 The Guidance only requires a record to be prepared where land is determined to be "contaminated"; it does not cover the situation where the opposite result is reached. However, local residents or amenity groups may be very concerned to know why land has been determined not to be "contaminated" within the statutory criteria, and may well request written reasons, having, they may well claim, a legitimate expectation that such reasons will be forthcoming. Similarly, persons whose land has been identified as contaminated may claim they have a valid interest in knowing why the land of others—which may have been subject to similar past contamination—has not been so identified. In either case, the absence of reasons may make it difficult to challenge the decision effectively. Accordingly, there may well be strong legal arguments that reasons should be given for a determination that land is not "contaminated".[35] Certainly, to give reasons can be regarded as contributing to good administration, and may in due course be a statutory requirement.[36] Any refusal to give reasons could also have implications under Article 6 of the European Convention on Human Rights and, consequently, under the Human Rights Act 1998.[37]

---

[35] See generally, Michael Fordham [1998] J.R. 158.
[36] See the Written Answer of the Solicitor-General of January 12, 1998 noted at [1998] J.R. 92.
[37] See *Stefan v. General Medical Council, The Times*, March 11, 1999.

Chapter 6

Special Sites

Summary

Certain types of contaminated land are required to be designated as "special sites"; the effect being that the Environment Agency or SEPA, rather than the local authority, is the enforcing authority.

The descriptions of contaminated land which are required to be designated as special sites are set out in the Contaminated Land (England) Regulations 1999. These categories relate to the presence of certain types of substance, or to certain types of use or occupation, or to certain types of effects, such as serious water pollution. The presumption is that these types of site are likely to present particularly difficult or serious problems.

Where the local authority and the Agency disagree on whether a site should be "special" or not, there is a dispute resolution procedure before the Secretary of State. After designation by the local authority or Secretary of State, the appropriate Agency takes responsibility for further enforcement, including the service of remediation notices. Appeals against remediation notices concerning special sites lie to the Secretary of State rather than to magistrates.

The Guidance requires the local authority to consider whether a site may be a special site at an early stage, and to make arrangements with the appropriate Agency to carry out inspection on behalf of local authority where this is the case.

The concept of "special sites"

6.01 Whilst all contaminated land within the meaning of Part IIA involves, of its nature, significant harm or risk, some types of contaminated land may present particular risks or technical problems making them more suitable to be dealt with by the Environment Agency or SEPA, rather than local authorities. An example is land on which an IPC-prescribed process has been or is being carried on, where the relevant expertise is likely to reside within the Agencies. As initially drafted, the Environment Bill proposed the creation of a special class of "closed landfills", so as to provide a "more tailored approach" to this type of site.[1] The distinction was ultimately dropped however, on the basis that other types of site could present equally serious problems.[2] In fact, landfill sites or other land used for waste activity are not as such designated as special sites, on the basis that Part II of the 1990 Act already contains wide powers for the Agency to tackle the problems presented by such sites.[3] It should not be assumed that a local authority will be reluctant to designate land a special site: the authority may be very happy to see responsibility for enforcement pass to the Environment Agency or SEPA.

Special sites defined

6.02 Although the legislation contains a definition of a "special site" it brings us no nearer to what exactly the legislature has in mind for their designation. Section 78A(3) defines a special site as:

> "any contaminated land—(a) which has been designated as such a site by virtue of sections 78C(7) or 78D(6) below; and (b) whose designation as such has not been terminated by the appropriate Agency under section 78Q(4)."

6.03 It follows from this definition that a special site must be contaminated land within section 78A(2), that is, in such a condition by reason of substances in, on or under the land, that significant harm is being caused, or there is significant possibility of such harm being caused, or pollution of controlled waters is being caused or likely to be caused. The key to what constitutes a special site lies in the Regulations. By section 78C(8) land is required to be designated as a special site if, but only if, it is land of a description prescribed for that purpose. The Regulations may make different provision for different cases, circumstances, areas or localities, and may describe land by reference to its area or locality.[4]

6.04 By section 78C(10), in prescribing descriptions of land as special sites, the Secretary of State may have regard to whether the harm or pollution involved are likely to be serious, and whether the appropriate Agency is likely to have expertise in dealing with the kind of pollution in question. However, this is expressly stated to be without prejudice to the generality of the power to prescribe such descriptions of land. Annex 4 to the Circular, which constitutes a Guide to the Regulations, indicates three main groups of cases where descriptions of land have been prescribed as special sites[5]:

---

[1] *Hansard* H.L. Vol. 560, col. 1432.
[2] *Hansard* H.L. Vol. 562, cols. 156, 158, 160.
[3] DETR Circular, *Contaminated Land,* Annex 4, para. 14.
[4] Section 78C(9).
[5] DETR Circular, Annex 4, paras 11 and 12.

(a) water pollution cases, in particular those where the Agency will have concerns under other legislation, on drinking water, surface water quality, and groundwater;
(b) industrial cases, where the land in question is being or has been used for specific types of activity that either pose particular remediation problems or are subject to other statutory regimes, for example, oil refining, explosives, integrated pollution control, nuclear sites;
(c) defence cases, designed to ensure that the Agency deals with most cases where the land involves the Ministry of Defence Estate, on the basis that the Agency is best placed to ensure uniformity and appropriate liaison with the MoD across the country.

The Regulations on special sites

6.05 The Regulations prescribe those descriptions of land which are required, if contaminated, to be designated as a special site. These are[6]:

(a) Land to which regulation 3 applies (as to which, see below).
(b) Land which is contaminated by virtue of waste acid tars. These are defined as tars which contain sulphuric acid, which were produced as a result of the refining of benzole, used lubricants or petroleum, and which are or were stored on land used as a retention basin for the disposal of such tars. The Guidance to the Regulations states that the retention basins (or lagoons) involved are typically those where waste arose from the use of concentrated sulphuric acid to produce lubricating oils or greases, or to reclaim base lubricants from mineral oil residues; the description is not intended to cover cases where the tars resulted from the manufacture of coal products, or where these tars were placed in pits or wells.[7]
(c) Land on which at any time there has been carried out either:
  (i) the purification (including refining) of crude petroleum or of oil from petroleum, shale or any bituminious substance except coal, or
  (ii) the manufacture or processing of explosives.
(d) Land on which an IPC-prescribed process (Part A) has been or is being carried on under an authorisation under Part I of the 1990 Act. It does not cover cases where the process comprises solely things being done that are required by way of remediation, that is where an authorisation has had to be obtained to carry out the

---

[6] Regulation 2.
[7] DETR Circular, Annex 4, para. 11(a).

remediation process itself. The Guidance points out that it also does not cover land where the activity was of the nature of a prescribed process, but ceased before the application of the IPC regime to it.[8] This category of special site may require amending in due course to reflect the new regime of pollution prevention and control under the Pollution Prevention and Control Act 1999.

(e) Land within a nuclear site, that is any site in respect of all or part of which a nuclear site licence under the Nuclear Installations Act 1965 is in force, or where the nuclear site licence has terminated, but where the licensee's period of responsibility under the 1965 Act still applies.[9] In any event, any designation of such a site under Part IIA can only be in respect of non-radioactive contamination, since section 78YC excludes radioactive contamination from that regime.

(f) Land owned or occupied by or on behalf of the Secretary of State for Defence, by the Defence Council, or by the relevant bodies under the Visiting Armed Forces Act 1952 or the International Headquarters and Defence Organisations Act 1964, being land used for naval, military or defence purposes. Land used for residential purposes or by the NAAFI is to be treated as such land only if it forms part of a base occupied for naval, military or air force purposes.[10] This category covers only land in current military use, and will not extend to land which has been disposed of to civil ownership or occupation; nor does it cover training areas or ranges which are not owned by the MoD but are subject to occasional military use.[11]

(g) Land on which the manufacture of certain military weapons has been carried on at any time. These are: chemical weapons, biological agents or toxins and associated weapons or delivery systems.[12] The current ownership or occupation is irrelevant in this case.

(h) Land designated under the Atomic Weapons Establishment Act 1991.

(i) Land held for the benefit of Greenwich Hospital to which section 30 of the Armed Forces Act 1996 applies.

(j) Land which is adjoining or adjacent to land falling within categories (b)–(i) above, and which is contaminated land by virtue of substances which appear to have escaped from land within those categories. The intention here is that the Agency will be the enforcing authority for both sites, thereby avoiding the splitting of regulatory control.[13]

---

[8] *ibid.*, para. 11(d).
[9] Regulation 2(4).
[10] Regulation 2(5).
[11] DETR Circular, Annex 4, para. 12.
[12] See further the Biological Weapons Act 1974 and the Chemical Weapons Act 1996.
[13] DETR Circular, Annex 3, para. 13.

6.06 Regulation 3 deals with special sites which relate to water pollution and covers the following categories of land:

(a) Where controlled waters used or intended to be used for the supply of drinking water for human consumption are being affected by the land, and as a result require treatment or additional treatment before use in order to be regarded as "wholesome" under the Water Industry Act 1991, Part III. The category therefore relates essentially to where the contamination impacts on the standards of wholesomeness currently set out in the Water Supply (Water Quality) Regulations 1989 (S.I. No. 1147) as amended, or the Private Water Supplies Regulations 1991 (S.I. No. 2790). The Guidance suggests that an intention to use water for drinking purposes would be demonstrated by the existence of an abstraction licence for that purpose, or an application for such a licence.[14]

(b) Where controlled waters are being affected so that they do not meet or are not likely to meet the criteria for water quality classification for waters of their relevant description. Again, it is standards set under other legislation which are critical: in this case, the Surface Waters (Dangerous Substances) (Classification) Regulations (1989 S.I. No. 2286, 1992 S.I. No. 337, 1997 S.I. No. 2560 and 1998 S.I. No 389).

(c) This category covers impacts on major aquifers. It applies where controlled waters are being affected by substances falling within a number of families or groups listed in Schedule 1 to the Regulations, and the waters or any part of them are contained within underground strata comprising or including rock formations listed in Schedule 1. The substances involved include organohalogens, mercury, cadmium, mineral oils and hydrocarbons, cyanides, and those possessing carcinogenic, mutagenic or teratogenic properties in the aquatic environment. The list corresponds to List I of the Groundwater Directive 80/86/EEC. The formations of rock include various chalk, sandstone and limestone formations, and are essentially the major aquifers. The fact that the land lies over one of these aquifers does not of itself make it a special site: pollution of the controlled waters within the relevant strata by the relevant substances must be occurring or likely to occur.[15]

Investigation and designation of special sites

6.07 Normally the procedure under Part IIA involves simply identification of land as contaminated, following which the remediation provisions apply. However, for special sites, the process is identification, followed by

[14] *ibid.*, para. 9(a).
[15] *ibid.*, para. 10.

designation, then remediation. Identification can be by either the local authority (section 78C(1)) or by the Secretary of State (section 78D(5)) or can be initiated by the Environment Agency (section 78C(4)).

6.08 The question of whether a site falls within one of the special site categories must be considered before any inspection is authorised or takes place, since if it appears that the land would be a special site, or this is a reasonable probability, the authority should seek to make arrangements with the Environment Agency or SEPA for that Agency to carry out the inspection on behalf of the authority.[16]

6.09 Where contaminated land has been identified by the local authority, a decision has to be taken as to whether or not the land is required to be designated as a special site, by reference to the Regulations and in accordance with any statutory guidance.[17] This may be a committee decision or one delegated under proper procedures to an officer. Before making the decision, the local authority must request the advice of the appropriate Agency, and have regard to it.[18] Having taken the advice of the appropriate Agency, either for or against designation as a special site, the local authority has a discretion not to follow it, provided it has regard to it. Where it is decided that land is required to be designated as a special site then notice must be given to those persons set out in section 78C(2).[19] For the purposes of section 78C the relevant persons are[20]:

(a) the appropriate Agency;
(b) the owner of the land;
(c) any person who appears to the local authority concerned to be in occupation of the whole or any part of the land; and
(d) each person who appears to that authority to be an appropriate person.

6.10 Where the appropriate Agency disagrees with a decision that land should be designated a special site, it may, by giving a counter-notice and a statement of reasons before the expiry of 21 days from the statutory notice being given, require the authority to refer the matter to the Secretary of State.[21]

Agency notice as to special site designation

6.11 In parallel with the procedure set out above where the local authority identifies and designates a special site, the appropriate Agency can also give notice to the local authority in whose area the land is situated that it

---

[16] See Chapter 4.
[17] Sections 78C(8) and 78B(2).
[18] Section 78C(3).
[19] Section 78C(1)(b).
[20] Section 78C(2). See Chapter 5.
[21] Section 78D(1).

considers contaminated land should be designated as a special site.[22] The local authority are then obliged to consider the Agency's notice and decide whether the land is required to be designated as a special site or otherwise, and having made their decision shall give notice to the Agency accordingly and to all the other relevant persons.[23] In other words, it is still the local authority which designates the site; the role of the Agency is to initiate the procedure which may lead to designation.

6.12 So what happens if the local authority disagrees with the Agency in this instance? Under section 78D, if a local authority gives notice of its decision to the Environmental Agency that the land is not to be designated as a special site, then before the expiry of the period of 21 days beginning on the day the notice was given, the Agency may serve a counter-notice that it disagrees with the decision, together with a statement of its reasons for disagreeing; the local authority must then refer the decision to the Secretary of State and send to him a statement of its reasons for the decision it has reached.[24] At the same time that the appropriate Agency gives the local authority notice that it disagrees with its decision, it must also send to the Secretary of State a copy of the notice of disagreement, together with a statement of reasons why it disagrees with the local authority.[25]

6.13 Where a local authority refers the decision to the Secretary of State under these provisions, it must give notice of the fact to the relevant persons.[26]

Disagreement between local authority and Environmental Agency: Secretary of State's decision

6.14 Basically, there are two ways in which disputes between the local authority and Agency may be referred to the Secretary of State. First where the authority decides that a site is a special site and the Agency disagrees; secondly, where the Agency gives notice that it considers the site to be a special site and the local authority disagrees.

6.15 Where a dispute has arisen between the local authority and the Environmental Agency as to whether a site should or should not be designated, then the Secretary of State, by section 78D(4), on reference to him by the authority, in accordance with the procedure in the previous paragraph:

(a) may confirm or reverse the decision of the local authority in respect of the whole or any part of the land to which it relates; and

[22] Section 78C(4).
[23] Section 78C(5).
[24] Section 78D(1).
[25] Section 78D(2).
[26] Section 78D(3).

(b) shall give notice of his decision on the referral:

(i) to the relevant persons, and
(ii) to the local authority.

6.16 No procedure is laid down for how the Secretary of State is to arrive at his decision, and it seems to be envisaged that he will deal with the matter on the basis of the statement of reasons given by each party under sections 78D(1) and (2). No doubt the Secretary of State may request further information if he so chooses, and may consider representations submitted by third parties and other Government departments and agencies. The Secretary of State's decision here seems essentially an administrative rather than a judicial one; accordingly it may be subject to challenge by a party with sufficient interest on ordinary administrative law principles.

Notice of decision: when designation as special site takes effect

6.17 There are a number of somewhat complex statutory provisions as to when a decision that a site shall be designated as a special site takes effect. A distinction is drawn between the decision that a site is required to be designated and the effective date of designation. Generally, where a local authority makes a decision of its own volition that land shall be designated a special site,[27] or accepts the decision of the Environment Agency to that effect,[28] the decision takes effect on the day after whichever of the following events first occurs[29]:

(a) the expiry of twenty-one days beginning with the date on which notice of the local authority's decision is given to the appropriate Agency; or
(b) if the Agency gives notice that it agrees with the decision of the local authority, the giving of that notice.

6.18 Where the decision takes effect as above, the local authority must give notice of that fact to the relevant persons.[30] The point of the 21 day period at (a) above is that under section 78D(1) the Agency may within that period give notice that it disagrees with the decision, thereby triggering a reference to the Secretary of State.

6.19 There is a distinction between the time when a decision that a site is required to be designated as "special" takes effect, as set out above, and the time when the designation itself shall take effect. It is the notice of the

[27] Section 78C(1)(b).
[28] Section 78C(5)(a).
[29] Section 78C(6).
[30] *ibid.*

decision given to the relevant persons which has effect as the designation, and it does so as from the time the decision takes effect.[31] However, where a decision of a local authority is referred to the Secretary of State because of a disagreement with the Environment Agency under section 78D, the decision does not take effect until the day after that on which the Secretary of State gives notice of his decision: it then takes effect as confirmed or reversed by him.[32] Where this is the case the notice given by the Secretary of State of his decision to the local authority and to the relevant persons takes effect as a designation as from the same time as the decision takes effect.[33]

6.20 These provisions need to be read in conjunction with the restrictions on serving a remediation notice until a three month period has expired, as set out in section 78H(3). This period runs from the date on which the notices of decision are given to the "relevant persons", including the owner of the land, any apparent occupiers, and each apparent appropriate person.[34] There are therefore no less than three relevant dates envisaged by the legislation for special sites:

(1) the date that the decision takes effect;
(2) the effective date of designation; and
(3) the date of service of notices on relevant persons (from which the three month period is calculated).

Special sites: adoption by Agency of remediation notice

6.21 We have already seen that one of the factors in framing the Regulations on special sites turns on the relative ability of local authorities and the Environment Agency/SEPA to deal with the problem in question, and in particular which of them has the necessary expertise.[35] It is therefore not surprising to find provisions under which the appropriate Agency may adopt a remediation notice where it later becomes apparent that the land in question should be treated as a special site.[36] If it does so, it must give notice of that decision to the appropriate person and to the local authority,[37] and the remediation notice then takes effect henceforth as a remediation notice given by that Agency.[38] It is provided that this process shall not affect the validity of the original remediation notice.[39] The effect

---

[31] Section 78C(7).
[32] Section 78D(5).
[33] Section 78D(5).
[34] See Chapter 9.
[35] Section 78C(10)(b).
[36] Section 78Q(1).
[37] Section 78Q(1)(a).
[38] Section 78Q(1)(b).
[39] Section 78Q(1)(c).

of adoption is that the local authority ceases to have jurisdiction as the enforcing authority, this passing to the Agency. However, where the local authority has already begun to act under section 78N to carry out remediation itself, the authority may continue with that action, and will still be able to use section 78P to seek to recover its costs.[40] This ability extends only to the "thing, or series of things", which the local authority has begun to do prior to adoption of the notice by the Agency; it does not extend to the remediation process in its entirety.

Special sites: termination of designation

6.22 If it appears to the appropriate Agency that a special site is no longer land which is required to be designated as such, they may give notice to the Secretary of State and to the local authority concerned terminating the designation as from the date specified in the notice.[41] If such a notice is given, this does not prevent all or any part of the land being designated as a special site on a subsequent occasion.[42] The Agency is required in exercising this function to act in accordance with any guidance for the purpose given by the Secretary of State.[43] Unlike the original designation of a special site, no procedure is provided for resolving disputes between the Agency and local authority, which may not necessarily want to see the designation terminated. The Agency's action under section 78Q(4) is presumably subject to judicial review on the normal grounds, but the local authority may also seek to persuade the Secretary of State to give site-specific guidance under Section 78W to the Agency on the issue.

Special sites: review

6.23 If and so long as any land is a special site, the appropriate Agency may from time to time inspect that land for the purpose of keeping its condition under review.[44] The purpose of this provision is not entirely clear; the review might lead either to further remediation requirements, or to possible withdrawal of special site status. The function is expressed in terms of a power rather than a duty. However, in exercising this function the appropriate Agency must act in accordance with government guidance.[45]

---

[40] Section 78Q(2).
[41] Section 78Q(4).
[42] Section 78Q(5).
[43] Section 78Q(6).
[44] Section 78Q(3).
[45] Section 78Q(6).

Chapter 7

Relationship with Other Powers

Summary

The application of the statutory duties on contaminated land should be seen in the context of other provisions which may have potential application to contaminated land situations. In particular, these are powers in relation to water pollution, statutory nuisance, and land use planning.

In some cases (described in Chapter 2) the existence of other powers has legal effects in terms of excluding or restricting the use of Part IIA; for example, integrated pollution control and waste management licensing. Even where this is not the case, the enforcing authority will need to be aware of other possibly applicable powers, as these may affect its stance. Some of these aspects are addressed by non-prescriptive guidance.

Introduction

7.01 As described in Chapter 2, there are certain types of site or situations where the contaminated land provisions are not applicable. In addition to these "no go areas", there are other occasions where there is a potential overlap between the contaminated land and other provisions, which are considered in this Chapter.

Water pollution powers: works notices

7.02 Contaminated land may be a source of pollution of both surface water and groundwater. Powers available to the Environment Agency or SEPA to prevent, remedy or mitigate pollution of controlled waters have been contained in section 161 of the Water Resources Act 1991 and earlier provisions for a number of years.[1] Those provisions include powers to

[1] See Stephen Tromans and Robert Turrall-Clarke, *Contaminated Land* (London, Sweet & Maxwell, 1994) pp. 81–83.

recover the expenses reasonably incurred in taking such action from the person who caused or knowingly permitted the polluting matter to be present in the controlled waters, or to be present at the place from which it is likely to enter such waters.[2] As we pointed out in *Contaminated Land*,[3] these powers present a potentially heavy threat to the owner or occupier of contaminated land, but in practice the power is not attractive to the enforcing authority, involving as it does the expenditure of potentially large sums with no certainty of ultimate recovery. Perhaps for that reason, the section 161 powers were little used in practice.

7.03 This main deficiency in section 161 was recognised and rectified by the new powers contained in sections 161A–D of the Water Resources Act 1991[4] and, for Scotland, sections 46A–D of the Control of Pollution Act 1974.[5] These provisions create a new type of notice, the "works notice", which may be served where it appears to the Environment Agency or SEPA that any poisonous, noxious or polluting matter, or any solid waste matter, is likely to enter, or to be or have been present in, any controlled waters. The notice is to be served on any person who, as the case may be:

(a) caused or knowingly permitted the matter in question to be present at the place from which it is likely, in the opinion of the Agency/SEPA. to enter controlled waters; or
(b) caused or knowingly permitted the matter in question to be present in any controlled waters.

7.04 The legislation states that the works notice may require the carrying out of specified works or operations for the purpose of preventing the polluting matter entering controlled waters, or where this has already happened, removing the matter, remedying or mitigating the pollution, and restoring the waters and any dependent flora or fauna, so far as is reasonably practicable.[6]

7.05 The more detailed requirements of works notices are provided for in section 161A and in Regulations made thereunder.[7] These include a requirement to reasonably endeavour to consult with the intended recipient of a works notice before serving it.[8] However, failure to consult in this way is not of itself a ground for regarding a works notice as invalid.[9] Other sections deal with the grant of rights of entry necessary to comply with a works notice,[10] appeals against notices,[11] and with a works notice.[12]

---

[2] Section 161(3).
[3] Note 1 above, p. 82.
[4] Inserted by the Environment Act 1995, Sched. 22, para. 162.
[5] Inserted by the Environment Act 1995, Sched. 22, para. 29(22).
[6] Water Resources Act 1991, s.161A(2); Control of Pollution Act 1974 s.46A(2).
[7] Water Resources Act 1991, s.161A(3), (5), (7)–(9); Control of Pollution Act 1974, s.46A(3), (5), (7)–(9). See the Anti-Pollution Works Regulations 1999 (S.I. 1999 No. 1006).
[8] Water Resources Act 1991, s.161A(4); Control of Pollution Act 1974, s.46A(4).
[9] Water Resources Act 1991, s.161A(6); Control of Pollution Act 1974, s.46A(6).
[10] Water Resources Act 1991, s.161B; Control of Pollution Act 1974, s.46B.
[11] Water Resources Act 1991, s.161C; Control of Pollution Act 1974, s.46C.
[12] Water Resources Act 1991, s.161D; Control of Pollution Act 1974, s.46D.

Works notices and remediation notices compared

7.06 There may be situations where it is clear that a remediation notice cannot be served, for example, where all contamination has already passed from the soil to controlled waters, but in many cases the circumstances which require the service of a remediation notice will also be those which would allow service of a works notice. The provisions on works notices in the 1991 Act and the accompanying regulations are terse when compared with those on remediation notices, and do not contain anything equivalent to the sophisticated provisions on allocation and apportionment of liability of the latter; nor is there any provision for detailed Government guidance, though there is a general power of direction by the Secretary of State. Service of a works notice could therefore be a more attractive option in terms of administrative simplicity to the regulator.

Relationship of contaminated land and water protection provisions

7.07 As explained in the previous paragraph, there may be cases where although land is not to be regarded as "contaminated", there may still be the possibility of clean-up requirements being imposed in relation to water pollution under sections 161 or 161A–D of the Water Resources Act 1991 or the equivalent Scottish provisions. Also, there may be cases where the two regimes overlap, raising the question as to which one should be used to deal with a specific problem. Situations can be foreseen where a chicken and egg situation could arise, with subsequent uncertainty for enforcing authorities and potentially liable parties alike.

7.08 The issue is addressed at Annex 1 of the Circular, paras 64–68, which provides a procedural description of the system. This refers to the policy statement of the Agency, *Policy and Guidance on the Use of Anti-Pollution Works Notices*, which sets out how the Agency intends to use the works notice powers in cases where there is overlap with the Part IIA regime. The effect of this policy, taken together with the legislation, is summarised as follows:

(a) the local authority should consult the Agency before determining that land is contaminated under Part IIA;

(b) where the authority has identified contaminated land that appears to be affecting controlled waters, it is required by the Guidance to consult the Agency, and to take into account any comments of the Agency with regard to remediation requirements. This allows the Agency to indicate the type of measures it would require in any works notice if such notice were served;

(c) where the Agency identifies any land which is affected by contamination and is causing actual or potential water pollution, it will notify the local authority, to enable the authority to identify the land as contaminated under Part IIA (if it is appropriate to do so); and

(d) in any case where contaminated land is identified under Part IIA, the Part IIA remediation notice procedure should normally be used rather than the works notice procedures. This follows from the fact that Part IIA involves a duty to serve a remediation notice, whereas service of a works notice is discretionary.

7.09 This advice does not address the situation where the Environment Agency is already taking enforcement action under the Water Resources Act 1991 before the local authority becomes involved. If such action is in progress, then the authority should consider what standard of remediation would be achieved by it. If the authority is satisfied that the standard would be such as to deal with all the relevant pollutant linkages, then subsection 78H(5)(b) will preclude service of a remediation notice, and the person who is carrying out, or will carry out, the action is required to prepare and publish a remediation statement.

7.10 The situation can therefore perhaps be summarised as follows:

1. If the Agency is already using its works notice powers, the local authority should not serve a remediation notice if satisfied that all the problems which arise from the land being contaminated will thereby be resolved.
2. If the local authority is using the contaminated land provisions, there is nothing as a matter of law to prevent the Agency using its works notice powers. However, its own policy should preclude the duplication of regulatory effort, and therefore it is likely to confine itself to giving site specific guidance to the local authority.
3. If the problem relates to controlled waters which are not subject to further contamination from the land in question, then the only remedy for such historic pollution will be through the works notice powers.
4. Difficulties could arise in the "chicken and egg" type of situation where neither the Agency nor the local authority has yet taken action, each perhaps hoping the other may do so first. Here it will be relevant that the local authority is under a duty to act, whereas the Agency only has power to act. The Agency's own guidance suggests that Part IIA should be used in such cases.

The Groundwater Regulations

7.11 The Groundwater Regulations 1998[13] came into force on April 1, 1999 and complete the transposition into U.K. law of the E.C. Groundwater Directive.[14] The Regulations do not apply to activities for which a waste

---

[13] S.I. 1998 No. 2746.

[14] [1980] O.J. L20, 80/68/EEC. Draft Guidance on the 1998 Regulations was issued for consultation in November 1999.

management licence is required.[15] One of the aspects of the Regulations is the need to prevent or control the disposal or tipping of List I or List II substances[16] in circumstances which might lead to an indirect discharge of those substances to groundwater, as well as other activities which might lead to such discharges. The Regulations deal with this issue as follows:

1. Causing or knowingly permitting such disposal or tipping without authorisation is treated as an offence under section 85 of the Water Resources Act 1991 or section 30F of the Control of Pollution Act 1974.[17]
2. Authorisation for these purposes may be granted by the appropriate Agency and may be subject to conditions.[18] Contravention of the conditions of such authorisation is an offence.[19]
3. Where a person is carrying on, or is proposing to carry on, any activity in or on the ground which might lead to an indirect discharge to groundwater of a List I substance, or to pollution of groundwater, by a List II substance, the appropriate Agency may by notice in writing either prohibit the carrying on of that activity, or may authorise it subject to conditions.[20] Contravention of any such authorisation is an offence.[21]

7.12 The Groundwater Regulations may obviously apply to situations where contaminated land is involved, including situations where it is being remediated or investigated. However, the main thrust of the Regulations is the *prevention* of groundwater pollution rather than its clean-up. Accordingly, they should be seen as complementing rather than supplementing the procedures of Part IIA. So, where for example, a remediation notice requires the excavation of contaminated material, and this activity might lead to listed substances being disturbed and so contaminating groundwater, the Agency may wish to impose conditions on the carrying out of those works by way of notice under the Regulations. To that extent, the requirements of the Regulations are no different from any other of the permits or consents which may be needed to carry out the remedial works.

7.13 A different situation might arise where Listed substances are continuing to leach into groundwater. Because the pollution is ongoing, this may lead to service of a Part IIA remediation notice, or to a works notice under section 161A. The Groundwater Regulations only allow control to be exercised over the tipping or disposal of Listed substances, or other activities in or on land. They cannot therefore be used to prohibit or

---

[15] Regulation 2(1)(d).

[16] See the Schedule to the Regulations as to these substances, which include many organics, metals and pesticides.

[17] Regulation 14(1)(a).

[18] Regulation 18.

[19] Regulation 14(1)(b).

[20] Regulation 19.

[21] Regulation 14(1)(b).

control passive discharges, where the activities in question have ceased. This limits their applicability in contaminated land situations.

Relationship to statutory nuisance

7.14 The provisions on contaminated land were to a great extent modelled on the existing law of statutory nuisance, which was already capable of applying to some of the effects caused by contaminated land. Having formulated a regime of the complexity of Part IIA, it could clearly be thought inappropriate to allow the less precisely defined system of statutory nuisance to remain as an alternative course of action.[22] Accordingly, the 1995 Act amends section 79 of the Environmental Protection Act 1990 to provide that no matter shall constitute a statutory nuisance to the extent that it consists of, or is caused by, any land being in a contaminated state.[23]

7.15 "Contaminated state" is defined more widely than the statutory definition in Part IIA, as the presence of substances causing harm, or the possibility of harm, or actual or likely pollution of controlled waters.[24] The key difference is that the concept of significance is omitted from the definition of "contaminated state". Accordingly, if land is in a condition such that contaminants in, on or under the land are causing harm, but that harm is not "significant", statutory nuisance will not provide a remedy. However, matters which would be a statutory nuisance, but do not arise from the "contaminated state" of the land as defined in section 79(1B), will still be subject to statutory nuisance procedures. Effectively, these are matters which do not involve "harm", as defined in section 78A(4), or water pollution. The Circular gives the example of stenches or other offences to human senses caused by the deposits of substances on land.[25]

7.16 The idea behind these provisions is clearly to avoid a situation where the carefully crafted scheme of Part IIA could easily be circumvented by a simple statutory nuisance abatement notice, though there is little evidence that these provisions were widely used hitherto in the context of contaminated land. Another important consequence is that it will not be open to private individuals to seek to make a direct complaint in relation to the "contaminated state" of land as "persons aggrieved"—a remedy which is of course a possibility for statutory nuisance under section 82 of the 1990 Act.

7.17 The saving provisions of the Environment Act 1995 (Commencement No. 16 and Saving Provision) (England) Order 2000 (S.I. No. 340) provide that in relation to abatement notices served or section 82 complaints or orders

---

22 DETR Circular, Annex 1, para. 59.

23 Environment Act 1995, Sched. 22, para. 89, inserting subss. 79(1A) and (1B). The provisions of Pt III of the 1990 Act on contaminated land are extended to Scotland by s.107 and Sched. 17 of the 1995 Act.

24 Environment Protection Act 1990, s.79(1B).

25 DETR Circular, Annex 1, para. 63.

made before April 1, 2000, such notices, complaints or orders (and any proceedings consequent upon them) shall continue to have effect despite the amendments made to section 79.[26] Any enforcement action already in hand under Part III can therefore continue uninterrupted by the Part IIA regime.

Planning law

7.18 Planning law presents one way in which problems of contaminated land can be addressed, but only if the development of land is in prospect. In such circumstances planning conditions or section 106 agreements can provide legally-binding mechanisms to ensure that risks from contamination are effectively addressed at the expense of the landowner or developer applicant.[27] It is likely that revised planning guidance will amplify PPG 23, *Planning and Pollution Control*, on the relationship between Part IIA and planning powers but until then the guidance in PPG 23 remains valid.[28]

7.19 The situation may arise where contamination comes to the attention of the district council as enforcing authority when development is in prospect—perhaps as a result of information supplied as part of a planning application. The fact that planning powers *may* be able to secure decontamination will not of itself discharge the enforcing authority's duty to serve a remediation notice. The appropriate course is rather to consider whether the appropriate things will in any event be done by way of remediation without service of a notice—if so then by section 78H(5)(b) service of a remediation notice will be precluded.[29] The authority should however bear in mind that whilst planning powers may be able to secure remediation in the event that development goes forward, the fact that planning permission will be implemented is not a foregone conclusion. This is a situation that may therefore need to be kept under review on a regular basis.

7.20 Local planning authorities may need to resist the temptation to elide the two regimes, which have different purposes. Planning law is concerned with ensuring that the risks consequent on developing and changing the use of contaminated land are properly identified and addressed. Part IIA is concerned with ensuring that unacceptable risks arising from the land in its current use are removed and allocating and apportioning the liability for the costs of doing so. To try to deal with complex liability issues in the context of a planning application or section 106 agreement risks going

---

[26] Article 3.

[27] See Stephen Tromans and Robert Turrall-Clarke, *Contaminated Land* (London, Sweet & Maxwell, 1994) Chapter 8; also Stephen Tromans and Robert Turrall-Clarke, *Planning Law, Practice and Precedents* (London, Sweet & Maxwell, 1991) Chapters 3 and 16.

[28] DETR Circular, Annex 1, para. 47.

[29] See Chapter 10.

beyond the proper bounds of land use planning. It may also result in an adverse costs award against the authority, as in one case involving an application for 80 dwellings on an old gasworks site in Leamington Spa, where the planning authority tried to impose detailed arrangements for long-term liability on the developer and its consultants.[30]

[30] *Warwick District Council v. Tom Pettifer Homes Ltd* (1995) 10 P.A.D. 665.

Chapter 8

REMEDIATION NOTICES—INTRODUCTION

Summary

At the heart of the system under Part IIA is the statutory duty to require the remediation of land which has been identified as contaminated, by service of a remediation notice. It is the remediation notice which states what should be done, and who should do it. The provisions are complex, and are discussed in the following chapters. The key issues are the determination of the remediation requirements and the "appropriate person" or persons to be responsible for complying with the notice. Remediation can include assessment and monitoring actions, as well as actual clean-up. Government guidance plays an important role in setting the remediation requirements, and a determinative role in deciding who should be liable and to what extent.

The basic scheme

8.01 The essential idea behind the Guidance is that the significant harm (or pollution of controlled waters) must be cleaned up or otherwise dealt with by those responsible for it. Because different present or past owners or occupiers, or indeed third-party perpetrators, may be responsible for substances in or under the land during a site's history, the principle is that they should only be responsible for the significant harm (or pollution of controlled waters) that those substances caused. The way in which the Guidance seeks to achieve this result is to make use of the same pollutant linkage concept as is central to the identification of contaminated land; that is, a harmful relationship between a contaminant, a pathway, and a receptor (or target).

8.02 A remediation notice will contain a "remediation action", which is the action to be taken by way of remediation, or a "remediation package"; that is a set or sequence of remediation actions, perhaps with associated timescales. "A liability group", that is a group of "appropriate persons" who are responsible for the linkage in question, bear responsibility to carry out the remediation action or package required to remediate that problem.

Each member of the group gets effectively the same notice, save that the apportionment of liability may differ as between them.

8.03 It can also be the case, where a number of distinct pollutant linkages exist on the same site, that separate remediation notices will be served, on different liability groups.

Duty to serve remediation notice

8.04 By section 78E(1) where land has been identified as contaminated land or designated as a special site, the enforcing authority shall serve on each person who is an "appropriate person" a notice (known as a remediation notice) specifying what that person is to do by way of remediation and the periods within which he is required to do each of the specified things.

"Pollutant linkage"

8.05 The term "pollutant linkage", though it appears nowhere in Part IIA itself, is an important concept in the guidance dealing with the definition of contaminated land.[1] The term refers to the relationship between the contaminant, a pathway which it can take, and a receptor which is susceptible to damage by the contaminant. The concept is also relevant for the purposes of the Guidance on who should be the appropriate person, and on how the remediation requirements should be framed, as will be seen from the next paragraph.

"Remediation" and associated terminology

8.06 Remediation is a defined term by section 78A(7), meaning:

(a) the doing of anything for the purpose of assessing the condition of:

(i) the contaminated land in question;
(ii) any controlled waters affected by that land; or
(iii) any land adjoining or adjacent to that land;

(b) the doing of any works, the carrying out of any operations or the taking of any steps in relation to any such land or waters for the purpose:

(i) of preventing or minimising , or remedying or mitigating the effects of, any significant harm, or any pollution of controlled

---

[1] See Chapter 2.

waters, by reason of which the contaminated land is such land; or
(ii) of restoring the land or waters to their former state; or

(c) the making of subsequent inspections from time to time for the purpose of keeping under review the condition of the land or waters.

8.07 It will be seen therefore that the concept of remediation extends through the assessment of land which is found to be contaminated (but not the original assessment as to whether it is contaminated) to the carrying out of preventive, remedial or restorative measures, and to subsequent inspection and review.[2] The term therefore encompasses the whole spectrum of clean-up techniques for contaminated land, from basic removal of contaminated material, to encapsulation or in-situ remediation, or the treatment of contaminated groundwater. The definition is not inherently limited to practicable or reasonable measures; rather the protection against unreasonable requirements is to be found in the provisions relating to the content and procedures for remediation notices, discussed below.

8.08 The Guidance contains its own terminology on remediation, using the following definitions[3]:

(a) "remediation action": any individual thing which is being, or is to be done by way of remediation (for example, excavating and removing material, installing a gas control system, etc.);
(b) "remediation package": the full set or sequence of remediation actions, within a remediation scheme, which are referable to a particular significant pollutant linkage (*i.e.* the totality of actions relating to a pollutant linkage);
(c) "remediation scheme": the complete set or sequence of remediation actions (referable to one or more significant pollutant linkages) to be carried out with respect to the relevant land or waters (*i.e.* the totality of actions relating to the land as a whole);
(d) "relevant land or waters": the contaminated land in question, any controlled waters affected by that land, and any land adjoining or adjacent to the contaminated land on which remediation might be required as a consequence of the contamination. It clearly therefore goes potentially much wider than the contaminated land itself[4];
(e) "assessment action": an action falling within section 78A(7)(a) (*i.e.* anything for assessing the condition of the land or controlled waters);

---

[2] In that sense it is wider than the meaning sometimes associated with remediation of just the clean-up stage: see DETR Circular, *Contaminated Land*, Annex 3, para. C.7.
[3] *ibid.*, para. C.8.
[4] See the definition at section 78G(7) in the different context of rights of entry and compensation

(f) "remedial treatment action": an action falling within section 78A(7)(b) (*i.e.* works or operations for preventing, remedying or mitigating harm or pollution);

(g) "monitoring action": an action within section 78A(7)(c) (*i.e.* the making of subsequent inspections for keeping the condition of the land or waters under review).

"Appropriate person"

8.09 Remediation notices are to be served on the "appropriate person" or, if more than one exists, on each of them. "Appropriate person" is defined by section 78A(9) to mean any person who is an appropriate person, determined in accordance with section 78F, to bear responsibility for any thing which is to be done by way of remediation in any particular case. The crucial words are therefore the reference to section 78F; it is this section which is decisive in allocating or channelling liability for remediation.

Chapter 9

Consultation before Service of Remediation Notice

Summary

Before serving a remediation notice the enforcing authority must reasonably endeavour to consult those who may be responsible for remediation. This requirement is coupled with a moratorium of at least three months following (in broad terms) the decision that the land is contaminated, intended to allow time for such consultation to take place. No remediation notice can generally be served within that period. Consultation and the three-month moratorium may be dispensed with where it appears that there is imminent danger of serious harm or pollution.

Consultation with affected parties and prescribed persons

9.01 By section 78H(1), before serving an enforcement notice, the enforcing authority must reasonably endeavour to consult:

(a) the person on whom the notice is to be served;
(b) the owner of any land to which the notice relates;
(c) any person who appears to be in occupation of the whole or any part of the land; and
(d) any person of any such other description as may be prescribed.

9.02 The consultation should concern what is to be done by way of remediation. The question of which additional persons need to be consulted, and the steps to be taken for the purpose of consultation, may be provided for by Regulations—see section 78H(2). In fact, the Regulations for England do not deal with this issue.

Consultation relating to possible grant of rights

9.03 Section 78G relates to the grant of rights which may be necessary to allow the recipient of a remediation notice to carry out the works concerned, whether on the land subject to the notice, or on adjacent or adjoining land. The relevant consent is required to be given by section 78G(2), and compensation may be claimed for the grant of such rights under section 78G(5).

9.04 Under Section 78G(3) an entirely separate consultation requirement arises before serving a remediation notice. The enforcing authority must reasonably endeavour to consult every person who appears to the authority:

(a) to be the owner or occupier of any of the relevant land or waters; and
(b) to be a person who might be required under section 78G(2) to grant, or join in granting, the necessary rights.

9.05 Here the consultation should concern the rights which the relevant person may be required to grant. A couple of examples may help to illustrate how these provisions will work and how they relate to the consultation requirements under section 78H.

Example 1

9.06 The contaminated land in question comprises a gassing landfill site. The remedial measures involved will include the provision of extraction and monitoring boreholes on adjacent land. As well as consulting the appropriate person, owner, occupier, etc., in relation to the landfill site, the authority would need to consult under section 78G the adjacent landowners on whose property the boreholes would need to be placed, concerning the rights they would be required to grant.

Example 2

9.07 The contaminated land question is an old industrial site which used to be operated by X Co. Ltd, which is still in existence. The site is now owned by Y plc. X Co. Ltd is a potential appropriate person to be made responsible for clean-up and as such would have to be consulted as to what is to be done by way of remediation under section 78H. Y plc would have to be consulted in two capacities: (1) under section 78H as the current owner of the site concerning what is to be done by way of remediation; and (2) under section 78G as someone who may well be required to give rights to entry to the site to X Co. Ltd in order to carry out the works, concerning the rights it would be required to grant.

The three-month period

9.08 Section 78H(3) provides that no remediation notice is to be served within a period of at least three months beginning and ending on the dates specified in paras 78H(3) (a)–(c). Two points need to be kept in mind with regard to this provision. First, it is subject to being overridden in cases of imminent danger of serious harm (see below). Secondly, whilst the intention may well be to allow time for the consultation processes required by sections 78G(3) and 78H(1) to take place, the period is not expressly a consultation period as such. The period is not linked to when consultation commences, and applies even if consultation is concluded satisfactorily within that period. In practice, in many cases the consultation process is likely to require a longer period than three months, in order to reach a satisfactory conclusion, or at least to offer reasonable prospects of such a conclusion.

When does the "moratorium" period begin and end?

9.09 Section 78H(3) contemplates three distinct situations:

(a) Where the land is not a special site and is simply identified by the local authority. Here the period begins with the date on which the land is identified as contaminated, and ends with the expiration of the period of three months beginning with the day on which notice was given as required by sections 78B(3)(d) or (4) to each person who appears to the authority to be an "appropriate person".
(b) Where the land is determined by the local authority to be a special site, the period begins with the making of that decision and ends with the expiration of the period of three months beginning either with the day on which notice of the decision taking effect was given under section 78C(6) to the relevant persons. Alternatively, if the decision has been referred to the Secretary of State to resolve, the three month period begins with the day on which the Secretary of State gave notice of his decision under section 78D(4)(b) to the relevant persons and to the local authority.
(c) Where the land is considered by the Agency to be a special site and notice has accordingly been given by the Agency to the local authority under section 78C(4), then the period begins with the day on which that notice was given and ends with the expiration of the period of three months beginning with:
    (i) where the local authority decides the land is indeed a special site, the date on which the authority gives notice under section 78C(6) to the relevant persons of the decision taking effect;
    (ii) where the local authority decides the land is not a special site, and the Agency does not give notice within the prescribed 21-day period that it disagrees, the day following the expiration of the 21-day period. In other words, the three month period does not begin running until the period for possible dispute as to special site status has passed;
    (iii) where the local authority decides the land is not a special site, the Agency disagrees within the 21-day period, and the

decision is then referred to the Secretary of State, the day on which the Secretary of State gives notice of his decision under section 78D(4)(b) to relevant persons and to the local authority.

9.10 As will be appreciated, the period will in total usually be more than three months—the principle is that the three months does not begin to elapse until:

(a) it is clear that no dispute exists as to whether the site is a special site, or if there is a dispute, it is settled; and
(b) the statutory notices of the decision have been given.

Calculating the period

9.11 In calculating the end date for the period, after which a remediation notice can be served, the starting point is to ascertain the correct day on which time begins to run under section 78H(3). The period ends with the expiry of three months beginning with that date.

9.12 The starting point will in many cases be the date on which the statutory notice was "given". This will probably be treated as being the date on which the notice reached its relevant destination[1] rather than the date on which it actually came to the attention of the intended recipient.[2] This will usually be the date on which the notice was personally delivered,[3] or was received through the post. If posted, the notice will be presumed to have been delivered in the ordinary course of the post, but this may be rebutted by proof of late delivery, or no delivery at all.[4] It should also be possible to give the notice by fax, but the legal uncertainties involved mean that service personally or by post will be the safest means to use.[5]

9.13 The next question is what is meant by "three months". "Month" here means calendar month, by reference to the Interpretation Act 1978, ss.5 and 22(1), and Schedule 1. The courts will presumably follow the "corres-

---

[1] See section 160 of the 1990 Act, and the further discussion in Chapter 12.

[2] *Papillon v. Brunton* (1860) 5 H. & N. 518; *Price v. West London Investment Building Society* [1964] 1 W.L.R. 616; *Lord Newborough v. Jones* [1975] Ch. 90; *Sun Alliance and London Insurance Co. Ltd v. Hayman* [1975] 1 W.L.R. 177, at 183, 185, *Re: 88 Berkeley Road, NW9* [1971] Ch. 648.

[3] Putting it through the letter box of the relevant premises may be sufficient: *Lambeth London Borough Council v. Mullings* [1990] R.V.R. 259.

[4] Interpretation Act 1978, s.7; *R. v. London Quarter Sessions Appeal Committee, ex p. Rossi* [1956] 1 Q.B. 682; *Beer v. Davies* [1958] 2 Q.B. 187; *Hewitt v. Leicester City Council* [1969] 1 W.L.R. 855.

[5] The fax must be shown to have reached its destination in a legible form: see *Ralux NV/SA v. Spencer Mason*, *The Times*, May 18, 1989; *Hastie and Jenkerson v. McMahon* [1990] 1 W.L.R. 1575; *Pearshouse v. Birmingham City Council* [1999] E.H.L.R. 140.

ponding date rule" so that the period will expire on the same number day in the month as the start date.[6] Thus if the statutory notice was given on, say, February 13, the period would end on May 13. If there is no corresponding day in the end month, because it is a short month, then the period will end on the last day of the month. So if notice was given on November 30, the period would end on February 28 (or 29 if a leap year).

The consequences of inadequate notification on the period

9.14 As mentioned above, the three month embargo on service of a remediation notice is calculated by reference to the date on which the notices required by statute were given under sections 78B(3)(d), 78B(4), 78C(1)(b), 78C(6) or 78D(4)(b) as the case may be.

9.15 In relation to the simplest cases of sections 78B(3)(d) or 78B(4), where there is no question of the land being a special site, the notices are required to be given simply to each person who appears to the authority to be an appropriate person. In the other cases, where the issue arises as to whether the land may be a special site, the notices must be given to "the relevant persons", defined as[7]:

(a) the appropriate Agency;
(b) the owner of the land;
(c) any person appearing to the local authority to be in occupation of the whole or part of the land; and
(d) each person who appears to the authority to be an appropriate person.

9.16 There is no obvious reason why section 78H(3)(a)–(c) should distinguish between putative "appropriate persons" and "relevant persons" in this way.

9.17 In any event, failure to give the necessary notices to all those who should receive them could have serious consequences; until the notices have been given, the three month moratorium period cannot begin to run, and until it expires "no remediation notice shall be served on any person" in relation to the contaminated land. This is therefore a matter that goes to the *vires* of the authority to serve a notice, and it seems is an issue that could be raised by a person on whom notice had in fact been served, if he can show that there are others to whom notice should have been given, irrespective of whether he himself was prejudiced.[8] Also the prohibition is on serving any notice in relation to the land identified as contaminated, not in

[6] *Dodds v. Walker* [1981] 1 W.L.R. 1027.
[7] Sections 78C(2) and 78D(7).
[8] Contrast *O'Brien v. London Borough of Croydon* [1999] J.P.L. 47 at p. 52.

relation to the particular pollutant linkage[9]; this means that the authority will have to consider who are the appropriate persons for all significant pollutant linkages and serve notices accordingly if the problem is to be avoided. On the other hand, the requirement to give notice relates to those persons who appear to the authority to be appropriate persons—a subjective test based on knowledge at the relevant time, rather than hindsight. Should other appropriate persons come to the attention of the local authority subsequently, then a further notice can be given under section 78B(4), and the three months will be calculated from that later notice.[10]

Cases of imminent serious harm

9.18 Neither the duty to consult under section 78G(3) and section 78H(1), nor the three month moratorium under section 78H(3), preclude the service of a remediation notice in any case where it appears to the enforcing authority that the contaminated land in question is in such a condition, by reason of substances in, on or under the land, that there is imminent danger of serious harm, or serious pollution of controlled waters, being caused—see sections 78G(4) and 78H(4). The use of the words "imminent danger" and "serious harm" obviously indicate a situation going beyond the criteria of "significant harm" and "significant possibility" of such harm, which are the test for contaminated land.

Failure to consult

9.19 The imposition of an express duty to consult avoids the difficulties that might otherwise arise in deciding whether any duty would arise on grounds of fairness or natural justice, and the scope and content of any such duty.[11] The fact that the statute expressly indicates who is to be consulted probably precludes success by others in arguing that they should have been consulted.[12]

9.20 The case law on consultation requirements is extensive,[13] but it can be anticipated that the context of the consultation will be relevant, and that

---

[9] This is not surprising, since the pollutant linkage concept is found only in the Guidance and appears nowhere in the legislation.

[10] Curiously, there is no similar provision in relation to the notices required to be given in relation to Special Sites under sections 78C and 78D.

[11] For example, in *R. v. Falmouth and Truro Port Health Authority, ex p. South West Water Ltd* [1999] Env. L.R. 833 it was held that there was no duty to consult the prospective recipient of an abatement notice, this being a matter of discretion. See also *R. v. Birmingham City Council, ex p. Terrero* [1993] 1 All E.R. 530.

[12] See *Bates v. Lord Hailsham* [1972] 2 W.L.R. 1373; *R v. Secretary of State for Education and Employment, ex p. Morris*, *The Times*, December 15, 1995.

[13] For example, see Michael Supperstone Q.C. and James Goudie Q.C., *Judicial Review* (London, Butterworths, 1997), Chapter 7.

the courts are likely to view seriously any failure to comply with the statutory requirements here, given the serious consequences of a remediation notice. Whereas the consultation requirements in relation to works notices for water pollution state specifically that failure to comply with the requirement to consult will not render the notice invalid, or invalidly served,[14] there is no such provision for remediation notices. On the other hand, the requirement is to "reasonably endeavour" to consult, and as such is not an absolute requirement. It will be a question of fact whether reasonable endeavours have been made.

9.21 A court will no doubt be concerned to see that the consultation followed a fair process, involving[15]:

(a) consultation while the proposals for the notice are still formative;
(b) the provision of adequate information on which a response can be based (including disclosure of any technical data);
(c) allowing adequate time in which to respond, bearing in mind the complexity of the matters concerned; and
(d) conscientious and open-minded consideration of the response.

9.22 One other factor also needs to be borne in mind given the nature of the matters being consulted on here: the parties consulted may have very different interests in whether, and on what terms, a remediation notice is served. The authority may therefore well find itself in receipt of conflicting representations. The key principle to follow would seem to be that of fairness. Clearly the authority should not act so as to favour one consultee over another. The issue will also arise as to whether the authority should give parties the chance to comment on representations made by other consultees.[16] Provided the authority treats all parties fairly in this respect, it might be argued that the best course would be to allow that opportunity, subject to the time constraints imposed by the legislative guidance.

9.23 However, it ought not to be overlooked that the authority's functions are being exercised in the public interest to remedy or avoid harm or pollution. It is possible to envisage circumstances where allowing each party to comment on the other's representation might impact adversely on the effective discharge of those public functions. One solution which may be helpful is to convene a meeting (rather like a "time and place" meeting in planning law) which consultees and their experts could attend. Such a meeting would have the elements of an informal hearing, *i.e.* the chair

---

[14] Water Resources Act 1991, s.161A(6); Control of Pollution Act 1974, s.46A(6).

[15] For example, see *R. v. Warwickshire County Council, ex p. Bailey* [1991] C.O.D. 284; *R. v. Gwent County Council and Secretary of State for Wales, ex. p. Bryant* [1988] C.O.D. 19; *R. v. Brent London Borough Council, ex p. Gunning* (1986) 84 L.G.R. 168; *R. v. North and East Devon Health Authority, ex p. Coughlan* [2000] L.G.L.R. 1, 38; *R. v. Wrexham Borough Council, ex p. Wall and Berry* [2000] J.P.L. 32, 50.

[16] For cases suggesting there is generally no duty to consult further unless the proposal changes radically, see *R. v. Islington L.B.C., ex p. East* [1996] I.L.R. 74; *R. v. Secretary of State for Wales, ex p. Williams* [1996] C.O.D. 127.

could identify issues from the representations made and invite comment and discussion. The practicalities of the consultation exercise are discussed in the following paragraph.

The practicalities of consultation

9.24 The starting point for considering how consultation prior to serving a remediation notice is to be conducted are the words of sections 78G and 78H themselves, together with any Regulations made under section 78H(2). Issues to be considered are:

1. How to initiate consultation.
2. How much time to allow.
3. How to draw up the list of consultees.
4. Whether to consult on a proposed remediation notice in draft, or on some other basis.
5. Whether to make the process public, and if so, to what extent?.
6. How to deal with those who do not wish to cooperate.
7. Whether to invite further comment on responses received.
8. How and when to draw the process to a conclusion.
9. Whether to invite all consultees to a meeting.

9.25 The difficulty of some of these issues, and the serious likelihood of challenge in the event that the process is defective indicate the importance of establishing a considered and clear set of ground rules for consultation in advance of the process. The local authority's written strategy may be the right place for such ground rules.

9.26 In particular, whilst the interests of transparency might suggest that a draft remediation notice should form part of the consultation process, this may create potential problems for the authority. Information may be incomplete at the time of consultation, and the use of a draft notice may invite comparison with the final version on appeal, with the authority either being required to justify changes, or to explain why changes have not been made. If the draft does not change substantially, the argument may arise that the authority has effectively pre-judged the issue by producing a draft notice. It is suggested that it will be preferable therefore to consult on the basis of a general document covering:

1. The condition of the land which is regarded as rendering it contaminated
2. The reasons why harm or the possibility of harm are regarded as significant.
3. The range of remediation measures which at this juncture appear appropriate.
4. Whether there is the prospect of voluntary remediation.
5. Who appears to be the appropriate persons to take responsibility for the various aspects of remediation, and whether there are

circumstances which would bring the case within one of the exclusion tests in the Guidance.

9.27 This approach also has the advantage of being consistent with the non-prescriptive guidance as to the type of discussion and consultation which the authority may find helpful.[17]

[17] DETR Circular, *Contaminated Land*, Annex 2, paras 6.10–6.17.

Chapter 10

Restrictions on Serving Remediation Notices and Procedure where a Notice Cannot be Served

Summary

There are a number of instances where the enforcing authority is precluded from serving a remediation notice. These are set out in section 78H(5) and are discussed in this Chapter. Where these circumstances apply, then the legislation provides for either a remediation statement to be prepared and published by the responsible person (indicating what is going to be done by way of remediation) or a remediation declaration to be prepared and published by the enforcing authority (indicating why the authority is precluded from serving a notice).

Cases where no remediation notice may be served

10.01 Under section 78H(5) the enforcing authority shall not serve a remediation notice if one or more of the conditions set out in that section apply:

(a) The authority is satisfied there is nothing by way of remediation they could specify in a notice, taking into account sections 78E(4) and (5). These subsections deal with the only things that may be contained in a remediation notice, namely those that are reasonable having regard to the costs involved, the seriousness of the harm or of the pollution of controlled waters, etc., and having regard to Government guidance. In other words, this situation will arise where nothing could be required to deal with the contamination which would not involve unreasonable expense—a somewhat unlikely situation.[1]

[1] It should be remembered that at the very least the notice may require further investigation, or monitoring so as to keep the condition of the site under review.

(b) The authority is satisfied that appropriate things are being or will be done by way of remediation without the service of a notice. This is a very important provision in practice because it enables those who would be the appropriate persons to receive a remediation notice to give undertakings to carry out remediation as part of the consultation process preceding service of a notice. Anybody likely to be served with a remediation notice may well be wise to consider offering a programme which they can afford, over a given time-span, if the alternative is to become embroiled in proceedings.[2] They may also point to redevelopment proposed for the land and a section 106 agreement or obligation requiring clean up as part of the redevelopment.

(c) It appears to the authority that the person on whom the notice would be served is the authority itself. This is another extremely important provision and is discussed separately below.[3]

(d) The authority is satisfied that it can exercise the powers conferred on it by section 78N to do what is appropriate by way of remediation. The cross reference to section 78N is awkward drafting and does little to aid intelligibility. However, it will cover, for example, cases where action is necessary to prevent serious harm or pollution of which there is imminent danger[4]; or where the appropriate person has entered into a written agreement with the authority for the authority to carry out works at his expense[5]; or where after reasonable inquiry, no person has been found on whom responsibility for a particular remediation action can be fixed.[6] Perhaps most importantly, it will also cover those cases where the authority considers that if it were to carry out remediation itself, it would not seek to recover all of its costs on grounds of hardship.[7] It will also cover the situation where the special provisions of section 78J on water pollution or 78K on escapes of contamination prevent some particular remediation requirement.[8]

Remediation declarations

10.02 Under section 78H(6), where sections 78E(4) or (5) apply, so that service of a notice is precluded by the unreasonableness of the costs that would be involved, the authority must serve a remediation declaration stating why

[2] See further Chapter 11.
[3] Paragraph 10.09.
[4] Section 78N(3)(a).
[5] Section 78N(3)(b).
[6] Section 78N(3)(f).
[7] Section 78N(3)(e).
[8] Section 78N(3)(d). As to these provisions see Chapter 15.

they consider themselves precluded from specifying the relevant things in the remediation notice. This declaration must record the reasons why the authority would have specified those things, and the grounds on which it is satisfied it is precluded from doing so. The possible reasons for preclusion are that the proposed requirements are not considered to be reasonable, having regard to the likely cost involved and to the seriousness of the harm in question, or that the proposed requirements are contrary to guidance issued by the Secretary of State. The declaration must be prepared and published by the authority. Presumably, this means published in the local press by way of advertisement. The declaration must also be placed on the public register under section 78R(1)(c).

10.03 For the owner of land, the remediation declaration is something of a two-edged sword; on the one hand the polluter or owner may welcome the public recognition that the authority cannot require a particular type of remediation. On the other hand, the notice will make it clear that but for considerations of cost or Government guidance, such action would be required. A prospective purchaser may be understandably nervous that the circumstances which led the authority to be satisfied that the action cannot be required may change. The seriousness of the risk of harm may increase, possibly due to factors outside the owner's control; the cost of remedial techniques may fall, or cheaper techniques may become available; or Government guidance may change. Such factors, and others, may lead the authority to reconsider its decision under sections 78E(4) or (5).

10.04 Where an owner does not consider that the land should be the subject of a remediation declaration (*e.g.* because it is not within the definition of contaminated land at all) there is no statutory mechanism for appeal, though the declaration could no doubt be challenged by way of judicial review.

Remediation statements

10.05 Remediation statements represent a practical and important alternative to service of a remediation notice. They are issued under section 78H(7) in circumstances where there is something that can be done by way of remediation but as a matter of law, a remediation notice cannot be served. Those circumstances, as indicated above, are where:

(a) The authority is satisfied that the appropriate things are being or will be done by way of remediation by a person served with a remediation notice; this might be because of assurances that the works will be carried out voluntarily, or because, for example, a works notice under the Water Resources Act 1991 has been served requiring the appropriate action, or because such works are required by planning obligations.

(b) It appears to the authority that the person on whom a remediation notice should be served is the authority itself, *i.e.* because the authority is itself the appropriate person.

(c) The authority is satisfied that the powers conferred on it by section 78N (that is powers enabling them to remediate the contaminated land concerned) are exercisable.

10.06 Whereas a remediation declaration is prepared and published by the enforcing authority, it is the "responsible person" who must prepare and publish the remediation statement. This term is defined by section 78H(8). In situation (a) above, it is the person who is responsible for carrying out the relevant works. In cases (b) and (c) it is the authority. The sanction for failure to prepare and publish the statement within a reasonable time in case (a) is that under section 78H(9) the authority may do so itself and recover its reasonable costs. The statement must be placed on the public register under section 78R(1)(c).

10.07 The remediation statement shall include:

(a) the things which are being, have been, or are expected to be done by way of remediation;
(b) the name and address of the person who is doing, or has done, or is expected to do the things in question; and
(c) the period or periods within which these things are being or are expected to be done.

10.08 Where there are a number of persons who have agreed to carry out remediation jointly, then it appears that they will jointly be responsible for preparing and publishing the statement. The question arises as to what happens when the responsible person fails to prepare a remediation statement, or fails to comply with a remediation statement they have prepared. As mentioned above, failure to publish a remediation statement at all within a reasonable time means that the authority itself may prepare and publish the statement and recover its reasonable costs of so doing—see section 78H(9). This begs the question of what then happens if the responsible person, having produced a statement, fails to comply with it. The answer appears to lie in section 78H(10). If the authority was precluded from serving a remediation notice because it believed assurances that the relevant works would be carried out without a remediation notice being served, then if it appears that those assurances will not in fact be honoured, a remediation notice must be served under subsection 78H(10). This raises the question of why an authority would wish to follow the procedure of section 78H(9). Indeed, where there is a failure to publish the remediation statement, there seems no reason why the authority could not proceed to serve a remediation notice on the basis that it is not now satisfied that the appropriate things will be done.

Where the local authority is itself the appropriate person

10.09 In practice, 78H(5)(c) is a very important provision. Here it is the authority itself which is responsible for the contamination as original polluter or as current owner or occupier. In these circumstances the authority does not have to serve a remediation notice on itself. The question arises, and it may well be a frequent one, of what next is to happen in circumstances where the neighbouring landowners would expect the authority to serve notice upon itself but for subsection 78H(5)(c). The answer is that the authority is required by section 78H(7) to publish a remediation statement and should then take the appropriate steps to comply with it. The legislation presumes that the authority will act responsibly, and no doubt failure to do so could be the subject of judicial review proceedings.

10.10 Another difficulty relates to the situation where there are a number of appropriate persons, of whom the authority is one. The language of section 78H(5)(c) refers to the authority as "the person" (not "a person") on whom the notice would be served, and therefore does not expressly contemplate the position where the authority is one among a number of appropriate persons. In such circumstances, by virtue of section 78H(5)(c) the local authority would not serve a remediation notice on itself; instead it would prepare and publish a remediation statement. However, in respect of the others who are responsible it would serve a remediation notice. The situation will raise important issues of fairness and possible bias in terms of allocation and apportionment of liability. These are considered at Chapter 19, below.

Remediation notice where circumstances change

10.11 The circumstances under section 78H(1) where a remediation notice should not be served apply "if and so long as any one or more of the following conditions is for the time being satisfied in the particular case". It will be recalled that those conditions include the cases where there is nothing by way of remediation that can be specified in a notice, or the authority is satisfied that the appropriate things are being done, or the authority itself is the person on whom a notice should be served, or they have the power to carry out the works themselves. However, there may come a time when the "if and so long as" proviso ceases to apply. Where those circumstances were the only reason why a notice could not be served then, if those circumstances change, the authority is required by section 78H(10) to serve a remediation notice. Section 78H(10) expressly provides that such notice may be served without any further endeavours to consult the relevant persons, if and to the extent they have already been consulted pursuant to section 78H(1) concerning the things to be specified in the notice.

10.12 Even if the remediation actions described in the remediation statement are being carried out as planned, the enforcing authority may consider that further remediation is necessary, for example, where the action is phased, and a further phase can now be seen to be necessary, or where further significant pollutant linkages have been identified. In this situation, the authority will need to repeat the consultation procedures, and revisit the question of whether those additional measures will be carried out without a remediation notice being served.[9]

[9] DETR Circular, *Contaminated Land*, Annex 2, para. 8.28.

Chapter 11

Avoiding Service of a Remediation Notice

Summary

The person who is alerted to the fact that a remediation notice may be served on them would be well-advised to consider the means by which it may be possible to avoid the notice being served, either by drawing relevant circumstances to the attention of the enforcing authority, or by putting forward a voluntary scheme for remediation.

Using the consultation period

11.01 As indicated earlier, the owner, occupier or other "appropriate person" in relation to contaminated land is entitled to be consulted prior to the service of a remediation notice. Such a person will certainly have been informed of the fact that the land has been identified as contaminated under section 78B(3). However, in practice, it is likely that the fact that land is under consideration as potentially contaminated land will have been known some months previously, and indeed representations may have been put to the committee of the authority making the determination.

11.02 The period of consultation represents a valuable opportunity for those who are likely to be faced with service of a remediation notice. Once a notice has been served, an adversarial position will have been created which may then make further constructive dialogue more difficult. It may be possible to avoid that situation arising by providing the authority with relevant information, or by putting forward acceptable proposals for remediation.

11.03 The formal consultation process which is required by section 78H(1) relates only to "what is to be done by way of remediation." This will be an important focus for discussion, but the owner, occupier or other appropriate person may well wish to put to the authority other points, as discussed below.

Points for consideration

11.04 The owner, occupier or other appropriate person ("the consultee") may wish to raise with the authority a number of points which may preclude service of a notice, or the inclusion of certain requirements within it, or which may suggest that some other person would be the appropriate recipient of the notice:

(a) Arguments that the land is not "contaminated land" at all. The consultee may wish to put forward their own data, reports or risk assessments to suggest that the land does not in fact meet the statutory criteria for identification as contaminated, or for designation as a special site.

(b) Arguments that the consultee is not an appropriate person to receive a remediation notice at all, or is only an appropriate person with regard to certain aspects of the contamination, or should benefit from one of the exclusion tests contained in the statutory guidance. This will involve bringing to the attention of the authority relevant factual circumstances, and quite possibly the terms of past transactions or commercial agreements that may have a bearing on the allocation of responsibility.

(c) Information relevant as to how responsibility for a particular remediation action should be apportioned between the consultee and other appropriate persons. This may involve both the disclosure of past events and agreements, and also any current agreements reached as to how liability should be shared.

(d) Representations concerning the proposed remediation requirements of the authority, in particular, whether those requirements are reasonable having regard to the statutory criteria of section 78E(4) and to the Guidance. This may involve the consultee putting forward an alternative scheme, which meets the relevant criteria but at a lower cost. Clearly if more than one consultee is involved, such proposals are likely to carry more weight if they are put forward on a united basis.

(e) Facts relevant to whether the service and enforcement of a remediation notice would entail hardship to the person from whom the cost would be recoverable. Again, the guidance will be relevant to the discussion on this issue—see section 78P(2).

(f) Any circumstances which may restrict the requirements of a remediation notice under section 78J where water pollution is involved.

(g) Any circumstances relating to the origin and mobility of the substances where they have been subject to migration, which may affect the allocation of liability under section 78K.

(h) The timescale for any proposed remediation will be relevant if remediation can be achieved over a reasonable period of time, integrated into current use of the land so as to avoid, for example, loss of income, or so that the scheme is tax effective, then this may be a major consideration.

(i) Any proposed immediate action which will alleviate the most serious consequences and give time for longer term solutions to be evaluated. An example might be the creation of a temporary barrier or layer to prevent immediate exposure to pollutants.

Proposals for voluntary clean-up

11.05 By section 78H(5)(b), no remediation notice can be served so long as the enforcing authority remains satisfied that appropriate things are being done, or will be done, by way of remediation without the service of a notice.[1] The consultee wishing to avoid service of a remediation notice on this basis will need to satisfy the authority:

(a) that what is proposed is appropriate to address the harm, pollution or risk arising from the contamination; and

(b) that those measures either are already in progress, or will take place, and (presumably, though not expressly stated) that they will be satisfactorily completed on an acceptable timescale.

11.06 The first of these issues is essentially a technical one, where clearly views may differ as to the preferred solution. Here it is important to note that the test is whether the measures are "appropriate", not whether they are the optimal measures or constitute the best practicable environmental option, or use the best available technology, or some similar test. The consultee may well have in mind a relatively cheap solution, such as excavation and consignment to landfill, to which the authority may have objections on environmental grounds; however, provided the course proposed is appropriate to deal with the threat presented by the contaminated land, such objections ought not to be relevant.

11.07 Other questions may arise where what is proposed is a new or experimental technique, which may be relatively untried. The authority will have to consider here whether doubts as to its ultimate effectiveness mean that it is not an appropriate technique in the context in which it is to be used: this may involve a risk assessment as to what would be the consequences of the technique not working. It will be relevant to take into account here those aspects of the Guidance dealing with the content of remediation notices, which bear on similar issues.[2]

11.08 The second issue is a factual one, namely whether the authority can be satisfied that the remediation will be carried out without the need to serve

---

[1] See DETR Circular, *Contaminated Land*, Annex 2, paras 8.3–8.8 on voluntary clean-up.

[2] See Chapter 12.

a remediation notice. The following issues may require consideration in that respect:

(a) Have the works already commenced, and if so at what stage are they? Do they appear to have been carried out satisfactorily so far, and what is the timescale for their conclusion?
(b) If the works have not yet been commenced, is there a clear scheme, designed by reputable consultants? Have engineers or contractors been appointed or have tenders been invited?
(c) Are all necessary permits or licences for the scheme (*e.g.* planning permission, waste management licences) in place or have they been applied for, and if so are they likely to be granted? Where there is a planning application for development which may be objectionable in itself on grounds of planning policy, a view will have to be taken of the likelihood of its being successful.[3] The outcome of the application, or any ensuing appeal, may be uncertain, and the owner may argue that clean-up under Part IIA should not be enforced until it is known whether the larger scheme will proceed: in those circumstances, it will no doubt be relevant to consider the urgency for remediation, and how that relates to the likely period of uncertainty, as well as the costs involved and possible hardship issues.
(d) Is the scheme adequately financed, or are there otherwise clear commercial incentives for its completion within a reasonable timescale, *e.g.* the completion of proposed development?
(e) What are the risks associated with a partly-implemented scheme which, for one reason or another, is not completed?
(f) Are the grant of rights or the permission of third parties necessary for the implementation of the scheme? If so, have those rights been obtained or appropriate arrangements made? If not, service of a remediation notice may be necessary in order to serve the grant of such rights under section 78G.
(g) Where a number of potential appropriate persons is involved, is there sufficient agreement between them as to what is to be done, by whom and at whose cost, to ensure implementation of the scheme?
(h) The authority may well wish to see a draft remediation statement, recording these matters.

11.09 It will be appreciated that there may well be difficult issues of judgment as to what degree of assurance the enforcing authority will need on these matters. The consultees might point to the fact that the authority can always keep the situation under review and serve a remediation notice if it becomes apparent that matters are not progressing as they should. On the

[3] See S. Tromans and R. Turrall-Clarke, *Contaminated Land* (Sweet & Maxwell, 1994), Chapter 8.

other hand, the authority will no doubt be mindful that a partly remediated site may leave just as great a problem (if not a greater one) than the site in its original condition.

Written agreement with appropriate person

11.10 Section 78N(2)(b) provides for the situation where an appropriate person has entered into a written agreement for the enforcing authority to undertake the necessary remediation at the expense of the appropriate person. In such circumstances, section 78N applies and accordingly by section 78H(5)(d) no remediation notice may be served. It seems debatable whether most local authorities would wish to enter into such arrangements, with the scope for potential contractual liability if things go wrong, unless they are themselves partly responsible for remediation.[4] However, there may be attractions in the authority undertaking the work where other statutory powers might be useful, for example, if the matter involves highway works or highway closure. The other key attraction is perhaps simply that the authority gets the job done as it would like. Clearly if the authority does undertake such works it will need to be insured for contractors' liability and will probably enter into an engineering works contract through its environmental services department.[5] The authority will also have to comply with applicable rules on procurement and with its statutory "best value" duties in relation to any contract for the works.

11.11 The agreement can only relate to those works which the appropriate person would have been obliged or is obliged to carry out; it cannot bind a third party. Accordingly, such agreements are perhaps more likely where there is a single appropriate person and a clearly defined set of remediation actions in contemplation. Clearly, the appropriate person will be concerned to have some control over the total costs for which they are responsible. The contract between the appropriate person and the authority will therefore probably contain provisions for the appropriate person to be consulted over any proposed variations in the works or in the price during the course of remediation.

---

[4] See Chapter 19.

[5] To avoid disputes at later cost-recovery stage, more than one estimate should be obtained and put to the appropriate person for his consideration.

Chapter 12

Drafting and Serving Remediation Notices

Summary

Unless one of the circumstances set out in section 78H(9) applies, the enforcing authority is under a duty to serve a remediation notice on each "appropriate person". The notice must specify what the recipient is to do by way of remediation and the period within which he is to do each of the things specified. The authority is subject to constraints under sections 78E(4) and (5) as to what a remediation notice may require. By section 78E(6), the requirements as to form, content and procedure are otherwise as set out in Regulations.

Duty to serve notice

12.01 When land has been identified as contaminated land or designated as a special site then the local authority or (as the case may be) the Environment Agency or SEPA is under a duty to serve a remediation notice on each person who is an appropriate person.[1] Identification and designation are in this sense formal legal processes which are described earlier.[2] As has already been explained, "identification" is not in this context to be equated with mere suspicion that land may be contaminated.[3] The duty to serve a notice is one that may be enforced by judicial review proceedings: it is not a matter for the general discretion of the authority.

The nature of remediation notices

12.02 The key purpose of a remediation notice is to specify what the recipient is to do by way of remediation and the periods within which he is to do

[1] Section 78E(1).
[2] See Chapters 5 and 6.
[3] See Chapter 5.

each of the things specified.[4] As described above, remediation is a defined term.[5]

12.03 The other vital point to note is that a remediation notice is personal in nature—it does not describe what is to be done in general with regard to a contaminated site, but rather what the individual recipient is to do. This is where the principle of referability arises.[6] Only certain remedial actions may be properly referable to an individual appropriate person, and accordingly the requirements of an individual remediation notice may be narrower than the works needed on the site as a whole.

12.04 The drafting of remediation notices presents many potential difficulties, some of which are addressed in this Chapter and some in the Chapters dealing with the identification of the "appropriate person" or persons to receive notices.[7] It is conceivable that the words of Harman L. J. in *Britt v. Buckinghamshire County Council*[8] may come to be applied to remediation notices:

> "Local authorities . . . have had practically to employ conveyancing counsel to settle these notices which they serve . . . instead of trying to make this thing simple, lawyers succeeded day by day in making it more difficult and less comprehensible until it has reached a stage where it is very much like the state of the land which their plaintiff has brought about by his operations—an eyesore, a wilderness and a scandal".

Jurisdiction

12.05 The enforcing authority will need to be satisfied that it has jurisdiction to serve a remediation notice. In the case of the Environment Agency or SEPA, the land will need to have been designated as a special site under sections 78C(7) or 78D(6).[9] In the case of a local authority, the land must have been formally identified as contaminated, and must be in the authority's area,[10] except where the authority relies on section 78X(2) to serve notice in respect of land outside its area. The authority must also be satisfied that it is not precluded from serving a notice by section 78H.[11]

---

[4] Section 78E(1).
[5] See Chapter 8.
[6] See section 78F(3).
[7] See Chapters 13–18.
[8] [1964] 1 Q.B. 77 at 87. Cited in the *Encyclopedia of Planning Law*, para. P171A.01.
[9] Section 78E(1)(a).
[10] Section 78E(1)(b).
[11] See Chapter 10.

Land covered by remediation notice

12.06 The notice must make clear to which land it relates. The primary legislation is silent on the issue, simply stating at section 78E that where land is identified as contaminated, a remediation notice shall be served on every appropriate person. However, the Regulations on form and content of remediation notices state that the notice must specify the location and extent of the contaminated land, sufficient to enable it to be identified.[12] This issue is both important and potentially difficult.

12.07 The area covered by an individual remediation notice will not necessarily be the same area as the whole of the land identified as contaminated: this potential difference stems from the fact that it is the presence of substances which determines the extent of the land to be identified as contaminated, whereas it is the involvement of the individual appropriate person that determines the land to be included within the remediation notice served on them. It is also possible that remediation notices requiring the doing of different things may be served in respect of the same land, if different pollution linkages arise on that land. This is clear from section 78(E)(2), which provides for the situation where different remediation actions are necessary to deal with the presence of different substances. The authority will have had to take a decision as to the extent of the land to be identified as contaminated at the earlier stage of identification; a further decision will be required when drafting the individual notice or notices. The extent of the notice clearly cannot be wider than the extent of the land identified as contaminated.

12.08 Whilst the Regulations do not state that a plan must be annexed to the notice showing the boundaries of the land covered by the notice, in many cases this will be advisable. The requirement of regulation 4(1)(b) is that the land can be identified, "whether by reference to a plan or otherwise". The question is whether it is possible to specify the location and extent of the land, sufficient to enable it to be identified without a plan. In this respect, the requirements for remediation notices may be compared with those applying to planning enforcement notices under section 173 of the Town and Country Planning Act 1990. Here the notice must specify "the precise boundaries of the land to which the notice relates, whether by reference to a plan or otherwise"[13]: even this more exacting requirement has been held to be capable of being satisfied by giving an address.[14]

12.09 It is also possible to argue by extrapolation from planning enforcement cases that there should be a degree of discretion as to the area covered by

---

[12] Regulation 4(1)(b).

[13] Town and Country Planning (Enforcement Notices and Appeals) (Amendment) Regulations 1992 No. 1904, reg. 3.

[14] *Wiesenfeld v. Secretary of State for the Environment* [1992] 1 P.L.R. 32.

the notice: the concept of the "planning unit"[15] may perhaps be translated in this connection into the "pollutant linkage unit".

Multiple remediation notices

12.10 The legislation envisages the possibility of several remediation notices requiring different works; they may well be served on different persons because they relate to different substances for which the persons in question are responsible.[16] This situation needs to be distinguished from the position where more than one person is responsible for the same works in respect of the same substance. Here the notice is served on each appropriate person, but must state the proportion of the cost, determined under section 78F(7), which each of them is to bear.[17]

12.11 In practical terms, the use of multiple remediation notices would seem to be fraught with difficulty. Where one site with a range of inter-linked pollution problems is involved the remediation works will also often in practice be inter-linked. The problem with multiple notices is that it may be difficult to determine appeals against such notices on an individual, isolated basis: any appeal hearing may well need to be on a joint basis to operate satisfactorily.

Consecutive remediation notices

12.12 It is clear that the remediation of contaminated land may involve a phased approach, and there is nothing to prevent this being reflected in a series of consecutive remediation notices which might require sampling, trials and feasibility studies prior to actual remediation, and operational monitoring thereafter. If after one remediation notice has been served the land continues to be "contaminated" then another notice should be served. This might occur if, for example, an authority determined that land was contaminated, but required monitoring or sampling to establish what should be done by way of remediation. Similarly, if it transpires that remediation has been inadequate or ineffective, and the land remains contaminated, another notice will have to be served.

12.13 The Guidance refers to the fact that a phased approach may be necessary, with different actions being carried out in sequence.[18] For example, the land may have been identified as contaminated, but further

---

[15] See, *e.g. Gregory v. Secretary of State for the Environment* [1990] 1 P.L.R. 100; *Richmond upon Thames London Borough Council v. Secretary of State for the Environment* [1988] J.P.L. 396; *Ralls v. Secretary of State for the Environment* [1998] J.P.L. 444.

[16] Section 78E(2).

[17] Section 78E(3).

[18] DETR Circular, *Contaminated Land,* Annex 3, para. C.12.

assessment may be necessary to determine what type of remedial action is appropriate. The authority will have to consider what phasing is appropriate in the individual circumstances of each case: the first stage of remediation actions may, for example, yield information which indicates that assessment is desirable before any further phase of remediation is commenced.[19] It may not be possible in such phased cases to include all the requirements in a single notice, so that further notices will be served (assuming of course that the land still meets the definition of "contaminated land").[20]

Restrictions on requirements of remediation notice: reasonableness

12.14 Sections 78E(4) and (5) provide restrictions on what may be required by a remediation notice. Under section 78E(4) the authority may not require more than it considers reasonable having regard to:

(a) the costs likely to be involved; and
(b) the seriousness of the harm or pollution involved.

12.15 Clearly, this issue is likely to be a major focus of discussion between the authority and potential recipients of a notice. The issue is not dissimilar to that arising under section 39 of the 1995 Act in relation to action by the Agency, in the sense of balancing costs and benefits of the exercise or non-exercise of powers. The more serious the harm which may be averted, the greater the cost that may be justified. However, given that land will not be "contaminated" at all unless the harm or risk involved is significant, the issue will perhaps most often be one of incremental benefit, *i.e.* comparing different techniques which will have different costs and which may offer different marginal benefits.

12.16 The second restriction is that by section 78E(5), in determining (a) what is to be done by way of remediation, (b) to what standard, and (c) what should or should not be regarded as reasonable in terms of cost versus harm, the authority must have regard to the Guidance issued by the Secretary of State. These issues are discussed in the following paragraphs.

Identifying a remediation scheme

12.17 The procedural description of Part IIA in Annex 2 of the Circular[21] outlines the approach to identifying a remediation scheme. This involves identifying the remedial treatment action or actions which will ensure the relevant

---

[19] *ibid.*, paras C.12 and C.13.
[20] *ibid.*, paras C.14 and C.15.
[21] See para. 6.17.

land or waters are remediated to the appropriate standard (see below). Where assessment is necessary before decisions can be taken on the remedial treatment aspect of the scheme, then the first step will be to identify the appropriate assessment actions. At all stages, the authority should review whether urgent action is necessary. In the most straightforward cases there will be only one significant pollutant linkage involved. However, where there is more than one, then the authority should be aware that to consider each linkage in isolation may result in conflicting or overlapping requirements. A remediation scheme should therefore deal with the relevant land or waters as a whole, which may involve subsuming what would otherwise have been separate remediation actions within a "remediation package". In other words, in considering what is the appropriate remediation action for a particular pollutant linkage, the authority cannot ignore any practical limitations which are imposed by other problems on the same site.

Guidance on the standard of remediation

12.18 A specific issue on which guidance is provided is the standard to which remediation should be required—see section 78E(5)(b). The authority is required to have regard to this Guidance, but not necessarily to act in accordance with it. Part 4 of Chapter C of the Guidance deals with this issue.

12.19 Just as the "suitable for use" concept underlies the issue of what constitutes "contaminated land", the intention of remediation is that the land should be brought into such a condition that, in its current use, it is no longer "contaminated land".[22] The standard should be established by considering separately each significant pollutant linkage: the appropriate standard is "that which would be achieved by the use of a remediation package which forms the best practicable techniques of remediation" for[23]:

(a) ensuring that the linkage is no longer a significant pollutant linkage, by removing or treating the source, breaking or removing the pathway, or protecting or removing the receptor; and
(b) remedying the effect of any significant harm or pollution of controlled waters which is resulting, or has already resulted from, the pollutant linkage.

12.20 The issue is therefore to consider what means might be employed to address the pollutant linkage, whether by source, pathway or receptor and to determine what measure or combination of measures would constitute the "best practicable techniques". The use of the word "practicable"

[22] DETR Circular, Annex 2, para. C.17.
[23] *ibid.*, para. C.18.

implies that considerations of cost and convenience will be involved, and that the exercise will involve striking a balance. The Guidance addresses the question of what represents the best practicable technique by stating that the authority should look for the method of achieving the desired result which, in the light of the nature and volume of the pollutant concerned, and the timescale for remediation is[24]:

(a) reasonable taking account of Guidance on that issue; and
(b) represents the best combination of the following qualities:

(i) practicability, both in general and in the specific circumstances of the case;
(ii) effectiveness in achieving the aims above; and
(iii) durability in maintaining that effectiveness over the timescale within which the significant harm or pollution of controlled waters may occur.

12.21 The balancing exercise involved is both crucial and potentially highly controversial, with possibly large discrepancies between the cost of different solutions. Robustness, certainty of outcome and durability will often come at a heavy price. Different solutions may also have different impacts in terms of disruption on those caught up in the problem: a solution which benefits the person responsible for clean-up by lower costs may be highly inconvenient for the occupier of the land affected. The Guidance states that in considering these questions, the authority should work on the basis of authoritative scientific and technical advice, considering what comparable techniques have recently been carried out successfully on other land, and also any technological advances or changes in scientific knowledge and understanding.[25]

12.22 Where there is "established good practice" for the remediation of a particular type of linkage, this should in general be assumed to represent the best practicable technique,[26] though the Guidance does not indicate the criteria for judging whether something is established good practice. In any event, the authority should satisfy itself that the practice is appropriate to the circumstances in hand, and that the costs would be reasonable, having regard to the seriousness of the relevant harm.

12.23 The Guidance also acknowledges the limitations of the remediation measures which may be available, in that they may not be able fully to ensure termination of the pollutant linkage. In such cases, the required standard is that which comes as close as practicable to that objective, and which remedies the adverse consequences of the harm or pollution caused and puts in place arrangements to remedy such future effects which may be caused by the continuing existence of the pollutant linkage.[27] Similarly,

---

[24] *ibid.*, para. C.19.
[25] *ibid.*, para. C.21.
[26] *ibid.*, para. C.22.
[27] *ibid.*, para. C.23.

total rectification of the effects of past harm or pollution may not be possible, in which case the requirement is to mitigate such harm or pollution so far as is practicable.[28]

12.24 In all cases, remediation should be implemented in accordance with best practice, including any precautions necessary to prevent damage to the environment and any other appropriate quality assurance procedures.[29] This aspect of the Guidance will itself however be subject to the test of reasonableness in terms of cost, since quality assurance procedures may themselves constitute a significant component of the cost of some schemes.

Multiple pollutant linkages

12.25 The position where there are multiple pollutant linkages on a given piece of land has been mentioned briefly above.[30] The statutory Guidance states that it may be possible to arrive at the necessary overall standard of remediation for the land by considering what would constitute the best practicable techniques for each linkage in isolation, and implementing each separately.[31] However, the authority must also consider whether there is an alternative scheme which, by dealing with the linkages together, would be cheaper or more practicable to implement: such a scheme should be preferred if it can be identified.[32]

Practicability, effectiveness and durability

12.26 As mentioned above, the practicability, effectiveness and durability of possible remediation schemes are factors which are required to be considered. Part 6 of Chapter C deals with those issues. The Guidance acknowledges that in some cases there may be little to go on in assessing particular remediation actions or packages.[33] Field-scale testing may not have been carried out. Here the authority will have to consider the issues on the basis of the information which it has at the relevant time. This may involve allowing a person who wishes to use innovative techniques to do so, and requiring further remediation if they prove ineffective.[34] The authority should not however, force an unwilling person to use innovative techniques for the purpose of establishing their effectiveness in general.[35]

---

[28] *ibid.*, para. C.24.
[29] *ibid.*, para. C.25.
[30] See paragraph 12.17 above.
[31] DETR Circular, Annex 3, para. C.26.
[32] *ibid.*, para. C.27.
[33] *ibid.*, para. C.45.
[34] *ibid.*, para. C.46.
[35] *ibid.*, para. C.47.

Practicability

12.27 The key question for practicability, according to the Guidance, is whether the remediation can be carried out in the circumstances of the land and waters under consideration.[36] Relevant factors should include[37]:

(a) Technical constraints, such as the commercial availability of the relevant technologies on the scale required, and constraint or problems imposed by the inter-relationship with other remedial action required.
(b) Site constraints, such as access, the presence of buidings or other structures, and the condition of the land and water concerned.
(c) Time constraints, bearing in mind the need to obtain any necessary regulatory permits, and to design and implement the relevant actions.
(d) Regulatory constraints, such as statutory requirements on health and safety, whether any necessary permits are likely to be forthcoming, whether the conditions likely to be attached to such permits would affect practicability. The authority should also consider the possibility of any adverse environmental impacts in this context (discussed below)

Adverse impacts of remediation

12.28 This aspect is considered at paragraphs C.51–C.57 of the Guidance. The process of remediation may itself create adverse environmental impacts, for example, dust, noise, odours, the possibility of water pollution, and movements of heavy vehicles. In some instances, the process of remediation may require a permit, for example, for a prescribed process under Part I of the 1990 Act, or a Part II waste management licence, or a discharge consent or abstraction licence under the Water Resources Act 1991. In such cases the authority should assume that the conditions attached will provide adequate levels of environmental protection. However, where no such permit is needed, the authority should consider whether the proposed remediation package can be carried out without damaging the environment, and in particular:

(a) without risk to water, air, soil, plants and animals;
(b) without causing a nuisance through noise or odours;
(c) without adversely affecting the countryside or places of special interest; and
(d) without affecting buildings of special architectural or historic interest.

---

[36] *ibid.*, para. C.48.
[37] *ibid.*, para. C.49.

12.29 These criteria correspond (though the Guidance does not expressly say so) to the objectives of the EC Waste Framework Directive 91/156/EEC, and the intention is clearly to ensure that remediation which could arguably be said to involve the disposal of waste does not fall foul of the Directive. Where the authority considers that there is such risk involved, it should address the question of whether the risk is sufficiently great to tip the balance of advantage towards a different remediation scheme, even though this might be less effective when judged against the test of effectiveness; it should also consider whether it may be possible to reduce that risk by incorporating particular precautions into the remediation scheme.[38] This requirement is in some respects akin to considering what constitutes the best practicable environmental option in terms of risk. For example, a site may be surrounded by housing, which would mean that some types of remediation measures may be particularly harmful in terms of nuisance. The impacts of remediation on groundwater will also need careful consideration and consultation with the Environment Agency.[39] The concept of attaching what may effectively be requirements akin to waste licensing conditions to a remediation notice is an intriguing one, which merits careful consideration. There will be no equivalent detailed enforcement procedures for such requirements, akin to those found in Part II of the 1990 Act. It may well be preferable to deal with such matters by planning conditions or section 106 agreements, where the nature of the remediation is such that planning permission is needed.

Effectiveness

12.30 Effectiveness is to be considered in the sense not only of how far the relevant objectives would be achieved, but also in terms of timescale before the remediation becomes effective.[40] Hence the authority will need to balance the speed of reaching a particular result against the longer timescales which may be involved in obtaining a higher level of effectiveness.

Durability

12.31 On durability, the authority will have to consider how long the remediation package will need to remain effective in order to control and resolve the problem, taking into account normal maintenance and repair.[41] For example, if a site is producing gas or leachate which need to be managed,

---

[38] *ibid.*, para. C.55.
[39] *ibid.*, paras C.56 and C.57.
[40] *ibid.*, paras C.58 and C.59.
[41] *ibid.*, para. C.61.

for how long is that production likely to continue? If development is in prospect within a short timescale, which may resolve the problem, then this should be taken into account so as to justify a shorter period being conisidered.[42]

12.32 Should it not be possible to ensure that the remediation measures will remain effective during the whole period of the problem, then the authority should require such measures as will be effective for as long as is practicable.[43] Additional monitoring may be required, and when the measures cease to be effective it may be necessary to consider whether another remediation notice should be served, since new methods of remediation may by then have emerged. If the remediation method chosen requires on-going maintenance and management, then this should be specified in the remediation notice; additional monitoring actions may also be required.[44]

Reasonableness criteria

12.33 Reasonableness of the remediation action, having regard to cost, is a factor in its own right under section 78E(4), and Part 5 of Chapter C provides the Guidance on this. The status of this Guidance under section 78E(5)(c) is again that the authority must have regard to it. The key issue is whether the anticipated benefits justify the likely costs.[45] The authority should therefore prepare an estimate of the costs of the proposed course and a statement of the benefits (which need not necessarily have financial values ascribed).[46] By "benefits" the Guidance means the benefits of reducing the seriousness of the harm or pollution involved.[47] In carrying out the assessment the authority should make allowance for timing issues, of which three examples are given[48]:

(a) Expenditure at a later date will have a lesser impact on the person liable than immediate expenditure than would an equivalent cash sum to be spent immediately.
(b) The gain from achieving earlier remediation should be set against any such higher costs from immediate expenditure. Allowance should also be made for the fact that natural processes may have a beneficial effect over that longer timescale.
(c) The same benefits may be achievable in future at a significantly lower cost, possibly through the development of new techniques or as part of a wider redevelopment scheme.

---

[42] *ibid.*
[43] *ibid.*, para. C.62.
[44] *ibid.*, para. C.63.
[45] *ibid.*, para. C.30.
[46] *ibid.*
[47] *ibid.*, para. C.31.
[48] *ibid.*, para. C.32.

12.34 What is not relevant to the issue of reasonableness of costs is the identity or financial standing of the person responsible for remediation.[49] This may however be relevant at the stage of considering hardship.

Remediation costs

12.35 The authority in considering the cost of the proposed remediation package should take into account[50]:

(a) all initial costs (including tax) of the actions, including feasibility studies, design, specification, management, and making good;
(b) ongoing management and maintenance costs;
(c) any "disruption costs", that is depreciation in the value of land, or other loss or damage likely to result from carrying out the remediation. Specific mention is made in this context of any statutory compensation payable to others in order to carry out the works.[51] However, it appears that this category could include other forms of consequential loss, such as loss of rental income, disruption to business, or relocation costs.

12.36 For the purpose of assessing these costs, it is irrelevant whether they are carried out by contractors, or "in-house".[52] The evaluation of the cost in this sense is not affected either by the identity of the person carrying out the work, or the internal resources available to them. Evaluation is therefore an objective exercise, which may result in valuing costs at greater than their actual cost in the specific circumstances.

12.37 In looking at the costs of a proposed remediation action or package, an overriding test of reasonableness is whether there is any alternative that would achieve the same purpose, to the same standard, for a lower costs.[53] This may be particularly important when considering how to deal overall with a number of pollutant linkages.[54]

Assessment of seriousness of harm

12.38 Assessment of the benefits of the remediation action will involve looking at the seriousness of the significant harm which needs to be addressed. This involves the following factors[55]:

---

[49] *ibid.*, para. C.33.
[50] *ibid.*, para. C.34.
[51] *ibid.*, para. C.35.
[52] *ibid.*, para. C.36.
[53] *ibid.*, para. C.38.
[54] *ibid.*, para. C.37.
[55] *ibid.*, para. C.39.

(a) whether it is already being caused;
(b) the degree of the possibility of it being caused;
(c) the nature of the harm with respect to the type and importance of the receptor, the extent and type of effects, the number of receptors and whether the effects would be irreversible; and
(d) the context in which the effects might occur, in particular whether the receptor has already been damaged by other means and, if so, whether further effects from the contaminated land would materially affect its condition. Secondly, this element involves considering the relative risk associated with the harm or pollution in the context of wider environmental risks. This issue of "context" might in practice be an important moderating factor, in areas of widespread environmental degradation, but authorities are unlikely to accept the argument when pushed to extremes, that harm from a badly contaminated site ought to be discounted because it is an already polluted area.

12.39 Separate and essentially similar advice is given in relation to water pollution.[56] The consideration of effects on water quality will involve advice from the appropriate Agency,[57] and where harm to ecological systems are involved, advice from the relevant nature conservation body will be needed.[58]

What is to be done by way of remediation

12.40 To complete the picture, the Guidance also covers "what is to be done by way of remediation" (section 78E(5)(a)). This is dealt with at Part 7 of Chapter C. This covers the three aspects of assessment, remedial treatment and monitoring. Assessment should only be required where it falls within one of three purposes and constitutes a reasonable means of achieving them[59]:

(a) characterising the pollutant linkage which has already been identified for the purpose of establishing the appropriate remedial action;
(b) enabling the technical design or specifications of treatment to be established; and
(c) evaluating the ongoing condition of land which remains contaminated after remedial treatment, to support future decisions on whether further remediation might be required.

---

[56] *ibid.*, para. C.41.
[57] *ibid.*, para. C.42.
[58] *ibid.*, para. C.40.
[59] *ibid.*, para. C.66.

The main point of this aspect of the Guidance is to make it clear, or to reinforce the point, that a remediation notice cannot be used for the purpose of assessing whether the land should be identified as contaminated.

12.41 For remedial treatment actions, the test is whether the action is necessary to achieve the standard of remediation required by the Guidance, but no more.[60] Such action can include complementary assessment or monitoring measures to evaluate its implementation, effectiveness and durability. It should include appropriate verification measures to provide assurance that the treatment has been properly carried out.

12.42 Monitoring should be limited to providing information on changes to aspects of an already identified pollutant linkage, where the authority needs to consider whether any further remediation should be required in consequence of such changes.[61] It should not be used as a means of gathering general information to satisfy the authority's duty of inspecting its area.

12.43 Finally, perhaps from an abundance of caution, paragraph C.70 emphasises that remediation may not be required for dealing with matters that do not in themselves form part of a significant pollutant linkage, or to make the land suitable for any use going beyond its current use. These wider objectives are a matter for voluntary remediation.

Summary and checklist on remediation requirements

12.44 It will be appreciated from the previous paragraphs that the Guidance involves a complex and sophisticated exercise in determining what remediation should be required. It may therefore be helpful to summarise the elements involved in that exercise (references are to Parts and paragraph numbers in Chapter C of the Guidance).

1. Has each significant pollutant linkage been identified? (para. C.18).
2. What standard would be achieved for each significant linkage by use of a package forming the best practicable remediation techniques? (para. C.18). Consider here:
   (a) reasonableness (Part 5, see point 5 below);
   (b) practicability, including any adverse environmental impacts (Part 6; paras C.48–C.51);
   (c) effectiveness (Part 6; paras C.58–C.60);
   (d) durability (Part 6; paras C.61–C.63);
   (e) striking a balance between these factors (para. C.20).
3. Also take into account on what is the best practicable technique: (Part 4)

[60] *ibid.*, para. C.67.
[61] *ibid.*, para. C.68.

(a) nature and volume of pollutants;
(b) timescale needed for remediation;
(c) comparable techniques that have been successful;
(d) technological advances and changes in knowledge;
(e) appropriate and authoritative technical advice;
(f) established good practice.

4. For multiple pollutant linkages, consider whether practicability favours addressing each pollutant linkage separately or as an overall scheme (paras C.26–C.27).
5. On the issue of reasonableness (see 2 above) consider:

(a) costs involved (Part 5; paras C.34–C.38); and
(b) seriousness of harm or pollution (Part 5; paras C.39–C.43).

6. Keep in mind the general tests on what can and cannot be required by way of remediation for:

(a) assessment (Part 7; paras C.65–C.67);
(b) remedial treatment (Part 7; para. C.67);
(c) monitoring (Part 7; paras C.68 and C.69).

Restrictions on requirements of remediation notice: pollution of controlled waters

12.45 Where land is identified as contaminated by virtue of its effects in polluting controlled waters, consideration must also be given by the enforcing authority to the restrictions on what a remediation notice may require which are imposed by section 78J. This section is discussed in Chapter 15.

Remediation notices and certainty

12.46 The issue may well arise as to whether a remediation notice is defective for failing to specify the remediation actions to be carried out sufficiently clearly. This has been a frequent argument in relation to abatement notices in the case of statutory nuisance and enforcement notices in relation to planning, and can also be expected to arise in the context of Part IIA.

12.47 The wording of section 78E(1) itself, referring to "each of the things so specified", is indicative that some degree of precision is required. On the other hand, in debates the Government indicated their intention that remediation notices "should generally be phrased in terms of objectives to be achieved rather than specific works which have to be undertaken".[62] The

[62] *Hansard*, H.L. Vol. 562, col. 1047.

Regulations on remediation notices do not add any extra requirements with regard to the remediation actions to be specified in the notice. The overall intention is, however, that the notice be clear and self-contained; and that it gives a clear indication of what is to be done to whom, where and by when.[63]

12.48 The problem with specifying works as opposed to objectives is where to stop in terms of detail, given that the success or otherwise of a remediation scheme may well lie in the detail of how the scheme is executed. In the context of planning and waste management licensing conditions, one possible solution to this type of problem is to require a detailed scheme to be prepared, and submitted to the authority for approval: the question is whether this approach, which might be eminently sensible, is sanctioned by section 78E(1). Given the need for the recipient of the notice to know what they must do, and the short timetable allowed for appeals against notices, it seems unlikely that such an approach would be acceptable.

12.49 It also needs to be remembered, however, that a remediation notice should have been preceded by a period of consultation as to its requirements with the recipients, during which time the authority's requirements may have been explained and clarified. In practice, the authority should seek to agree the works if possible before serving the notice. The debate may in some cases centre on the cost of the works rather than their nature.

12.50 The problem of clarity has arisen frequently in cases relating to abatement notices for statutory nuisance. On *Kirklees Metropolitan Council v. Field*,[64] interpreting section 80 of the 1990 Act, the Court held that an abatement notice must inform the recipient of what he did wrong, and must also ensure that he knows what he has to do to abate the nuisance.[65] In some cases where it is clear what is to be done it will be sufficient simply to require the nuisance to be abated,[66] but in many cases it will be necessary to state the works to be carried out.[67]

12.51 Quite clearly, remediation notices are likely to fall into this second category in that it will not be at all obvious simply from the fact that the land is contaminated as to what type and degree of work will be necessary to satisfy the authority. Indeed, section 78E(1) makes it clear that the notice must specify what is to be done by way of remediation. In that sense, the decisions on statutory nuisance do not take matters much further forward. What the *Kirklees* case does make clear is that lack of clarity is likely to be a material defect which cannot be remedied by a covering letter[68] or rectified on appeal.[69]

---

[63] See DETR Circular, Annex 4, para. 17.

[64] [1998] Env. L.R. 337.

[65] *ibid.*, p. 342. See also *R. v. Wheatley* [1885] 16 Q.B.D. 34; *Millard v. Wastall* [1898] 1 Q.B. 342.

[66] For example, cases of noise from barking dogs (*Myatt v. Teignbridge D.C.* [1995] Env. L.R. 78; *Budd v. Colchester Borough Council* [1999] J.P.L. 739) or amplified music (*SFI Group plc v. Gosport Borough Council* [1999] Env. L.R. 750).

[67] *Network Housing Association v. Birmingham City Council* [1996] Env. L.R. 121.

[68] [1998] Env. L.R. 337 at 340.

[69] [1998] Env. L.R. 337 at 342.

12.52 However, in the last resort this issue will turn on the width of the statutory discretion to correct or vary notices on appeal, and in particular whether this involves injustice to any party.[70] It would therefore be unwise to seek to draw firm conclusions from the *Kirklees* case.

12.53 Where the nature of the work required is specified by the authority the temptation to qualify this by adding words such as "or other equivalent work", "other equally effective work" or similar phrases, should be resisted. This may have the effect of vitiating the notice by referring to other, unspecified measures.[71]

Form and content of remediation: Regulations

12.54 Section 78E allows Regulations to make provision for or in connection with the form or content of remediation notices and with procedural steps in connection with service. Regulation 4(1) of the Regulations sets out the matters which a remediation notice must specify (in addition to those required by sections 78E(1) and (3)):

(a) the name and address of the person on whom the notice is served;
(b) the location and extent of the land, sufficient to enable it to be identified, whether by reference to a plan or otherwise;
(c) the date of any notice under section 78B identifying the land as contaminated;
(d) whether the authority considers the person on whom the notice is served to be an appropriate person because they are a causer or knowing permitter or because they are owner or occupier of the land;
(e) particulars of the significant harm or pollution by reason of which the land is contaminated land;
(f) the substances by reason of which the land is contaminated and, if they have escaped from other land, the location of that land;
(g) brief particulars of the reasons for the remediation action required, showing how the Guidance has been applied;
(h) where there are two or more persons in relation to the contaminated land:

 (i) a statement that this is the case;
 (ii) their names and addresses; and
 (iii) the thing by way of remediation for which each is responsible.

(i) where the authority has applied the tests in the Guidance to exclude an appropriate person from liability, the reasons for the authority's determination and how the Guidance has been applied;

---

[70] See Chapter 23.
[71] *Perry v. Garner* [1953] 1 All E.R. 285.

(j) where the notice involves apportionment of responsibility, the reasons for the apportionment, and how the Guidance on that issue has been applied;
(k) where known to the authority, the name and address of the owner and of any person who appears to be in occupation of the whole or any part of the land;
(l) where known, the name and address of any person whose consent is required under section 78G before anything required by the notice may be done;
(m) where appropriate, that it appears the land is in a condition of imminent danger of serious harm or serious pollution to controlled waters;
(n) a statement that a person on whom such a notice is served may be guilty of an offence for failure, without reasonable excuse, to comply with any of the requirements of the notice;
(o) for such an offence the penalties which may be applied on conviction;
(p) the name and address of the enforcing authority serving the notice; and
(q) the date of the notice.

12.55 The notice must also inform the recipient of the right of appeal under section 78L and of how, where, within what period and on what grounds such appeals may be made, and that the notice will be suspended, where an appeal is duly made, while an appeal is pending until final determination or abandonment of the appeal.[72] The actual form (as opposed to the content) of remediation notices is not prescribed. However, it appears from the Circular that the DETR and the Environment Agency intend to draw up a model form of notice which can be used to assist consistency and minimise preparatory work.

Apportionment of costs for the same remediation works

12.56 Where the situation is that two or more persons are responsible for the same remediation works, then, although subsection 78E(3) does not expressly say so, each is presumably liable to secure compliance with the notice; but as to cost, the notices in question must specify that proportion of the cost of remediation that each of them has to bear separately. It is the cost of the works, not the works themselves, which are apportioned. It follows that it would be a defence to the non-completion of the works in question that the other party served with the notice is unable to, or has otherwise failed to, fulfil his part of the obligation and that it was therefore impossible from a practical point of view to start with remediation, the party in question being unable to bear the whole of the costs.

---

[72] Regulation 4(2).

12.57 This is effectively the statutory position under section 78M(2), which provides a defence that the only reason that the defendant has not complied with the notice is that one or more of the other persons who are liable to bear a proportion of the cost refused or was not able to comply with the requirement. In such circumstances, it is presumed that the local authority would do the works itself and charge the parties according to the proportions already determined and stated in the notices concerned. As to how precisely the apportionment is to be done, section 78F(7) requires the authority to act in accordance with Government guidance. Apportionment is dealt with in Chapter 18.

Giving reasons for notices

12.58 As will be appreciated many of the decisions on which a remediation notice is based may well be controversial; in particular, what is required, who is determined to be an appropriate person, and how the costs are apportioned where there are a number of appropriate persons. Service of the notice may well have been preceded by dialogue on these issues, and may well be followed by an appeal or other legal challenge. The question therefore arises as to whether the enforcing authority, when serving the notice, is under any obligation to explain the reasoning behind its decisions.

12.59 As mentioned above, the Regulations on form and content of notices require reasons to be given as to the remediation action required, the allocation of liability and the apportionment of costs (if any). This is clearly a mandatory requirement.

12.60 The question is whether general legal principles require any further reasons. The current principles relating to the giving of reasons for decisions were distilled and summarised by the Divisional Court in *R. v. Ministry of Defence, ex p. Murray*[73] as follows:

(a) The law does not at present recognise a general duty to give reasons.

(b) Where a statute has confirmed the power to make decisions concerning individuals, the court will not only require the statutory procedure to be followed, but will readily imply as much (and no more) to be introduced by way of additional procedural standards as will ensure fairness.

(c) In the absence of a requirement to give reasons, the person seeking to argue that reasons should have been given must show that the procedure adopted of not giving reasons is unfair.

---

[73] [1998] C.O.D. 134, see also *R. v. Civil Service Appeal Board, ex p. Cunningham*; [1991] 4 All E.R. 310; *R. v. Secretary of State for the Home Department, ex p. Doody*; [1994] 1 A.C. 531; *R. v. Higher Education Funding Council, ex p. Institute of Dental Surgery* [1994] 1 W.L.R. 241.

(d) There is a perceptible trend towards an insistence on greater openness or transparency in the making of administrative decisions.
(e) In deciding whether fairness requires reasons, regard will be had to the nature of any further remedy. The absence of a right of appeal strengthens the need for reasons, and if the remedy is judicial review, then reasons may be necessary in order to detect the type of error which may justify intervention.
(f) The fact that a tribunal is carrying out a judicial function (which the local authority is not) is a consideration in favour of requiring reasons, particularly where personal liberty is concerned.
(g) If the giving of a decision without reasons is insufficient to achieve justice, then reasons should be required.
(h) Giving of reasons is helpful in concentrating the decision maker's mind on the right questions and in demonstrating to the recipient that this was so.
(i) Against giving reasons is the fact that it may place an undue burden on decision makers; demand an appearance of unanimity where there is diversity; call for articulation of sometimes inexpressible value judgments; and offer an invitation to nit-pick the reasons to discover grounds of challenge.

12.61 Clearly, some of these criteria are more relevant than others to remediation notices. However, based on them, a respectable case can be mounted that, irrespective of the Regulations, remediation notices should be accompanied by reasons. Such notices will have considerable financial and legal consequences for those in receipt of them; they also have penal consequences. Although they can be subject to appeal, no recipient is going to embark on a potentially lengthy and expensive legal process lightly, and in the absence of reasons it may be difficult or impossible to gauge the likelihood of such a challenge being successful. Particularly compelling is the fact that the decisions in question are such as ought to be taken on a rational basis, rather than "value judgments". The matters will have been the subject of a reasoned report by officers, often after discussion with potential recipients, and debate in committee: the formulation of reasons should therefore not constitute an undue administrative burden.

12.62 However, there is also the issue of the standard and detail of the reasons required. Some aspects of the process are more susceptible to detailed reasons than others: matters based on scientific considerations may well be capable of being supported by detailed reasons. The same is less true of decisions on apportionment or relief on hardship grounds, which involve value judgments. Whilst reasons will be required in both cases, the reasons involving scientific facts may need to be given in greater depth than those involving value judgments.

12.63 At heart, the question is one of fairness. Parliament has seen fit to provide a formal consultation process for potential recipients of notices to express their views and it may be seen as unfair for those who have such rights to have an inadequate indication as to the thought processes of the authority or to find their representations rejected out of hand, without reasons. There may also be arguments in favour of reasons based on the right to an effective remedy under Article 6 or Article 13 of the European Convention on Human Rights,[74] now incorporated into U.K. law by the Human Rights Act 1998.[75]

Delegation of the service of notices

12.64 Normal principles of local government law will apply to the internal procedures for authorising service of remediation notices. By section 101 of the Local Government Act 1972 the function of authorising the issue of a notice may be delegated to a committee, sub-committee or officer; similarly the function of preparing and serving the notice may be delegated in this way. On the basis of *Albon v. Railtrack plc*,[76] there seems to be nothing to prevent a senior officer who is empowered to serve a notice authorising a member of his or her staff to do so.[77] However, to reduce so far as possible the risk of challenge to the notice, the authority should ensure that the arrangements for delegation are clearly evidenced by specific authority or by standing orders.[78] Whilst it may be possible for a decision to serve a notice to be subsequently ratified, no sensible authority would wish to enter such uncharted waters.[79]

Signature of notices

12.65 On general principles, the notice must be properly authenticated to be valid.[80] Neither the Act or Regulations requires a remediation notice to be signed on behalf of the authority. Given the absence of any statutory requirement of signature, the notice probably does not require to be

---

[74] Compare *R. v. Secretary of State for the Environment, Transport and the Regions* and *Parcelforce, ex p. Marson* [1998] J.P.L. 869 where the applicant was held to have a "fundamental right" to reasons as to why an environmental assessment was not required, The reasoning in that case does not appear particularly apt to the situation discussed above.

[75] See also the comments of Laws J. at *Chesterfield Properties plc v. Secretary of State for the Environment* [1998] J.P.L. 568, p. 579 on the importance of justifying interference with fundamental or constitutional rights.

[76] [1998] E.H.L.R. 83 (a case on the Prevention of Damage by Pests Act 1949).

[77] In that case, authorisation was by the Assistant Chief Environmental Health Officer to a Principal Environmental Health Officer. See also *Fitzpatrick v. Secretary of State for the Environment* [1990] P.L.R. 8.

[78] See *Cheshire County Council v. Secretary of State for the Environment* [1988] J.P.L. 300.

[79] *Cooperative Retail Services v. Taff-Ely Borough Council* (1981) 42 P & C.R. 1, HL. By section 100G(2) of the 1972 Act, a list of powers delegated to officers must be kept open for public inspection.

[80] See Carter, Pengelly and Saunders, *Local Authority Notices* (Sweet & Maxwell, London, 1999) pp. 54–55.

signed: the essential point is that it should be possible for the recipient to see where it emanated from and verify its authenticity.[81] Reference to the name of the officer with primary authority to issue the notice should be sufficient for this.

12.66 However, by section 234 of the Local Government Act 1972 a remediation notice (being a document which the local authority are required to issue) may be signed on behalf of the authority by the proper officer; that is, an officer appointed for that purpose by the authority.[82] The "signature" may be a facsimile of the signature, though it is questionable whether remediation notices will be issued in such numbers as to make this necessary.[83] The benefit of signature under this provision is that the notice is deemed by section 234(2) to have been duly issued, unless the contrary is proved.[84]

Copies of notices

12.67 Under the Regulations, when serving a remediation notice the authority must send a copy of it to[85]:

(a) any person who was required to be consulted under section 78G (being a person whose consent would be needed for the remediation actions to be carried out);
(b) any person who was required to be consulted under section 78H(1) before serving the notice;
(c) where the local authority is the enforcing authority, the appropriate Agency; and
(d) where the appropriate Agency is the enforcing authority, the local authority in whose area the land is situated.

12.68 It will be good practice to notify such persons of the capacity in which they are being sent a copy of the notice.[86] Where there is imminent danger of serious harm or pollution, the authority must send copies of the notice to those persons as soon as practicable after service of the notice[87]; in other words the requirements for sending copies should not delay service in urgent cases. What constitutes a "copy" of a notice was considered in *Ralls*

---

[81] *Albon v. Railtrack plc* [1998] E.H.L.R. 83 (see n. 76); *Pampian v. Gorman* [1980] R.V.R. 54.
[82] Section 234(2).
[83] *FitzPatrick v. Secretary of State for the Environment* [1990] P.L.R. 8.
[84] Section 234 will probably be treated as an enabling rather than a mandatory provision in this respect: *Tennant v. London County Council* [1957] 121 J.P. 428; *Albon v. Railtrack plc* (above). These are cases on section 284 of the Public Health Act 1936.
[85] Regulation 5(1).
[86] DETR Circular, Annex 4, para. 19.
[87] Regulation 5(2).

v. Secretary of State for the Environment.[88] The addition to the copy of the name and address of the person on whom the copy is being served will not make it a bad copy, but may give rise to confusion and is probably best avoided. The Courts could possibly construe the requirement to serve copies of notices as directory only, so that failure to do so will not necessarily be fatal to the original notice, provided that no-one is prejudiced by the breach.[89]

Service of notices

12.69 By section 78E(6)(b), Regulations may be made with regard to procedures in connection with, or in consequence of, service of a remediation notice. In fact, the Regulations do not make such provision, other than that in relation to copies of notices.

12.70 Since the remediation notice is one required to be served under the 1990 Act, section 160 of that Act will apply to it. This provides that the notice may be served by delivering it to the intended person, or by leaving it at his proper address, or by sending it by post to him at that address.[90]

12.71 Special provision is made for companies and partnerships.[91] In the case of a company or other body corporate, the notice may be served on or given to the secretary or clerk; in the case of a partnership, it may be served on a person having the control or management of the partnership business.[92]

12.72 As to the time of service, by section 7 of the Interpretation Act 1978, where service is effected by post, then it will be deemed to be effected by properly addressing, prepaying and posting a letter containing the notice, and unless the contrary is proved, shall be deemed to have been effected at the time when the letter would be delivered in the ordinary course of post.

12.73 A key question is therefore what constitutes the "proper address" for service. This can be the address specified by the person to be served as one at which he or some other person will accept notices of that description.[93] Otherwise, it is the last known address of the person, except that[94]:

---

[88] (1998) J.P.L. 444.

[89] See *O'Brien v. London Borough of Croydon* [1999] J.P.L. 47 at p. 52; *Nahlis, Dickey and Morris v. Secretary of State for the Environment* (1996) 71 P. & C.R. 553.

[90] Section 160(2).

[91] Section 160(3).

[92] In *Leeds v. London Borough of Islington* [1998] Env. L.R. 655, it was held by the Divisional Court that section 160(3) is mandatory as to the person to whom the notice must be addressed. There a notice addressed to "the Senior Estate Manager" was held not sufficient. Compare the more relaxed approach taken in *Pearshouse v. Birmingham City Council* [1999] L.G.R. 169 and *Hall v. Kingston-upon-Hull City Council* [1999] L.G.R. 184 (but on the bais of procedures intended to be operated by private individuals).

[93] Section 160(5). The question is whether they have accepted service for that particular purpose—accepting service for general property notices may not be sufficient—see *Leeds v. London Borough of Islington* [1998] Env. L.R. 655 at p. 661. Also the person who specifies the address for service need not be the secretary or clerk—see *Hall v. Kingston-upon-Hull City Council* [1999] L.G.R. 184.

[94] Section 160(4).

(a) in the case of a body corporate or its secretary or clerk, it shall be the address of the registered or principal office of that body; and
(b) in the case of a partnership or person having control or management of the business, it shall be the principal office of the partnership.

12.74 For this purpose, the principal office of a non-U.K. registered company or of a partnership carrying on business outside the U.K. shall be their principal office within the U.K.[95] Given that the end result of a remediation notice may be criminal proceedings, the courts can be expected to take a relatively unforgiving approach to errors in service.[96] So, getting the name of a company wrong, or specifying the wrong company within a group of companies, might give rise to serious difficulties.[97]

12.75 For a company, its registered office will be ascertainable, in that every company must have a registered office,[98] to which as a matter of company law communications, notices and proceedings can be addressed.[99] That address must be shown on the company's business letters[1] and service will be valid for a 14 day transitional period at the old office where a change of office is registered.[2] The authority will be entitled to proceed on the basis of the registered office as disclosed by the official file.[3] Thus the registered office will be the safest address for service.

12.76 The "principal office" of a company is the place at which the business of the company is controlled and managed.[4] "Principal" in this sense means chief, or most important.[5] There is a question of statutory construction arising from section 160(4) as to whether the "principal office" is an alternative address for service in all cases, or whether it is only relevant for companies registered in Scotland or outside the U.K. The better view seems to be that the "principal office" is a true alternative address for service in all cases. The concept of a principal office is clearly distinct from the registered office as a matter of company law.[6] References to service on companies at their principal office go back to the Public Health Act 1936,[7]

---

[95] *ibid.*
[96] *Leeds v. London Borough of Islington* [1998] Env. L.R. 655 at p. 661.
[97] See *Amec Building Ltd v. London Borough of Camden* [1997] Env. L.R. 330; but compare *Mulkins Bank Estates Ltd v. Kirkham* (1966) 64 L.G.R. 361.
[98] Companies Act 1985, section 287(1).
[99] *ibid.*, section 725(1).
[1] *ibid.*, section 351(1).
[2] *ibid.*, section 287(4).
[3] *Ross v. Invegordon Distillers Ltd* 1961 S.L.T. 358.
[4] *Palmer v. Caledonian Railway Co.* [1982] 1 Q.B. 823, pp. 827–8.
[5] *The Rewia* [1991] 2 Lloyd's Rep. 325. A company can have more than one place of business (*Davies v. British Geon Ltd* [1956] All E.R. 389) but only one principal place.
[6] *Palmer's Company Law* (Sweet & Maxwell) para. 2.504.
[7] The Public Health Act 1936, s.285 refers to service at the registered office or principal office. Earlier public health legislation (*e.g.* the Public Health Act 1875) referred to notices being serviced at the residence of the recipient.

by which time the separate concept of a registered office had emerged,[8] and Parliament may be assumed to have been aware of the difference between the two terms. Nor does the language of section 160(4) justify confining "principal office" to the more limited circumstances mentioned there of companies registered outside the U.K.

12.77 Be that as it may there may well be uncertainties as to what is a company's principal office, and as mentioned above, for practical purposes the safest course will be service at the registered office, addressed to the company secretary.[9]

12.78 In the case of service of notices on an owner or occupier of land, if that person cannot after reasonable inquiry be ascertained, the notice may be served either by leaving it in the hands of a person who is or appears to be resident or employed on the land or by leaving it conspicuously affixed to some building or object on the land.[10] The authority can, and should, before doing so, consider whether it could gain that information by use of its statutory powers to obtain particulars of persons interested in land under section 16 of the Local Government (Miscellaneous Provisions) Act 1976, or whether it can obtain the necessary information direct from the Land Registry.

Withdrawal of notices

12.79 Neither Part IIA nor the Regulations makes express provision for withdrawal of a remediation notice; unlike planning enforcement notices, where section 173A of the Town and Country Planning Act 1990 states that they may be withdrawn whether or not they have taken effect. Nonetheless, there may well be circumstances where the appropriate course is to withdraw a remediation notice. Examples might include:

1. Cases where the authority is now satisfied that the land is no longer "contaminated land" (for example, because it has been cleaned up voluntarily).
2. Cases where the authority is satisfied that things will now be done within an appropriate timescale, which although not in compliance with the terms of the notice, will in fact effect satisfactory remediation.

---

[8] The Companies Act 1929, s.370 stated that service may be effected by leaving notices at the registered office. Section 62 of the Companies (Consolidation) Act 1908 required every company to have a registered office to which all communications and notices could be addressed, as did section 39 of the Companies & c. Act 1862. The concept of principal office goes back further: the Companies Clauses Consolidation Act 1845, s.135 (and associated railways legislation) provided for service at the principal office of the company, and the Companies Act 1844, s.7, required the certificate of registration to include the principal or only place for carrying on business, and any branch office.

[9] *Leeds v. London Borough of Islington* [1998] Env. L.R. 655.

[10] Local Government Act 1972, s.233(7).

3. Cases where the authority has now concluded that the notice was served on the wrong person, or stated the wrong apportionment of liability.
4. Cases where the authority simply "got it wrong" so that the notice is defective or is open to challenge, and the authority wants to start again with a clean slate.

12.80 Notwithstanding any express statutory power, the authority almost certainly has inherent powers to withdraw a notice it has served by analogy with *R. v. Bristol City Council, ex p. Everett*,[11] on statutory nuisance. It would be sensible to give notice of withdrawal to every person who should have originally been notified. Alternatively, the authority may simply decide to take no further action on the notice, *i.e.* not to enforce it. Again, notice of that decision should be given to all the relevant parties.

12.81 The authority should also bear in mind that it is under a duty to serve a remediation notice or notices where contaminated land has been identified. Clearly, therefore, withdrawing a notice without sound reasons for doing so, and without replacing it with a further notice, may be subject to challenge by third parties by way of judicial review; as may a decision to take no further action on a notice.

12.82 Unlike planning enforcement notices, remediation notices do not have to state a period after which they take effect; the ability to withdraw a notice is not therefore a question as to whether the notice has taken effect. It may well be advisable to obtain the consent of the appropriate person to withdrawal of the notice. Apart from anything else, unilateral withdrawal where the recipient has expended money on compliance or on legal or other advice may result in a claim for losses incurred in tort, or a complaint to the local government ombudsman for alleged maladministration.

[11] [1999] 1 W.L.R. 1170, applying the first instance decision at [1999] 1 W.L.R. 92.

Chapter 13

The "Appropriate Person"

Summary

A remediation notice must be served on each "appropriate person", *i.e.* any person who should bear responsibility for any thing which is to be done by way of remediation in relation to the land in question. Who is an appropriate person is to be determined in accordance with section 78F, which contemplates two main categories of appropriate persons.

The first category (described in the Guidance as "Class A") are the persons who have caused or knowingly permitted the relevant contaminated substances to be in, on or under the land. The second category ("Class B") comprises the current owners and occupiers; this category is only responsible where no person falling within Class A can be found.

Once the appropriate persons have been ascertained then, if there is more than one such person, the enforcing authority will have to apply the rules on exclusion of persons from liability and on apportionment. These rules are discussed in Chapters 17 and 18.

"The appropriate person"

13.01 The term "appropriate person" in Part IIA is defined by section 78(A)(9) as the person "who is the appropriate person to bear responsibility for any thing which is to be done by way of remediation in any particular case". This question naturally lies at the heart of the statutory regime. Who is an appropriate person is to be determined in accordance with section 78F, but it may be noted from the definition that the crucial issue is the relationship between the person and the particular thing which is to be done by way of remediation.

13.02 In relation to section 78F, government guidance plays a central role in channelling and apportioning liability in the situation where two or more persons are potentially appropriate persons. However, the starting point is subsections 78F(2), (3) and (4), all of which deal with the primary question of who are appropriate persons. Essentially, there are two ranks, tiers, or groups of appropriate persons: the first defined in subsections 78F(2) and (3) and the second in subsection 78F(4). The Guidance refers to these respectively as "Class A" and "Class B". Class B only comes into play where no person in Class A can be found. This effectively continues a system of liability for statutory nuisances which goes back at least as far as the Nuisances Removal Act 1855[1] in channelling liability first to the perpetrator or person responsible and only secondarily to the owner or occupier.

13.03 By Schedule 1 of the Interpretation Act 1978, the word "person" includes a body of persons, corporate or unincorporate. It could therefore include an association, club or partnership. As a matter of general law, the term "person" denotes an entity which is the subject of rights and duties,[2] and has been held to include, for example, a fund,[3] and a Hindu Temple.[4] It would clearly include a local government body, statutory body, or corporation. The Crown has been held to be a "person".[5]

The first rank of appropriate persons: "Class A"

13.04 The ambit of Class A is defined by sections 78F(2) and (3) "as any person, or any of the persons, who caused or knowingly permitted the substances, or any of the substances, by reason of which the contaminated land in question is such land to be in, on or under that land is an appropriate person". This approach is qualified by the concept of referability, discussed in the next paragraph.

13.05 It is important to note that the terminology is in the past tense, which might arguably suggest that sections 78F(2) and (3) are directed toward the original contaminator who either introduced the substances in question into the land or who caused them to be introduced by instructing servants, agents or independent contractors to put them there, or who stood by and watched them being introduced into the land. Although the use of the term "knowingly permitted" indicates that passive conduct can be sufficient, nevertheless it might on that construction denote only passive conduct in respect of the past event or events giving rise to the original contamination. On this construction it would be wrong to suggest that the terminology would catch someone who merely came to the land, who knew or ought to have known of the existence of contamination caused by a predecessor in title, and nevertheless did nothing about it. It could be said that if the legislation was intended to catch him, then the word would be in the present tense, namely "permits".

---

[1] Statutes 18 & 19 Vict. C. 121, section 12.

[2] See L.S. Sealy, *Cases & Materials in Company Law* (Butterworths, 6th edition, London, 1996) p. 37.

[3] *Arab Monetary Fund v. Hashim (No. 3)* [1991] 2 A.C. 114.

[4] *Bumper Development Corporation Ltd v. Metropolitan Police Commissioner* [1991] 4 All E.R. 638.

[5] *Boarland v. Madras Electrical Supply Co. Ltd* [1954] 1 W.L.R. 87.

13.06 However, "knowingly permitted" might also plausibly be read as denoting not only those concerned with the original introduction of the substances to the land, but also those who were subsequently in control of the land so as to be in a position to remove or render harmless the substances, but who, knowing of its presence, did not take steps within their power to do so: on that reading they knowingly permitted the presence of the substance during their period of control. This difficulty is discussed below in the context of subsequent owners and occupiers.

The referability exclusion

13.07 Subsection 78F(3) is conceptually complicated. Instead of saying that a person shall only be responsible for contamination arising from substances for which he is responsible the subsection says a person shall only be an appropriate person within Class A in relation to things which are to be done by way of remediation which are to any extent referable to the substances which he caused or knowingly permitted to be present. This is a very important provision because of what follows. The effect of subsection 78F(3) is to take certain persons out of the category of "appropriate persons" altogether and thus there could in certain circumstances be no "appropriate person" in the first rank in Class A. If the remediation works are not referable to substances which he caused or knowingly permitted to be present in, on or under the contaminated land, then he is not an appropriate person at all. The enforcing authority is then entitled, in such circumstances, to look to the second rank of Class B.

13.08 The effect of this referability exclusion is to negative what would otherwise be the wide words of section 78F(2), referring as they do to "any of the substances", and the effect of joint and several liability which those words would otherwise potentially create. If a number of substances are causing land to be contaminated, it will be necessary to establish to which of them the remediation action is referable. The range of people who may be held responsible for those requirements will be limited accordingly.

13.09 However, it is equally important to note some limitations on the referability exclusion. First, a person may be liable if there is referability "to any extent" (section 78F(3)). This implies that the referability need not be total, *i.e.* not all of the measures in question need to be referable to the contamination for which the person is responsible. Secondly, the position on referability may be affected by specific provisions on chemical reactions between substances (section 78F(9)) and on escapes of substances between different sites (section 78K). Finally, section 78F(10) provides that a thing to be done by way of remediation may be regarded as referable to the presence of a substance notwithstanding that it would not have to be done in consequence of the presence of *that substance* alone, or in consequence of the presence of the quantity for which the person was responsible. This makes it clear that where a number of people have contributed the same substance or different substances so as to result in contamination which requires remediation, it will not be open to any of them to argue that their contribution alone would not justify such remediation measures. In such cases, were it not for the requirement under section 78E(7) that the remediation notice must apportion liability for the costs of remediation, the end result would be joint and several liability.

13.10 Essentially, referability will be a question of fact and of science. It stresses the critical importance of a causal link between the substances in respect of which the person is responsible and the remedial action necessary to deal with the harm or risk which makes the land "contaminated".

Caused

13.11 One way to be a first tier appropriate person is to have caused the contaminating substances to be present in, on or under the land. This will generally be a question of fact, and by its nature the concept of causation seems most naturally to relate to the entry of the substances in question on to or into the land. Case law in the field of water pollution offences suggests that the courts are likely to approach the issue in a robust, common sense way, without knowledge, fault or negligence needing to be shown.

13.12 There have of course been many recent cases interpreting the word "caused" in the context of water pollution offences.[6] The leading authority is the decision of the House of Lords in *Environment Agency v. Empress Car Co. (Abertillery) Ltd.*[7] Lord Hoffman there laid down a series of principles to be considered in cases on causing pollution. Many of these will be equally relevant to the contaminated land regime and section 78F(2) in particular:

1. What is it that the defendant did to cause pollution? This is equally valid as a starting point under section 78F: what is the action or activity which forms the basis of a person having caused the presence of contaminating substances?
2. That thing need not have been the immediate cause of the pollution. Maintaining tanks, lagoons or storage systems containing polluting substances is "doing something" for this purpose, even if the immediate cause of the pollution was lack of maintenance, a natural event, or the act of a third party. This point seems highly relevant to many contaminated land situations.
3. Did that thing cause the pollution? This is not the same as asking whether it was the sole cause, or whether something else was also a cause. The fact that something else (*e.g.* vandalism or natural

---

[6] See for example *Attorney General's Reference No. 1 of 1994* [1995] Env. L.R. 356 setting out general principles.

[7] [1999] 2 A.C. 22; [1998] 2 W.L.R. 350.

events) was also a cause does not mean that the defendant's conduct was not a cause. Again, this has relevance to contamination of land.

4. Where a third party act or natural event was involved in the escape of the pollution, it should be considered whether that act or event was a normal fact of life, or something extraordinary. Ordinary occurrences will not negate the causal effect of the defendant's conduct, even if the intervention was not foreseeable at all or in the form in which it occurred. This distinction is one of fact and degree, to which common sense and knowledge of what happens in the locality should be applied. In that case, vandalism was regarded as an ordinary normal fact of life, whereas an act of terrorism might well not be. In a subsequent case the failure of a seal on a pipe, though unforeseeable, was held to be an ordinary rather than extraordinary event.[8]

13.13 These principles would appear to constitute a sound basis for decisions on whether a person or company be present within section 78F(2). Contamination will often come about because pipes leak, tanks corrode, vandals open taps, etc. In such circumstances, responsibility seems likely to follow. In particular, earlier cases which suggested that some positive act was required, have been disapproved as too restrictive.[9] This approach seems to accord with the Government's view that the test of "causing" will require that the person concerned was involved in some active operation, or series of operations, to which the presence of the pollutant is attributable, and that such involvement may also take the form of a failure to act in certain circumstances.[10]

Causing and consignment to landfill

13.14 In some cases, the contaminated land in question may comprise a former tip, landfill or similar facility to which many companies (or for that matter local authorities) may have consigned material over the years. The person or company which operated the site may long ago have ceased to exist, but some or all of those whose waste was sent there may still be in existence. In such circumstances it is important to decide whether or not the consignors of the waste may be regarded as having "caused" the material to be present. The exact circumstances of the deposit may now be lost in

---

[8] *Environment Agency v. Brock Plc* [1998] J.P.L. 968. See also *CPC (U.K.) Limited v. National Rivers Authority* [1995] Env. L.R. 131 (pollution arising from a latent defect in pipework at the defendant's factory held to be caused by defendant).

[9] *Price v. Cromack* [1975] 1 W.L.R. 988 and *Wychavon District Council v. National Rivers Authority* [1993] 1 W.L.R. 125, disapproved in *Environment Agency v. Empress Car Co. (Abertillery) Ltd.*

[10] DETR Circular, *Contaminated Land*, Annex 2, para. 9.9.

the mists of history—for example, it may not be clear whether the consignor's own employees transported and tipped the waste themselves, or whether the material was collected by the operator of the tip or by a carrier (who, likewise, may or may not still be in existence and, if so, may be a candidate to be regarded as an appropriate person).

13.15 Ideally, the primary legislation would have made the law clear on what is, essentially an issue of policy.[11] In the event, it is left to the Guidance to deal with this issue. This is an important matter not least for the current owner and occupier of the site, who if the waste consignors are not regarded as falling within Class A, may well find themselves held responsible for the cost of remediation. Applying authority on the meaning of "caused", the starting point would appear to be to identify what it is said the consignor of the waste did that could constitute causing the contaminants to be present in the land.[12] Obviously if the consignor's own employees or agents carried out the tipping there is such an act. But what of the situation where the consignee simply had the waste collected? In the days before the duty of care on waste producers under section 34 of the Environmental Protection Act 1990, a waste consignor may well not have known what the destination of the waste would be[13]; though good practice would suggest even at that time that he should have enquired.

13.16 The issue was one which arose in debate on the Environment Bill. The Government expressly indicated that it did not intend the words "caused or knowingly permitted" to be construed as including persons "merely on the grounds that they had consigned materials to an authorised waste stream".[14] The Government believed that this was already the effect of the relevant words without the need for any amendment. However, comments were also made at the Commons Standing Committee Stage indicating that the Government was not necessarily as sympathetic to those transporting waste to sites:

> "When individual lorry drivers or companies deliver material, they bear some responsibility for the quality of the material."[15]

13.17 This would appear to reinforce the distinction, suggested above, between those who actively participated in bringing material on to the land in question, and those whose involvement was less direct. In any event, the position is clarified in practice by the Guidance on exclusion discussed in Chapter 17.

---

[11] The liability of waste producers is a key feature and complicating factor in the U.S. "Superfund" legislation.

[12] See paragraph 13.13 above.

[13] See Stephen Tromans and Robert Turrall-Clarke, *Contaminated Land*, (London Sweet & Maxwell, 1994) para. 7.31 *et seq.*

[14] *Hansard*, H.L. Vol. 562, col. 182.

[15] *Hansard*, H.C. Standing Committee B, May 23, 1995, col. 341.

Knowingly permitted

13.18 The other basis for a person falling within the first tier of appropriate persons is having knowingly permitted the contaminating substances to be in, on or under the land. In *Alphacell Ltd v. Woodward*,[16] knowingly permitting was said to involve a failure to prevent the relevant situation accompanied by knowledge. Thus under Part IIA, the test of "knowingly permitting" would require both knowledge that the substances in question were in, on or under the land and the possession of a power to prevent their presence there.[17]

13.19 Whilst it is difficult to see how someone could be said to have permitted something they were unaware of, in that knowledge is already inherent within the concept of permitting, it is helpful for the purposes of analysis to consider the two issues separately. The following paragraphs attempt such analysis.

13.20 A preliminary and vital question is, however, what precisely "knowingly permitted" relates to. As mentioned above,[18] it could be interpreted as relating to the original entry of the contaminants to the land, *i.e.* a specific event or series of events; or to the continued presence of the substances in the land, *i.e.* a state of affairs. The question obviously has fundamental implications for subsequent owners, amongst others.

13.21 Comments during the passage of the Environment Bill suggest that the Government intended the words to have the latter, wider, effect. In particular, references were made to the responsibility of those having control over the sites where pre-existing contaminants remain[19];

> "We believe that it would be reasonable for somebody who has had active control over contaminants on a site, for example, when redeveloping it, to become responsible for any harm to health or the environment that may result, even if he did not originally cause or knowingly permit the site to become contaminated."

13.22 Another indication of Parliamentary intention arises when at one point an amendment was proposed to replace the words in section 78F(2) "*be* in, on or under the land" with "*come* into, onto or under the land".[20] This amendment, which would obviously have excluded subsequent owners, was rejected by the Government on the basis that it would ignore the responsibility of those who ". . . genuinely and actively permit the continued presence of contaminating substances in land".[21]

---

[16] [1972] A.C. 824.
[17] See also DETR Circular, Annex 2, para. 9.10 to similar effect.
[18] Pararaph 13.05.
[19] *Hansard*, H.L. Vol. 560, col. 1461.
[20] *Hansard*, H.L. Vol. 562, col. 189.
[21] *ibid.*

13.23 It therefore seems clear from these statements, and the retention of the word "be" in section 78F(2), that the concept of permitting can refer to the continued presence of the contaminants. Whilst the Government's subsequent guidance cannot be conclusive as to the intention behind the statute, it seems apparent from various of the exclusion tests that the Government envisages that subsequent owners may fall within the first tier of appropriate persons.[22]

Knowingly permitted: what knowledge?

13.24 The issue of what must be known in order to be said to knowingly permit may be a difficult one; it raises the question of whether there must be knowledge simply of the entry or presence of the substances, or also of their contaminating nature or potential. The only interpretation which would make the Act workable would appear to be the former. If the enforcing authority had to demonstrate knowledge of the contaminating effect of substances—perhaps in relation to activities many years ago—this would be a difficult task, indeed impossible in many cases. A decision which supports the former interpretation is *Shanks & McEwan (Teeside) Ltd v. Environment Agency*[23] where it was held that in the context of a waste licensing offence, the word "knowingly" related only to the fact of the deposit, and not to the other elements of the offence.

13.25 Another issue relates to knowledge which is supplied by the enforcing authority itself: a subsequent owner may not know of the actual presence of contaminating substances until he is informed by the authority that they have been discovered. The question is whether, by virtue of that knowledge, he then becomes a knowing permitter if he fails to take appropriate action. The difficulty with treating such a person as a knowing permitter is that it would potentially result in the statutory distinction drawn between knowing permitters in Class A and current owners or occupiers in Class B being almost entirely eroded. For this reason the Government's view in the Guidance is that knowledge resulting from notification under section 78B or consultation under section 78H is not sufficient to trigger the "knowingly permit" test.[24] However, this in turn involves drawing a distinction between knowledge deriving from notification and from other sources; a distinction which does not appear in the Act itself. Also, it might be argued that if Parliament had intended to draw such a distinction then it could have done so expressly, by providing that only knowledge prior to the notification of contaminated land being identified was relevant, or that knowledge arising from notification should be disregarded. In any event, it is quite likely in practice that formal

---

[22] As to the Guidance, see Chapter 17.
[23] [1997] J.P.L. 824; see also *Ashcroft v. Cambro Ltd.* [1981] 1 W.L.R. 1349.
[24] DETR Circular, Annex 2, para. 9.14.

notification under section 78B may be preceded by a period of dialogue between the authority and owner, during which the owner may acquire the relevant knowledge.

13.26 Another issue is whether actual knowledge is required, or whether constructive knowledge may suffice. There is authority to suggest that a person may be held to know that which in all the circumstances they could reasonably be expected to have known. Shutting one's eyes to the obvious may therefore not assist.[25] Such cases may be relevant where, for example, the person or company in question was aware of circumstances which might well have resulted in the land being contaminated (for example, a spill, a defective pipe, or a contaminative previous use), yet refrained from making any further enquiries.

Knowingly permitting: knowledge by companies

13.27 Where it is a company which is said to have knowingly permitted the presence of contaminants, the question arises as to what knowledge is to be attributed by the company. This will be determined on ordinary principles of corporate law, which were discussed by the Privy Council in *Meridian Global Funds Management Asia Ltd v. Securities Commission*.[26] In that case, Lord Hoffman indicated that it will first be necessary to look at the rules of attribution of knowledge (if any) contained in the company's constitution and in general company law. These would seem unlikely to provide much help in the specific case here. A second line of approach is to look at the general rules of attribution contained in the principles of agency, *i.e.* the extent to which the knowledge of a servant or agent is to be attributed to the company as employer or principal. The question is then essentially one of construction as to whether section 78F(2) intends that such knowledge shall be attributed. This is a question of the language, content and policy of the statute, and given the very serious consequences inherent in being held to have knowingly permitted under section 78F(2) it seems somewhat unlikely that the draftsman intended knowledge by every employee or agent to be attributed.[27]

13.28 Lord Hoffman in the *Meridian* case went on to suggest that if applying these primary and secondary rules of attribution did not produce a satisfactory result, then the court "must fashion a special rule of attribution for the particular substantive rule". This will be a matter of statutory interpretation, bearing in mind the purpose, content and policy of the

---

[25] See *Westminster City Council v. Croyalgrange* [1986] 2 All E.R. 353; *Schulmans Incorporated v. National Rivers Authority* [1993] Env. L.R. DI; *Roper v. Taylor's Central Garages (Exeter) Ltd.* [1951] 2 T.L.R. 294; *Vehicle Inspectorate v. Nuttall* [1999] 1 W.L.R. 629.

[26] [1995] 3 W.L.R. 413.

[27] But see also the discussion of *Shanks & McEwan (Teeside) Ltd v. Environment Agency* (paragraph 13.30 below).

relevant provision. The answer will not necessarily be confined to knowledge by persons who could be described as representing the "directing mind and will" of the company. In the context of the securities legislation under consideration in the *Meridian* case, it was held that the relevant knowledge was that of the person who had the authority of the company to acquire the relevant interest. In a different context, in *R. v. Rozeik*,[28] the relevant knowledge was held to be that of an employee who had authority to draw a cheque and sign it, but not those whose job was simply to type it out.

13.29 It may therefore be that in the context of contaminated land, knowledge by someone who was not a director, yet who had authority to take decisions involving the presence of contaminants, might well be attributed to a company. Such persons might include the environmental manager or safety officer, for example.

13.30 The question of the level at which there must be knowledge within a company arose in the environmental context in *Shanks & McEwan (Teeside) Ltd v. Environment Agency*.[29] The issue was whether the company could be said to have knowingly permitted the deposit of waste at one of its sites in breach of a licence condition. It was held that if knowledge by senior management was required for the offence, then it was sufficient that the management knew that the site was generally used for receiving waste of the relevant type. This approach may be appropriate for some, but not all, types of contaminated land situations. The Court also suggested, without deciding, that a wide rule of attribution might be appropriate for that purpose under section 33 of the Environmental Protection Act 1990, "where the company would be treated through its employees at the site as knowingly causing or knowingly permitting the deposit of any controlled waste in respect of which instructions or authorisation for off-loading onto the site were given by such employees".[30] However, whereas in the context of section 33 the strictness of such an approach is tempered by the availability of a defence of all reasonable precautions and all due diligence, such a defence would not be available in the context of section 78F.

13.31 The position here is somewhat different to that of the "wilful blindness" described in the previous paragraph: it is rather the situation where a company had within its organisation the knowledge of facts relevant to the condition of the land as contaminated, yet that information never reached the senior management of the company. Bearing in mind that the condition of the land must be significant in terms of harm before Part IIA can apply at all, a court may not be sympathetic to a company which claims not to have had the relevant knowledge because of the absence or inadequacy of its own internal management systems, or a company which had in place "paper systems" but did not implement them adequately.

---

[28] [1996] 2 All E.R. 281.
[29] [1997] J.P.L. 824.
[30] *ibid.*, p. 833.

Knowingly permitting: what constitutes permitting?

13.32 "Permitted" is a difficult word to pin down to a single precise meaning.[31] It could mean giving permission, leave or licence for something to be done. In this sense it relates most aptly to permitting the original entry of contaminants into or onto the land and in this context might include, for example, the owner or land who grants a licence for waste to be tipped on it. It could also in this sense be taken to cover someone who has the power to prevent the entry of the contaminants, yet knowing of their entry fails to exercise that power. It is this interpretation which is potentially of concern to parties such as landlords and lenders. The safeguard for such persons lies in the test provided in the statutory guidance, which has the effect of excluding a range of persons such as landlords, lenders, insurers and statutory permitting authorities, who fall within Group A only by reason of the capacity in which they stand in relation to the polluter. This test is discussed in more detail in Chapter 17, but it should be noted that the inclusion of a particular description of persons within the test does not imply that they will necessarily fall within Group A. This must be correct, since clearly the subsequent guidance cannot alter the effect of, or place a gloss on, the words used by Parliament.

13.33 The second aspect of "permitting" relates to the permitting of a state of affairs, *i.e.* the continued presence of the contaminants in the land. Whether "permit" is used in the broader sense will depend on the context and the objective of the legislation.[32] As discussed above, it appears to be the case that Parliament intended this aspect to be covered by section 78F(2). If this is correct, the question will then turn on what the relevant person could have done, at the relevant time, in relation to the presence of the relevant substances. This would seem to involve consideration of what opportunity the person had to remove the contaminants and whether it was reasonable to expect them to take advantage of that opportunity. Clearly, a freehold owner who is developing land and knows of the presence of contaminants will have the opportunity to remove them: whether it was reasonable to expect him to do so may depend on many factors, including the use for which the land is being developed. On the other hand, a tenant of property may not have the opportunity to carry out works to remove contaminants; it would not generally be reasonable to expect a tenant of, say, an office building to excavate the subsoil of the building to remove contaminants which are present. Indeed, he may be barred from so doing under the terms of the lease.

Substances: in, on or under the land

13.34 A substance may be present on land before it makes its way into or under the land. An example would be chemicals which are stored in a surface installation at an industrial plant, but then are accidentally

---

[31] *Vehicle Inspectorate v. Nuttall* [1999] 1 W.L.R. 629.
[32] *ibid.*

released and contaminate the subsurface: the chemicals were originally "on" the land, but are now "in" or "under" the land. The person (A) who caused or knowingly permitted the chemicals to be placed originally on the land may be a different person to the person (B) who caused or knowingly permitted them to escape and thus to be in or under the land. Since the chemical which was in the tank on the surface is the same substance by reason of which the land is now contaminated, a strict and literal reading of section 78F(2) would indicate that A, as well as B, is an "appropriate person". Such a result could obviously be unjust to A, and might well be avoided by construing the subsection so that the presence of the substance in, on or under the land is read as referable only to the point at which the land is in such a condition that harm or pollution of controlled waters (or the possibility or likelihood thereof) occur. On that basis, the presence of the substances on land in a proper storage installation would not be relevant so as to fix A with liability. Such a reading is not easy to reconcile with the wording of the subsection however, and in any event would not benefit A if he had placed the chemicals in a defective storage installation which presented the significant possibility of a harmful escape. In that case, A would have caused the substances to be on land and the land thereby to come within the definition of contaminated land in section 78A. The issue may therefore be essentially one of apportionment as between A and B.

13.35 The point was not addressed directly in debate, though at one point an amendment was proposed to replace the words "be in, on or under land" with "come into, onto or under land" as defining more accurately the polluter: such an amendment would also have limited the scope for subsequent owners to be liable, and as pointed out above was rejected by the Government.[33]

13.36 One practical implication of the storage installation, example given above, is that a seller of land on which potentially contaminative substances are stored should consider emptying such installations on sale or seeking indemnities from any purchaser whose activities or failure to maintain such installations might result in the escape of substances originally brought on to the land by the seller.

The second rank of appropriate persons: "Class B"

13.37 Subsection 78F(4) provides that if no person has, after reasonable enquiry, been found who is under subsection 78F(2) an appropriate person to bear responsibility for the things which are to be done by way of remediation, then the owner or occupier for the time being of the contaminated land in question is an appropriate person. The subsection is not readily intelligible, and raises a number of questions which are

---

[33] See paragraph 10.06.

considered in the following paragraphs. In particular, does it mean that the person concerned has not been "found" physically, or does it mean "identified"?[34]

13.38 The question has a bearing on individuals who are known, but who have died, and on companies or other corporate entities which have ceased to exist. The previous Government's view, expressed in debates, was that the circumstances where a polluter cannot be found would include cases where the company concerned had gone into liquidation.[35] On that basis, the "reasonable enquiry" will be not only into whether the person who caused or knowingly permitted the contaminating event or events can be identified, but also whether they still exist as a person or legal entity.

13.39 What does "reasonable enquiry" mean in this context? It is an important question because the enforcing authority cannot refer to the second group of appropriate persons until they have satisfied themselves as to the absence of anyone in the first rank. How long, for example, does this "reasonable enquiry" have to go on?[36] The authority which seeks to fix liability on a Class B party may well find itself faced with the argument either that it should have looked harder for a Class A party, or possibly that it should have pursued the estate of a deceased Class A party.

13.40 The problem does not end there. Section 78F(4) contemplates not just a reasonable inquiry into whether there is a person who is an "appropriate person"; it refers to an "appropriate person to bear responsibility for the things which are to be done by way of remediation". Those additional words might be thought to cause a further difficulty in suggesting that even where a person within section 78F(2) is found, they may not be appropriate to bear responsibility, for example, because they lack the funds. However, similar words appear in the definition of "appropriate person" itself in section 78A(9). Thus analysed, the words add nothing and accordingly the subsection does not mean that the appropriate person in the front rank (the actual contaminator) would have to be solvent and financially capable of complying with the notice in order to be an appropriate person. To hold otherwise would be to encourage those with contaminated land to divest themselves of adequate funding.

13.41 One other issue is what happens if a person or company within Class A has been found, but dies or ceases to exist before a remediation notice is served, for example during the statutory period of consultation. It is submitted that in such circumstances the party has been "found" and accordingly section 78F(4) cannot then operate to make the current owner or occupier liable. The consequence is that remediation will then be at the cost of the authority. The situation where the person dies or ceases to exist *after* the remediation notice has been served is considered in Chapter 14.

---

[34] See paragraph 13.42 for further discussion.
[35] *Hansard*, H.L. Vol. 562, Col. 209.
[36] See paragraph 13.45 for further discussion.

When is no Class A person found?

13.42 As stated above, the meaning and construction of the word "found" in section 78F(4) is crucial to the operation of Part IIA and will in particular in many cases determine whether or not the current owner or occupier can be held responsible for remediation. As indicated above,[37] the view of the Conservative Government which promoted Part IIA was that a company which no longer exists could not be said to be "found". This approach is followed by the Guidance, which suggests that a person who is no longer in existence cannot meet the description of "found".[38] Earlier versions of the Guidance referred to the Oxford English Dictionary definition of the word "found" as meaning "Discovered, met with, ascertained, etc." Unfortunately, this definition actually serves only to highlight the possible ambiguity of the word. Clearly, a person or company which is no longer in existence cannot be "met with", but their identity can be ascertained. Reference in the OED to the verb "find" demonstrates the richness of meanings attributable to the word, some of which imply a physical encounter or determination of the whereabouts of a person or object; but others of which could certainly be applied to the process of discovering the identity of a person who no longer exists.[39]

13.43 There are authorities which have construed "found" as denoting physical presence, but it may be difficult to extrapolate from these as they relate to matters such as the offence of being "found" drunk in a public place,[40] or being "found within the jurisdiction" for enforcement purposes.[41] Nor does such case law as exists on similar wording under earlier public health legislation help greatly.[42] In construing the provisions under trust legislation as to powers when a trustee cannot be "found", there are two cases which suggest that a company which has been dissolved cannot be "found".[43] However, there is also a case to the opposite effect, that where a trustee company is known to have been dissolved, it is not correct to say it "cannot be found": ". . . I am not ignorant where it is, I know that it is not anywhere; it does not exist."[44]

13.44 The best way of approaching the matter may be to look at the structure and purpose of Part IIA as a whole. The point of section 78F is to set out a scheme with the object of getting contaminated land cleaned up in the public interest.[45] If someone is an "appropriate person", then the expecta-

---

[37] Paragraph 13.38.

[38] DETR Circular, Annex 2, para. 9.17.

[39] For example, "to discover, come to the knowledge of"; "to discover or attain by search or efforts".

[40] *Palmer-Brown v. Police* [1985] 1 N.Z.L.R. 365, 369.

[41] *R. v. Lopez, R. v. Sattler*, 27 L.J.M.C. 48.

[42] *The Conservators of the River Thames v. The Port Sanitary Authority of London* [1894] 1 Q.B. 647.

[43] *Re General Accident Assurance Co. Ltd* [1904] 1 Ch. 147; *Re Richard Mills & Co. (Brierley Hill) Limited* [1905] W.N. 36.

[44] *Re Taylors Investment Trusts* [1904] 2 Ch. 737 at 739, Buckley J.

[45] Compare *Rhymney Iron Co. v Gelligaer District Council* [1917] 1 K.B. 589 at 594, Ridley J.

tion is that they will be served with a remediation notice and that legal sanctions and consequences will follow if they do not comply. If there is no-one against whom such action can be taken then the statutory scheme is that the owner and occupier are responsible. To hold that a person or company which no longer exists can be "found" would mean that the legislation would be ineffective in those circumstances, and would thereby leave a potentially large gap in the scheme of channelled liability envisaged by section 78F. Coupled with the *Hansard* statement of intention referred to at paragraph 13.38 above, this gives robust support to the view that "found" implies physical existence of the person in question. The issue is, however, undoubtedly capable of argument the other way.

"After reasonable enquiry"

13.45 The question is not simply whether a Class A person can be found, but whether such person can be found "after reasonable enquiry". In *Rhymney Iron Co. v. Gelligaer District Council*,[46] the question arose of what steps a local authority had to take under the Public Health Act 1875, s.94, to establish the cause of a blocked sewer before serving notice on the owner of the house in question. It was stressed by the Court that the aim and purpose of the 1894 Act in protecting the public had to be kept in mind, and any duty on the local authority to carry out what might be a lengthy or difficult inspection process was rejected. As Ridley J. put it, the object of section 94 is[47]:

> ". . . not to put upon the local authority the duty of finding out who in the ultimate result is the person responsible for the nuisance; the object is, as it appears to me, that they should by all means possible procure that such a nuisance should be abated."

13.46 However, given the express pre-condition of reasonable enquiry in Part IIA and the magnitude of the potential liabilities that may fall on the owner or occupier if persons in Class A cannot be found, the courts seem likely to take a more rigorous approach to what constitutes reasonable inquiry under section 78F(4). At the same time, the enforcing authority is fulfilling a public function in which it needs to balance the cost and delay of possibly fruitless inquiries against the public interest in getting the land cleaned up.

13.47 What is critical is that the authority should be able to demonstrate that it has approached its task of inquiry logically and consistently. The relevant steps are likely to include the following:

1. Careful review of data collected on the initial investigation of the land to establish how far this helps to identify a relevant person or persons.

---

[46] [1917] 1 K.B. 589.
[47] *ibid.*, p. 595.

2. Further enquiries as necessary as to ownership and occupation of the land during the relevant period, including appropriate Land Registry searches and enquiries.
3. Enquiries as to the nature of activities and substances used on the site during that period. These may possibly be by way of reviewing local sources of information, by direct enquiries of those involved, or former employees, or neighbours. It is possible that the authority may wish to use consultants or enquiry agents to undertake some or all of this work.
4. Searches at the Companies Register, or other sources of information on companies, to establish the current status of any companies involved, and the identities of their directors and senior officers.
5. Service of a requisition for information on all relevant parties, using where necessary its statutory powers to obtain information under section 16 of the Local Government (Miscellaneous Provisions) Act 1976.

13.48 This aspect of local authority and Agency duties under Part IIA is likely to be one of the more difficult in practice. Whilst there is considerable technical experience of tracing those responsible for recent pollution incidents—often successfully—the attribution of responsibility for contamination of soil or groundwater which may have occurred many decades ago, and over the course of many years, is a different matter entirely. The nature of the inquiry will differ depending on the circumstances, and will, for example, be very different if the contamination is the result of a single, recent and identifiable incident.

13.49 The problem is that if an owner or occupier receives a remediation notice in circumstances where a Class A person cannot be found, they will be able to argue that if reasonable enquiry has not been made by the local authority, and thus the precondition for the notice being served on them has not been fulfilled. The notice would in fact be *ultra vires*.[48] This problem may arise in many cases. The owner or occupier may argue either that the original polluter is or should be traceable; or where they are themselves in Class A as a knowing permitter, that the original polluter is traceable and should be in Class A as well.

Companies in liquidation

13.50 It follows from the discussion above that if a company is in liquidation, which may either be voluntary or compulsory, then such a company could be an appropriate person. A company in liquidation still exists for the time

[48] This of course brings us back to the question of whether the owner/occupier can be said to be a "knowing permitter" and so within Class A themselves.

being, and there is nothing in subsections 78E(4) or 78H(5) which suggests that insolvency, receivership or liquidation necessarily bars the service of a remediation notice.[49] However, if a company has been struck off the Companies Register it ceases to exist as a legal entity. In any liquidation situation, recovery of the costs of clean-up would be tenuous. The enforcing authority carrying out the works themselves under section 78N would not be a preferred creditor. Any charging notice served under section 78P is registrable as a charge but may not necessarily take priority.[50] The enforcing authority has the same powers and remedies under the Law of Property Act 1925 as if it were a mortgagee by deed having power of sale and lease.[51]

Privatised utilities and reorganised local authorities

13.51 Class A appropriate persons may well include utility companies such as gas, electricity and water undertakers, which may have gone through a serious of transitions, possibly from private to nationalised companies, then privatisation. In the course of this process the entity which caused or knowingly permitted the relevant contamination may have ceased to exist, to be replaced by a successor. The same may well be true of local authorities which have been subject to reorganisation, and of NHS trusts. The question is whether the successor can be regarded as the Class A appropriate person. On a strict application of the rules of corporate personality, the two are separate entities and there is no reason why, for example, Railtrack plc should be treated as the appropriate person in respect of contamination caused by British Railways or, for that matter, Great Western Railways before it.

13.52 However, one additional factor is that in many of these situations, there will have been a statutory transfer scheme passing assets, rights and liabilities. The question is whether such a scheme could have the effect of bringing the successor company or authority within Class A. The difficulty with this argument is that at the time of the transfer the liability in question would not have existed. Indeed, the legislation creating that liability would not have existed. It is difficult to see how a transfer scheme could properly be said to have the effect claimed in such circumstances.

Groups of companies

13.53 Another practical issue which may frequently arise is identifying which one of a group of companies actually operated a site or process so as to be the appropriate person within Class A. It may well be common knowledge

[49] The issue is discussed more fully in Chapter 14.
[50] This issue is discussed at Chapter 22.
[51] Section 78P(11).

that a particular contaminated site was for many years operated by "X Co." which may have comprised a group of several companies, all of which are still in existence. In the absence of cooperation from X Co. the authority may not be able to identify with certainty on which company in the group a notice should be served.

13.54 A practical first step is possibly for the local authority to serve a requisition for information on the companies. If this does not elicit any helpful response, then one possible pragmatic solution may be to serve the holding company and each of the subsidiaries with a remediation notice, stating that some of the notices may be withdrawn when liability is clarified. The authority will however have to attempt to apportion liability between the companies, which may be far from straightforward.

Foreign companies

13.55 Another potential problem may occur where the company, which is the Class A appropriate person, is now outside the jurisdiction: it may for example be a foreign company registered outside the U.K., which had operated for a time in the U.K. The company may have been registered in the Channel Isles or elsewhere, and may have operated in the U.K. direct, or through a subsidiary. If such a company can be found, there would seem to be no mandate in section 78F(4) for turning to the persons in Group B, simply because it would be complex, costly and difficult to seek to enforce against it.

Directors of companies

13.56 One other important issue is whether officers of a company can be responsible as Class A appropriate persons in their own right. There seems no reason why not in principle, if the director can be said to have caused the contamination, or knowingly permitted it through decisions which they took or the way in which they ran the company. At a practical level, the smaller the company, the easier it will probably be to find such circumstances.

13.57 The issue has practical significance, not only the obvious concern for the individuals involved, but also for the current owners and occupiers. If the company which caused contamination has ceased to exist, its directors may still be alive and as such potential Class A persons. This in turn may mean that the current owner or occupier will not be liable.

13.58 At one level, there seems no reason why a director should not be held responsible in this way. A director who undertakes wrongful trading prior to a liquidation may be liable personally under statutory provisions, so why not a director who knowingly operates a contaminating process, then winds up the company? It should be noted, however, that the practical consequence may be that although liability will fall on the director, he or she may be able to argue hardship. In that case, the authority will not be able to turn to the owner and occupier as Class B persons, so the clean-up may have to be in whole or in part at the public expense.

"Owner"

13.59 As discussed above, the second group of appropriate persons only come into contention when no appropriate person from the first rank has been found. The second group comprises the owner or occupier for the time being of the contaminated land. There is a definition of "owner" in section 78A(9), namely:

> "a person (other than a mortgagee not in possession) who, whether in his own right or as a trustee for any other person, is entitled to receive the rack rent of the land, or, where the land is not let at a rack rent, would be so entitled if it were so let."

13.60 For Scotland, the definition is different:

> "a person (other than a creditor in a heritable security not in possession of the security subjects) for the time being entitled to receive or who would, if the land were let, be entitled to receive, the rents of the land in connection with which the word is used and includes a trustee, factor, guardian or curator and in the case of public or municipal land includes the persons to whom the management of the land is entrusted."

13.61 Taking the long leaseholder at a ground rent as an example, he may well have sub-let and will not therefore be an occupier. He will not be the freeholder either, but he will fall within the definition of "owner", as one entitled to the "rack rent". The freeholder will not, if the leaseholder is paying only a ground rent, be the "owner".

13.62 A person who holds an option or other contractual interest in the land (for example, a contract to purchase) would not appear to fall within the definition of "owner".

13.63 Other cases, such as trustees and managing agents, are discussed in Chapter 14.

Proving ownership

13.64 Ownership can be a difficult issue to prove conclusively in some cases. It seems likely that the Courts will not require an enforcing authority to prove the exact nature of the legal interest held by the appropriate person, and that evidence such as letters or applications for certificates of lawful use and the like may be sufficient. This was the position in *Walton v.*

Sedgefield Borough Council[52] where it was said that what is sufficient to establish ownership depends on the circumstances of the particular case. It was also accepted in that decision that if ownership is proven at a particular date, there will be a presumption of its continuance.

"Occupier"

13.65 The term "occupier" is not defined in Part IIA and the question of whether a person is in occupation will have to be determined on the facts of each case. The test is that of the degree of control exercised over the land rather than exclusively of rights of occupation: a licence entitling a person to possession may make someone an "occupier".[53] Similarly, it appears that a statutory tenant is an "occupier"[54] but there is authority to suggest that a person who entered premises forcibly and unlawfully is not.[55] In the context of the Wildlife and Countryside Act 1981, it has been held that a contractor coming onto land on a transient basis to carry out work is not an "occupier".[56]

13.66 In Scotland, it has been established that receivers may become occupiers.[57] Thus, those receivers acting in a management capacity may conceivably find themselves regarded as occupiers for the purposes of contaminated land provisions, although the specific statutory protection afforded to them should be noted.[58]

Liability of owner and occupier: generally

13.67 Government policy is that the owner and occupier of land, even if "innocent" in relation to the presence of the contaminating substances, should in some cases bear responsibility for the condition of their land. Lord Northbourne, in debate, put forward extreme examples of cases where it would be unjust to regard an innocent owner as liable: for example, an owner whose land was contaminated by a crashed tanker, or by migrating dust or particles.[59] The Government's response was that it was not justifiable to relieve owners of liabilities which they might already incur under existing legislation, and that it was reasonable for owners to bear responsibility for their property and its effects on others and the wider environment in cases where no original polluter can be found.[60]

---

[52] [1999] J.P.L. 541.
[53] *Stevens v. Bromley London Borough Council* [1972] 1 All E.R. 712, CA.
[54] *Brown v. Minister of Housing and Local Government* [1953] 2 All E.R. 1385.
[55] *Woodcock v. South West Electricity Board* [1975] 2 All E.R. 545.
[56] *Southern Water Authority v. Nature Conservancy Council* [1992] 3 All E.R. 481, HL.
[57] *Lord Advocate v. Aero Technologies (in receivership)* 1991 S.L.T. 134.
[58] See further, Chapter 14.
[59] *Hansard*, H.L. Vol. 562, cols. 1052.
[60] *Hansard*, H.L. Vol. 562, cols. 1052.

13.68 As originally drafted, there were three circumstances where the current owner or occupier would be the appropriate person to bear responsibility for remediation. Two of these circumstances were dropped in the course of the passage of the Bill; namely cases where the owner/occupier refused consent for remediation works to be carried out on their land, and cases where the liability of the original "polluter" had been directly or indirectly transferred. The original provisions on transfer of liability were particularly difficult to understand, though they were basically prompted by the Government's wish to respect contractual provisions.[61] On looking at the matter more closely, however, the Government concluded that it would be more practical to leave the question to be dealt with through the normal contractual means of guarantees and indemnities rather than detailed statutory provisions.[62]

13.69 Two circumstances now remain where the owner or occupier is liable under this provision; namely where after "reasonable enquiry" no appropriate person has been found under subsection (2), or where something to be done by way of remediation cannot be regarded as "referable" to anyone under subsection (3).

Liability of owner or occupier where original polluter still in existence

13.70 What happens if there are still in existence original contaminators but they are not appropriate persons because what is to be done by way of remediation is not referable to the substances which they caused or knowingly permitted to be present, etc., in the land in question? This is dealt with by subsection 78F(5). In those circumstances it is the owner or occupier for the time being who is the appropriate person in relation to the things in question. Thus there could be a situation where a number of remedial actions are required as a result of land having been contaminated in various ways. It may be possible to find Group A appropriate persons to whom some but not all of the remedial actions are referable. The owner and occupier will in such circumstances be appropriate persons with respect to those things which are not referable to the Class A appropriate persons. The general concept of referability is discussed above.[63]

Protection of owner and occupier

13.71 Persons liable under subsections (4) and (5) are in a better position than persons liable under subsection (2) in one respect. This relates to the situation where the contamination results or is likely to result in pollution of controlled waters—see Chapter 15.

---

[61] *Hansard*, H.L. Vol. 562, cols. 1048–1051.
[62] *Hansard*, H.L. Vol. 562, cols. 1498.
[63] See paragraph 13.07.

The innocent fly-tipped owner

13.72 One situation for which the Government did have evident sympathy was the plight of the owner or occupier who suffers from fly-tipping on their land. Imposing liability for clean-up of such contamination on the owner or occupier on the basis that the fly-tipper could not be found would have resulted in significantly harsher liabilities than under section 59 of the 1990 Act which deals with the removal of unlawfully deposited waste. The Government's view was that the best way of dealing with the problem was to disapply Part IIA in cases where section 59 could be used.[64] This was achieved by an amendment at third reading stage in the House of Lords. Full responsibility for dealing with unlawfully deposited waste would thereby be placed with the Environment Agency/SEPA as the waste regulation authority and the exemption for innocent victims of fly-tipping was retained.[65] Certainly in relation to contamination occurring after controls over waste deposits on land were introduced, the availability of section 59 as a remedy might turn out to be a significant restriction on the use of Part IIA powers. However, it must be remembered that section 59 is a very limited remedy compared with Part IIA. It only deals with removal of the waste, not the consequences of the deposit of the waste. Also, it applies only to controlled waste, that is to say "household, industrial and commercial waste or any such waste."[66] Not every type of contaminant dumped on land will necessarily fall within the relevant definitions.

The role of Guidance

13.73 This Chapter has concentrated on the statutory provisions of section 78F, which as will be appreciated, are complex enough in their own right. They are, however, not the whole story.

13.74 Section 78F(6) applies where there are two or more appropriate persons in relation to any particular remediation action. Here the enforcing authority must determine in accordance with guidance from the Secretary of State, whether any of them is to be treated as not being an appropriate person. In other words, an appropriate person may be entitled to be excluded from the liability group if the guidance so dictates. Exclusion is dealt with in Chapter 17.

13.75 Secondly, by section 78F(7), where two or more persons are appropriate persons in relation to a particular remediation action, they shall be liable to bear the cost of that action in proportions determined in accordance with guidance issued by the Secretary of State. Again the guidance is mandatory in effect. Apportionment is covered in Chapter 18.

---

[64] *Hansard*, H.L. Vol. 562, col. 182.
[65] *Hansard*, H.L. Vol. 562, cols. 1048–1051.
[66] Environmental Protection Act 1990, s.75(4).

Chapter 14

LIABILITY—PARTICULAR CASES

Summary

Certain commercial situations may give rise to particular concerns under Part IIA as to who is an appropriate person. In particular, there is the position of banks, insolvency practitioners, trustees, and parent companies. Other issues of concern are the position on the dissolution of companies and the death of individuals, and the position of the Crown. These issues will often require consideration of other areas of law, such as company law, insolvency, and succession.

Banks and other lenders

14.01 The question of the potential liability of lenders under the new provisions aroused considerable interest in debate. The Government did not regard it as likely that the act of lending money to a polluter would of itself constitute causing or knowingly permitting contamination.[1] This view is reflected and supported in the statutory Guidance on the exclusion of lenders from liability.[2] Lending money in itself is, however, not really the issue: what is more important for most lenders is the security for the loan provided by the land.

14.02 The definition of "owner" adopted in the Act expressly excludes a mortgagee not in possession and the Scottish equivalent. The Government expressed the view that banks should not be treated as "deep pockets", that the simple act of lending should not result in liability, and that the lender should retain the right to walk away from security without taking possession if the costs of remediation appeared to exceed its ultimate value.[3] However, the banks and other institutions such as the Council of

---

[1] *Hansard*, H.L. Vol. 565, col. 1497; see also DETR Circular, *Contaminated Land*, Annex 2, para. 9.11.

[2] See Chapter 17.

[3] *Hansard*, H.L. Vol. 560, col. 1448.

Mortgage Lenders were still concerned that a lender might find itself in possession by default where a borrower simply abandoned the property.[4] Although it agreed to look into the issue, the Government concluded that no changes were necessary to protect the lenders in this situation; they were already exposed to similar liabilities under existing legislation on public health, highways and building standards.[5] Similarly, the Government were not sympathetic to the argument that special provision was needed to protect lenders who took possession to a limited extent to secure property or deal with obvious hazards.[6]

Insolvency practitioners and similar persons

14.03 Persons acting in certain capacities in relation to insolvency enjoy specific protection by virtue of sections 78X(3) and (4). Subsection (3) is intended to protect persons acting in various capacities (specified in subsection (4)) in insolvency situations. The provisions cover liquidators, administrators, administrative receivers, supervisors of voluntary arrangements, trustees in bankruptcy, the official receiver, and any person acting as a receiver under an enactment (*e.g.* the Law of Property Act 1925) or appointed as such by a court, or by any instrument. The protection is twofold:

(a) the person is not liable in a personal capacity for remediation costs unless the remediation requirement is referable (see section 78F(10) for the meaning of this) to substances whose presence in, on or under the land is a result of any act done or omission made by him, which was unreasonable for a person acting in his capacity; and
(b) he shall not be guilty of an offence of failing to comply with a remediation notice unless the relevant requirement relates to a thing for which he is personally responsible under (a).

14.04 This wording is potentially difficult. It seems unlikely that the presence of contamination is likely to be the result of the positive act of an insolvency practitioner or similar person, but its presence might well be said to result from an omission on his part, *i.e.* failure to remove it, or failure to take steps which would have prevented it occurring. An obvious example might be the case where a receiver is appointed and becomes aware that leaking drums of chemicals are being stored on the property, in such conditions that soil contamination is continuing to occur.

---

[4] *Hansard*, H.L. Vol. 560, col. 1445; Vol. 562, col. 1040.
[5] *Hansard*, H.L. Vol. 562, col. 1042–1053.
[6] *Hansard*, H.L. Vol. 562, cols. 1042–1053.

14.05 The concept of an unreasonable omission is a difficult one, raising issues of what it is or is not reasonable to expect an insolvency practitioner to do as regards land which is or might be contaminated. Where there is an obvious problem such as corroded drums, it might well be said to be unreasonable to take no action to prevent further contamination occurring, for example by arranging for disposal by a specialist contractor. In less obvious cases, however, it might be questioned to what extent an insolvency practitioner should investigate for possible contamination, or should expend money on remedial measures, in order to avoid being held to have acted (or failed to act) unreasonably. In this respect, the case of *John Willment (Ashford) Limited*[7] may be relevant, in that it was suggested there that a receiver could not exercise his discretion in such a way as to lead the company to act unlawfully. It will also be relevant to consider the terms of appointment or the debenture under which appointment takes place, to establish exactly what powers the receiver has. A receiver can hardly be said to be acting unreasonably in failing to do what he has no powers to do.

Managing agents

14.06 The definition of "owner" in section 78A(9) includes a person receiving the rack rent of the land, whether in their own capacity or as a trustee for another person. This means that a professional managing agent could find themselves falling within the category of Class B appropriate persons.[8] The agent in such circumstances will not have the benefit of the statutory protection accorded to insolvency practitioners, as outlined above.

Trustees

14.07 The definitions of owner at section 78A(9) include a trustee, and in Scotland also factors, guardians and curators. The possible hardship of the provisions applying to trustees was drawn to the attention of the Government in debate.[9] The response was that to provide an exemption for trustees would be to open up an easy route for evasion.[10] Therefore there is no provision in the legislation to protect trustees as such, for example by reference to reasonableness of conduct (as in the case of insolvency practitioners) or by limiting liability to the extent of assets held. The trustee will however have a lien over the trust funds to be indemnified for work and expenses properly incurred in relation to liabilities under Part IIA.[11]

---

[7] [1979] 2 All E.R. 615.
[8] See *London Borough of Camden v. Gunby* [2000] 1 W.L.R. 465 (a case on statutory nuisance); also see *Midland Bank Ltd v. Conway Corporation* [1965] 1 W.L.R. 1165.
[9] *Hansard*, H.L. Vol. 562, col. 163.
[10] *Hansard*, H.L. Vol. 562, col. 165; Vol. 560, col. 1448.
[11] See *X v. A and others* [2000] Env. L.R. 104.

Unincorporated associations

14.08 Where property belongs to a club or unincorporated association, it will be a question of fact who is the owner or occupier.[12] Where the property is held by trustees, then they will be the owner.[13] Otherwise, consideration will have to be given as to whether liability rests with the members as a whole or with a committee which exercises control over the use of the property.[14]

Parent companies

14.09 Part IIA does not deal expressly with the issue of parent and subsidiary companies, but it is undoubtedly an issue which will arise in practice, particularly in cases where a subsidiary which caused or knowingly permitted contamination either no longer exists or has insufficient assets to meet its remediation obligations.

14.10 The question here is whether it may ever (and if so, in what circumstances) be appropriate for an enforcing authority to treat the parent company as an appropriate person. It is submitted that this raises three distinct questions, as follows, which are considered in subsequent paragraphs:

1. Did the parent company itself cause or knowingly permit the contamination, *i.e.* is there some direct relationship between the parent and the contamination?
2. Did the parent company cause or knowingly permit the contamination indirectly, through the activities of the subsidiary, for example by funding the contaminating activity, or by failing to exercise control or supervision so as to prevent it?
3. Are there broader policy reasons for attributing the act of a subsidiary to its parent, for example to prevent evasion of the liability scheme of the Act?

14.11 At a practical level, local government or Environment Agency officers may have no wish to become embroiled in complex issues of corporate personality and control. If during the period of consultation prior to service of a notice neither the parent or subsidiary has accepted responsibility, the pragmatic answer may be to serve notice on both companies, and leave it to them as to how they respond in terms of compliance or any appeal.

Direct activity by parent

14.12 Each case here will depend on its own facts, and close attention will need to be paid to which company actually carried out the activities in question which gave rise to contamination: who provided the employees, engaged

---

[12] *Bolton v. Stone* [1951] A.C. 850, 858.

[13] *Verrall v. Hackney London Borough Council* [1983] Q.B. 445.

[14] See *Evans v. Waitmeta District Pony Club, East Coast Bays Branch* [1972] N.Z.L.R. 773; *affirmed* [1974] 1 N.Z.L.R. 28.

the contractors, gave the instructions, owned the relevant assets? On the basis of the case-law on "causing" it will not necessarily be a matter of choosing whether one or other company caused the contamination: it may well be possible for both to be said to have caused it.[15] It may, depending on the circumstances, be possible to find that a relationship of agency existed between a company and one of its members, or between companies in the same group, so that (for example) one company might have been acting as another's agent in carrying on the business ostensibly carried on by the principal company.[16]

14.13 One potentially important issue will be whether the parent's own directors took an active management role at the facility, or if there were common directors, whether they were acting on behalf of the parent or the subsidiary. The identity of the directors should be readily ascertainable from the Register of Companies, but the capacity in which they acted will require examination of the facts. Here, norms of corporate behaviour may well be relevant in deciding what conduct might normally be expected in the parent/subsidiary relationship.[17]

Indirect activity by parent

14.14 Here the issue is not whether the parent itself directly caused the contamination, but whether it might be said to have caused or knowingly permitted it by reference to the relationship between parent and subsidiary. This might take the form of active facilitation (for example, the parent providing resources to a subsidiary which is carrying on a contaminating activity) or failure to prevent (for example, knowing that a subsidiary is causing contamination and failing to prevent it). Another possible scenario would be the situation where a parent exercises strict financial control over a subsidiary, so that the subsidiary is denied the resources to deal with contamination that is known to be present.

14.15 Again, each set of circumstances will require careful consideration. The nature and degree of control of the parent over the subsidiary will be relevant. Certain cases have considered, in the context of company law, whether one company can be a "shadow director" of another by exerting management, strategic or financial control.[18] However, it is questionable

---

[15] See Chapter 13.

[16] *Smith Stone & Knight Limited v. Birmingham Corporation* [1939] 4 All E.R. 116; *Re FG (Films) Limited* [1953] 1 W.L.R. 483; *Firestone Tyre and Rubber Co. Ltd v. Lewellin* [1957] 1 W.L.R. 464. Compare however, *William Cory & Son Ltd v. Dorman Long & Co.* [1936] 2 All E.R. 386.

[17] See the decision of the U.S. Supreme Court in *United States v. Bestfoods* [1998] U.S. Lexis 3733; 118 S. Ct. 1876.

[18] The concept is found in section 741 of the Companies Act 1985. See *Re Unisoft Group Ltd (No. 2)* [1994] B.C.L.C. 766; *Re Hydrodam (Corby) Ltd* [1994] 2 B.C.L.C. 180; *R. v. Secretary of State for Trade and Industry, ex p. Laing* [1996] 2 B.C.L.C. 324; *Standard Chartered Bank of Australia v. Antico* [1995] 13 A.C.L.C 1381.

whether the concept of shadow director adds very much to the environmental law tools of causing and knowingly permitting.

14.16 A parent company may have all sorts of reserve powers over a subsidiary. These may not necessarily be exercised, but their existence may be relevant to the question of whether the parent could have prevented a particular course of action and thus may be held to have knowingly permitted it.[19]

Attributing liability to parent on policy grounds

14.17 On ordinary principles of corporate personality a subsidiary company is a separate legal entity, with rights and liabilities distinct from those of its parent or other shareholders.[20] It is common for these rules to be utilised, often for entirely legitimate reasons, so as to isolate potential liabilities within particular companies in a group structure.

14.18 The provisions of Part IIA are silent on the issue of liabilities of parent companies, though it would have been possible to insert provisions making the position clear. In the context of the U.S. "Superfund" legislation, the Supreme Court held in *United States v. Bestfoods*[21] that the relevant U.S. statute does not affect the fundamental principle of corporate law that a parent is not liable for the acts of its subsidiaries simply by being a parent. The same position is likely to apply under U.K. law; namely that in view of the reasoning behind the principle of separate corporate personality, and its long-standing recognition, any Parliamentary intention to pierce the corporate veil would need to be expressed in clear and unequivocal language.[22]

14.19 Nonetheless, under U.K. law there is certainly authority to suggest that a court may look beyond corporate structures which are a device to circumvent legal restrictions,[23] to perpetuate fraud,[24] to conceal the true facts,[25] or to avoid liability to another.[26] Such principles may apply to

---

[19] See *Unilever plc v. Gillette (U.K.) Limited* [1989] R.P.C. 583, CA (reserve powers of U.S. parent over U.K. subsidiary were relevant to application for patent action against subsidiary).

[20] *Salomon v. A Salomon & Co. Ltd* [1987] A.C. 22.

[21] [1998] U.S. Lexis 3733; 118 S. Ct. 1876.

[22] *Dimbleby & Sons Ltd v. National Union of Journalists* [1984] 1 W.L.R. 427, HL, at 435.

[23] *Merchandise Transport Ltd v. British Transport Commission* [1962] 2 Q.B. 173, 206–207; *Adams v. Cape Industries plc* [1990] Ch.433; compare *Ord v. Belhaven Cut Ltd* [1998] N.L.J. 938.

[24] *Aveling Barford Ltd v. Perion Ltd* [1989] B.C.L.C. 626; *Re H and Others* [1996] 2 All E.R. 391; *Re Darby, ex p. Brougham* [1911] 1 K.B. 95.

[25] *Woolfson v. Strathclyde Regional Council* (1978) 38 P & CR 521; *Re FG (Films) Ltd* [1953] 1 W.L.R. 483.

[26] *Creasey v. Beachwood Motors Ltd* [1993] B.C.L.C. 480; *Gilford Motor Co. Ltd v. Horne* [1933] Ch. 935; *Yukong Line Ltd of Korea v. Rendsburg Investments Corp of Liberia and others (No. 2)* [1988] 4 All E.R. 82.

property transactions.[27] However, the use of a company as a single purpose vehicle may be an entirely legitimate approach which is more than a "mere facade"; nor is a court free to disregard the principle of separate corporate personality merely because it considers that justice so requires.[28] Nor should the closeness of economic and commercial relationships between companies in a group justify, in general, treating them as a single entity.[29]

14.20 Particular problems may arise in relation to loose inter-company arrangements where one company in a group occupies property belonging to another on an informal basis, which confers no rights of exclusive occupation. This may fall short of giving rights of occupation which are recognised as having legal consequences for the purposes of Part IIA.[30] The question is perhaps most likely to arise where a subsidiary company is liable as owner of a contaminated site and has insufficient assets to carry out the necessary remediation. Much will no doubt depend on particular circumstances[31]: a court may view very differently a parent which has conducted a particular aspect of its business through a trading subsidiary for many years, from a company which, knowing it is to be faced with a remediation notice for a site of which it is owner, transfers that site to a subsidiary company with no assets, created for that purpose. There is much to be said for retaining some flexibility in this area, provided the courts are aware of the dangers of too readily upsetting matters of property or contract.[32] In the circumstances where a parent transfers a site knowing of its contaminated condition, there may well be an argument that the parent is responsible in its own right for knowingly permitting. Those exclusion tests described in Chapter 17 which relate to transfers of property will also require consideration in this context.

Dissolution of companies

14.21 Liability of a company under Part IIA may arise from its past acts as a causer or knowing permitter of contamination, or from its ownership or occupation of contaminated land. Both situations need to be considered in

---

[27] *Gisborne v. Burton* [1989] 1 Q.B. 390; *MacFarlane v. Salfield Investments Ltd* [1997] S.C.L.R.

[28] *Re Polly Peck International plc (in administration)* [1996] 1 All E.R. 433, 447–8.

[29] *Lonrho Ltd v Shell Petroleum Co. Ltd* [1980] Q.B. 358; affirmed [1980] 1 W.L.R. 627 (HL); *The Albazero* [1977] A.C. 774 at 807; *Bank of Tokyo Ltd v. Karoon* [1987] A.C. 45n at 64.

[30] *Butcher Robinson & Staples v. London Regional Transport* [1999] 36 E.G. 165. Compare the more liberal approach taken in *DHN Food Distributors Ltd v. Tower Hamlets London Borough Council* [1976] Q.B. 852.

[31] Sealy points out that commentators have failed to discern any set pattern in the decided cases: ". . . indeed, in many instances they seem to contradict each other in the most baffling way". (L.S. Sealy, *Cases and Materials in Company Law*, Sixth edition, 1996, p. 56.)

[32] Sealy, *op. cit.*, at p. 56.

the situation where a company is dissolved. Dissolution can occur under various provisions of the Insolvency Act 1986, Chapter IX, after winding up[33] and under the Companies Act 1985, Part XX, Chapter VI.[34]

14.22 Once the company is dissolved it ceases to exist as a legal entity and, it is submitted, can no longer be "found" for the purposes of section 78F.[35] It may however be possible for an application to be made for the dissolution to be declared void, or for the company to be restored to the register.[36] The enforcing authority would need to consider carefully whether any purpose would be served by such an application.

Winding-up and dissolution of companies: effect on property

14.23 One of the assets of the company being wound up may be contaminated land. The issue may arise in two ways: first, on disclaimer by the liquidator during winding up, and, secondly, the effect of dissolution of the company if by that time the property has neither been disclaimed nor disposed of.

14.24 By section 178 of the Insolvency Act 1986, where a company is being wound up in England or Wales the liquidator may disclaim onerous property, even if he has already taken possession, attempted to sell, or otherwise exercised rights of ownership.[37] The definition of "onerous property" under section 178(3) includes property which may give rise to an obligation to pay money or perform an onerous act or which may well not be saleable readily or at all. Contaminated land may well fall within this definition, whether or not a remediation notice has yet been served.[38]

14.25 The effect of disclaimer under section 178(4) is to determine, as from the date of disclaimer, the rights, interests and liabilities of the company "in or in respect of the property disclaimed". However, the subsection expressly provides that it does not affect the rights or liabilities of any other person except insofar as is necessary for the purpose of releasing the company from liability. It is not totally clear how this provision will apply to the situation where the company may be subject to liabilities either in its capacity as the original perpetrator of contamination, or in its capacity as continuing owner or occupier. Where the company's liability depends on its continuing ownership of the property, then disclaimer will terminate

---

[33] See sections 201 (voluntary winding up), section 202 (order of court), section 204 (Scotland), and section 205 (final meeting of creditors or winding up by court).

[34] Section 652 (defunct companies), section 652A (private companies).

[35] See Chapter 13.

[36] Companies Act 1985, ss.651 and 653.

[37] The liquidator will need to follow the appropriate procedures of section 178(2) by giving the prescribed notice.

[38] For the application of section 178 to a waste management licence as onerous property, see *Official Receiver (as liquidator of Celtic Extraction Ltd. and Bluestone Chemicals Ltd.) v. Environment Agency* [2000] Env. L.R. 86.

that liability. In respect of liability for past acts as a perpetrator of the contamination, the question is perhaps whether this type of liability constitutes liability "in respect of the property disclaimed". Whilst it could be argued that the liability stems from the past actions of the company rather than its current ownership or occupation of the property, and so is not affected by section 178(4), the case law on the section and its predecessors does not address this specific issue.[39]

Position of parties affected by disclaimer

14.26 By section 178(6) any person who has sustained loss or damage in consequence of disclaimer is deemed to be a creditor of the company to the extent of their loss or damage, and can prove for it in the winding up as ordinary creditors. This would potentially apply to an enforcing authority, which claims to have suffered loss in having to take responsibility for dealing with disclaimed property. This course may have its own problems for the enforcing authority.[40] However the specific statutory procedures for charging notices[41] will in general make this unnecessary. Section 178(6) might also potentially apply to another appropriate person who claims that their share of liability has been increased by the effect of disclaimer in extinguishing the liability otherwise attributable to the company. Such a third party is unlikely to be able to argue that their own share is extinguished by the effect of the disclaimer, in view of the proviso to section 178(4) that disclaimer should not in general affect the liabilities of other persons.

Effect of dissolution on property: *bona vacantia*

14.27 By section 654 of the Companies Act 1985 the effect of dissolution of a company is that all property and rights vested in the company (excluding those held on trust for others) are deemed to be *bona vacantia* and to belong to and vest in the Crown, the Duchy of Lancaster or the Duke of Cornwall, as appropriate.[42] It is possible for the Crown's title to be disclaimed within 12 months by notice under section 656, signed by the Treasury Solicitor or by the Queen's and Lord Treasurer's Remembrancer as representative of the Crown. By section 657 the effect of such disclaimer is that the property is deemed not to have vested in the Crown, so determining the relevant

---

[39] See especially *Stacey v. Hill* [1901] 1 K.B. 660; *Warnford Investments Ltd v. Duckworth* [1979] 1 Ch. 127.
[40] See the discussion in *Re Mineral Resources Ltd* [1999] 1 All E.R. 756.
[41] See Chapter 22.
[42] For consideration of section 654 in the different context of a waste management licence, see *Wilmott Trading Co. Ltd (No. 2)* [2000] Env. L.R. 54.

interests, rights and liabilities in the same way as disclaimer by a liquidator. However, this may be a pointless exercise, since in such circumstances the property will revert to the Crown on general principles of escheat, as the owner of the ultimate reversion in the property (assuming there is no mesne lord).[43]

Effect of death of individual

14.28 In the same way as for a company, an individual may incur liability for their own acts or omissions with regard to contamination, irrespective of ownership or occupation of land. Death of the individual will mean that they can no longer be found,[44] and the question will be whether there is any right to proceed against their estate. The following paragraphs consider the principles of testate and intestate succession.

Effect on property of individual dying testate

14.29 Where the estate of a deceased person includes contaminated land, consideration will need to be given to the effect of the rules of testate and intestate succession.

14.30 As a trustee, an executor can incur personal liability in their capacity as owner of the land.[45] The executor may quite understandably in those circumstances wish to consider renouncing probate, the effect of which under section 5 of the Administration of Estates Act 1925 will be that his rights will cease and that the administration of the estate will devolve as if he had not been appointed. Renunciation is equally an option open to trust corporations, normally by an instrument under seal or by an official duly appointed for that purpose. Similarly, it is a course open to a person appointed by an order of the Court of Protection or an attorney acting under an enduring power or attorney, on behalf of the incapable person. What is vital is that the executor should not have intermeddled in the estate of the deceased.

14.31 A distinction needs to be drawn between the liabilities of the deceased, and the ongoing liabilities which as it were pass with the land. To the extent that the deceased had liabilities as an "appropriate person" under Part IIA, his real and personal estate will constitute assets to meet these liabilities, by section 32 of the Administration of Estates Act 1925. If there was a subsisting cause of action against the deceased before his death then this will survive against his estate by section 1(1) of the Law Reform

---

[43] See Ing, *Bona Vacantia* (Butterworths, London 1971) p. 5; Enever, *Bona Vacantia* (HMSO, 1927) p. 15.

[44] Chapter 13.

[45] See paragraph 14.07 above.

(Miscellaneous Provisions) Act 1934. Since the section refers to all causes of action "whether in contract or in tort or otherwise", it appears wide enough to cover statutory liabilities under Part IIA. Whilst the personal representatives should pay such liabilities with due diligence having regard to all the circumstances of the case, they should also consider whether any defence is available, and rely on it if appropriate.[46]

14.32 Where the land remains in a contaminated condition, liability as knowing permitter or owner could obviously pass to the beneficiary as the next owner. He may elect not to take under the will, or who may disclaim the legacy, or subject to the normal principles of succession, the onerous property. As it was colourfully put in one case: a man "cannot have an estate put into him in spite of his teeth."[47]

Effect on property of individual dying intestate

14.33 By section 6 of the Administration of Estates Act 1925 the estate of an intestate person vests in the President of the Family Division of the High Court as the Probate Judge until an administrator of the estate is appointed. By section 21 of the 1925 Act the administrator will be liable in the same manner as an executor. The administrator may, in the same way as an executor, renounce his right to grant of letters of administration.

14.34 The distinction between the liabilities of the deceased for previous actions and those liabilities which follow the ownership or occupation land needs to be borne in mind. With respect to the first type of liability the principles set out in paragraph 14.31 above will apply with respect to claims against the estate. In relation to the second type of liability which passes with the land, the view of commentators is that a beneficiary in an intestacy can disclaim his entitlement.[48]

14.35 Subject to the possibility of such disclaimer, the property will pass according to the rules of succession contained in section 46 of the 1925 Act, and by section 46(1)(vi) in default of any person being entitled to take an absolute interest, the residuary estate will belong to the Crown, Duchy of Lancaster or Duke of Cornwall as *bona vacantia*, and in lieu of any right to escheat.[49] It is the practice of the Treasury Solicitor not to administer an estate which is prima facie insolvent, or the solvency of which is doubtful, but to leave it to the creditor to do so: this is on the basis that in such cases there is no *bona vacantia* to collect.[50] This may be a relevant principle where

---

[46] *Re Rownson* (1885) 29 Ch. D. 358.

[47] *Townson v. Tickell* (1819) B. & Ald 31 at 37.

[48] Sherrin & Bonehill, *The Law and Practice of Intestate Succession* (London, Sweet & Maxwell, 1994), pp. 104, 358. But compare Williams, Mortimer & Sunnucks, *Executors, Administrators and Probate* (London, Sweet & Maxwell, 1993) p. 940.

[49] As to procedural issues, see Ing, *Bona Vacantia* (Butterworths, London, 1971), Part II.

[50] *Op. cit.*, p. 36.

the estate includes contaminated land or associated liabilities. To what extent the Crown can be liable (bearing in mind that *bona vacantia* was originally an aspect of the Royal Prerogative)[51] is doubtful and is considered below.

Crown land and Crown liability

14.36 The Crown may be an appropriate person to be liable under Part IIA either because its own activities (*e.g.* those of the armed forces) have caused the contamination, or in its capacity as an owner or occupier of land. Such ownership may arise in respect of operational land, or land which devolves on the Crown under the principles of *bona vacantia*, as described above. It is therefore necessary to consider section 159 of the Environmental Protection Act 1990, which deals with its application to the Crown.

14.37 By section 159(1) the 1990 Act and regulations under it bind the Crown, but by section 159(5) do not affect Her Majesty in her private capacity. The Crown can therefore be an appropriate person under Class A or B in the same way as any other person. But by section 159(2) the Crown cannot be made criminally liable for non-compliance with a remediation notice. However, it will be possible to obtain a declaration from the High Court or Court of Session that the Crown is acting unlawfully in failing to respond to a remediation notice, if that be the case.

---

[51] Enever, *Bona Vacantia* (HMSO, 1927) pp. 13, 87.

Chapter 15

Restrictions on Liability Relating to Water Pollution

Summary

Section 78J places certain restrictions on the possible requirements of remediation notices in cases where the contaminated land in question has been identified as such wholly or partly as a result of water pollution. The section must therefore be considered carefully by the enforcing authority before serving a remediation notice. There is one restriction of a general nature (section 78J(2)), and one relating specifically to water from abandoned mines (section 78J(3)).

When section 78J applies

15.01 The section is stated to apply where any land is contaminated land by virtue of section 78A(2)(b) of the Act, whether or not the land is also contaminated by virtue of paragraph (a) of that subsection. In other words, if the condition of the land is such that pollution of controlled waters is being caused, or is likely to be caused, then section 78J applies. It does not matter that significant harm, or the significant possibility of such harm, are also involved.

The general restriction

15.02 Section 78J(2) provides that no remediation notice shall require a person who is an appropriate person by virtue by section 78F(4) or (5) (*i.e.* a Class B owner or occupier where a causer or knowing permitter cannot be found) to do anything by way of remediation which he could not have been required to do if the water pollution limb of the definition of "contaminated land" in section 78A(2) did not exist. In other words, the remediation required may relate only to the "significant harm" limb of the definition, that is, harm to the health of living organisms, interference with ecological systems, or harm to property. The restriction relates not only to the remediation required on the contaminated land itself, but also to remediation of other land or of any waters.

15.03 The object behind this somewhat convoluted wording is to avoid any additional liability accruing to an owner or occupier of land beyond that which could accrue under the provisions of the Water Resources Act 1991 (or, in Scotland, the Control of Pollution Act 1974), dealing with the clean-up of polluted waters or the prevention of such pollution. These provisions (including the "works notice" procedures introduced by the Environment Act 1995) are based on liability for causing or knowingly permitting, hence the restriction of the section 78J protection to owners and occupiers.

15.04 Two points may be made on this restriction. First, its effectiveness as a means of protection for owners and occupiers will depend on the interpretation given to "knowingly permitting."[1] If an owner or occupier who knows of the presence of contamination can be regarded as knowingly permitting its continued presence, then they will fall into the first tier of appropriate persons and will not be able to take advantage of section 78J(2). They will be an appropriate person by virtue of section 78F(2) rather than sections 78F(4) or (5).

15.05 Secondly, pollution of controlled waters may well involve one of the forms of "significant harm" under section 78A(4), allowing remediation requirements to be imposed on that basis. Examples would be harm to fish or plant life, and possibly, in the case of groundwater contamination, rendering the water unfit for the purpose for which it is abstracted.)[2] Where remediation cannot be required in consequence of section 78J(2) the authority may still use its own clean-up powers under section 78N, but will not be able to recover its costs under section 78P.[3]

Pollution from abandoned mines

15.06 Sections 78J(3)–(6) provides a further restriction on the requirements of remediation notices, relating to water pollution from abandoned mines. As with section 78J(2), the notice may not go further than could be required were the contamination related only to significant harm or the significant risk of such harm. The provision was introduced to prevent the potential problem that the Part IIA provisions might be used to impose liability in the three year period of grace given by Parliament under the Environment Act 1995 before the withdrawal at the end of 1999 of the defence in relation to water pollution from abandoned mines.[4] In the event, the provisions of Part IIA were not introduced during that period.

---

[1] See Chapter 13.

[2] This could arguably constitute damage to property, in the sense that abstraction is a natural right incidental to the ownership of land.

[3] Section 78J(7).

[4] See *Hansard*, H.L. vol. 562, cols. 144–5; *Hansard*, H.L. Vol. 562, col. 151. For detailed explanation of the statutory regime relating to abandoned mines; see Stephen Tromans, *The Environment Acts 1990–95: Text and Commentary* (London, Sweet & Maxwell, 1996, p. 569).

15.07 The provision applies to any person who permits, has permitted, or might permit water from an abandoned mine or part of such a mine to enter controlled waters, or to reach a place from which it is or was likely to enter controlled waters. The effect is to mirror the position as to liability for pollution from abandoned mines contained in the Water Resources Act 1991, as amended by the Environment Act 1995 (or in Scotland, the Control of Pollution Act 1974, as amended). As with those provisions, by section 78J(4) the restriction does not apply to owners or former operators of mines which become abandoned after December 31, 1999. The restriction only applies to those permitting (as opposed to causing) such pollution or potential pollution.[5]

15.08 The question of what is a "mine" is dealt with by section 78J(8) which applies the definition contained in section 180 of the Mines and Quarries Act 1954. This refers to:

> ". . . an excavation or system of excavations, including all such excavations to which a common system of ventilation is provided, made for the purpose of, or in connection with, the getting, wholly or substantially by means involving the employment of persons below ground, of minerals (whether in their natural state or suspension) or products of minerals."

15.09 This definition relates to the original purpose of the excavations in question. In some cases, mine workings may have been used for the disposal of waste, but would still appear to fall within the definition, given their original purpose. An opencast mine or quarry is an "excavation", but would not fall within the definition since the getting of opencast minerals does not involve the employment of persons below ground (as opposed to below the pre-existing ground level). The 1954 Act draws a distinction between mines and quarries, and opencast working will be a quarry rather than a mine. The point is one of some practical significance, since many open excavations will subsequently have been used for waste tipping or landfill.

15.10 The vital question of when abandonment occurs is dealt with by sections 78J(5) and (6). "Abandonment" is not defined by section 78J, but receives a detailed definition as to what it does and does not include by sections 91(A)(1) of the Water Resources Act 1991 and 30Y(1) of the Control of Pollution Act 1974, inserted by sections 58 and 59 of the 1995 Act.

15.11 The question arises as to what should happen when water pollution is occurring from a mine abandoned before the relevant date, but is not causing significant harm or a significant risk of such harm. By section 78F(7) the enforcing authority can still carry out remediation itself under section 78N, but will be precluded from recovering its expenses under section 78P in relation to any thing which it could not have included in a remediation notice.

---

[5] See Chapter 13 above as to "planning".

15.12 Section 91B(7) of the Water Resources Act (section 30Z(7) of the Control of Pollution Act 1974 in Scotland) provides a link between the new requirements on abandonment and the contaminated land provisions. Where the Environment Agency/SEPA receive notice of abandonment from the operator, or otherwise learn of proposed abandonment, if they consider that any land (including land beyond the mine) has or is likely to become "contaminated land" in consequence of the abandonment, then they must inform the local authority for the area. It is therefore clear that the local authority is intended to use its contaminated land powers where appropriate in relation to mines abandoned after December 31, 1999.

Chapter 16

Migration of Contamination and Chemical Changes

Summary

It will frequently be the case that contamination present in the soil or groundwater undergoes chemical or other transformations over time, or in some cases spreads to other land or groundwater. Provisions of Part IIA deal with these specific eventualities, and will require careful application to the facts where migration or transformation has occurred.

Chemical reactions or biological process

16.01 Where as a result of chemical or biological changes a substance which a person caused or knowingly permitted to be in the ground becomes or generates a contaminant, which it may not have been initially, the person who caused or knowingly permitted the original substance to be present bears responsibility for the resulting contamination.[1] This is so whether or not he knew or ought to have known that the reaction or process was likely to happen.[2] The provision will thus cover substances which degrade so as to produce more toxic or harmful products, or substances which react together to create harmful synergistic effects.

Escape of contaminating substances to other land

16.02 Section 78K deals with liability in respect of contaminating substances which escape to other land. Who is the "appropriate person" in such circumstances? A number of different situations can arise—section 78K distinguishes four, which are considered separately below.

---

[1] Section 78F(9).

[2] See, however, the exclusion test relating to changes to substances, explained in Chapter 17.

Liability of contaminator for escapes to other land

16.03 The original contaminator, *i.e.* the person who caused or knowingly permitted the substances to be in, on or under the land, is taken, by section 78K(1), to have caused or, as the case may be, knowingly permitted, the substances to be in, on or under any other land to which they appear to have escaped.

16.04 "Appear" here means appears to the local authority; determining where contaminants may have originated from can be a highly complex technical issue. The word "appear" indicates that the authority must have some factual basis for its view, but that 100 per cent certainty is not necessary.[3] It will of course be open to adduce evidence to the contrary on appeal.

16.05 Responsibility therefore follows the event of the escape, making the contaminator a first tier appropriate person in relation to that other land. There is no requirement that the escape should have been foreseeable, or that it must have been caused by the original contaminator. The original contaminator stands in the same position to the escaped substances as he does in relation to their original presence on the land from which they escaped: this will have relevance for the purposes of exclusion and apportionment tests.[4]

Persons onto whose land substances escape

16.06 Section 78K(3) applies where it appears to the enforcing authority that substances are present in, on, or under land as a result of their escape, directly or indirectly, from other land where some person caused or knowingly permitted them to be. Its effect is to prevent any remediation notice requiring the owner or occupier of the land to which the substances have escaped to carry out remediation to any land or waters other than that of which he is the owner or occupier. By inference, the owner or occupier can be required to carry out remediation to his own land, although provided the original contaminator of the land from which the escape occurred can be found, they ought to be a first tier appropriate person under section 78K(1).

16.07 The section applies only to innocent owners, in that it excludes those who caused or knowingly permitted the substance to be in, on, or under their land. In that respect, there appears to be no difference between A, who knows his predecessor X dumped chemicals on the property, and B who knows his neighbour Y dumped chemicals which have migrated onto his land.

---

[3] See *Ferris v. Secretary of State for the Environment* [1988] J.P.L. 777 in the context of planning enforcement.

[4] See Chapters 17 and 18.

16.08 The net effect is therefore that someone onto whose land contaminants have escaped and who is not a knowing permitter should only be liable in respect of his own land, and then only if the original contaminator cannot be found. The potential flaw in this logical scheme lies in the use of the words "knowingly permitted" at section 78K(3)(b). If the person to whose land the substances have escaped knows of their presence and fails to take action, can they be regarded as knowingly permitting their presence? If so, they will lose any protection conferred by section 78K(3).

16.09 As a matter of policy, it could be asked whether there is any logical distinction between a person who comes to a site, discovers it is contaminated, and does nothing, as against a person whom comes to a site which has been contaminated by substances which have escaped from elsewhere, and does nothing?

Escapes to further land

16.10 Section 78K(4) complements and reinforces (arguably, duplicates) section 78K(3). It refers to the situation where contaminants escape from their original location where some person caused or knowingly permitted them to be, to land A, and then escape to further land, (land B). This situation may occur where, for example, a plume of contamination passes through land which is in a number of ownerships. The effect of section 78K(4) is that the owner or occupier of land A—provided he cannot be said to have caused or knowingly permitted the substances to be in, on or under land A—will not be responsible for the remediation of land B unless he is also the owner or occupier of that land. In such circumstances, the enforcing authority should be able to pursue the original contaminator under section 78K(1), or if he cannot be found, the owner or occupier of land B.

Subsequent owners or occupiers of the land from which escape takes place

16.11 The situation can obviously arise where a person A contaminates land and the contamination escapes to adjoining land; person B at some stage becomes the owner or occupier of the land from which the escape took place. Is B liable to receive a remediation notice? Section 78K(5) states that no remediation notice may require B to remediate the other land to which the substances escaped, except where he himself caused or knowingly permitted the escape. The wording of section 78K(5) is somewhat curious, stating that B should not be required to do anything in consequence of the apparent acts or omissions of A. But there is no reason in any event why B should be held liable for A's acts or omissions unless he himself is an appropriate person on the basis of causing or knowingly permitting.

16.12 The subsection could potentially apply in two cases. First, where the escape took place entirely before B became owner or occupier. In that situation it seems difficult to see on what possible basis B could be liable, unless he had some prior separate connection with the escape (*e.g.* as a contractor); in which case his liability would not arise as owner or occupier in any event.

16.13 The second situation is where the escape occurs or continues after B becomes the owner or occupier. Here, B may be liable as occupier or owner of the land from which the escape took place, if he caused or knowingly permitted the escape. It is not the presence of contamination on his land which makes him liable to receive a remediation notice, but rather causing or allowing the escape, which may have happened some time later. B could still be responsible for remediation of the land he acquired, to clean up contaminants remaining there, or to prevent future escapes. He will not be responsible however for past escapes, save to the extent he caused or knowingly permitted them.

16.14 There is a possible conflict between the specific drafting of section 78K(5) and the general principle at section 78K(1), which could apply where person B knowingly permitted the continuing presence of the substances in his land, but did not knowingly permit their escape. It appears fairly clear that the more specific wording of section 78K(5) should prevail in this respect.

Use of authority's own clean-up powers in case of escapes

16.15 Nothing in subsections (3), (4) or (5) of section 78K prevents the enforcing authority from carrying out remediation itself under section 78N, but the authority will not be entitled under section 78P to recover from any person any part of the cost incurred by the local authority in carrying out the remediation in circumstances where it is precluded from requiring the person concerned to carry out the remediation himself—see section 78K(6). In other words, the use of the cost recovery powers in section 78P cannot circumvent the scheme of liability created by section 78K.

Example of provisions on escape

16.16 A is the original contaminator of land X. Land X is sold to B. An escape of contaminants to land Y (owned by C) occurred both before and after this sale. The contaminants also escaped from land Y to land Z (owned by D). The position under section 78K is as follows (leaving aside the ensuing application of rules on exclusion and apportionment):

1. A is an appropriate person (Group A) for land X, land Y and land Z (section 78K(1)).
2. B can be an appropriate person (Group A or B depending on the knowingly permitting issue) in relation to land X.

3. B will not be an appropriate person in relation to land Y and Z for substances which escaped prior to his acquiring the land (section 78K(5)).
4. B may be an appropriate person (Group A) in relation to land Y and Z to the extent only that he caused or knowingly permitted the escape from land X (section 78K(5)).
5. C may be an appropriate person (Group B or possibly Group A) in relation to his own land Y (section 78K(3).
6. C will not be an appropriate person in relation to land Z (section 78K(4)) unless he caused or knowingly permitted the escape from land Y to land Z (section 78K(5)).
7. D will be an appropriate person (Group B or possibly Group A) in relation to land Z (section 78K(3)).

Relevance of Guidance

16.17 These provisions also need to be read in conjunction with the Guidance on exclusion from liability, discussed in Chapter 17. In particular, Test 4 deals with "Changes to substances" and Test 5 deals with "Escaped substances". Both may have the effect of excluding from liability some of the parties involved where substances have interacted or escaped.

Chapter 17

Exclusion of Parties from Liability

Summary

Where two or more persons are appropriate persons to be served with a remediation notice in relation to the same remediation action, section 78F(6) requires the enforcing authority to determine, in accordance with statutory guidance, whether any of them should be treated as not being an appropriate person in relation to that thing. The Guidance provides six tests for this purpose for Group A persons and a single test for Group B persons.

The general approach to exclusion

17.01 Before considering in detail the content of the Guidance on the exclusion of appropriate persons from liability, a number of general points should be restated:

1. The starting point is in fact not the Guidance, but the words of section 78F and the related sections 78J and 78K. The first step is to establish who, under those sections, are the appropriate persons in relation to each relevant remediation action[1] ("the liability group", which may be a "Class A" or a "Class B" group).[2]
2. The Guidance on exclusion is only relevant where there are two or more appropriate persons. It therefore cannot help a person who is the only appropriate person in relation to a particular remediation action.[3]
3. By the same token, if there is a group of appropriate persons, it does not appear that all of them can be excluded. If some of them can benefit from an exclusion test and once they have been excluded only one member of the group is left, then the situation

---

[1] DETR Circular, *Contaminated Land,* Annex 3, paras D.7–D.10.
[2] *ibid.*, paras D.11–D.14.
[3] *ibid.*, para. D.28.

will be as at (2) above and that person will not be able to claim exclusion.[4]

Terminology

17.02 Chapter D of the statutory Guidance, dealing with exclusion from and apportionment of liability, creates its own inner world of terminology, in addition to the statutory definitions of Part IIA. This is set out at paragraph D.5:

(a) appropriate persons who caused or knowingly permitted are "Class A persons";
(b) appropriate persons who are unfortunate enough to be the current owner or occupier where no Class A person can be found are "Class B persons";
(c) appropriate persons in relation to a particular pollutant linkage form a "Class A liability group" or "Class B liability group", as the case may be;
(d) determination that a person is not to be treated as an appropriate person under section 78F(6) is "exclusion";
(e) determination under section 78F(7) dividing the cost between appropriate persons is "apportionment". The process of dividing the cost between liability groups is "attribution";
(f) each individual thing to be done as remediation is a "remediation action";
(g) all the remediation actions referable to a particular pollutant linkage is a "remediation and package"; and
(h) the complete set or sequence of remediation actions to be carried out on the land in question is a "remediation scheme".

The procedure by stages

17.03 Part 3 of Chapter D deals with the discrete stages involved in the procedure for determining liabilities, and as such sets the exclusion process in context. Not all stages will necessarily be relevant in each case.

17.04 *Stage 1—identifying appropriate persons and liability groups* (paragraphs D.9–D.19)—The starting point is the pollutant linkage or linkages by virtue of which the land is contaminated land. The authority should first make reasonable enquiries to find all Class A persons to whose activities the linkage is referable. If no Class A persons can be found, and the linkage

[4] *ibid.*, para. D.41(C).

relates solely to water pollution, then by section 78J there can be no liability on Class B persons, and the linkage will be an "orphan linkage".[5] In other cases, the authority seeks to identify all Class B persons for the land in question. This process is repeated for each pollutant linkage until all liability groups have been identified, bearing in mind that a person may be a member of more than one liability group for different linkages. The authority should then consider whether any member of a group can benefit from the statutory exemptions of sections 78J, 78K or 78X(4), it should be noted that hardship does not enter into the exercise at this point.

17.05 *Stage 2—characterising remediation actions* (paragraphs D.20–D.23)—This stage is only relevant where there is more than one significant pollutant linkage on the land involved. The authority should establish whether each remediation action is referable to a single linkage or to more than one linkage, in which case it is termed a "shared action", such shared actions may be either "common actions" or "collective actions". This issue is of primary relevance to the apportionment exercise, and is discussed in Chapter 18.

17.06 *Stage 3—attributing responsibility between groups* (paragraphs D.24–D.27—Again this stage is only relevant where there are multiple pollutant linkages. The exercise may result in a Class B group not having to bear any liability. Again, it is discussed in Chapter 18.

17.07 *Stage 4—excluding members of a group* (paragraph D.28)—This involves applying the Guidance on exclusion, which forms the substance of this Chapter.

17.07 *Stage 5—apportioning liability between members of a group* (paragraphs D.29–D.31)—This is covered in Chapter 18.

General guidance and information for appropriate persons

17.08 Part 4 of Chapter D contains guidance on some general considerations applying to the exclusion, and apportionment processes. It requires that enforcing authorities should ensure that any person who might benefit from an exclusion, apportionment or attribution rule is aware of the Guidance, so that they may make appropriate representations to the authority. This raises some interesting questions of how far the authority must go to comply with this aspect of the Guidance. At the very least, the authority should presumably tell the person that there are rules which may benefit them, and that those rules may be found in the Guidance. Many

[5] See Chapter 20.

appropriate persons will no doubt have access to sophisticated legal advice, and nothing more than this will be needed. Other appropriate persons may find it difficult to understand the Guidance and how it may apply to their particular circumstances. All parties involved in the more complex multiple linkage cases will find it difficult to take sensible decisions until they know how the authority is proposing to analyse the situation in terms of the remediation package. The authority will need to be careful not to leave itself open to the accusation of unfairness by giving one party more help than another. Authorities may wish to consider producing a standard form letter which summarises the main points, and directs the reader to the relevant parts of the Guidance for the definitive position. Clearly the person must be given an adequate time to take legal advice and make representations. Authorities faced with complex multi-linkage situations might also need to consider adopting an iterative process, whereby further representations are invited at particular stages, so that the impact of decisions taken at the earlier stages can be assessed.

Financial circumstances

17.09 Another general factor is financial circumstances. The Guidance states that, in relation to exclusion, apportionment and attribution, the financial circumstances of individual members of the liability group should have no bearing on the application of the tests and procedures involved.[6] In particular, it should be irrelevant:

(a) whether the person would benefit from any limitation on the recovery of costs under the hardship provisions of section 78P(2) or the guidance on cost recovery; or
(b) whether they would benefit from any insurance or other means of transferring their responsibilities to any other person, for example, an indemnity.

17.10 The intention behind this aspect of the Guidance is clearly to avoid a "deep pocket" approach of making those with substantial resources responsible for a proportionately greater responsibility for remediation.

Information available and decisions

17.11 The application of the principles of exclusion and apportionment may require much information about past activities and current circumstances which may be impossible or very difficult to obtain. The authority may also

[6] DETR Circular, Annex 3, para. D.35.

be faced with conflicting accounts of such activities or circumstances. The Guidance recognises the difficulties which may arise, by stating that the authority should seek to obtain only such information as it is reasonable to seek, having regard to[7]:

(a) how the information might be obtained;
(b) the cost of obtaining the information for all parties involved; and
(c) the potential significance of the information for any decision.

17.12 The authority should however at least make reasonable endeavours to consult those who may be affected by any exclusion or apportionment.[8]

17.13 The authority can only make judgments on the information available to it at a particular time, and should do so on the basis of the balance of probabilities.[9] The burden of providing additional information relevant to an exclusion or apportionment rests with the person wishing to benefit from it.[10] Where such information is provided, any other person who may be affected by decisions based on that information should be given a reasonable chance to comment on it before the decision is made.[11]

17.14 One can readily see how all too easily an authority could become embroiled in a dispute between a number of potentially liable parties, each seeking to refute or comment on information provided by the others. There will have to come a point where the authority draws such exchanges to a conclusion and makes its determination. It will be for the authority to judge when that point has been reached. However, it is important that the authority seeks to establish a clear framework or set of ground rules for this process, so that those affected know where they stand. Otherwise, arguments of procedural unfairness may well arise.

Agreement on liabilities

17.15 Where two or more parties are or may be responsible for remediation it is quite conceivable that they may reach agreement as to the basis on which they wish to share or to allocate the costs. Such an approach may be a sensible alternative to lengthy litigation and in those circumstances it would be a waste of time and effort for the authority to form its own view on exclusion or apportionment. There may be legitimate tax advantages arising from one company rather than another funding the work. The Guidance therefore provides that in such cases, where a copy of the agreement is provided to the authority, and none of the parties has informed the authority that it challenges the application of the agree-

---

[7] *ibid.*, para. D.36.
[8] *ibid.*
[9] *ibid.*, para. D.37.
[10] *ibid.*
[11] *ibid.*

ment,[12] then the authority should generally make such determinations on exclusion or apportionment as are needed to give effect to the agreement.[13] This principle can naturally only apply as between the parties to the agreement, and the normal rules on exclusion or apportionment should still be applied as between those parties and any other appropriate persons who are not parties to the agreement.[14]

17.16 An important exception to this aspect of the Guidance is that such agreements should not be allowed to be used for the purpose of evading liabilities. So, where giving effect to such an agreement would increase the share of the costs by a person who would be able to claim the benefit of the hardship provisions of the Act and Guidance, the authority should disregard the agreement.[15] It is therefore not possible to agree to pass liability to a "man of straw". It should be noted, however, that this provision of the Guidance does not apply only to agreements entered into with that purpose in mind, but to any agreement which would in practice have that effect. The authority may, therefore, be obliged to disregard a bona fide agreement, where one party has later become insolvent or otherwise subject to hardship factors. Parties entering into such agreements should bear this in mind.

Exclusion of members of Class A groups

17.17 The Guidance lays down some six tests which can have the effect of excluding one or more persons from a class A liability group (*i.e.* those who are responsible for causing or knowingly permitting). The tests are stated to be based on the notion of fairness[16] and are subject to specific "overriding guidance"[17]:

(a) The exclusions are to be applied to each specific pollutant linkage separately. Accordingly, it is perfectly possible for a person to be excluded with respect to one pollutant linkage, but not another.
(b) The tests are to be applied in the order in which they are set out. This is important since clearly different members of the group may be able to benefit from different tests—the order of application may therefore be critical. The Guidance does not explain the reasoning behind the order in which the tests are set out, but presumably this is again a reflection of what the Government regards as fair.

---

[12] The challenge might relate to the existence, enforceability or terms of the agreement. The agreement will need to be sufficiently precise as to the intended outcome if the authority is to be able to give effect to it.
[13] DETR Circular, Annex 3, para. D.38.
[14] *ibid.*
[15] *ibid.*, para. D.39.
[16] *ibid.*, para. D.40.
[17] *ibid.*, para. D.41.

(c) No test should be applied if it would result in the exclusion of all the remaining members of the group. In other words, there must always be at least one member of the group left to bear responsibility for each pollutant linkage.

Effect of applying tests

17.18 Paragraphs D.42 and D.43 draw an important distinction between the effect of applying different tests. For Tests 1, 4, 5 and 6, the effect is that the relevant person is completely excluded. It is as if he had never been a member of the liability group at all, and any further exclusion or apportionment process proceeds accordingly. For Tests 2 and 3 (respectively, "payments made for remediation" and "sold with information") the position is different. The point about these Tests is that they are based on an inferred transactional transfer of responsibility for contamination. accordingly the person who it is inferred accepted responsibility (either for payment or through buying with knowledge) should bear the responsibility of the notional transferee of responsibility as well as his own. Accordingly, the authority is required to assume that the person excluded under these Tests remains part of the liability group for the purposes of applying further Tests and apportioning liability. The notional or hypothetical share of the excluded person, determined on this basis, is then allocated to the recipient of the payment or the buyer, as the case may be. This is so irrespective of whether the recipient or buyer would be excluded in respect of their own acts by another exclusion test.

17.19 To give an example of how the Guidance might work in this regard, assume there are three Class A appropriate persons, X, Y and Z. X sold to Y with a payment for remediation. X is excluded from liability, but is assumed still to be a member of the group for apportionment, which is determined to be in equal shares. The final result is that Y bears two-thirds and Z one-third. Taking the same parties, now assume X sold to Y with information, and Z introduced a later pathway or receptor (Test 6). X is excluded by Test 3, but is assumed to remain a member of the group for applying Test 6. If the application of Test X is to exclude both X and Y, then 6 has no ultimate share for Y to bear. But if for some reason that Test did not apply to X, but only to Y, then Y would assume X's residual share.

"Related companies"

17.20 The Guidance also considers exclusion tests by reference to the situation where "related companies" are involved. Where the application of an exclusion test involves the relationship between, or relative positions of, "related companies", the authority should not apply a Test so as to exclude any of the related companies.[18] The objective of the Guidance in this

[18] *ibid.*, paras D.45–D.46.

respect is probably to prevent group companies evading their proper liabilities by structuring themselves to place liability on a subsidiary without resources.[19] Whilst this may be the intention, the same result will apply whether the relevant structure was created for such reasons, or for entirely legitimate fiscal, business or other reasons.

17.21 The question is whether at the "relevant date" the companies are or were "related" for the purposes of the Guidance. The "relevant date" is the date on which the authority first served on anyone a notice under section 78B(3) identifying the land as contaminated land.[20] The critical date is, therefore, service of the notice identifying the land as contaminated. If the companies were "related" at that date, their position before and after is irrelevant. Once a notice has been received, it is therefore too late to try to create a non-group relationship.

17.22 The rule is predicated on both companies being members of the same liability group for the pollutant linkage in question. It does not allow (for example) liability to be attributed to a parent company if that parent is not a member of the liability group into which its subsidiary falls. The circumstances in which a parent company may be held liable in such circumstances is a different question, discussed elsewhere.[21]

17.23 It may be noted that the approach required by the Guidance is simply to preclude the operation of exclusion in relation to any related company by virtue of transactions or relative circumstances within the group. Earlier drafts of the Guidance applied a different approach of requiring the authority to treat the companies as a single person. This would have created a number of uncertainties and, it is submitted, has wisely not been pursued as an approach in the final Guidance.

Test 1—Excluded activities

17.24 This is a complex test, set out at paras D.47–D.50, the general purpose of which is to exclude those who have been identified as having caused or knowingly permitted contamination solely through having carried out certain activities, which are such as to carry limited responsibility, in the Government's view.[22] It is by no means certain that any of the listed activities would amount to "causing or knowingly permitting" in the first place and the Guidance does not imply that this would be the case.[23]

17.25 The test operates by listing a number of activities. Where the person in question is within the liability group solely by reason of those activities (not including any associated activity falling outside the descriptions), the

---

[19] The terms "holding" and "subsidiary" companies have the same meaning as in Section 736 of the Companies Act 1985: see para. D.46.

[20] DETR Circular, Annex 3, para. D.46.

[21] See Chapters 13 and 14.

[22] DETR Circular, Annex 3, para. D.47.

[23] *ibid.*

person should be excluded from the Group. The list of activities is detailed, and subject to detailed qualifications in some cases. As well as the commentary below, reference should accordingly be made to the actual wording of the Test. The activities are:

17.26 (a) Providing (or withholding) financial assistance to another person (whether or not a member of the liability group) in one or more of the following ways:

(i) making a grant;
(ii) making a loan or providing credit in any other form, including leasing arrangements and mortgages;
(iii) guaranteeing the performance of a person's obligations;
(iv) indemnifying a person against loss, liability or damage;
(v) investing in a company by acquiring share or loan capital, but without acquiring control[24]; or
(vi) providing any other financial benefit, including the remission in whole or in part of any financial liability or obligation.

17.27 (b) Underwriting an insurance policy under which another person was insured in respect of matters by reason of which that person has been held to have caused or knowingly permitted contamination.[25] It is irrelevant whether or not that person can now be found.

17.28 (c) Carrying out any action for the purpose of deciding whether to provide such financial assistance or underwrite such an insurance policy.[26]

17.29 (d) Consigning the substance to another person as waste under a contract[27] whereby he knowingly took over responsibility for its proper disposal or other management on a site note under control of the person seeking exclusion. The thinking behind this category appears to be to exclude from liability waste producers who contracted for others to dispose of or otherwise manage their wastes. It would appear to be irrelevant whether the waste producer in so doing complied with their statutory responsibilities under the "duty of care" provisions of section 34 of the Environ-

---

24 "Control" is by reference to the test of section 736 of the Companies Act 1985 as to whether a holding company has control over a subsidiary.

25 Underwriting a policy for this purpose includes imposing any conditions on the insured, for example relating to the manner in which he carries out the insured activity.

26 The exclusion will not however include any intrusive investigation in respect of the land for the purposes of indemnity, where the investigation itself is a cause of the existence, nature or continuance of the pollutant linkage and the person who applied for the financial assistance or insurance is not a member of the liability group.

27 The contract need not be in writing. It is quite conceivable that waste may have been collected for disposal in the past without any written contract prior to modern legislation.

mental Protection Act 1990. It is irrelevant whether or not the person to whom the waste was consigned can now be found.

(e) Creating at any time a tenancy over the land in question in favour of another person who subsequently caused or knowingly permitted contamination (whether or not they can now be found). This category protects a landlord who finds their tenant has contaminated the land, provided their presence in the group of category A persons is due simply to their identity as landlord. The test may not assist where they are implicated in the contamination in other ways than having granted the lease (*e.g.* by failing to take steps within their power to prevent contamination). **17.30**

(f) As owner, granting a licence to occupy to another person who has subsequently caused or knowingly permitted contamination (whether or not they can now be found). Importantly, this does not include a case where the person granting the licence operated the land as a site for waste disposal or storage at the time of the grant of the licence. The wording is slightly curious. The exclusion does not apply if the licensor was already operating the site for waste disposal prior to the licence, but would apply if the licence related, as it were, to a greenfield site to be used for waste disposal. **17.31**

(g) Issuing any statutory permission, licence or consent required for any action or omission by which some other person causes or knowingly permits contamination, whether or not that person can now be found. The Test does not apply to statutory undertakers who grant permission to contractors to carry out work. **17.32**

(h) Taking, or not taking, any statutory enforcement action with respect to the land, or against some person who has caused or knowingly permitted contamination, whether or not that person can now be found. This avoids arguments either that some enforcement authority such as the Environment Agency has failed to take steps to prevent land becoming contaminated (for example, by failing to exercise its waste regulation functions), or that a planning authority has failed to enforce against some unlawful contaminative use and as such in either case has knowingly permitted contamination. **17.33**

(i) Providing legal, financial engineering scientific or technical advice[28] to a client: **17.34**

 (i) in relation to an action or omission by reason of which the client causes or knowingly permits the presence of the pollutant; or
 (ii) for the purpose of assessing the condition of the land; or

[28] This includes other design, contract management or works management services.

(iii) for the purpose of establishing what might be done to the land by way of remediation.[29]

17.35 (j) Carrying out any intrusive investigation in respect of the land as a person providing advice or providing services within (i) above except where:

(i) the investigation is itself a cause of the existence, nature or continuance of the pollutant linkage; and

(ii) the client is not himself a member of the liability group.[30]

17.36 (k) Performing any contract (whether of employment or for the provision of goods or services) where the contract is made with another member of the liability group, whether the service is provided on a direct subcontracting basis. The exclusion does not apply if the activity falls within another part of Test 1, or if the act or omission in question was not in accordance with the terms of the contract, or in a situation where the contractual service was provided to a company by a director or similar officer, and the actions or omissions resulting in the company being liable were carried out with the consent or connivance of the officer, or were attributable to any neglect on his part. This Test can operate at different levels. For example, a company which supplies vehicles or equipment to another company, which then uses them to commit acts causing contamination, would be excluded from any liability they might thereby have. The same would apply to a company which supplies fuel or chemicals which its customer spills, or which installs tanks or pipework from which contaminants escape. The contractor's acts must however be in accordance with the contract—so, for example, a supplier of defective pipework which causes a contaminating release would not be excluded. Employees who act in accordance with their conditions of employment will also be able to rely on this Test; but not senior corporate officers whose consent, connivance or neglect is involved. The wording of the Test on this latter point is a somewhat strange mix of the principles of criminal liability for senior managers contained in provisions such as section 157 of the 1990 Act; the essential idea is that senior officers should not

---

[29] The purpose is presumably to ensure that knowledge gained by a consultant or other advisor is not treated as constituting "knowingly permitting". But in any event, even with such knowledge, it is difficult to see how such an advisor would have the power autonomously to deal with the contamination so as to be said to permit its continued presence.

[30] The exception here relates to the situation where in carrying out intrusive investigation, the consultant might be said to have caused a pollutant linkage. The consultant will not be excluded from liability if the investigation is itself the cause of the problem and the client cannot be held liable. Where the client is liable he may of course have a cause of action against the consultant.

escape personal liability where they were personally implicated. So, for example, a director of a company who was aware, or ought to have been aware, that the company was allowing seepage of contaminants from its equipment to occur, yet did nothing, could not rely on the fact that he was acting under a service contract with the company to obtain exclusion from liability as a causer or knowing permitter.

17.37 As will be appreciated, the application of Test 1 may vary in complexity in that some of the categories are quite straightforward (and are clearly intended to allay concerns of commercial sectors such as banks, insurers and consultants) whereas others are much more complex and will involve detailed consideration of the relevant circumstances. The following paragraph contains some further thoughts on the application of Test 1.

Test 1: Further remarks

17.38 Two further points merit consideration before passing on to Test 2. First, it is presumably the case that the various activities listed are to be applied in terms of exclusion in the order in which they appear. Thus a bank would be excluded before a guarantor, who would be excluded before an investor, who would be excluded before an insurer, who in turn would be excluded before a lawyer or consultant, etc. So long as there is at least one other person in the group left, this should make no difference, but the logic behind the order is not apparent.

17.39 Secondly, there remains the somewhat uneasy relationship between the excluded activities and the concept of "causing or knowingly permitting". If the person did not cause or knowingly permit they will not be in category A at all, and the exclusion test will be irrelevant. The excluded activities focus for the most part on the status of the person concerned: a lender should not be liable just because they provided credit, an insurer because they wrote a policy, a landlord because they granted a lease, a planning authority just because they gave permission, etc. But this leaves open the possibility of liability stemming from later activities associated with the main activity.

17.40 Indeed, the Guidance is explicit on this point; paragraph D.48 removes from the exclusion "any associated activity outside these descriptions". So a landlord who having granted a lease then acts as a landlord in such a way as to knowingly permit contamination, may not be able to rely on the exclusion test. Indeed, if all he had done was to grant a lease, it is difficult to see how he could be within Group A at all. Test 1, therefore, for all its length and complexity, may remain in reality something of a "belt and braces" exercise.

Test 2: Payments made for remediation

17.41 The purpose behind this Test (paragraphs D.51–D.56) is to exclude from liability those who have already effectively met their clean-up responsibilities by having made a payment, sufficient to provide for remediation, to some other member of the liability group.[31] For past transactions this avoids the unjust enrichment which might otherwise occur, and for future transactions it offers a way of managing risk.

17.42 The Guidance requires the authority to consider whether all of the following circumstances exist:

(a) one member of the Group has made a payment to another member for the purpose of carrying out particular remediation on the land in question. In this respect, only certain types of payments are to be considered (see below);
(b) the payment would have been sufficient at the date when it was made to pay for the remediation;
(c) had that remediation been carried out effectively, the land would not now be regarded as "contaminated land"; and
(d) the remediation in question was not carried out or was not carried out effectively.

17.43 It will be appreciated that the key to this test is payment made to cover a particular remediation action: an assessment then has to be made as to whether that action would have been sufficient to deal with the problem if it had been carried out effectively. This may prove to be quite a difficult technical question. The test precludes payments made simply to reflect the fact that the land is or may be contaminated, with no particular plan of remediation in mind.

17.44 Only three types of payment[32] may be taken into account for the purpose of this test. These are[33]:

(a) payments made in response to a claim for the cost of particular remediation, whether voluntarily or to meet a contractual obligation;
(b) payments made in the course of civil proceedings, arbitration, mediation or other dispute resolution procedures, whether paid as part of an out of court settlement, or under the terms of a court order; or
(c) payments as part of a contract[34] for the transfer of ownership of the land, which is either specifically provided for in the contract to meet the relevant cost or which consists of a reduction in purchase price explicitly stated to be for that purpose.

17.45 One important qualification is that the benefit of the test will be lost where the person making the payment retained any subsequent control over the condition of the land in question.[35] Holding interests, such as

[31] DETR Circular, Annex 3, para. D.51.
[32] "Payments" includes consideration of any form: see para. D.54.
[33] *ibid.*, para. D.53
[34] This includes "a group of interlinked contracts".
[35] DETR Circular, Annex 3, para. D.55.

easements, restrictive covenants, reversions expectant on long leases, similar statutory agreements (such as section 106 agreements) or statutory undertakers' rights to install equipment, is not "control" for this purpose. Nor is "holding contractual rights to ensure the proper carrying out of the remediation for which the payment was made". The application of this principle may be a trap for those who make such payments yet wish to assure themselves that the money has been properly spent. Requiring proof of remediation activity ought not to be a problem, or imposing a contractual obligation to carry out the works; but when the person making the payment reserves rights of direct supervision or intervention over the remediation activity, that may be another matter.

17.46 Another possible trap lies within the test for innocent parties whose land is contaminated by migration or escape of substances from elsewhere. If the original polluter agrees to fund clean-up measures which are properly carried out but turn out to be ineffective, the application of the test effectively places the risk in those circumstances on the person receiving the payment. This should be borne in mind by that person's professional advisors.

17.47 The end result of applying Test 2 is that where the conditions are met, the authority will exclude the person who made the payment in respect of the remediation action in question. Subsequent tests on exclusion and apportionment should however be made as if the exclusion had not occurred and the person in receipt of the payment ultimately bears that excluded person's share.[36]

Test 3: "Sold with information"

17.48 The purpose of this important test (paragraphs D.57–D.61) is to exclude from liability those who sell or grant long leases of land, and who in doing so ensure that the purchaser or tenant had sufficient information as to the presence of the pollutant and thus (inferentially) the opportunity so take that into account in agreeing the price. The test will only operate at all where both vendor and purchaser are within Group A, and the test does not presuppose that the purchaser will be a "knowing permitter" simply by having such information: this will require consideration of all the circumstances.

17.49 The authority is required to consider whether all the following circumstances exist[37]:

(a) one of the members of the liability group has sold the land to a person who is another member of the group;
(b) the sale took place at arm's length, (that is, on terms which could be expected in a sale on the open market between a willing seller and a willing purchaser);

---

[36] See paragraphs 17.18 and 17.19 above.
[37] DETR Circular, Annex 3, para. D.58.

(c) before the sale became binding, the purchaser had information[38] that would reasonably allow that particular person[39] to be aware of the presence of the relevant pollutant "and the broad measure of that presence"[40]; and the seller has done nothing material to misrepresent the implications of that presence to the purchaser[41];

(d) after the date of the sale the seller retained no interest in the land or any rights to use or occupy the land.[42]

17.50 The test applies in the equivalent circumstances in relation to the grant or assignment of a long lease as well as a sale of the freehold.[43] For these purposes, a long lease is a lease or sublease granted for a period of more than 21 years under which the lessee satisfies the definition of "owner" in section 78A.[44]

17.51 Where the Test applies the seller or lessor is to be excluded. As with Test 2 however, they are treated as remaining liable for the purposes of future exclusion and apportionment, the buyer then becoming responsible for their share.[45]

Test 3 in practice and commercial transactions

17.52 Test 3 seems likely to be one of the most difficult to operate in practice, because of the uncertainty inherent in deciding whether sufficient information was given to activate it—particularly where the transaction may have taken place some years ago. As indicated above, the question is whether "the buyer had information that would reasonably allow that particular buyer to be aware of the presence on the land of the pollutant identified in the significant pollutant linkage in question, and the broad measure of that presence". Thus it would appear that reports which simply give a generalised indication of potential contamination (for example, a desk survey relating to past uses) will not of themselves suffice, in that they will not provide details of the actual substances involved.

17.53 Another possibility is that the buyers obtain the information for themselves, rather than having it provided by the seller. In this context it has become more common for buyers in commercial transactions to carry out

---

[38] Such information may be generated by the purchaser's own investigations, or by the provision of reports or data by the seller.

[39] The test is thus a subjective one, depending on the buyer's characteristics.

[40] The meaning of this phrase is not totally clear, but presumably it refers to the physical extent of contamination.

[41] In this respect the purchaser will presumably be taken to know the legal position and the consequent risks of liability.

[42] Easements, rights of statutory undertakers, reversions expectant on long leases and restrictive covenants are disregarded for this purpose.

[43] DETR Circular, Annex 3, para. D.59(a).

[44] See Chapter 13.

[45] See paragraphs 17.18 and 17.19 above.

their own intrusive investigations. The Guidance recognises this by formulating a rule (at paragraph D.59(d)) that in transactions since January 1, 1990 where the buyer is (here it presumably means to say "was") a "large commercial organisation" or public body, permission from the seller to carry out investigations on the condition of the land should "normally" be taken as a sufficient indication that the buyer had the requisite information to activate the Test. This is problematic Guidance in certain respects. What constitutes a "large commercial organisation" is not defined, and will have to be decided by the authority or the appellate tribunal. Also, what may constitute permission to carry out investigations "of the condition of the land" is far from clear. The standard reply to preliminary enquiries, "The purchaser should rely on his own investigations" would not of itself seem to amount to such permission, though the point is arguable. Would access to carry out a normal structural survey constitute permission to investigate the condition of the land for this purpose? Such permission would clearly not extend to sinking boreholes or taking soil or groundwater samples. Unless there is specific evidence of an offer of access for this purpose being made, the matter may effectively come down to the seller saying, "If you had asked me for access for investigations, I would have given it". This, however, is not the same as giving permission. If permission was given, but the investigations were inadequate to discover the contamination, the seller will be excluded, and the buyer will need to consider action against his consultants.

17.54 Finally, paragraph D.59(c) states that where there is a group of transactions or wider agreement (such as the sale of a business) which includes a sale of land, the sale should be taken to have been at arm's length where the person relying on the test can show the net effect of the overall deal was a sale at arm's length. The classic example is probably the situation where assets are transferred within a group of companies, to place them in a corporate vehicle (often "Newco") which is then sold. The transfer to Newco may not, seen in isolation, be at arm's length; but the transaction as a whole is. Therefore, if the other relevant criteria of Test 3 are met, the transfer of land to Newco may have the effect of excluding from liability the company within the seller's group which transferred it. It does not follow from this that the purchaser of the shares in Newco will become liable, since at that point there is no sale of land.

Test 4: "Changes to substances"

17.55 The purpose of this test is to protect by exclusion those who find themselves within a liability group having caused or knowingly permitted the presence of a substance which in itself would not have resulted in a pollutant linkage, but has only done so because of the later actions of others in adding another substance which has interacted with the earlier substance; or otherwise causing a change to that substance.[46]

---

[46] DETR Circular, Annex 3, para. D.62.

17.56 In applying the Test, the authority must consider whether all these circumstances set out at paragraph D.63 exist:

(a) the substance in question is only present, or is only a significant pollutant, because of a chemical, biological or other change (called "the intervening change") involving both a substance previously present which would not have given rise to a pollutant linkage in itself ("the earlier substance") and a "later substance" (which might or might not have given rise to a pollutant linkage of itself);
(b) the intervening change would not have occurred in the absence of the later substance;
(c) a person (A) is a Class A member of the liability group in relation to the first substance, but not the second;
(d) one or more persons are Class A members of the liability group in relation to the second substance;
(e) before the date when the later substances were introduced, A either:
  (i) could not reasonably have foreseen their introduction; or
  (ii) could not reasonably have foreseen the intervening change; or
  (iii) took what were at the time reasonabie precautions to prevent these consequences; and
(f) after that date, A did not cause or knowingly permit any further earlier substances to be introduced, or do anything to contribute to the conditions that brought about the intervening change, or fail to do something that he could reasonably be expected to do to prevent the intervening change happening.

17.57 If conditions (a)–(f) are satisfied, person A is excluded. The requirements at (f), effectively to mitigate and minimise the harm after the intervening change has commenced, could be onerous—at least where person A retains control of the land and has the power to act.

Test 5: "Escaped substances"

17.58 Escapes and migration of contaminants from one piece of land to another is a complex issue in its own right. The purpose of Test 5 is to exclude those who would otherwise be liable for contamination caused by the escape of substances from other land, where it can be shown that another Class A person was responsible for their escape.[47]

17.59 Given the statutory rules on escapes, it is conceivable that a liability group might comprise the original polluter, those onto whose land the

[47] *ibid.*, para. D.65.

contamination has escaped, and the person who caused the escape of the pollutant. In applying this test, the authority is required to consider whether all the following circumstances exist:

(a) a significant pollutant is present on the land wholly or partly as a result of its escape from other land.
(b) a member of the liability group (person A) caused or knowingly permitted the pollutant to be present in the land from which it escaped and is liable only for that reason, and
(c) one or more other members of the group caused or knowingly permitted the pollutant to escape from that land and the escape would not have occurred but for their acts or omissions.

17.60 In these circumstances, person A is excluded. Unlike Test 4, there is no qualification as to the foreseeability of the escape or as to the precautions taken by B to prevent it. The Test might be applicable where trespassers or vandals cause the escape of substances from A's land. Provided the vandals can be found (and as such are members of the liability group) it seems that A could rely on the Test. In this respect it is perhaps curious that A would not need to show that his precautions to prevent vandalism had been reasonable.

Test 6: Introduction of pathways or receptors

17.61 The purpose of this Test is to exclude from liability those who would otherwise be liable solely because of the subsequent introduction by others of the relevant pathways or receptors.[48]

17.62 The Test uses the terms "relevant action" and "relevant omission". A relevant action is the carrying out at any time of building, engineering, mining or other operations in, on, over or under the land in question, or the making of any material change in use for which a specific application for planning permission was required at the time the change was made (as opposed to permission granted by a general development order, simplified planning zone, enterprise zone or the like).[49] A relevant omission is failing to take a step in the course of carrying out a relevant action which would have ensured that the pollutant linkage did not come into being; or unreasonably failing to maintain or operate a system installed for the purpose of reducing or managing a risk associated with the contamination (for example, gas venting systems).[50]

17.63 The authority must consider whether all the following circumstances exist:

---

[48] *ibid.*, para. D.68.
[49] *ibid.*, para. D.70(a).
[50] *ibid.*, para. D.70(b).

(a) one or more members of the liability group have carried out a relevant action or made a relevant omission ("the later actions") either as part of a series of actions or omissions which resulted in them being Class A persons in relation to the pollutant linkage, or in addition to that series of actions or omissions;
(b) the effect of the later actions was to introduce the pathway or the receptor forming part of the pollutant linkage;
(c) if those later actions had not occurred, there would not have been a significant pollutant linkage; and
(d) person (A) is a member of the liability group solely by reason of other earlier actions or omissions which were completed before the later actions were carried out.

17.64 In these circumstances, any person meeting the description of person A is excluded from liability. In legal terms, it is as if the later action which introduced the pathway or receptor broke the chain of causation from the original polluter. However, the Test has its limitations: it can only apply where the person carrying out the later action is himself a Class A person; that is, he caused or knowingly permitted the contamination to be present. Introducing a pathway or receptor in itself does not make someone a causer or knowing permitter of the presence of the contamination. One consequence of this is that the Test can only apply in respect to developments on the contaminated land itself—it does not apply where the relevant acts or omissions take place on other land, even if their effect is to introduce a receptor.[51] This is because the person carrying out development on neighbouring land cannot be said to be knowingly permitting the presence of contamination on the land which presents the threat—unless that is, he also owns the contaminated land (an issue which the Guidance does not address).

17.65 It is not immediately apparent why the introduction of a pathway or receptor under this test is limited by the definition of "relevant action" effectively to development which would require planning permission. The making of a material change of use does not constitute a relevant action where specific planning permission is not required because of an enterprise zone or simplified planning zone scheme. This could potentially exclude much development carried out during the 1980s and 1990s in the traditional heavy industrial areas. However, such development will in practice have involved building operations, and so will be caught by the other limb of "relevant action", covering all operational development. The concept of "relevant omission" is, by its nature, more likely to give rise to a pollution pathway than a receptor.

Exclusion of members of Class B liability group

17.66 Compared with the rules for exclusion from Class A, those from Class B are simple. This is because the members of Class B will simply be the owner and occupier of the land in question.

---

[51] *ibid.*, para. D.71.

17.67 The purpose of the single test for Class B is to exclude from liability those who do not have a capital interest in the land.[52] There are various precedents for what is perceived to be the justice of allocating responsibility to the person whose interest carries with it the actual or potential right to receive the rack rent.[53]

17.68 Where Class B comprises two or more people, the authority should exclude any of those persons who[54]:

(a) occupies the land under a licence or other agreement of a kind which has no marketable value or which he cannot legally assign or transfer to another person[55]; or

(b) is liable to pay a rent equivalent to the rack rent and who holds no beneficial interest in the land other than any tenancy to which such rack rent relates.[56]

17.69 The effect is therefore in general to channel liability from those with only personal interests or full market-rent leaseholds to the "owner", who may be either the freeholder or the holder of a long lease. The position of a tenant who is, for whatever reason, paying a rent less that the full market rent is problematic in that he may well be unable to rely on the exclusion test. What is the market rent due, however, needs to be considered in the context of all the terms of the lease, as a matter of valuation.

---

52 *ibid.*, para. D.87.

53 See *Pollway Nominees Ltd v. Croydon London Borough Council* [1986] 2 All E.R. 849, HL, at 853.

54 DETR Circular, Annex 3, para. D.89.

55 For these purposes, the fact that a licence may not actually attract a buyer in the market, or its actual marketable value, are irrelevant.

56 Thus a tenant with an option to purchase the freehold might not be able to rely on this exclusion. Where the rent is subject to periodic review, it should be considered to be the rack rent if, at the latest review, it was set at the current full market rent.

Chapter 18

Apportionment of Liability

Summary

Where two or more persons are appropriate persons in relation to any particular remediation action, they shall be liable to bear the cost of doing that thing in proportions determined by the enforcing authority (and stated in the remediation notice) in accordance with statutory guidance. The Guidance deals with three separate issues: apportionment within a Class A liability group; apportionment within a Class B liability group; and attribution of responsibility between different liability groups.

The general approach to apportionment

18.01 As with exclusion, discussed in the previous Chapter, it is important to keep in mind the statutory basis for apportionment before attempting to apply the Guidance. The starting point, again, is to establish who, under the primary legislation, are the appropriate persons with regard to each remediation action. The general principle of sections 78E and 78F is one of joint and several liability, modified by the principle of referability in relation to Group A appropriate persons,[1] and by the requirement of apportionment.[2] Having established who are the appropriate persons for each liability group and how, if necessary, responsibility is to be attributed between the groups as a whole, it should then be considered whether any of them should be treated under section 78F(6) as not being appropriate.[3] Only then, if there are two or more left, will the principles of apportionment under section 78F(7) come into play.[4]

---

[1] Section 78F(3).
[2] Sections 78E(3) and 78F(7).
[3] See Chapter 17.
[4] See the process at Part 3 of Chapter D of the Guidance, summarised in Chapter 17 above.

18.02 In practical terms, the authority may well begin by looking at the circumstances of the case in a "common sense" way, and applying the Guidance in the context of that initial evaluation. The rules are really no more than a sophisticated attempt to formulate principles of fairness.

Different types of apportionment

18.03 As the Guidance points out, apportionment operates at two levels. The first is apportionment between members of the liability group for the pollutant linkage in question. The Guidance considers this issue at Parts 6 and 8 of Chapter D, dealing with Class A and Class B groups separately. However, there is also the possibility that there may be more than one significant pollutant linkage, and therefore more than one liability group for a given area of land. In that case, it may be necessary for the costs of some remediation actions to be allocated between the various liability groups, and Part 9 of Chapter D deals with this issue, which it terms "attribution".[5]

Class A apportionment: general principles

18.04 Given that the history and circumstances of any piece of contaminated land may vary widely, as may the nature of the responsibility of different persons, the Guidance does not attempt to prescribe detailed rules for apportionment which would be "fair" in every case.[6]

18.05 Instead, the authority is required to follow the general principle that liability should be apportioned to reflect the relative responsibility of each member of the Class A group for creating or continuing the relevant risk.[7] This general principle is subject however to guidance on a number of "specific approaches", discussed below.

18.06 If appropriate information is not available and cannot reasonably be obtained to allow an apportionment based on individual responsibility to be made, then liability should be apportioned in equal shares, subject to a special rule for companies and their officers which is explained in the next paragraph.[8]

Companies and officers

18.07 Where a Class A liability group, after the application of all relevant exclusion tests, includes a company and one or more of its officers,[9] then the authority should by paragraph D.85:

---

[5] For the terminology used in Chapter D, see Chapter 17.

[6] DETR Circular, *Contaminated Land*, Annex 3, para. D.74.

[7] *ibid.*, para. D.75.

[8] *ibid.*, para. D.76.

[9] "Relevant officers", by DETR Circular, Annex 3, para. D.86, means any director, manager, secretary or similar officer, or any other person purporting to act in any such capacity.

(a) treat the company and its officers as a single unit for the purpose of applying the general principles of relative responsibility and apportionment in equal shares; and

(b) having thus determined the aggregate share of responsibility for the company and its officers together, that share should then be apportioned between them on a basis which takes into account the degree of personal responsibility of those directors and the relative levels of resources which may be available to them and the company to meet the liability.

18.08 Effectively, (b) means that hardship considerations for individual directors will apply at this stage (rather than at enforcement) so potentially increasing the share of liability of the company; this constitutes a specific exception to the normal principle of not targeting the "deep pocket". Of course, the principle could work the other way in the case of the wealthy director of an impecunious company.

18.09 When considering the relative responsibility of a company and its officers, as against that of the other members of the group, the company and the officers, under the principle at (a), should be treated as a single unit. In practice, this will mean that knowledge and fault amongst a number of directors can be aggregated with that of the company. Also it will avoid the risk of the company being disadvantaged by any element of "double-counting" of responsibility in the apportionment exercise.

Class A groups—specific approaches for apportionment

18.10 In applying the general principle of relative responsibility, the authority should apply a number of specific approaches where these are relevant. Effectively, they are attempts to refine the general principles of fairness and relative responsibility.

Partial applicability of an exclusion test

18.11 If, for any member of the liability Group the circumstances set out in one or more of the exclusion tests[10] applies to some extent, but not sufficiently to allow the test to apply, the authority should assess the person's degree of responsibility as being reduced to the extent which is appropriate in the light of all the circumstances, and the purpose of the test in question. The example is given where a payment made for remediation was sufficient to pay for only half of the necessary

---

[10] See Chapter 17.

remediation". In that case translating the shortfall into numerical terms is not difficult, but in other cases it may be less easy to quantify the extent to which the test has not been met; for example, in relation to the "sold with information" test or the test on the introduction of pathways and receptors. In such cases, a broad, equitable approach will be needed, involving judgments of reasonableness.[11]

Original contamination versus continued presence

18.12 Situations may well arise where a Class A liability group comprises a person (X) who caused or knowingly permitted the initial entry of a contaminating substance onto land, and one or more others (Y) who knowingly permitted its continued presence. Paragraph D.78 requires the authority, when assessing their relative responsibility, to consider the extent to which Y "had the means and a reasonable opportunity to deal with the presence of the pollutant in question or to reduce the seriousness of the implications of that presence." The assessment should then be on the following basis.

(a) if Y had the necessary means and opportunity, he should bear the same responsibility as Y;
(b) if Y did not have the necessary means and opportunity, his responsibility relative to that of X should be "substantially reduced"; and
(c) if Y had some, but insufficient, means or opportunity, his relative responsibility should be reduced "to an appropriate extent".

18.13 As originally drafted, this guidance contained the added complication of the forseeability of harm on the part of Y. This aspect has been dropped, but even without it the Guidance may still pose problems for enforcing authorities. It is not clear what is menat by the phrase "means and opportunity", but this could in principle embrace considerations such as the length of time that Y has had to deal with the problem since becoming aware of it; the technical feasibility of dealing with the problem; and the financial and other resources at Y's disposal. The concept of "knowingly permitting" (discussed at Chapter 13) itself inherently requires some opportunity or ability to deal with the problem. The responsibility of Y should, it appears, never be greater than that of X, but it can be reduced to an appropriate extent where there was less than full opportunity, and "substantially" where there was no opportunity. In this last case, the guidance appears to contemplate something less than complete exoneration, or reduction to nil; presumably if there really was no opportunity at all to remove or deal with the contamination, Y would not be a knowing permitter within the liability group in the first place.

---

[11] DETR Circular, Annex 3, para. D.77.

Original polluters

18.14 The specific Guidance also deals with the situation where within the liability group are a number of persons who all caused or knowingly permitted the entry of the pollutant into the land. This could be because they operated similar processes on the land at different times, or because they operated processes on different sites, each of which contributed to the contamination of soil or, perhaps more likely, groundwater.

18.15 The starting point, set out at paragraph D.80, is to consider whether the nature of the remediation action required points clearly to different members being responsible for particular circumstances. The example given in the Guidance is where different persons were in control of different areas of land, with no interrelationship between those areas. In that case, the authority should regard the persons in control of the different areas as being "separately responsible for the events which make necessary the remediation actions or part of actions referable to those areas of land". The wording of the Guidance is rather opaque on this issue and, in particular, it is not clear why in such circumstances it would not be possible to identify distinct pollutant linkages in relation to the different areas requiring remediation.

18.16 If the circumstances in paragraph D.80 do not apply, but the quantity of the pollutant is a major influence on the cost of remediation, then by paragraph D.81 the authority should regard the relative amounts of the pollutant involved as an appropriate basis for apportionment. An example might be the amount of oil or solvent in groundwater, which determines the length of time that pumping and treatment has to occur. The first step is to consider whether there is direct evidence of the relevant quantities involved—unless there are clear records of spillages or losses in known quantities this seems unlikely. "Surrogate measures" will then have to be used, which may include the relative periods during which broadly equivalent operations were carried out, the relative scale of such operations (which may be indicated by the quantities of a product manufactured), the relative areas of land on which operations were carried out and combinations of these factors.[12] The area of land may be a very rough factor, since the intensity of operations may vary widely.

18.17 If none of the above circumstances apply, then the authority should by paragraph D.84 consider the nature of the activities involved. If they are "broadly equivalent", then responsibility should be apportioned on the basis of the relative periods of time that the different persons were in control. This should be adjusted where the activities were not broadly equivalent, for example where they were on a different scale.

Knowing permitters

18.18 Paragraph D.84 deals with apportionment between persons who have knowingly permitted the continued presence of a contaminant over a period of time. Here the apportionment should be in proportion to the

---

[12] *ibid.*, para. D.87.

length of time each controlled the land, the area that they controlled, the extent to which each had the means and a reasonable opportunity to deal with the problem[13] and combinations of these factors.

18.19 Applying these principles seems unlikely to be a straightforward or simple exercise, and will involve the assimilation of potentially complex and ascertain facts coupled with the exercise of judgment. In reality, once the authority leaves the relative certainty of apportionment in equal shares, the Guidance cannot provide the answer: it can simply list the factors to be taken into account. Except in relation to the issue of "means and opportunity" for knowing permitters to deal with continuing pollution, the Guidance does not refer to what might be termed "behavioural factors". Such factors might include whether the pollution was foreseeable, the precautions actually taken by the appropriate person, contemporary standards and best practice at the relevant time, and so on. These matters might be thought to be relevant to the general test of "relative responsibility". However, their absence from the Guidance on specific issues might be taken to preclude their consideration: against this, it could be said that the overriding test is that of "relative responsibility" at paragraph D.75, and that these are material considerations in that determination.

Class A apportionment: examples

18.20 It may be helpful to give a few examples of how the principles of apportionment within a Class A liability group may apply; though given the complexity of the issues, they are offered with some trepidation. Of necessity the examples are simpler and clearer than the likely cases which will arise in practice. However, they do illustrate that the exercise may well prove contentious. They assume that Part IIA is fully applicable and is not ousted by IPC or some other regime.

Example 1

18.21 S, a company which has operated on the same site for sixty years, sold in 1989 a small portion of its site to B, a management buy-out company. B's site is now found to be contaminated both by organic chemicals released before its acquisition, but also by such chemicals spilled since 1989. When selling the site to B, a certain amount of information on contamination was provided by S, but this gave only a partial picture, insufficient to trigger the "sold with information" test.

The authority will have to consider the following issues as between S and B.

[13] See paragraph 18.13 above.

1. Assuming the organic substances are of the same nature and the amount of those substances is a determinative factor of the cost of remediation, can the relative contributions of S and B be quantified? Their respective periods of occupation are known, but the scale of operations may have been very different, as may the respective precautionary measures taken, so that time in itself may not necessarily be helpful as a guide.
2. As well as bearing responsibility for its own contamination, does B bear any responsibility for the contamination already present when it acquired the land, having knowingly permitted its continued presence? If B had the means and opportunity to prevent or reduce the risks then the Guidance suggests that B may bear equal responsibility with S.
3. To what extent should the responsibility which S would otherwise bear be reduced to reflect the fact that some information was provided?

Example 2

18.22 On the facts of Example 1, now assume that it is the whole of S's present and former site (including what is now B's land) which requires comprehensive remediation for a single pollutant linkage.

1. An initial issue is whether it is possible to draw a clear distinction in terms of allocating responsibility between the land of S and that of B and to distinguish clearly between the costs attributable to each area.
2. If so, then S would be wholly responsible for the costs relating to the S land, and the costs relating to the B land would be apportioned as in Example 1 above.
3. Otherwise, if the S and B land have to be treated as a whole, can the respective contributions of the polluting organic solvents by S and B be quantified or assessed? Such assessment may be by a combination of the areas controlled and length of control. However, that in itself would not attribute any responsibility to B as a knowing permitter, only as a causer.
4. An alternative approach would be to look at the respective areas as between the S land and the B land, to apportion the appropriate proportion of the total reflecting the area of the S land to S, then to apportion separately the portion attributable to the B land in accordance with the principles at Example 1 above.

Example 3

18.23 S, an industrial company, occupied the same site for thirty years, having acquired it from P (another industrial company) which had occupied it for fifty years before that. The land is now divided into light industrial units, which in 1985 were sold by S to X and Y. X still occupies its unit, whereas Y sold its unit to Z in 1993. The sale to Z was not "with information". The whole area is contaminated by heavy metals and requires clean-up.

1. Assuming there is a single pollutant linkage which must be dealt with together, there are five Class A parties: P (causer), S (causer and knowing permitter), X (knowing permitter), Y (knowing permitter) and Z (knowing permitter).
2. Assuming X, Y and Z all had knowledge of the harm in question and the means and opportunity to do something about it, the Guidance might suggest that they should all bear equal responsibility with the original causers, P and S. But does that mean each of the five bears 20 per cent, or that P and S together bear 50 per cent and X, Y and Z together bear 50 per cent?
3. Whichever is the case, if any of X, Y or Z did not have a reasonable opportunity to correct the problem, to what extent does that reduce their responsibility, and if so which of the other parties picks up that share?
4. The issue of apportionment as between P and S as causers will involve consideration of their respective contributions to the contamination, which may relate to periods of occupation, modified by the intensity of operations.
5. The issue of apportionment as between the knowing permitters (X, Y and Z) will depend on their respective areas of occupation and length of occupation, which ought in principle to be a straight-forward mathematical exercise. However, if there are differences in opportunity which they had to deal with the problem, this will also have to be factored in.
6. However, what of the position of S, who has "dual capacity" both as a causer, and as a long term "knowing permitter"? Is S effectively apportioned two shares, one as a causer for 30 years and the other as a knowing permitter of previous contamination over the same period? The Guidance does not expressly address that issue, yet it will have an important bearing on the ultimate outcome. If S is simply regarded as a causer, this will dilute the responsibility of P, but not that of X, Y and Z. Regarding S as both causer and knowing permitter will result in S bearing a proportionately heavier burden than P, and will greatly dilute the responsibilities of X, Y and Z, given the fact that S occupied the whole site for 30 years compared with their shorter occupations of different parts of the site. It would seem to be inequitable to X, Y and Z that S should effectively benefit from having not only knowingly permitted the continued problem, but also added to it.

18.24 These examples, brief and imperfect as they are, should illustrate how difficult it is in fact to adopt a totally prescriptive approach to the apportionment issue. What is perhaps most important in practice is that the authority should be able to explain the process by which it reached its decision. This may, however, be easier said than done, given the complexity and subjective nature of some of the tests. In practice, the authority may have received various conflicting representations from the parties, and it may be easier to state why a submission from a particular party is accepted or rejected rather than starting from scratch.

Apportionment within Class B

18.25 Part 8 of Chapter D deals with apportionment between members of a Class B liability group. The Guidance states that where the whole or part of a remediation action for which a Class B liability group is responsible clearly relates to a particular area of land within the larger area of that pollutant linkage, then liability for that remediation action (or the relevant part of it) should be apportioned among those owning or occupying that area of land.[14] This is the first question to be considered, and the key words are "clearly relates to a particular area within the land".

18.26 The authority is therefore directed to consider the relationship between the remediation action and the area of land.[15] It is submitted that this must mean something more than simply the action being physically carried out on that land. Take the example of a large area of land affected by mobile contaminants with a sensitive river down-gradient. The remediation consists of inserting a cut-off wall at the site boundary to form a barrier and protect the river. It would clearly be inequitable if the whole cost fell on the person who happened to own that part of the site where the wall was constructed.

18.27 It is submitted that what is required is a relationship between the need for the remediation action and the land, not the action itself and the land. This would mean, for example, that where a contaminated site contains isolated "hotspots" which require additional remediation measures, the cost of those measures will fall to the owner or occupier of that specific area or areas.

18.28 In any event, where these circumstances do not apply (and in many cases they will not) the authority should apportion liability for all the actions relating to the significant pollutant linkage amongst all members of the liability group.[16] The Guidance goes on to state that in so doing it should do so in proportion to capital values, including those of buildings and structures on the land.[17]

18.29 Where different areas of the land are in different occupation or ownership the apportionment is on the basis of the respective capital values of the various areas relative to the aggregate of all such values.[18] Where different interests exist in the same area of land, the apportionment should be based on the respective capital values of those interests relative to the aggregate of all such values.[19] Where both ownership or

---

[14] DETR Circular, Annex 3, para. D.92.

[15] The fact that apportionment is required reflects the fact that it is not possible to characterise that area as giving rise to a separate pollutant linkage.

[16] DETR Circular, Annex 3, para. D.93.

[17] *ibid.*, para. D.94.

[18] *ibid.*, para. D.94(a).

[19] *ibid.*, para. D94(b). It should be recalled, in this respect, that the effect of the relevant exclusion test should be to channel liability from the rack-rented tenant to the owner.

occupation of different areas and the holding of different interests are involved, the overall liability should first be apportioned between the different areas, then between the interests in each area.[20]

18.30 The capital values used for this purpose should be estimated by the authority on the basis of the available information and disregarding the existence of any contamination.[21] The date for valuation is the date immediately before the notice of identification of the land as being contaminated was served under section 78B.[22] Where the land in question is "reasonably uniform in nature and amenity", it can be "an acceptable approximation" to apportion on the basis of the area occupied by each.[23] If appropriate information is not available to enable an assessment of relative capital values, and such information cannot reasonably be obtained, then the authority is directed to apportion liability in equal shares.[24]

18.31 The Guidance also deals with the situation where no owner or occupier can be found for part of the land. Here the authority should deduct from the overall costs a sum reflecting the apportioned share for that land, based on its relative capital value, before apportioning between those owners or occupiers who can be found.[25] This is effectively an "orphan share" situation.[26]

Attribution between liability groups

18.32 Part 9 of Chapter D deals with apportionment in the situation where one remediation action is referable to two or more significant pollutant linkages. Such a remediation action is referred to as a "shared action".[27] This situation can arise either where both linkages require the same action, or where a particular action is part of the best combined remediation scheme for two or more linkages. This process of apportionment between the liability groups is termed "attribution of responsibility".

18.33 The authority will have to consider whether the remediation action in question is referable solely to the significant pollutant in a single pollutant linkage ("a single-linkage action"), or is referable to the significant pollutant in more than one pollutant linkage ("a shared action").[28] Apportionment applies only in the case of shared actions. An example of a shared action might be the installation of a cut-off barrier to prevent mobile pollutants from more than one source reaching a watercourse. The

---

[20] *ibid.*, para. D.94(c).
[21] *ibid.*, para. D.95.
[22] Thus the effect on value of the notice itself is also disregarded.
[23] DETR Circular, Annex 3, para. D.95.
[24] *ibid.*, para. D.97.
[25] *ibid.*, para. D.96.
[26] See Chapter 20.
[27] DETR Circular, Annex 3, para. D.98.
[28] *ibid.*, para. D.21.

analysis of whether a remediation action is a shared or single action should be carried out for each separate remediation action rather than the package as a whole.

18.34 The Guidance also requires the authority to consider whether the shared action is "a common action" or "a collective action". A common action is one which addresses together all of the relevant pollution linkages, and would still have been part of the remediation package for each of those linkages if each had been addressed separately.[29] A collective action is one which addresses together all of the relevant pollutant linkages, but which would not have been part of the remediation package for each of them if they had been addressed separately.[30] This might be because some different solution would have been more appropriate for each linkage in isolation, or because the action would not have been needed to the same extent for one or more of the linkages, or because the action is adopted as a more economical way of dealing with the linkages than requiring separate solutions for each linkage.

18.35 Taking the example of the remedial cut-off barrier given above, the authority is thus required to ask, effectively, "Is this the solution we would have adopted for each of the pollution linkages had we been considering each alone?" The answer will of course depend on the exact circumstances.

Common actions

18.36 For common actions, the apportionment between the liability groups should be on the following basis:

(a) If there is a single Class A group, then the full cost of the common action falls to that group, with no cost falling to any Class B group or groups.
(b) If there are two or more Class A groups, then an equal share of the cost of the common action should be apportioned to each of those groups, and no cost should be attributed to any Class B group.
(c) If there is no Class A group and there are two or more Class B groups, the cost of the common action should be apportioned among the members of the Class B groups as if they were members of a single group, that is, on the basis of the value of their interests.

18.37 Effectively therefore, the approach is one of preferential liability for Class A groups over Class B, otherwise equal shares to reflect the fact that each linkage would have needed that remediation action in its own right.

---

[29] *ibid.*, para. D.22(a).
[30] *ibid.*, para. D.22(b).

Liability for Class A rather than Class B reflects both the "polluter pays" principle and the general liability scheme of Part IIA.

Collective actions

18.38 For collective actions the process of apportionment is the same as for common actions, except that where the cost falls to be divided among a number of Class A groups, then rather than being divided equally, they should be divided on the following basis[31]:

(a) The authority should estimate the cost of the collective action, and the hypothetical costs of the actions which would have been necessary for each pollutant linkage considered separately: these are termed the "hypothetical estimates".

(b) The authority should then apportion the cost of the collective action among the liability groups in proportion to the hypothetical estimate for that group relative to the aggregated hypothetical estimates for all groups.

18.39 This attempts to reflect the fact that the collective action solution may be quite different to the common actions relative to each linkage and seeks to achieve fairness.

18.40 The Guidance also allows for any appropriate person to demonstrate that the result of an attribution on the above basis would be that his liability group bears a disproportionate burden, taking into account the overall relative responsibilities for the condition of the land of the person or persons concerned, so that the result as a whole would be unjust.[32] The authority should then reconsider the attribution, consulting the appropriate persons, and adjust it if necessary so as to make it fair in the light of all the circumstances. Such an adjustment, according to the Guidance, should be necessary only in very exceptional circumstances.[33] One possible case is where the same group of persons forms the liability groups in respect of several pollution linkages, which though numerous, are of relatively low significance compared with one or more of the other linkages. The risk is that the larger number of the less serious linkages might result in those responsible bearing a disproportionate share.

Orphan linkages and attribution

18.41 As discussed elsewhere[34] an orphan linkage may arise either where no owner or occupier can be found, or where those who would otherwise be liable are exempted by one of the relevant statutory provisions. The Guidance considers the attribution process in four separate situations here:

---

[31] *ibid.*, para. D.100.
[32] *ibid.*, para. D.101.
[33] *ibid.*, para. D.102.
[34] See Chapter 20.

(a) If a shared action is referable to an orphan linkage and to one Class A group, then the entire cost should be attributed to the Class A group.[35] Thus, the polluter pays, rather than the public purse.
(b) Similarly, if the shared action is referable to the orphan linkage and to a number of Class A groups, the entire cost is attributed between the Class A groups, ignoring the orphan linkage.[36]
(c) If a common action is referable to an orphan linkage and a Class B group, then the entire cost should be attributed to the Class B group.[37] This reflects the fact that the common action would have been necessary for the Class B linkage in any event.
(d) If a collective action is referable to an orphan linkage and a Class B group, then the attribution to the Class B group should not exceed the hypothetical cost of dealing with the Class B group's pollution linkage in isolation.[38] To the extent that the collective action involves a more expensive solution, the excess will fall to the public purse.

[35] DETR Circular, Annex 3, para. D.107.
[36] *ibid.*, para. D.108.
[37] *ibid.*, para. D.109(a).
[38] *ibid.*, para. D.109(b).

Chapter 19

When the Local Authority is Itself an Appropriate Person

Summary

A local authority may well find itself in the position of an appropriate person, either in relation to land which it has itself contaminated in the past (for example as a municipal waste disposal facility or transport depot) or as the current owner or occupier of land. Particular care is needed in such cases in relation to the legal framework, and in managing situations of potential bias or conflict of interest which may arise where the authority is also the enforcing authority for the purpose of Part IIA. Such notices may be subject to appeal by one or more of the recipients, which may in turn lead to the remediation declaration requiring reconsideration.

Local authorities as "appropriate persons"

19.01 There is no reason in principle why a local authority cannot be an "appropriate person", either as an original contaminator or as a current owner or occupier of land. An authority's regulatory role in issuing planning permission or other licences should not of itself lead to liability, since the authority will be able to rely on exclusion test 1 in the statutory guidance in this regard.[1] However, local authorities may well have operated facilities, such as highways maintenance depots, waste transfer stations or landfill sites, which may have caused contamination. Attention will need to be given here to whether liability has passed from an earlier authority to the current one where local authority reorganisation has occurred. The terms of the relevant orders and of any related agreements may vary in this respect and will need to be considered carefully.[2]

---

[1] DETR Circular, *Contaminated Land*, Annex 3, para. D.48(g). See Chapter 17 generally.

[2] Compare, for example, the Local Government Reorganisation (Property) Orders 1986 (S.I. Nos. 24 and 148), made under the Local Government Act 1985 and the Local Government Area Changes Regulations 1976 (S.I. No. 246), made under the Local Government Act 1972.

The legal framework

19.02 As is discussed elsewhere,[3] section 78H(5)(c) provides that if it appears to the enforcing authority that the person on whom a remediation notice would be served is the authority itself, it should not serve the notice on itself. Rather, by section 78F(7), the authority as the "responsible person" should prepare and publish a remediation statement, indicating what it will do and within what timescale.

19.03 This will be the position where the authority is the only appropriate person with regard to a particular remediation action. In circumstances where the authority is one among a number of appropriate persons, the authority will serve remediation notices on the other appropriate persons, stating the appropriate remediation action, and will publish a remediation statement with regard to its own responsibility. Naturally, depending on the circumstances, the authority may wish to take the lead in securing remediation, and one way forward may be for the authority and other parties involved to enter into an agreement under section 78N(3)(b) for the authority to carry out the work under powers in that section. In that case the authority will not need to serve a remediation notice on the other parties (indeed, it will be precluded from doing so by section 78H(5)(d)).

19.04 If remediation notices are to be served, the authority will have to apply the statutory guidance to determine the proportions of the cost which it and the other parties should bear, and state those proportions in the notices—see section 78E(3). Such notices may be subject to appeal by one or more of the recipients, which may in turn lead to the remediation declaration requiring reconsideration.

19.05 This represents the bare statutory framework under which local authorities will operate in such cases. However, there are other issues which must also be considered, which are canvassed below.

Bias

19.06 The problem is that where the enforcing authority is itself an actual or potential appropriate person, there is a danger that the authority may favour itself (or be perceived to do so) in the decisions which it takes as enforcing authority. There are a number of stages where this problem might potentially arise:

1. Considering whether the relevant land is "contaminated" at all.
2. Considering what should be the remediation requirements.
3. Applying the statutory provisions to determine who are appropriate persons.

---

[3] See Chapter 10.

4. Applying the tests on whether any appropriate persons should be excluded.
5. Applying the Guidance on apportionment as between appropriate persons.

19.07 It is not necessarily the case that any authority would consciously set out to apply these rules in such a way as to favour itself. However, any owner or other person served with a remediation notice is almost certainly going to be aggrieved if he feels either that the authority has taken a more lenient approach with regard to comparable local authority land, or has been the very body applying the rules which determine who is liable for what, in a way that could be seen as being to its own advantage.

Relevant legal principles

19.08 The legal principles of administrative law on bias are concerned not only with actual bias but also with the appearance or risk of bias: "Justice should not only be done, but should manifestly and undoubtedly be seen to be done."[4] The fact that the authority is involved in taking decisions which will be to its own financial benefit or detriment might be thought to raise a reasonable suspicion of bias in itself.[5]

19.09 However, it is hardly possible to treat the local authority as disqualified from fulfilling statutory duties expressly assigned to it by Parliament. The statute makes no provision for any other body to become the enforcing authority simply because the local authority is an "appropriate person": though, of course, in cases where the land is a special site this function will fall to the Environment Agency. Accordingly, the focus seems likely to rest not so much on the fact that the authority is in a situation of potential bias,[6] but in how it behaves in that situation and in particular whether there is any suggestion that it acted unfairly. Whilst there is a statutory safeguard open to aggrieved parties by the statutory appeal process against remediation notices, this does not in any sense absolve the authority from the overriding duty to act fairly.

19.10 Potential conflicts of interest where a local authority is fulfilling a statutory function are not new; for example, in the planning law context. The important point is that the authority should preserve its impartiality by keeping an open mind,[7] take account of all proper considerations, exclude all improper ones, and reach its decision fairly.[8]

---

[4] *R. v. Sussex Justices, ex p. McCarthy* [1924] 1 K.B. 256, 259 (Lord Hewart C.J.); see also *R. v. Bow Street Magistrates, ex p. Pinochet* [1999] 2 W.L.R. 272, 284 (Lord Browne-Wilkinson).

[5] For cases, see de Smith, Woolf & Jowell, *Judicial Review of Administrative Action* (London, Sweet & Maxwell, 1995) Fifth edition, p. 526.

[6] *ibid.*, p. 543.

[7] *Lower Hutt City Council v. Bank* [1974] 1 N.Z.L.R. 545, 550.

[8] *R. v. St. Edmundsbury Borough Council, ex p. Investors in Industry Commercial Properties Ltd* [1985] 3 All E.R. 234, 256.

Practical measures

19.11 The authority can best guard itself against allegations of bias by ensuring that it follows scrupulously the legal requirements on matters such as statutory consultation, that it takes great care in applying the statutory Guidance on exclusion and apportionment, and that it is in a position to give clear reasons for its decisions. It should ensure that it takes so far as possible a consistent approach as between its own sites and those of others, and that it is in a position to explain and justify any differences of approach. Where in doubt on technical issues, it might well consider seeking site specific guidance from the Environment Agency under section 78V.

19.12 The main problems are likely to arise where the authority is itself one of a group of appropriate persons. The authority will still have to carry out an apportionment exercise, but clearly they could be wise to call in an independent consultant to advise on the issue, whose report could be made public. Another possibility might be to separate out the role of dealing with enforcement from that of dealing with the authority's own liabilities and to allocate these responsibilities to different people.

Funding

19.13 Where the authority is liable for clean-up as an appropriate person, the question of funding will arise. This is discussed below in the context of "orphan sites".[9]

[9] See Chapter 20.

Chapter 20

"ORPHAN SITES" AND "ORPHAN LINKAGES"

Summary

The term "orphan site" has no precise meaning and can encompass a range of situations. The basic concept is of a site in respect of which the necessary remediation works, in whole or in part, cannot be attributed to any appropriate person, thereby leaving remediation in the hands of the enforcing authority. The Guidance uses the term "orphan linkage" to describe the situation where either no owner or occupier can be found for some land, or where those who would otherwise be liable are exempted by one of the statutory provisions. This Chapter considers the various situations in which orphan sites, orphan shares or orphan linkages may arise, and how the enforcing authority should address such situations.

Can there be orphan sites?

20.01 The basic scheme of the legislation would suggest that orphan site situations should not readily arise. If the original contaminator (the causer or knowingly permitter) cannot be found under section 78F(2), then the authority may turn to the owner or occupier. In most cases there will be both, and in all cases there should be an owner, even if it is the Crown as a matter of escheat.[1] Even in cases where contamination has migrated from one piece of land to another, in general under section 78K each owner or occupier will at least be responsible, in the last resort, for cleaning up their own land.[2]

20.02 The real possibility of an orphan site therefore rises not from the inability to find someone who is an appropriate person within the scheme of Part IIA, but rather from the fact that the authority may be precluded

---

[1] See Chapter 14.
[2] See Chapter 16.

from serving a remediation notice on that person under section 78J because the remediation relates solely to water pollution or to water pollution from abandoned mines[3]; or under section 78H(5)(d) because of hardship considerations[4]; or in respect of persons acting as insolvency practitioners or some other relevant capacity under section 78X(4).[5]

20.03 One particularly important situation in practice may be where a Class A appropriate person has been found, but they can successfully argue on hardship grounds that a remediation notice may not be served on them. Here an appropriate person under section 78F(2) has actually been found (albeit that service of a notice on them is precluded) and on that basis it is not possible to look to those in Group B under section 78F(5). In that case, the site is effectively an orphan site. Interestingly, the Guidance, in discussing orphan linkages at paragraph D.103, does not refer to the hardship scenario, but does refer to section 78K on escapes.

Can there be orphan actions?

20.04 Another possibility is that a series of different remediation actions, relating to different types of contamination or pollution linkages, may be required for a site. Some of those actions may not be referable to a Class A appropriate person under section 78F(3).[6] However, those actions would then naturally fall to the owner and occupier as Class B appropriate persons under section 78F(5).

20.05 The same question will then arise as described above, namely whether the authority is precluded from serving notice on those persons. If so, then the relevant actions will effectively be orphan actions. The Guidance generally treats the situation of orphan sites and orphan actions together, within the concept of "orphan linkages".[7]

Can there be orphan shares?

20.06 The situation that may potentially arise here is where a number of appropriate persons are found to be responsible for the same remediation action, the cost of which has been apportioned between them by the authority. The issue may be one of hardship, in that one or more of the recipients may be able to resist service of a notice on them on grounds of hardship with regard to their share. In that situation, it is the share of such

---

[3] See Chapter 15.
[4] See Chapter 22.
[5] See Chapter 14.
[6] See Chapter 13.
[7] See DETR Circular *Contaminated Land,* Annex 3, paras D.12, D.14, D.17 and D.103.

persons which is, effectively, an orphan share. The problem then is that other appropriate persons may seek to argue a defence under section 78M(2) that others involved refused, or were unable, to bear their share.[8] However, the answer to such an argument is that the person on whom the remediation notice could not be served is not a "person liable to bear a proportion of that cost": hence section 78M(2) will not apply.

Dealing with orphan sites

20.07 This is the most straightforward case, where there is no pollutant linkage other than an orphan linkage; in other words, there is no-one whom the authority could proceed against in relation to the whole site. The Guidance at paragraph D.34 states that in such circumstances the enforcing authority should itself bear the cost of any remediation which is carried out. The question is then whether the authority uses its section 78N powers to cany out remediation at its own cost. Doing nothing may be politically, and possibly legally, unacceptable. The issue of funding work in such circumstances is considered below.

Dealing with orphan actions

20.08 This situation is more complex. A site may be subject to a number of pollutant linkages involving different substances and different factual backgrounds. One or more of these linkages may be orphan linkages, on the principles described above. The authority will then need to consider each remediation action separately.[9] If the action in question is referable only to the orphan linkage, and no other, then the authority will itself have to bear the cost of carrying out that action.[10] In those circumstances the authority's decisions as to how and when it uses its powers to carry out work under section 78N will need to take into account the other remediation works being carried out on the same site

20.09 However, where the action is referable to both the orphan linkage and to one or more other linkages in respect of which there are laibility groups, then the cost of the shared action may fall on those liability groups. If the persons involved are Class A causers or knowing permitters, then the authority should attribute the entire cost to them, apportioning it if necessary between the separate liability groups, if there are more than one, as if the orphan linkage did not exist.[11] In other words, it is the polluters, rather than the public purse, which bear the cost of that part of the shared

---

[8] See Chapter 22.
[9] DETR Circular, Annex 3, para. D.105.
[10] *ibid.*, para. D.106.
[11] *ibid.*, paras D.107 and D.108.

action attributable to the orphan linkage. Class B owners and occupiers are treated somewhat more favourably in this respect. Such groups will only be exposed to bearing the cost of action attributable to an orphan linkage where there is a shared action which cannot be referred to a pollutant linkage involving a Class A group.[12] In such cases, the authority's approach will have to vary depending on whether the shared action involved is a common action or a collective action. For common actions (that is, those which would have been required if each pollutant linkage was considered individually) the entire cost should be attributed to the Class B Group.[13] For collective actions (those that would not have been part of the remediation package for individual linkages) the attribution of cost to the Class B liability group should not exceed the estimated hypothetical cost of dealing with their pollutant linkage separately[14]

Dealing with orphan shares

20.10 The Guidance does not deal explicitly with the orphan share situation where within the relevant liability group there are one or more individuals on whom the authority cannot serve a remediation notice, for example under section 78J or 78X, or for hardship reasons. The point to keep in mind is that such persons are not excluded from the liability group (unless, of course, they can rely on one of the exclusion tests). Rather, they are exempted from having a notice served on them. So long as there remains one or more persons in the group who do not benefit from such exemptions, the linkage will not be an orphan linkage.[15] The exempted persons must still be apportioned an appropriate share of the cost, and for this purpose it is irrelevant whether or to what extent they would benefit from the hardship provisions.[16] To the extent that the authority cannot recover the cost from them, it will be an "orphan share" falling to the authority.

Fairness

20.11 As will be appreciated, the way in which orphan linkages are treated may be highly controversial, since it may involve on the one hand substantial calls on the public purse, and on the other persons having to bear the cost of dealing with actions for which they do not feel fairly responsible. It is therefore vital that the authority acts fairly and transparently in applying the Guidance and in making the relevant decisions.[17]

---

[12] *ibid.*, paras D.108 and D.109.
[13] *ibid.*, para. D.109(a).
[14] *ibid.*, para. D.109(b).
[15] *ibid.*, para. D.17.
[16] *ibid.*, para. D.35.
[17] See Chapter 19 generally.

Funding

20.12 Substantial expenditure could be necessary to deal with actions for which the authority has to bear the cost, either as an orphan share or linkage, or as an appropriate person in its own right. On general principles of local government finance such expenditure is not subject to capital controls if funded from revenue or from reserves. However, if it is funded from borrowing, credit provisions or capital receipts, then credit approval will be necessary. Supplementary credit approvals (SCAs) can be issued by Ministers to cover a particular project or programme under section 54 of the Local Government and Housing Act 1989, and may then be used either to enter into credit arrangements or to capitalise expenditure in accounts, so increasing borrowing powers.[18] In practice, obtaining approval involves bidding for support from the programme of SCAs run by the DETR for dealing with contaminated land which local authorities own or for which they are otherwise responsible: applications are prioritised against environmental criteria.[19]

20.13 SCA support is not available for work needed to facilitate the development, redevelopment or sale of land.[20] Where clean-up is related to economic regeneration, funding may be available from Regional Development Agencies under the Regional Development Agencies Act 1998 or under the Single Regeneration Budget operated under Part IV of the Housing Grants, Construction and Regeneration Act 1996[21] or under the European Regional Development Fund in designated areas.

20.14 So far as the Environment Agency and SEPA are concerned, both have general statutory powers to carry out such engineering operations as they consider appropriate as facilitating, or conducive or incidental to, the carrying out of their functions.[22] In principle, the funding for such work as is necessary could come from Government grants,[23] or from borrowing,[24] or from Government loans.[25]

---

[18] See the Local Authorities (Capital Finance) Regulations 1990 No. 432; Circular 11/90; *A Guide to the Local Authority Capital Finance Systems* (DoE, 1997).
[19] DETR Circular, Annex 2, paras 16.12–16.14.
[20] *ibid.*, para. 16.13.
[21] See S. Tromans and R. Turrall-Clarke, *Contaminated Land—First Supplement* (1998) p. 101–107.
[22] Environment Act 1995, s.37(1)(b).
[23] *ibid.*, s.47.
[24] *ibid.*, s.48.
[25] *ibid.*, s.49.

Chapter 21

ACCESS FOR REMEDIATION AND COMPENSATION

Summary

The person on whom a remediation notice is served may not necessarily have the rights of access or other rights necessary in order to comply with the notice. For example, it may be necessary to install and maintain equipment on the land of a third party. Section 78G provides a procedure for obtaining the necessary rights and for the payment of compensation in respect of such rights.

The need for access or other rights

21.01 By section 78G(1) an enforcing authority is expressly permitted to include in a remediation notice things which the recipient is not entitled to do. These could include, for example, carrying out engineering works or conducting monitoring on adjacent land, obtaining access across the land of a third party or, in the case of an original contaminator, returning to a site which he no longer owns in order to carry out remediation.

Obligation to grant rights

21.02 By section 78G(2) any person whose consent is needed before the thing required by the remediation notice can be done is required to grant, or join in granting, such rights in relation to the "relevant land or matters" as will enable the appropriate person to comply with the requirements of the remediation notice. Although the word used in section 78G(2) is "granting", it does not follow that this need necessarily involve a grant by deed of legal rights. In most cases, a contractual licence may be all that is necessary. Where more permanent works are in prospect (for example the laying of pipework across land) then more formal rights may be appropriate. The rights to be granted relate to "relevant land or waters", defined by section 78G(7) to mean:

(a) the contaminated land in question;
(b) any controlled waters affected by that land; or
(c) any land adjoining or adjacent to that land or those waters.

21.03 It will be appreciated from this definition that the relevant land may be some considerable distance from the site of contamination in cases where controlled waters have been affected. The obligation under section 78G(2) might, for example, involve granting rights to pump and treat groundwater many kilometres downstream, or providing access to remove contaminated silt from the downstream bed of a river where it has been carried by the current.

21.04 The grant of the relevant rights may involve cost, disruption and inconvenience. The Act itself contains no sanction for failure to grant such rights, nor any mechanism by which the local authority or appropriate person may compel the grant of rights; nor is there any express dispute resolution mechanism, where the third party claims that the rights sought are unreasonable. It would appear therefore that the onus is on the appropriate person to take civil proceedings to secure the grant of the rights where necessary. The court would in those circumstances have to determine whether the rights were in fact needed to comply with the relevant requirement of the notice.

Consultation

21.05 Where the terms of a proposed remediation notice will involve the grant of rights under section 78G, the enforcing authority must reasonably endeavour to consult the relevant owners or occupiers—see section 78G(3). This requirement does not however preclude service of a notice where it appears there is imminent danger of serious harm or serious pollution being caused.[1]

Compensation

21.06 A person who grants, or joins in granting, rights under section 78G is entitled to make an application for compensation.[2]

Regulations may prescribe:

1. the time within which such applications shall be made;
2. the manner in which applications are made;
3. to whom they are made; and
4. how the amount of compensation is to be determined.

---

[1] Section 78G(4).
[2] Section 78G(5).

21.07 Regulation 6 of, and Schedule 2 to, the Regulations deal with these issues.

How applications for compensation are made

21.08 Any application for compensation under section 78G must be made within the prescribed period beginning with the date of the grant of the rights in question and ending on whichever is the latest of the following dates[3]:

(a) twelve months after the date of the grant;
(b) where operation of the remediation notice is suspended by an appeal, twelve months after the date of its final determination or abandonment; or
(c) six months after the date on which the rights were first exercised.

21.09 It must be made to the appropriate person to whom the right was granted and must be in writing and delivered at or sent by pre-paid post to the last known address for correspondence of the appropriate person.[4] It must contain, or be accompanied by, the following particulars[5]:

(a) a copy of the grant of right in respect of which the grantor is applying for compensation and of any plans attached to such grant;
(b) a description of the exact nature of any interest in land in respect of which compensation is applied for; and
(c) a statement of the amount of compensation applied for, distinguishing the amounts under the various prescribed headings, and showing how the amount applied for under each heading has been calculated.

Types of loss and damage which are compensated

21.10 Compensation is payable for five descriptions of loss and damage.[6] The five categories distinguish between effects on "relevant interests" (that is, those interests out of which the rights in question were granted) and other interests in land (which may include land other than that over which the rights were granted). They also distinguish between depreciation in the value of the interest and other loss or damage, disturbance or "injurious

---

[3] Regulations, Sched. 2, para. 2.
[4] *ibid.*, para. 3(1).
[5] *ibid.*, para. 3(2).
[6] *ibid.*, para. 4(1).

affection" (adopting the terminology of land compensation legislation). Finally, there is a distinction drawn between those effects stemming from the grant of the right in question, and those resulting from the exercise of the rights. It is not entirely straightforward to understand how those categories relate to each other, and there would appear to be some potential for overlap between them. The five categories of compensation are as follows:

(a) Depreciation in the value of any relevant interest to which the grantor is entitled which results from the grant of the rights. The "relevant interest" is defined by pararaph 1 of the Schedule to mean an interest out of which rights have been granted, and so will be that interest which enables the grantor to grant the rights in question. It may accordingly be freehold or leasehold. "Grantor" in this context is defined to include any person joining in the granting of rights. It should be noted that the depreciation in question for this category is that resulting from the grant of the right, not its exercise.

(b) Depreciation in the value of any other interest in land to which the grantor is entitled which results from the exercise of the rights. It therefore does not cover relevant interests falling within paragraph (a) above, and relates to depreciation caused by the exercise of the right rather than its grant. An example might be someone who is not the owner of the immediate interest in the land, but who has an interest which means that they need to join in the grant of the rights in question. Another example might be the situation where it can be shown that the exercise of the rights affects the value of neighbouring land owned by the grantor.

(c) Loss or damage in relation to any relevant interest to which the grantor is entitled, which is attributable either to the grant of the right or its exercise, but which does not consist of depreciation in its value, and is loss or damage for which he would have been entitled to compensation by way of compensation for disturbance if the interest had been acquired compulsorily under the Acquisition of Land Act 1981 (assuming the notice to treat to have been served on the day on which the rights were granted). The important point here is that the loss or damage in question must not be too remote, and must be a natural and probable consequence of the grant or exercise of the right, as the case may be.[7] Subject to this, as the Circular suggests, compensation might be payable where the land or things on it are damaged, or where there is a loss of income or profits on the part of the grantor.[8] This may be particularly relevant where remediation disrupts the ongoing business operations of the current owner or occupier.

[7] *Horn v. Sunderland Corporation* [1941] 2 K.B. 26; *Harvey v. Crawley Development Corporation* [1957] 1 Q.B. 485, 494.
[8] DETR Circular, Annex 4, para. 26(b).

(d) Damage to, or injurious affection of, any interest in land to which the grantor is entitled which is not a relevant interest, and which results from the grant of the rights or their exercise. Though the wording of the Regulations does not make this explicit, the intention of this category is to cover the situation where the works on the contaminated land involved had "some permanent adverse effect" on adjoining land.[9] Again the analogy is with compulsory purchase law, and on that basis the damage or adverse effects must not be too remote.[10] The reference to "permanent adverse effect" in the Circular is potentially confusing, however, since by definition the effects of the remediation works are unlikely to be permanent. However, it must also be recalled that what is being compensated is damage to an interest in land, not nuisance generally: accordingly it is probably the case that compensation will not be payable where the problems are temporary and do not affect the value of the land once the works are complete.[11] In any event, this head of compensation can only be helpful to those who have granted or joined in granting rights; those who are simply affected as neighbours will have to rely on their normal common law remedies in nuisance.

(e) Loss in respect of work carried out by or on behalf of the grantor which is rendered abortive by the grant of the rights or their exercise. The example given by the Circular is of a newly erected building which can no longer be accessed and used after the grant of the rights.[12] Less dramatic examples can be contemplated, for example, an access road or drainage system, the route of which needs to be altered.

Basis of assessing compensation

21.11 Schedule 2, paragraph 5 sets out a number of rules or principles for the assessment of compensation under section 78G. For the purposes of assessing valuations of land, the rules set out in section 5 of the Land Compensation Act 1961 have effect, so far as applicable and subject to any necessary modifications.[13] This involves valuation on the basis of sale in the open market by a willing seller. Where the interest in land is subject to a mortgage, the compensation is to be assessed as if it were not so subject,

---

[9] *ibid.*, para. 26(c).

[10] See *Metropolitan Board of Works v. McCarthy* (1874) L.R. 7 H.L. 243.

[11] See *Wildtree Hotels Ltd v. Harrow London Borough Council* [1998] 3 W.L.R. 1318; [1999] J.P.L. 136.

[12] DETR Circular, Annex 4, para. 26(d).

[13] The five rules contained in section 5 of the 1961 Act are discussed in S. Tromans and R. Turrall-Clarke, *Planning Law, Practice and Precedents*, paras 10.07 *ff.*

and no compensation is payable in respect of the interest of the mortgagee.[14]

Expenditure by grantor

21.12 In assessing compensation, no account is to be taken of expenditure which enhances the value of the land in question by carrying out works, improvements or erecting buildings, where that work was not reasonably necessary and was carried out with a view to obtaining enhanced compensation.[15] Where genuine abortive expenditure has been incurred on the land, this will be recoverable under (e) above, and can include expenditure on plans or similar preparatory matters,[16] such as obtaining planning permission or building approval.

Professional expenses

21.13 Compensation can include reasonable legal and valuation expenses.[17]

Mortgages

21.14 The position of any mortgagee will need careful consideration in relation to compensation. Since, as mentioned above the "grantor" for the purposes of compensation includes a person who joins in granting the relevant rights, this could include a mortgagee whose consent is required to the grant of an easement or similar right. Also, a mortgagee in possession might be required to grant rights of access. A mortgagee can therefore obtain compensation in his own right. However, by Schedule 2, para. 5(5) where the interest in which compensation is assessed is subject to a mortgage, then the mortgage is ignored for the purpose of assessing the compensation, and no compensation is payable separately in respect of the mortgagee's interest. This avoids the risk of double compensation being payable. The compensation in such circumstances must by Schedule 2 para. 6(1) be paid to the mortgagee, or to the first mortgagee if there is more than one; it must then be applied as if it were proceeds of sale. It is questionable how well thought out these provisions are: the intention, according to the Circular, is to ensure that the mortgagee and any subsequent mortgagee "will get any appropriate share".[18] However, it is

---

[14] Schedule 2, para. 5(5).
[15] Regulations, Sched. 2, para. 5(3).
[16] *ibid.*, para. 5(4).
[17] *ibid.*, para. 5(6).
[18] DETR Circular, Annex 4, para. 31.

possible to foresee disputes arising between the mortgagor and mortgagee, in particular where the compensation relates to matters such as disturbance to the mortgagor's business or to abortive expenditure by the mortgagor.

Dispute resolution

21.15 By Schedule 2, para. 6(3) disputes on compensation are to be referred to the Lands Tribunal for determination. The provisions in the Land Compensation Act 1961 are applicable as to the procedure and costs of such references.[19] As the Circular points out, under its own Rules, the Lands Tribunal may deal with the matter on the basis of written representations (rule 27), or by a simplified procedure, if appropriate (rule 28). One issue which may in practice lead to disputes, it is anticipated, is the relationship between the different categories of compensation, which do not appear to be entirely mutually exclusive. Care will be needed to ensure that the claimant is not compensated twice over in respect of what is essentially the same loss.

Date for payment and interest

21.16 Compensation is payable in accordance with the date or dates agreed between the parties, either in a single payment or in instalments; otherwise it is payable, subject to any direction of the Tribunal or Court, as soon as reasonably practicable after the amount is determined.[20] The Government proposes to apply the provisions for interest under the Planning and Compensation Act 1991 to compensation under section 78G.[21] Claims may be substantial in some cases, and may take some time to settle (particularly as novel points may arise). The question of interest may accordingly be an important one.

The authority's role

21.17 The enforcing authority is not involved directly in the question of compensation under section 78G, other than its obligation to consult them under section 78G(3) before serving a remediation notice and the requirement that it serve a copy of the notice on them. However, the Circular suggests that it is good practice for authorities to inform those they have

---

[19] Regulations, Sched. 2, para. 34.
[20] *ibid.*, para. 6(2).
[21] DETR Circular, Annex 4, para. 34.

consulted in this way of the final outcome of any appeal against the notice, so that they are alerted to the need to be ready to apply for compensation.[22] This raises the interesting question of whether an authority may be liable for failure to comply with this "good practice" if in consequence of not being notified, the grantor of rights misses the deadline for applying for compensation.

[22] *ibid.*, para. 38.

Chapter 22

ENFORCEMENT

Summary

By section 78M, failure to comply with a remediation notice without reasonable excuse constitutes an offence. As well as prosecuting, options open to the enforcing authority include the use of civil proceedings, and carrying out remediation itself with powers of cost recovery and charging orders. In seeking to recover its costs, the authority must take account of the Guidance on the issue of hardship to the appropriate person or persons concerned.

Offence of non-compliance with remediation notice

22.01 Section 78M(1) provides that a person on whom an enforcing authority serves a remediation notice who fails, without reasonable excuse, to comply with any of its requirements, shall be guilty of an offence.

22.02 These prosecution provisions seem weak in some respects. Under section 78M there is a general qualification, namely that only a person who fails to comply with any of the requirements of a remediation notice "without reasonable excuse" is guilty of an offence. This would appear to provide a great deal of latitude for debate whether or not an offence has been committed in the first place. Whilst lack of funds will not generally be a reasonable excuse,[1] the physical inability to carry out works, for example for lack of access, may be relevant. However, it would be difficult to rely on lack of access without first seeking to exercise the statutory rights of access given by section 78G. Care must be taken to distinguish between matters which constitute reasonable excuse, and those which go only to the issue of mitigation.[2] Nonetheless, it is not possible to provide an exhaustive and comprehensive definition of what may be a reasonable excuse, and reasons of personal hardship (*e.g.* illness) may be relevant.[3]

---

[1] *Saddleworth U.D.C. v. Aggregate and Sand* (1990) 114 S.J. 931; *Kent County Council v. Brockman* [1996] 1 P.L.R.1.

[2] *Wellingborough Borough Council v. Gordon* [1993] Env. L.R. 218.

[3] *Hope Butuyayu v. London Borough of Hammersmith and Fulham* [1997] Env. L.R. D13.

22.03 As with statutory nuisances, the question arises whether the prosecution have to produce evidence of the lack of a reasonable excuse. As with the provisions on statutory nuisance, the requirement of "without reasonable excuse" is an inherent part of the offence rather than a defence that has to be specifically raised as such. That being so, it is probably sufficient for the defendant to simply adduce some evidence of an excuse, and it will then be for the prosecution to satisfy the court on the criminal standard of beyond reasonable doubt that there was no reasonable excuse.[4]

22.04 Secondly, by section 78M(2) where the notice states that only a proportion of the cost is chargeable against the recipient, it is a defence for him to prove that the only reason why he has not complied with the requirements of the notice is that one or more other persons liable to pay a proportion of the cost have refused, or were not able, to comply with the requirement. If one of the other appropriate persons is in liquidation or bankrupt, insolvent or merely impecunious, the remediation works may well not have been carried out, since a responsible party will not wish to be committed contractually to the engineering or other costs involved if he is not satisfied that a contribution is likely to be forthcoming from the others responsible. The defence will be available to someone who can prove that this is the only reason for his non-compliance. The onus of proving the defence rests with the defendant, on the balance of probabilities.[5] It would therefore be prudent for the person seeking to rely on the defence to demonstrate that tenders have been obtained, that attempts have been made to invite the others involved to contribute, and that generally the defendant is willing and able to do his part.

22.05 The statutory provisions do not create an offence triable in the Crown Court, or either way, that is in either magistrates' or Crown Court at the election of the defendant. Section 78M(3) is concerned only with "summary conviction". This in itself seems extraordinary given that, by definition, the harm or risk of harm relating to the land will be of a serious nature. The summary-only nature of the offence has the consequence that any prosecution must be brought within six months from the time when the offence was committed.[6] By analogy with *Hodgetts v. Chiltern District Council*,[7] failure to comply with a remediation notice probably constitutes an offence which is complete once and for all when the period for compliance with the notice expires. The six-month period will therefore run from that date.[8]

---

[4] *Polychronakis v. Richards and Jerrom Ltd* [1998] J.P.L. 588, following *R. v. Clarke* [1969] 1 W.L.R. 1109. Compare the position where a defence is actually involved—see section 101 of the Magistrates' Courts Act 1980 and *R. v. Hunt* [1997] A.C. 392.

[5] *Islington Borough Council v. Panico* [1973] ER 485; *Neish v. Stevenson* 1969 SLT 229. See also *R. v. Hunt*, n. 4 above.

[6] Magistrates' Court Act 1980, section 127.

[7] [1983] A.C. 120. Like remediation notices, the case was concerned with a "do" notice, requiring positive steps.

[8] Compare *Camden London Borough Council v. Marshall* [1996] 1 W.L.R. 1345, where there was an express statutory provision (section 376(2) of the Housing Act 1985) that the obligation to carry out the works continued despite the expiry of the period. Part IIA contains no such provision.

22.06 The normal penalty is set at a maximum of level 5 on the standard scale, *i.e.* £5000. For each day on which failure to comply continues, until such time as the authority exercises its section 78N powers a further fine of one-tenth of that sum applies.[9]

22.07 In respect of an offence committed in circumstances where the contaminated land to which the remediation notice relates is industrial, trade or business premises the fine on summary conviction has an upper limit of £20,000 unless the Secretary of State by Order increases it.[10] The daily penalty is one-tenth of that sum (*i.e.* £2,000). "Industrial trade or business premises" are defined by section 78M(6) to mean premises used for any industrial, trade or business purposes, or premises not so used on which matter is burnt in connection with such premises; it specifically includes premises used for industrial purposes of any treatment or process as well as for manufacturing. The term, as so defined, refers at subsection 78M(4) to cases "where the contaminated land to which the remediation notice relates *is*" such premises. Where the contaminated land was once industrial, but has since been changed to residential use, subsection (4) will not apply.

Practical issues in prosecution

22.08 Prosecution for failure to comply with a remediation notice is, like other aspects of criminal enforcement, a discretionary matter. Where the Environment Agency is the enforcing authority, it will need to have regard to its own published prosecution policy.[11] However, having said that, prosecution will normally follow the situation where the recipient of a notice fails to comply with it. Issues of discretion may arise where the recipient of the notice—perhaps now taking the situation seriously for the first time—seeks additional time to comply, or partially complies with the notice. On the issue of time, the authority should be mindful of the fact that failure to comply with the notice is an offence triable only summarily, so that the six-month rule for bringing prosecutions applies.[12] The authority should therefore be careful not to prejudice its position by delay in proceeding. If the six-month period is missed, it will have to serve a new notice and start again.[13] Unless all the requirements of the notice are complied with by the time stated, an offence is committed; the authority

---

[9] Section 78M(3). This does not mean that the daily fine must be that figure. The general discretion under section 34 of the Magistrates' Courts Act 1980 allows a lesser figure to be imposed—see *Canterbury City Council v. Ferris* [1997] Env. L.R. D14.

[10] Section 78M(4).

[11] November 1998.

[12] See paragraph 22.01.

[13] See *Camden London Borough Council v. Marshall* [1996] 1 W.L.R. 1345, at 1350 (Henry L.J.).

may however wish to consider the practical extent of compliance in such circumstances, and whether prosecution would serve any useful purpose.

22.09 Another issue in prosecuting is whether there is sufficient evidence of non-compliance to sustain a conviction. Again, by analogy with statutory nuisance, the magistrates will only require proof that the notice has not been complied with, if the land remains harmful or potentially harmful so as to constitute "contaminated land".[14]

22.10 Finally, there is the situation where a number of parties are required to contribute to the cost of carrying out the same remediation action. One or more of them may conceivably have a defence under section 78M(2) as outlined in the previous paragraph. The question is whether the enforcing authority should prosecute all the group by a joint charge, or lay separate informations against each. As a matter of principle, each remediation notice is separate, and non-compliance with each notice is a separate offence. Each information can relate to one offence only, but a single document can contain a number of informations.[15] The best course is to charge separate offences, but to seek a joint trial on the basis that the facts are so interconnected that the interests of justice would be best served by a single hearing.[16]

Use of civil remedies

22.11 If the enforcing authority is of the opinion that proceedings for an offence under section 78M(1) would be an ineffectual remedy for non-compliance with the notice by section 78M(5) then they may take proceedings in the High Court (or any court of competent jurisdiction in Scotland)[17] for the purpose of securing compliance with the remediation notice. The question which the authority must ask is whether prosecution would be effectual against the person in default: the attitude, past conduct and resources of that person will therefore be relevant. On the basis of the decision in *Vale of White Horse District Council v. Allen & Partners,*[18] the power under section 78M to seek an injunction will probably be regarded as a self contained code, so that section 222(1) of the Local Government Act 1972 may not be used as a separate means of obtaining an injunction. Accordingly, the authority must have formed the opinion required by section 78M(5); it is not enough simply that injunctive proceedings are seen as more convenient or appropriate.[19]

22.12

[14] *AMEC Building Ltd and Squibb & Davies Ltd v. London Borough of Camden* [1997] Env. L.R. 330.
[15] Magistrates' Courts Rules 1981, rule 12.
[16] *Chief Constable of Norfolk v. Clayton* [1983] 2 A.C. 473; *R. v. Assim* [1996] 2 Q.B. 249.
[17] The appropriate sheriff court or the Court of Session.
[18] [1997] Env. L.R. 212 at 223–4.
[19] *ibid.*, at p. 224.

Given that a remediation notice will normally impose positive requirements, the appropriate remedy will be a mandatory injunction, or in Scotland, specific implement or *ad factum praestandum*. The wording of section 78M(5) would certainly seem wide enough to allow for a mandatory injunction to be sought.[20] There are precedents for the use of such injunctions in the environmental context, even where it is alleged that the defendant is impecunious.[21] The injunction will need to be as specific as possible, though the courts have accepted that where enforcement of public law is involved, if adequate protection cannot be given in any other way then an injunction may be granted in extensive terms.[22]

Remediation carried out by enforcing authority

22.13 By section 78N there are a number of situations in which the enforcing authority itself has power to carry out remediation works to the relevant land or waters. Those works will not always have been set out in a remediation notice because in some of the cases concerned no remediation notice will have been served, or if one has been served, it may not necessarily have included the works concerned. However, the structure of section 78P is such that the right to recover the costs is based on the person concerned being an "appropriate person" who could be required to do the works concerned in a remediation notice. Bearing in mind the Government's stated commitment that local authorities should not be involved in additional costs in implementing the legislation, it is perhaps curious to find them in a position where in certain cases they may have to carry out works without any legal right to reclaim the costs from any person. However, it should be remembered that section 78N confers a power to act, not a duty.

22.14 The situations where the enforcing authority may carry out remediation works are the following[23]:

(a) where the authority considers the works to be necessary for the purposes of preventing the occurrence of any serious harm, or serious pollution of controlled waters, of which there is imminent danger[24];

(b) where an appropriate person has entered into a written agreement with the enforcing authority for the authority to do the works that otherwise he would have to do[25];

---

[20] See, in the context of listed buildings, *South Hams District Council v. Halsey* [1996] J.P.L. 761, CA (injunction given although prosecution for breach of notice had failed in Crown Court because of a wrong decision on the law). See also *Runnymede Borough Council v. Harwood* [1994] 1 P.L.R. 22, CA.

[21] *Warrington Borough Council v. Hall* [1999] Env. L.R. 869, CA (mandatory injunction requiring £2 million expenditure).

[22] See *Kettering Borough Council v. Perkins* [1999] J.P.L. 166, p. 173.

[23] Section 78N(3).

[24] "Serious harm" or "serious pollution" are not defined in section 78(P) nor, it would appear, does the Secretary of State have powers to issue guidance or make regulations in respect of that aspect. See paragraph 22.17 below.

[25] See Chapter 11 generally.

(c) where a person on whom a remediation notice has been served fails to comply with any of its requirements;
(d) where the enforcing authority is precluded under sections 78J or 78K from including something by way of remediation in a remediation notice which nevertheless is required to be done[26];
(e) where the enforcing authority considers that if it did carry out works itself, it would decide not to seek to recover the costs of the work or to recover only a proportion of them under section 78P (this situation relates to hardship); and
(f) where no person after reasonable enquiry has been found to be an appropriate person.[27]

22.15 Only in case (c) does it follow that a remediation notice will have been served. Quite clearly, no notice will have been served in cases (d), (e) and (f).

22.16 By subsection 78N(2) the authority does not have power to do anything in circumstances where the contaminated land regime is excluded by section 78YB where the statutory regimes apply.[28]

Serious harm or serious pollution

22.17 There is no indication of what constitutes serious harm or serious pollution as the term is used in section 78N. It is therefore a matter of the judgment of the enforcing authority, and that judgment could not be challenged unless it was unreasonable in the *Wednesbury* sense. Further, there is no requirement to publish a decision that serious harm or serious pollution had or was likely to arise. Nevertheless, there would need to be a committee decision that the enforcing authority should carry out the works itself, or a decision to that effect by a responsible officer if the delegation agreement of the authority conferred delegated power to make that decision on him. In any event, the danger has to be "imminent". Bearing in mind that land is not contaminated at all for the purposes of the legislation unless and until "significant harm" is being caused, or there is a significant possibility of such harm being caused, the term "serious" in the present context would appear quite different from "significant". Clearly, harm could be "significant" without being "serious"; it could never be "serious" without also being "significant".

Failure to comply with notice

22.18 There are a number of practical points which are important in respect of an enforcing authority's decision to carry out works upon failure of the recipient to comply with a notice. First, it is prudent for notice to be given

[26] See chapters 15 and 16.
[27] See the discussion in Chapter 13 as to what is meant by "found".
[28] See Chapter 2.

to the appropriate person that this procedure is to be adopted and any such notice should contain a time limit. Secondly, the matter will have to go to committee, or a duly delegated officer, with estimates of the cost of the work, having in mind that full recovery may not be possible, even where legally permitted. Some experience of good practice in this field has been gained by local planning authorities carrying out works under section 178 of the Town and Country Planning Act 1990.[29] In particular, good practice may involve supplying the appropriate person with copies of a number of estimates for the cost of the works, giving an opportunity for comment.

22.19 There is not only the question of the nature of works that need to be done to remediate the harm (which may be the subject of an appeal against the notice) but the way in which those works shall be done. By section 78N(4)(c) the authority may do only those things which the recipient of the remediation notice was required to do. The terms of the notice therefore need to be considered carefully, since the authority's powers to act (and of cost recovery) may be jeopardised by departing from, or going beyond, what the remediation notice requires. In any event, the authority clearly should not exercise its powers before the statutory period for appeal has expired.

Works which cannot be included in a remediation notice

22.20 As we have seen,[30] an enforcing authority is precluded by sections 78J and 78K, from including certain things in a remediation notice which nevertheless ought to be done. Where that situation arises the enforcing authority can carry out the works, but it will not be able to recover the reasonable cost. This is because the power of cost recovery under section 78P(1) does not apply to the section 78N(3)(d) power to carry out remediation works in those circumstances; a position confirmed by sections 78J(7) and 78K(6).

What is the authority allowed to do?

22.21 Section 78N(1) allows the authority "to do what is appropriate by way of remediation to the relevant land or waters". The phrase "relevant land or waters" is defined to mean:

(a) the contaminated land in question;
(b) any controlled waters affecting that land; or
(c) any land adjoining or adjacent to those waters.

---

[29] See paras 9.133 *et seq.*, Tromans and Turrall-Clarke, *Planning Law Practice & Precedents*, (London, Sweet & Maxwell, 1991).
[30] See Chapters 15 and 16.

Thus the remediation may involve land some distance from the source of the contamination if controlled waters have been affected; for example, the construction of barriers or extraction wells. **22.22**

The power is not, however, as wide as this wording might indicate. The things which are "appropriate" are defined by section 78N(4) by reference to each of the grounds on which the action is based. The authority will need to refer carefully to paragraphs (a)–(e) of section 78N(4) in that regard, as some paragraphs are much wider than others. In relation to paragraphs (b) and (c) in particular, reference will have to be made to the terms of the agreement or of the remediation notice, as appropriate. **22.23**

Recovery of costs

By section 78P(1) the enforcing authority which exercises its powers of remediation under section 78N is entitled in some but not all cases to recover the reasonable costs incurred in carrying out the remediation from the appropriate person or persons. Where there is more than one such person, the costs are recoverable in the proportions determined under section 78F(7). Those cases where costs may be recovered are paragraphs (a), (c), (e) or (f) of section 78N(3), namely: **22.24**

(a) action to prevent serious harm or pollution where there is imminent danger;
(c) where a recipient of a remediation notice has failed to comply;
(e) hardship cases (where part of the costs may be recoverable);
(f) where no appropriate person has been found after reasonable enquiry.

In cases (c) and (e) it should be reasonably clear who is the appropriate person, but in cases under (a) urgent action may have been taken before any remediation notice was served, or before the identity of the appropriate person was established. In such cases, the phrase "appropriate person" must mean the person determined to be appropriate under section 78F, and the authority will have to make that determination before it can seek to recover its costs. **22.25**

With regard to case (f), it might seem somewhat curious to give the power to the enforcing authority to recover its costs from an appropriate person in cases where no appropriate person can be found. However, there is presumably the possibility that the authority might, having made enquiry and failed to find an appropriate person, commence the works under section 78N, then later find that an appropriate person exists. **22.26**

As to cases not falling within section 78P, in situations falling within section 78N(3)(b) (where there is a written agreement for the authority to carry out the work) the authority will be able to recover its costs by virtue of the agreement. The real situation where "orphan sites" or "orphan shares" may arise seems to be under section 78N(3)(d) (*i.e.* where the **22.27** authority finds itself precluded from serving notice under sections 78J or 78K), which is discussed at Chapter 20.

Hardship and discretion

22.28 Recovery of costs is not an automatic process. The enforcing authority has a discretion to be exercised in accordance with section 78P(2) which requires the authority to have regard to:

(a) any hardship which the recovery may cause to the person from whom the costs are recoverable; and
(b) any guidance issued by the Secretary of State.

22.29 In considering these issues, it should be remembered that, in the first place, the only things which the enforcing authority may require to be done by way of remediation are things which it considers reasonable having regard to the cost involved.[31] Also, the Courts will no doubt be concerned to ensure that impecuniosity should not be too readily accepted simply on the basis of the assertions of the appropriate person, or by the production of a bank statement showing an overdraft.[32] The Oxford English Dictionary definition of "hardship" refers to "hardness of fate or circumstance; severe suffering or privation". As the Circular points out, there is a certain amount of case law on the expression.[33] The cases suggest that a wide meaning will be given to the word, to cover any matter of appreciable detriment, whether financial, personal or otherwise.[34] The term would thus seem to embrace the hardship inherent in facing worrying legal proceedings in appropriate cases. Also, the test seems likely to be an objective one, based on what the reasonable person would view as hardship[35]:

> "The rich gourmet who because of financial stringency has to drink *vin ordinaire* with his grouse may well think that he is suffering a hardship; but sensible people would say he was not".

22.30 Whilst the authority can only judge these issues at the time before serving the notice, it may be that subsequent changes of circumstance in terms of hardship would be taken into account on any appeal.[36]

---

[31] Section 78E(4).
[32] *Kent County Council v. Brockman* [1996] 1 P.L.R. 1.
[33] DETR Circular, Annex 2, paras 10.8–10.10.
[34] See *Purser v. Bailey* [1967] 2 Q.B. 500; *F G O'Brief v. Elliott* [1965] N.S.W.R. 1473; *Director-General of Education v. Morrison* (1985) 2 N.Z.L.R. 431; *Re Kabulan* (1993) 113 A.L.R. 330.
[35] *Rukat v. Rukat* [1975] 1 All E.R. 343 at 351 (Lawton L.J.).
[36] *Leslie Maurice & Co. Ltd v. Willesden Corporation* [1953] 2 Q.B. 1.

Guidance on hardship

22.31 In deciding whether and to what extent to seek to recover costs, the authority must have regard to guidance of the Secretary of State.[37] Part 3 of Chapter E of Annex 3 of the Guidance deals with cost recovery decisions. It sets out a number of general principles. It should be kept in mind that this is "traditional" guidance to which the authority must have regard, not prescriptive advice which it must follow. The approach is to set out "principles and approaches, rather than detailed rules".[38] In particular, the authority should aim for an overall result which is as just, fair and equitable as possible to all those who have to meet the costs of remediation, including local and national taxpayers.[39] It should also have regard to the "polluter pays" principle, which will involve considering the nature and degree of responsibility of the persons concerned.[40] In general, this means that the authority's starting point is that it should seek to recover all its costs, subject to the considerations outlined below.[41]

Deferred recovery

22.32 As a general point, the authority should consider whether it could maximise recovery by deferring receipt and securing the costs with a charge on the land under section 78P.[42] In such a case costs may be recovered either in instalments or when the land is sold. This may mean that less hardship will be involved than would be the case if immediate recovery were demanded.

Information on hardship

22.33 As hardship is essentially a personal issue for the appropriate person concerned, it is to be expected that they rather than the authority will have access to the information to support any such claim. As such, the onus is on the person seeking a reduction in cost recovery on hardship grounds to present any information needed to support that request.[43] The authority should of course consider such information, but should also give thought to the possibility of obtaining relevant information itself where it is reasonable to do so, give the practicalities of how the information is to be

---

[37] Section 78P(2).
[38] DETR Circular, Annex 3, para E.10.
[39] *ibid.*, para. E.11(a).
[40] *ibid.*, para. E.11(b).
[41] *ibid.*, para. E.12.
[42] *ibid.*, para. E.13.
[43] *ibid.*, para. E.14.

obtained, the cost of doing so, and the potential significance of the information.[44] To take a practical example, where an authority is faced with an individual owner who is clearly of limited resources, the authority must ask itself what real purpose would be served in devoting time and money to investigating the issue of hardship further. On the other hand, where the remediation actions required involve very substantial cost, and the authority is met with a claim by a polluter who is a major company that they should pay only a small proportion of that cost, or indeed nothing at all, on hardship grounds, then the authority may well be justified in making its own enquiries into the matter, and taking professional advice as to the validity of the company's hardship case. If it did not, the authority could properly be criticised for failing to take proper steps to safeguard the public purse, which would otherwise have to bear the cost of remediation.

Cost recovery policies

22.34 The Guidance suggests the possibility that local authorities may wish to prepare and adopt public policy statements about the general approach to be adopted in making cost recovery decisions.[45] This, the Guidance suggests, would promote transparency, fairness and consistency, by outlining circumstances in which the authority would waive or reduce cost recovery, having had regard to hardship and the statutory guidance. It is questionable, however, how advisable such a policy really would be. An authority cannot fetter its statutory role of considering the circumstances of every case, and the application of the guidance to those circumstances. It is difficult to see what the authority could sensibly say about its general approach without simply paraphrasing the Guidance. An adopted policy on cost recovery might simply turn out to be a hostage to fortune, as the basis for arguments that a remediation notice should not have been served at all in any given case. It is also suggested in the Guidance that any cost recovery policy should be taken into account by the Environment Agency, when acting as enforcing authority in relation to a special site in the area; this of course means that the Agency may be faced with inconsistent or at least different policies when acting in relation to sites in different local authority areas.

General considerations on cost recovery

22.35 The Guidance sets out a number of general considerations on cost recovery which apply to both Class A and Class B persons. These are:

(a) General parity of approach for all commercial enterprises, whether public corporations, limited companies, partnerships or sole traders (para. E.20).

---

[44] *ibid.*, para. E.15.
[45] *ibid.*, para. E.17.

(b) In the case of small and medium-sized enterprises (defined by reference to E.C. state aid guidelines relating to numbers of employees, turnover or annual balance sheet), the authority should consider whether the imposition of full cost recovery would mean that the enterprise is likely to become insolvent and cease to exist. If so, then where the cost to the local economy of closure would exceed the cost to the authority of having to clean up the land itself, the authority should consider waiving or reducing its costs to the extent necessary to avoid insolvency (paras E.21, E.22, E.24).

(c) However, the authority should not waive or reduce cost recovery where it is clear that the enterprise has deliberately arranged matters to avoid liability, or where insolvency appears likely irrespective of the costs of remediation, or where the enterprise could remain in, or be returned to, business under different ownership (para. E.23).

(d) Where the enforcing authority is a local authority, it may wish to take into account any policies it may have for assisting enterprise or promoting economic development, for example, for granting financial assistance under section 33 of the Local Government and Housing Act 1989 (para. E.25). The Guidance is not explicit as to how these policies might be relevant, but "financial assistance" under section 33 of the 1989 Act can include the remission of any liability or obligation. Such policies might be seen as operating in addition to or in conjunction with the guidance on insolvency of small and medium-sized enterprises as set out above, in that particular enterprises might be seen as particularly strategic so far as the local economic situation is concerned.

(e) Where the Environment Agency is the enforcing authority, it should seek to be consistent with the approach of the local authority in whose area the land is situated, should consult that authority and take its views into consideration (para. E.26). The Agency's perspective may naturally be wider than the local one, since the threat posed by the contaminated land may be to resources of regional or even national importance—for example, where a major aquifer is involved.

(f) Where trustees are appropriate persons, the authority should assume that they will exercise all the powers they have or may reasonably obtain to make funds available from the trust, or from borrowing on behalf of the trust, to pay for remediation. The authority should consider waiving the recovery of costs to the extent that such costs exceed the amount that could be made available from these sources (para. E.27). However, the authority should not waive or reduce cost recovery where it is clear that the trust was formed for the purpose of avoiding payment for remediation, or to the extent that the trustees have personally benefited from, or will benefit from, the trust. The focus of this Guidance is on the effect of cost recovery on the trustees of the trust, rather than the beneficiaries. However, there appears to be nothing to prevent those beneficiaries seeking to raise general arguments of hardship flowing from depeletion or exhaustion of trust funds.

(g) Since charities are intended to operate for the benefit of the community in the broad sense, the authority should consider the extent to which cost recovery might jeopardise the charity's ability to continue to provide a benefit or amenity which is in the public interest. If this is the case, the authority should consider waiving or reducing its cost recovery accordingly (para. E.29).

(h) Similarly, the authority should consider waiving or reducing cost recovery in relation to bodies such as housing associations eligible for registration as social housing landlords under section 2 of the Housing Act 1996, where the liability relates to land used for social housing, and where full recovery would lead to financial difficulties, such that provision or upkeep of the social housing might be jeopardised (paras E.30 and E.31).

22.36 With the exception of the guidance on trustees (at (f) above), these aspects of the Guidance involve a utilitarian balancing exercise of considering the consequences of remediation falling on the public purse as against the adverse social consequences which would ensue from pursuing full cost recovery. It is a different issue from that of personal hardship on the part of the appropriate person. Determining which is the lesser of the two evils may not be an easy matter, and could be one which requires expert advice, for example, in relation to the likelihood of corporate insolvency, and the quantification of the wider economic consequences of such insolvency. In these circumstances, authorities may well see a charging order, with deferred cost recovery, or recovery by instalments, as a relatively attractive option.

Cost recovery against Class A persons

22.37 Part 5 of Chapter E on cost recovery deals with the specific issues on cost recovery relating to Class A appropriate persons. Essentially there are two principles. The first is that in general the authority should be less willing to waive or reduce costs where the Class A person caused or knowingly permitted the contamination in the course of carrying on a business, on the basis that he is likely to have earned profits from the activity in question (para. E.35). In practice, naturally, most if not all contamination will have been caused by those carrying on a business.

22.38 The other point made in the Guidance on Class A persons related to the position where one Class A person (X) can be found, but there are others who also caused or knowingly permitted who cannot now be found, for example, because they were a company which has now been dissolved. In that case, the authority should consider waiving or reducing cost recovery if

X can demonstrate to its satisfaction that if the other Class A person could in fact be found, then the tests in the Guidance on exclusion or apportionment would have operated to exclude X from liability altogether, or to reduce significantly its proportion of the cost (para. E.35). Firm evidence will be needed from X as to the actual identity of the other Class A person or persons involved (para. E.36). This aspect of the Guidance is therefore essentially based on fairness to a Class A party who has acted in ways recognised as exculpatory by the Guidance, but who cannot rely on those circumstances because he is now the only Class A party who can be found.

Cost recovery against Class B persons

22.39 The Guidance on cost recovery against Class B appropriate persons at Part 6 of Chapter E is potentially important in practice. It needs to be recalled that such parties are liable not because of their past actions, but because of their present ownership or occupation of land. There are three main categories of situation envisaged here.

22.40 The first situation is where the costs of remediation exceed the value of the land in its current use after remediation: here the authority should consider waiving or reducing cost recovery if the appropriate person can show this is the case (para. E.39). The value to be considered is that of the land on the open market, post clean-up, but disregarding any possible residual diminution in value from blight. In general, the authority should seek to ensure that the costs of remediation which it recovers do not exceed the value of the land. It may, however, take into account whether remediation would result in an increase in the value of other land that the relevant person owns. As will be appreciated, this principle could in some cases represent a very significant restriction on the ability to recover costs, or from the other side of the coin, protection for the appropriate person concerned. It emphasises the importance of two factors discussed in earlier chapters: first, determining whether a given person falls within Class A or Class B and, secondly, determining exactly what land should be included within the remediation notice.

22.41 Another issue canvassed in the Guidance is that of the precautions taken by the Class B person before acquiring the land to determine whether it was contaminated. The issue of due diligence is not relevant for the purposes of the primary legislation, and in terms of the prescriptive Guidance on exclusion is relevant primarily to Test 3 ("Sold with information") as between Class A persons. However, on cost recovery, the authority should consider reducing the costs recovered against a Class B person who can show that he took such steps as were reasonable at the time before acquiring his interest to check for the presence of contaminants, that despite those precautions he was not aware of the contamination and could not reasonably have been expected to be so aware, and that taking into account the interests of national and local taxpayers it is reasonable that he should not bear the entire cost of remediation (para. E.43). The authority should bear in mind that what constitutes reasonable precautions will differ as between different types of transactions and as between buyers of different types. The Guidance here appears to contemplate the reduction of costs, rather than complete waiver. There may be very considerable variations in practice as to the discount given for due diligence, as well as the inherent uncertainty as to what could reasonably have been expected at some point in time, by way of investigation on the specific transaction and by the specific purchaser. It is possible that authorities who are under financial pressure may not be particularly sympathetic to arguments that they should bear the partial cost of clean-up where there is a solvent land owner, notwithstanding the adequacy of the investigations undertaken by that person. Where there were investigations which failed to detect contamination, the question may arise as to why the landowner should not seek to recover the costs from the consultants who failed to detect the contamination, or their insurers.

22.42 The third category concerns Class B persons who own or occupy dwellings. It is suggested that where such a person did not know of the contamination at the time of purchase, and could not reasonably be expected to have known of it, then there should be waiver or reduction of costs to the extent necessary to ensure he bears no more than is reasonable, having regard to his income, capital and outgoings (paras E.44 and E.45). This approach is intended to be applied only to the dwelling and its curtilage, not to any more extensive land which the person in question owns. Again, this is a highly subjective issue, but the Guidance suggests the possibility of applying a means-test approach analogous to that used for applications for housing renovation grant (HRG) under the Housing Renewal Grants Regulations 1996 (S.I. No. 2890), which considers income, capital, outgoings and allowance for any special needs. How practical a suggestion this will be remains to be seen, but this is of course an area where the local authority will need to have regard to the political acceptability of being seen to recover substantial sums from private individuals who have the misfortune to have bought a property sited on contaminated land.

Reasons for decisions

22.43 In all cases, the enforcing authority, whether the local authority or Agency, should inform the appropriate person of any cost recovery decisions taken, and explain the reasons for those decisions (para. E.16). This constitutes a demanding requirement to give reasons for what will in many cases ultimately be value judgments as to what is fair and reasonable in the circumstances. The authority will have to pay close attention to the Guidance, and explain how it has been taken into account and applied, or not as the case may be.

Recovery of costs: charging notices

22.44 Sections 78P(3)–(14) deal with charging notices for the recovery of costs by enforcing authorities under section 78P(1) where the authority has carried out the work itself. The charging notice provisions do not extend to Scotland.[46] The powers only apply where the relevant costs are recoverable from a person who is the owner of any premises which consists of or includes the contaminated land in question and who caused or knowingly permitted the substance or any of the substances to be in the land concerned.[47] Thus the scope of recovery is relatively narrow. A notice cannot be served on an occupier, or on an original polluter who is no longer owner. The procedure is initiated by serving a charging notice, which must specify the amount of costs claimed and state the effect of subsections (4), (7) and (8).[48] A copy must on the same date be served on every other person who, to the knowledge of the authority, has an interest in the premises capable of being affected by the charge.[49]

22.45 The cost specified in the charging notice carries interest at such reasonable rate as the enforcing authority may determine (probably 2 per cent above base rate) from the date of service of the notice until the whole amount is paid.[50] More problematic from the chargee's point of view is that the costs and accrued interest are charged on the premises.[51] Such charge takes effect at the end of a period of 21 days from service of the notice, unless an appeal is made.[52] Where any cost is a charge on premises under the section, the enforcing authority may by order declare the costs to be payable with interest by instalments within the specified period until the whole amount is paid.[53] The enforcing authority has the same powers and remedies under the Law of Property Act 1925, and otherwise, as if it were a mortgagee by deed, having power of sale and lease, of accepting surrenders of leases and of appointing a receiver.[54]

22.46 A person served with a charging notice or a copy of a charging notice may appeal against the notice to a County Court within a period of 21 days beginning with the date of service.[55] Regulations will provide the grounds on which such an appeal may be made and the procedure for dealing with appeals of this sort.[56]

---

[46] Section 78P(14). See paragraph 22–50 below.
[47] Section 78P(3).
[48] Section 78P(5).
[49] Section 78P(6).
[50] Section 78P(4)(a).
[51] Section 78P(4)(b).
[52] Section 78P(7).
[53] Section 78P(12).
[54] Section 78P(11).
[55] Section 78P(8). See paragraph 22–47 below.
[56] Section 78(P)10.

Appeals against charging notices

22.47 As mentioned above, a right of appeal lies against a charging notice under section 78P(8), within the period of 21 days beginning with the date of service. The right of appeal extends to the person served with the notice and to anyone served with a copy pursuant to section 78P(6) who has an interest in the premises capable of being affected by the charge, for example, a mortgagee or tenant-in-common. Appeal is to the County Court. On appeal, the Court may confirm the notice without modification, substitute a different amount, or order that the notice is to be of no effect.[57] The grounds for appeal and procedures for appeal are to be dealt with by Regulations.[58]

Scope and priority of charge

22.48 As mentioned above, a statutory charge under section 78P may be enforced by the powers and remedies of sale, lease and receivership available to a mortgagee by deed under the Law of Property Act 1925.[59] The statutory charge is a charge on "the premises" (subs. (6)) "consisting of or including" (subs.(3)(a)(i)) the contaminated land. This raises the question of what constitutes "the premises" where these may be greater in extent than the contaminated area—a possibility which subs. (3)(a)(i) acknowledges. Where a charging notice may be served, there appears to be nothing to prevent the enforcing authority service a notice charging the whole of any premises in the ownership of the same person, provided that those premises include the previously contaminated land. Such premises may be far more valuable than the part of it which is contaminated; this is therefore an important protection for the enforcing authority, as well as a possible major concern for mortgagees.

22.49 The second question is how the statutory charge rates in terms of priority with existing mortgages. If the wording were a charge "on the land," then on the authority of *Westminster City Council v. Haymarket Publishing Ltd*[60] it could be said that the charge is on all the estates and interests in the land, including prior mortgages. The wording here is "premises", but in fact this phraseology was used in earlier statutes and was held to have the effect of charging all proprietary interests in a series of cases followed in the *Westminster City Council* case.[61] On this basis, there seems a strong argument that the statutory charge will affect, and take

---

[57] Section 78P(9).
[58] Section 78P(10).
[59] Section 78P(11).
[60] [1981] 2 All E.R. 555.
[61] *See Birmingham Corpn v. Baker* (1881) 17 Ch.D. 782, *Tendering Union Guardians v. Dowton* [1891] 3 Ch. 265, *Paddington Borough Council v. Finucane* [1928] Ch. 567.

priority over, all existing mortgages, charges, options and other legal or equitable estates or interests in land; another serious concern for mortgagees. The mortgagee will at least have the comfort that the works will have improved the condition of the property. In practice, the mortgagee would probably have had to carry out those works at its own expense in any event in order to be able to exercise its power of sale.

Scotland

22.50 Although the cost recovery provisions of sections 78P(1)–(2) extend to Scotland, the charging powers in relation to property do not. However, there seems in principle no reason why these provisions should not have been extended to Scotland and amendments were introduced in Parliament to that effect albeit unsuccessfully.[62] Explaining its refusal to accept these amendments, the Government stated that existing mechanisms for recovering sums due under Scots law were adequate and that the amendments would require Scotland to change its system of conveyancing.[63] This reasoning appears to be highly unsatisfactory, as statutory charging orders are by no means a new concept in Scots law, as was indeed pointed out in Parliament.[64] For example, the Building (Scotland) Act 1959, Sched. 6; the Sewerage (Scotland) Act 1968; the Water (Scotland) Act 1980, s.65; the Civic Government (Scotland) Act 1983, s.102; and the Housing (Scotland) Act 1987, Sched. 9, all make provision for charging orders. Primarily these charges may be created by local authorities in relation to works carried out as, for example, under the Buildings (Scotland) Act 1959 and the Housing (Scotland) Act 1987. Since their existence has not required the reform of the system of conveyancing in Scotland, it is not easy to understand the Government's refusal to give Scottish local authorities the power to make charging orders for remediation costs in relation to contaminated land.

22.51 It may as a result of this difference be significantly more difficult for enforcing authorities to recover their costs in Scotland than in England and Wales, although this would to a degree have been true in any event: unlike the English provisions which as noted above give the enforcing authority the same power as a mortgagee under the Law of Property Act 1925, the charging order provisions in the Scottish statutes do not put the enforcing authority in the same position as a heritable creditor under the Conveyancing and Feudal Reform (Scotland) Act 1970. They merely provide, where the charge is appropriately registered, for the burdening of the property concerned with an annuity to pay the amount due. In the absence of contaminated land charging order provisions in Scotland, local authorities

---

[62] *Hansard*, H.L. Vol. 562 col. 250.

[63] *Hansard*, H.L. Vol. 562, col. 22; see also Standing Committee B, May 23, 1995, col. 368; and *Hansard*, H.C. Vol. 262, col. 959.

[64] Standing Committee B, May 23, 1995, col. 364.

will be required to rely on standard court procedures and methods of diligence such as inhibition and adjudication, which are not ideal. For example, inhibition may only be used against the owner of the property and is merely a means of preventing its sale until payment of the debt secured. Furthermore, adjudication involves a lengthy and complex process.

22.52 What it means in practice may that Scottish authorities, lacking any reliable power to secure costs, may be more inclined to seek immediate recovery, even where this would give rise to a degree of hardship.

"Signing off"

22.53 One practical aspect of enforcement is how the authority signifies its view that the remediation notice has been properly and adequately complied with. The statute itself does not provide for this, but the procedural description of the system in Annex 2 of the Circular suggests a possible procedure. This involves the authority writing to the person concerned, stating that it currently sees no grounds, on the basis of available information, for further enforcement action (or where remediation has been carried out prior to the service of any notice, no grounds for serving such a notice). The limitations of such a system in practice are obvious: should it transpire that in fact the land remains contaminated, so that further action is necessary, any such letter (which is likely to have been heavily qualified in any event) could not bind the authority not to comply with its statutory duties or estop it from doing so (at least so far as the current owner is concerned). Nonetheless, it may be anticipated that owners of land will seek such letters in practice, if only on the basis that some reassurance is better than none. There might also arise the question of a person who acquires the land on the basis of any such letter by the authority, and whether such a party might claim any cause of action for negligent mis-statement, or perhaps argue some form of estoppel against the authority in relation to its cost recovery powers (having changed his position in reliance on the information). These considerations mean that any authority which is minded to "sign off" under the Circular should consider very carefully the wording of the letter.

Chapter 23

APPEALS AGAINST REMEDIATION NOTICES

Summary

A 21-day period is allowed by section 78L within which to appeal against a remediation notice. Appeals against notices served by a local authority go to the magistrates' court, or in Scotland to the sheriff by way of summary application. Appeals against notices served by the Environment Agency or SEPA go to the Secretary of State or the Scottish Ministers in Scotland. Chapter 24 considers other possible means of attacking remediation notices.

Appeals against remediation notices: the right of appeal

23.01 By section 78L(1) there is a right of appeal against a remediation notice within the period of 21 days beginning with the day on which the notice is served. This appears to be a remarkably short period, but it has to be recalled that there is a period for prior consultation under section 78H between the authority and the person on whom the notice is to be served. The recipient is therefore unlikely to be taken entirely unawares, and in any event at least three months will normally have passed since the contaminated land was identified.

23.02 The appeal lies to the magistrates' court (or the sheriff in Scotland) in circumstances where the notice was served by a local authority. If the notice was served by the appropriate Agency, appeal is to the Secretary of State or the Scottish Ministers in Scotland (referred to collectively as the Secretary of State in this Chapter). Whilst appeal to the magistrates' court follows the tradition of public health and other similar legislation, it must be questioned whether the magistrates' court is a suitable forum for resolving such appeals, and whether the civil procedures in the magistrates' courts are adequate for this purpose. It also seems strange that there should be two entirely different modes and forms of appeal for ordinary remediation notices and for those relating to special sites.

Appeals regulations

23.03 Regulations may deal with a considerable number of matters in relation to appeals,[1] including the grounds on which appeals may be made, the right of further appeal against the decision of a magistrates' court or sheriff court, and the relevant procedure. No express mention is made in this context of grounds on which an appeal may be made against a decision of the Secretary of State, and it is therefore assumed that, there being nothing in the primary legislation, judicial review will lie against such decisions.

23.04 The Regulations may also deal with the circumstances in which the remediation notice is to be suspended until the appeal has been decided.[2] The obvious problem is that of significant harm already being caused, which may continue or even worsen during any suspension period. Logically the enforcing authority should use its own powers under section 78N to deal with this problem, to the extent necessary, and seek recovery of its costs from liable persons.

23.05 There is also provision for the Regulations to prescribe cases in which the decision on appeal may in some respects be less favourable to the appellant than the original remediation notice.[3] The Regulations may also prescribe cases in which the appellant can claim that a remediation notice should have been served on some other person and the procedure to be followed in those cases.[4]

23.06 The Regulations made under the above powers for England are the Contaminated Land (England) Regulations 2000. The relevant provisions are regulations 7–14.

Grounds of appeal

23.07 The grounds on which a person on whom a remediation notice is served may appeal under section 78L(1) are set out at regulation 7 of the Regulations and are:

23.08 (a) That in determining whether any land to which the notice relates appears to be contaminated land, the authority either: (i) failed to act in accordance with the Guidance; or (ii) unreasonably identified all or any of the land as contaminated, whether by reason of such a failure or otherwise. This ground will be relevant where there is a dispute as to the proper extent of the contaminated land, or as to the harm, risk or pollution presented by the land, or

---

[1] Section 78L(4).
[2] Section 78L(b).
[3] Section 78L(c).
[4] Section 78L(5)(d).

where it is suggested that the authority otherwise made a wrong decision (for example, by incorrect analysis or identification of contaminants). It should be noted however that, in cases otherwise than where failure to act in accordance with the Guidance is alleged (sub-ground (i)), the ground is not that the authority "wrongly" identified the land as contaminated, but that they "unreasonably" did so. What is contemplated may be therefore something narrower than a completely open review on the merits, but at the same time not a ground as narrow as '*Wednesbury*' unreasonableness.

(b) That, in determining a requirement of the notice, the authority either: (i) failed to have regard to the Guidance; or (ii) whether by reason of such a failure or otherwise, unreasonably required the appellant to do anything by way of remediation. This is the appropriate ground where there is dispute as to the remediation action or package specified in the notice, though the Guidance here has advisory rather than prescriptive status. **23.09**

(c) That the authority unreasonably determined that the appellant was the appropriate person to bear responsibility for remediation. This ground may apply where there are factual disputes as to the involvement of the appellant, or legal dispute as to the proper interpretation of the legislation on who is an appropriate person. However, again it should be noted that the test is whether the determination was unreasonable, not whether it was wrong; presumably any determination based on a wrong interpretation of the law must inevitably be unreasonable. **23.10**

(d) That the authority unreasonably failed to determine that some person in addition to the appellant was an appropriate person. This covers the situation where the authority failed to find some other appropriate person who was in existence. It is qualified by regulation 7(2), which limits its application to cases where the appellant claims to have found some other appropriate person within the appropriate liability class. It cannot therefore assist the appropriate person who simply alleges generally that there must have been some other appropriate person, but who cannot identify them. **23.11**

(e) That the authority failed to act in accordance with the Guidance on exclusion of appropriate persons under section 78F(6). This is simply an issue of whether the authority understood the Guidance correctly and applied it to the facts. It does not explicitly address the situation where it is alleged that the authority got the facts wrong; in those circumstances the appeal may have to lie under ground (c). **23.12**

(f) That the authority, where two or more appropriate persons were involved, either: (i) failed to act in accordance with the Guidance on apportionment under section 78F(7); or (ii) by reason of that failure or otherwise, unreasonably determined the proportion of the cost to be borne by the appellant. Apportionment disputes seem likely to be fertile sources of appeals. **23.13**

23.14 (g) That service of the notice contravened the requirements of section 78H(1)–(3) for the three-month moratorium, and for consultation before service.

23.15 (h) In cases where the authority served notice without consultation and within the three-month moratorium, that the authority could not reasonably have taken the view that there was imminent danger of serious harm or pollution falling within section 78H(4). The way in which this ground is worded looks more like a *Wednesbury* test—which may well be appropriate, given that the authority will have had to act urgently, balancing the interests of proper consultation against the public interest in the face of serious risk.

23.16 (i) That the authority has unreasonably failed to be satisfied, in accordance with section 78H(5)(b), that reasonable things are being or will be done by way of remediation, without any notice being served. This will be relevant where the appellant has put forward a proposed scheme which has been rejected as inadequate by the authority.

23.17 (j) That anything required by the notice was in contravention of the restrictions on liability applying in cases of water pollution under section 78J.

23.18 (k) That anything required by the notice was in contravention of the provisions of section 78K dealing with the escapes of substances between land.

23.19 (l) That the enforcing authority itself has power under section 78N(3)(b) to do what is appropriate. This applies where the appropriate person has entered into an agreement with the authority for it to carry out the work at his expense.

23.20 (m) That the authority has power under section 78N(3)(e) to carry out the work itself. This covers cases where it is alleged that a notice ought not to have been served on hardship grounds.

23.21 (n) That in considering hardship and cost recovery issues, the authority either: (i) failed to have regard to hardship, or to Guidance; or (ii) whether by reason of that failure or otherwise, unreasonably determined that it would seek to recover all its costs.

23.22 (o) That the authority failed to have regard to any guidance issued by the Environment Agency under section 78V.

23.23 (p) That any period specified in the notice for doing anything is not reasonably sufficient.

23.24 (q) That the notice imposes requirements on an insolvency practitioner in his personal capacity, contrary to section 78X(3)(a).

23.25 (r) That service of the notice contravened the provisions of section 78YB, dealing with the interaction of Part IIA and other regimes, and where appropriate it ought reasonably to have appeared to the authority that the relevant other powers might be exercised.

23.26 (s) That there has been some informality, defect or error in, or in connection with the notice, in respect of which there is no other ground of appeal. By regulation 7(3) where an appeal is made on

> the basis of informality, defect or error in, or in connection with, the notice, the appeal must be dismissed if the appellate authority is satisfied that the informality, defect or error was not material.

23.27 As will be appreciated the number and diversity of the grounds of appeal reflect the complexity of the Part IIA regime itself, with its restrictions on service of notices, exclusion and apportionment provisions. The grounds of appeal will need to be considered carefully in each case against the detailed facts to determine which of them may be applicable. The question of whether a remediation notice can be a nullity, in which case there is nothing to appeal against, is discussed in the next Chapter. However, the comprehensive nature of the grounds, especially the "catch-all" ground (s), mean that the scope for judicial review of notice is likely to be limited in practice.

Who can appeal?

23.28 By section 78L(1), an appeal can be brought only by a person on whom the remediation notice is served. A person on whom a copy of the notice was served has no statutory right of appeal. An example of such a person would be the current owner or occupier of the land, where the notice has been served on the original polluter. The current owner or occupier, if they are aggrieved by the land being subject to a remediation notice, or by its terms, and do not wish to leave conduct of any appeal exclusively in the hands of the original polluter, will have to rely on challenging the notice by way of judicial review.

Appeal to magistrates' court

23.29 Appeals in relation to sites other than special sites lie to the magistrates' courts (to the Sheriff Court in Scotland). Regulation 8 of the Appeals Regulations deals with the process of making an appeal to a magistrates' court. Such an appeal shall be by way of complaint for an order and (subject to specific provisions on quashing, confirming and correcting notices) the Magistrate's Court Act 1980 shall apply to such proceedings. The procedure is therefore one of complaint and summons under section 51 of that Act.

23.30 By Rule 4 of the Magistrate's Court Procedure Rules 1981[5] a complaint may be made by the complainant in person or by his counsel or solicitor or any other person authorised on his behalf; and subject to the provisions of any other Act, need not be in writing or on oath. In practice, however, a

---

[5] S.I. 1981 No. 552, as amended.

complaint instituting an appeal under Part IIA is likely to be in writing, and should follow the form of complaint at Form 98 of the Magistrates' Court (Forms) Rules 1981.[6]

23.31 As well as making a complaint in the normal way, the appellant must comply with additional requirements under regulations 8(2) and (3) of the Regulations as follows:

(1) At the same time as making the complaint, the appellant must file a "notice of appeal" which states his name and address and the grounds on which the appeal is made.
(2) One copy of the remediation notice must be filed with the notice of appeal. By regulation 8(6), in this context, "file" means to deposit with the justices' clerk.
(3) As well as filing the notice of appeal the appellant must serve a copy of it on:

   (a) the enforcing authority;
   (b) any person named on the notice of appeal as an appropriate person;
   (c) any person named in the notice of appeal as an appropriate person; and
   (d) any person named in the remediation notice as the owner or occupier of the whole or any part of the land.

23.32 As a matter of normal practice the justices or clerk will have to apply their mind to the complaint and make a judicial decision as to whether to issue a summons. This will involve considering whether, prima facie, there is power to make the order sought,[7] whether the court has jurisdiction, and whether any applicable time limits have been complied with.

Appeal to sheriff

23.33 In Scotland, appeals in relation to sites other than special sites lie to the sheriff court by way of summary application. Summary applications are governed by sections 3(p) and 50 of the Sheriff Court (Scotland) Act 1907 and the Act of Sederunt (Sheriff Court Summary Application Rules) 1993 (SCSAR 1993).[8] However, the 1907 Act provides that nothing contained within its provisions affects any right of appeal provided by any Act of Parliament under which a summary application is brought. Therefore, the

---

[6] S.I. 1981 No. 553. The Form is Form 98 at *Stones Justice Manual*, Pt IX.
[7] This should not be a problem since jurisdiction to determine the appeal and to quash or confirm the notice stems from section 78L and from the Appeals Regulations.
[8] S.I. 1993 No. 3240.

Appeals Regulations (when these are made for Scotland) will take priority over the 1907 Act.

23.34 An appeal should be made by initial writ in the form of Rule 4 and Form 1 of the SCSAR 1993. It must be signed by the appellant or his solicitor (if any).[9] The title and section of the statute under which the appeal is brought should be stated in the heading of the writ.[10] Even though the person making the application is an appellant, he is to be known as the pursuer and the respondent is known as the defender.[11] The writ must contain an article of condescendence setting out the ground of jurisdiction and the facts upon which such ground is based.[12] Where the respondent's residence, registered office or place of business is not known and cannot reasonably be ascertained, the appellant must set out in the instance that the whereabouts of the respondent are not known and make averments in the condescendence indicating the steps that have been taken to ascertain his current whereabouts.[13] The writ should contain a crave which sets out the decree or order the appellant is seeking from the court, articles of condescendence setting out the facts which form the ground of action and pleas in law which would contain the legal basis (drawn from the Appeals Regulations) for the appeal.[14] The writ must contain averments about those persons who appear to the appellant to have an interest in the application and in respect of whom a warrant for citation is sought.[15]

23.35 The sheriff clerk or sheriff may sign a warrant for citation.[16] The forms of the warrant and the citation itself are prescribed by the SCSAR 1993.[17] Where the sheriff clerk refuses to sign a warrant, the appellant may apply to the sheriff to sign the warrant.[18] In deciding whether or not to sign a warrant, the sheriff clerk or sheriff will consider whether the court has jurisdiction and whether applicable time limits have been complied with.[19]

Appeal to Secretary of State

23.36 By regulation 9 of the Regulations, an appeal in respect of a special site made to the Secretary of State must be made by notice of appeal, which shall state[20]:

---

[9] SCSAR 1993, rule 4(7).
[10] SCSAR 1993, rule 4, App, Form 1. *Hutcheon v. Hamilton District Licensing Board*, 1979 S.L.T. (Sh.Ct) 44.
[11] SCSAR 1993, rule 2(1).
[12] SCSAR 1993, rule 4(5). However, for the sake of convenience, they are referred to as the appellant and respondent hereafter.
[13] SCSAR 1993, rule 4(6).
[14] SCSAR 1993, App., Form 1.
[15] SCSAR 1993, rule 4(8).
[16] SCSAR 1993, rule 7(1).
[17] SCSAR 1993, rule 7(4). App., Forms 2 and 3.
[18] SCSAR 1993, rule 7(2).
[19] SCSAR 1993, rule 4(5).
[20] Regulation 9(1).

(a) the name and address of the appellant;
(b) the grounds on which the appeal is made; and
(c) whether the appellant wishes the appeal to be in the form of a hearing or to be decided on the basis of written representations.

23.37 The requirements as to lodging a copy of the remediation notice and as to serving a copy of the notice of appeal on the enforcing authority and on relevant third parties, which apply in the case of appeals to the magistrates, are also applied to appeals to the Secretary of State, with the appropriate modifications.[21] Additionally, the appellant must include with the notice of appeal a statement of the names and addresses of all persons (other than the Agency) on whom he is serving a copy of the notice and must serve a copy of the remediation notice on the Secretary of State and on any person named in the appeal as an appropriate person who is not named in the remediation notice.[22]

23.38 The Regulations make provision for abandonment of appeals to the Secretary of State.[23] Notification to that effect is to be given to the Secretary of State. The appeal is then treated as abandoned on the date the Secretary of State receives the notification. Where the appeal is abandoned, the Secretary of State shall give notice to any person on whom the notice of appeal was required to be served.[24] However, the Secretary of State may refuse to allow abandonment of the appeal where he has already notified the appellant of a proposed modification to the notice.[25]

Suspension of notices pending appeal

23.39 Where an appeal is duly made against a remediation notice the notice is of no effect pending the final determination or abandonment of the appeal.[26] The effect is therefore that no offence is committed by failing to comply with the suspended notice, nor can the authority take other enforcement action. In order to have this effect, the appeal must be "duly made"; that is to say it must comply with the appropriate procedural requirements.[27] Perhaps, surprisingly, there is no exception to this rule in cases where it appears there is imminent danger of serious harm or pollution, and the notice recites those facts.[28]

---

[21] Regulation 9(2).
[22] Regulation 9(2)(b).
[23] Regulation 9(3).
[24] Regulation 9(5).
[25] Regulation 9(4).
[26] Regulation 14(1).
[27] Regulation 14(2).
[28] Earlier drafts of the Regulations did contain such provision.

Procedure for joining third parties

23.40 One of the potential shortcomings of the appeals procedures is the lack of any clear mechanism for involving third parties. Obviously, if the appellant is arguing that someone else other than the appellant is the appropriate person to be responsible for remediation, or that the appellant is entitled to benefit from some exclusion of liability to the detriment of another, or that the appellant's apportioned share should be less and some other person's commensurately greater, then the other person or persons involved have a direct interest in the outcome of the appeal. It would be contrary to natural justice, and quite probably the Human Rights Act 1998, for the issues to be determined without giving them adequate notice, time to prepare, the opportunity to submit evidence, cross-examine and present a case.

23.41 As explained above, the Regulations at least contain procedures whereby such persons are notified that an appeal has been made, and of the grounds of the appeal.[29] For appeals to the Secretary of State there is a right for such persons to be heard at a hearing[30] (and presumably to make written representations). For appeals to magistrates, there is a requirement for such persons to be given notice of the hearing, and of any hearing for directions, and an opportunity to be heard at such hearings.[31]

23.42 Beyond this there are no clear procedural rules (in England and Wales at least) governing the involvement of such third parties, either at the preliminary stages of the appeals process, or at the appeal hearing itself. In practice, therefore, this will be a matter for the magistrates, the sheriff, the inspector or reporter (as appropriate) to determine as part of the procedure of the relevant tribunal. In principle, the procedures should be capable of accommodating such third parties though complex issues may well arise as to how to structure the appeal. Structurally, there may be a situation where there are two or more appeals which are heard jointly, or where there is a single appeal in which a third party participates.

23.43 The position in Scotland may be somewhat clearer in that the Sheriff Court Summary Application Rules[32] make provision for:

(a) the initial writ to include averments about those persons who appear to the pursuer to have an interest in the application and in respect of whom a warrant for citation is sought[33]; and
(b) orders by the sheriff for intimation to any person who appears to him to have an interest in the summary application.[34]

---

[29] Regulation 8(2) and 9(2).
[30] Regulation 5(2).
[31] Regulation 8(3)(b).
[32] S.I. 1993 No. 3240.
[33] Rule 4(8).
[34] Rule 5.

Joined appeals

23.44 Where a number of persons have been served with remediation notices in respect of the same land, they may well all wish to appeal on the same ground, for example that the land was not contaminated land at all, or that the requirements of the notice were excessive.

Although the Regulations do not make any express provision for appeals being joined and determined together, this must in many cases be the sensible course, and there seems no reason why it cannot be done.

Powers of appellate authority

23.45 By section 78L(2) on any appeal against a remediation notice the appellate authority is required to quash the notice, if satisfied that there is a material defect in it; but subject to that duty, has a discretion to confirm the notice, with or without modification, or to quash it. Where the notice is confirmed, the appellate authority may extend the period for complying with the notice.[35]

23.46 Before modifying the notice in any way which would be less favourable to the appellant or any other person on whom the notice was served, the appellate authority must:

(a) notify the appellant and other persons involved in the proposed modification;
(b) permit those notified to make representations; and
(c) permit the appellant or any other person on whom the notice was served to be heard if they so request.[36] The enforcing authority is also entitled to be heard in such circumstances.[37]

23.47 In so far as an appeal against a remediation notice is based on the ground of some informality, defect or error in the notice, or in connection with it, the appellate authority shall dismiss the appeal if satisfied that the informality, error or defect was not a material one.[38] The reference to material defects is similar language to that used in planning legislation prior to 1981 in relation to enforcement notices.[39] Applying case law on that wording, the issue may well be whether the defect is such as to result in injustice.[40]

---

[35] Section 78L(3).
[36] Regulation 12(1).
[37] Regulation 12(2).
[38] Regulation 7(3).
[39] See S. Tromans and R. Turrall-Clarke, *Planning Law, Practice and Precedents*, para. 9.30.
[40] *Miller-Mead v. Minister for Housing and Local Government* [1963] 2 Q.B. 196, considered in *Simms v. Secretary of State for the Environment and Broxtowe Borough Council* [1998] P.L.C.R. 24.

Magistrates' court jurisdiction

23.48 In accordance with ordinary principles, a magistrates' court has no jurisdiction except in so far as it is conferred expressly by statute, and the territorial jurisdiction of a magistrate will extend only to the Commission area for which they are appointed. In the absence of any express provision in Part IIA or the Regulations as to which magistrates' court shall have jurisdiction, a court will have jurisdiction under section 52 of the Magistrates' Courts Act 1980 if the complaint relates to anything done or left undone within the commission area for which the court is appointed, or relates to any other matter arising within that area. The key issue will therefore be within which area the contaminated land in question is situated.

Magistrates' court procedures

23.49 An appeal against a remediation notice will be part of the civil jurisdiction of the court under Part II of the 1980 Act. As mentioned above, the procedure is therefore one of complaint and summons under section 51 of the Act. Regulation 8(4)(a) gives a general power to the magistrates and the justices' clerk to give directions for the conduct of the proceedings, including the timetable, service of documents, submission of evidence and order of speeches.[41]

23.50 Section 53 of the 1980 Act and Rule 14 of the Magistrates' Courts Procedure Rules 1981 provide for the procedure, order of evidence and submissions on hearing complaints. Whilst these provisions are no doubt subject to directions given under the specific discretion referred to in the previous paragraph, they can be summarised as follows:

23.51

1. The court shall, if the defendant appears, state to him the substance of the complaint (section 53(1)).
2. The court may determine under section 53(3) of the 1980 Act to make the order with the consent of the defendant without hearing evidence. However, this power appears to apply only to sums recoverable summarily or to other matters prescribed: it therefore will not apply to statutory appeals under Part IIA. Evidence must therefore be heard.
3. Otherwise, the complainant/appellant shall call his evidence, and before doing to may address the court. Witnesses may be called in whatever sequence is chosen by the advocate.
4. At the conclusion of the evidence for the complainant/appellant, the defendant/enforcing authority may address the court, whether

[41] Delegation of these functions is dealt with by regulation 8(5), albeit in a rather oblique fashion.

or not they afterwards call evidence. The court may stop the case at the close of the complainant's evidence, whether upon a submission to that effect or on their own motion.[42]

5. At the conclusion of the evidence, if any, for the defendant/enforcing authority, the complainant/appellant may call evidence to rebut that evidence.[43]
6. At the conclusion of the evidence for the defence/enforcing authority, and the evidence (if any) in rebuttal, the defendant/enforcing authority may address the court if they have not already done so.
7. Either party may, with the leave of the court, address the court a second time, but where the court grants leave to one party it shall not refuse leave to the other.
8. Where the defendant/enforcing authority obtains leave to address the court for a second time, its second address shall be made before the second address, if any, of the complainant/appellant.
9. The court, after hearing the evidence and the parties, shall make the order for which the complaint is made or dismiss the complaint (section 53(2)).
10. The form of order is later drawn up in writing in one of the prescribed forms as is appropriate to the case.[44]

Appeal procedures: sheriff court

23.52 The procedure is principally governed by section 50 of the Sheriff Courts (Scotland) Act 1907 ("the 1907 Act") which provides that:

> "In summary applications (where a hearing is necessary) the sheriff shall appoint the application to be heard at a diet to be fixed by him, and at that or any subsequent diet (without record of evidence unless the sheriff shall order a record) shall summarily dispose of the matter and give his judgment in writing: . . ."

23.53 The warrant of citation on the initial writ will specify the date and time when the respondent is to appear in the sheriff court.[45] The sheriff has a very wide discretion as to the nature of the procedure prior to and at the hearing.[46] In particular, he may make any order he thinks fit for the

---

[42] *Mayes v. Mayes* [1971] 1 W.L.R. 679.

[43] The use of this provision to call rebuttal evidence could be problematic. The object, presumably, is to allow the complainant/appellant to deal with specific issues on which it has been taken by surprise, not to completely reorganise its case and have "a second bite at the cherry".

[44] Magistrates' Court Rules 1981, r.16(1).

[45] Act of Sederunt (Sheriff Court Summary Application Rules) 1993 (S.I. 1993 No. 3240) (SCSAR 1993), rule 7(4)(a), App., Form 2.

[46] Sheriff Court (Scotland) Act 1907, s.50; *O'Donnell v. Wilson*, 1910 S.C. 799.

progress of the application as long as it is not inconsistent with section 50 of the 1907 Act.[47]

23.54 The case will normally call first on either the summary applications roll or the procedure roll.[48] If the appellant is neither present nor represented at the first calling the application should not be dismissed but simply dropped from the roll.[49] A motion may then be lodged by the appellant to enrol the application for further procedure. If the respondent is neither present nor represented at the first calling, although it is open to the appellant to move for decree, a summary application which is an appeal cannot generally succeed by default.[50]

23.55 Assuming both parties are present or represented at the first calling, the sheriff will usually require the respondent to lodge answers to the application within a specified time. He will also normally allow the parties to adjust the pleadings for a reasonable period, usually up to 14 days, before the hearing.[51] Where this is done, the appellant will be ordained to lodge a record of the pleadings as adjusted by a specified time, normally 7 days, prior to the hearing. At the first calling the sheriff should ascertain from the parties if the hearing is to take the form of a debate or proof.[52] It is likely that evidence will be heard in all appeals under Part IIA although it should be noted that any proof will be restricted to the averments made by the parties in the pleadings.[53] No record of evidence will be made unless the sheriff so orders.[54] However, where there is a right of appeal on a question of fact from the sheriff court, the evidence should generally be recorded.[55] If the evidence is not recorded, the sheriff's findings in fact will not be open to review[56] and any appeal will be restricted to questions of law alone.[57]

23.56 Although section 50 of the 1907 Act requires the sheriff to dispose of the application summarily this does not mean that normal rules of evidence, recovery of documents or normal principles of civil litigation can be set aside. Witnesses may be called in whatever sequence is chosen by the

---

47 SCSAR 1993, rule 32.

48 Nicolson & Stewart, MacPhail's Sheriff Court Practice (2nd ed, 1998), para. 25–20.

49 This approach was approved in *Saleem v. Hamilton District Licensing Board*, 1993 S.L.T. 1092.

50 See, *e.g. Muir v. Glasgow Corporation*, 1955 S.L.T. (Sh.Ct) 23; *Ladbrokes the Bookmakers v. Hamilton District Council*, 1977 S.L.T. (Sh.Ct) 86; *Charles B. Watson (Scotland Ltd v. Glasgow District Licensing Board*, 1980 S.L.T. (Sh.Ct) 37 and *Saleem v. Hamilton District Licensing Board, supra*, note 49.

51 Nicolson & Stewart, MacPhail's Sheriff Court Practice (2nd ed, 1998), paras 25–23.

52 Nicolson & Stewart, *op. cit.*, paras 25–24.

53 *Rank Leisure Services Ltd v. Motherwell and Wishaw Burgh Licensing Court*, 1976 S.L.T. (Sh.Ct) 70, *per* Sh.Pr. Reid at p. 73.

54 Sheriff Court (Scotland) Act 1907, s.50.

55 *Director-General of Fair Trading v. Boswell*, 1979 S.L.T. (Sh.Ct) 9.

56 *Sinclair v. Spence* (1883) 10 R. 1077.

57 *United Creameries Co. v. Boyd & Co.*, 1912 S.C. 617, at 623.

solicitor or advocate and may be examined, cross-examined and, where appropriate re-examined. At the close of evidence, legal submissions will be made by the appellant and respondent. Section 50 of the 1907 Act envisages that the hearing may be adjourned to a subsequent diet.

23.57 Section 50 of the 1907 Act requires the sheriff to give his decision in writing. This means that the sheriff will need to make avizandum although given that section 50 also requires the sheriff to dispose of the application summarily, it also implies that his written judgment must be produced as soon as possible. Although the form of the judgment is at the discretion of the sheriff it is recommended that where an application is contested or where there is a right of appeal, the sheriff should append a note of the grounds of the judgment to his interlocutor[58] and, where evidence has been led, the sheriff should include findings in fact and law in his judgment.[59] In all cases the sheriff's interlocutor must specify which plea in law is sustained as the ground of judgment.[60] In all appeals which seek an order in terms of a statutory provision, the sheriff must use the words of the statute since paraphrase or misquotation may lead to an appeal on the ground of error in law.[61]

Appeal procedures: Secretary of State

23.58 Appeals to the Secretary of State are by section 78L(6) subject to the provisions of section 114 of the Environment Act 1995. This allows the Secretary of State to appoint any person to exercise the appellate functions on his behalf. Appeals may therefore be referred to the Planning Inspectorate or in Scotland to the Scottish Office Inquiry Reporters' Unit, either for report to the Secretary of State, or to determine the appeal. Schedule 20 of the 1995 Act applies to such appeals.[62] This deals with the appointment process (which must be in writing),[63] the powers of appointed persons,[64] the holding of local inquiries and hearings,[65] and evidence and costs.[66]

23.59 The Guidance on the Regulations contained at Annex 4 to the Circular indicates (paras 60–63) that most cases will be delegated to inspectors, but that each "special site" case will be looked at individually to see if jurisdiction should be recovered by the Secretary of State. Cases which involve sites of major importance, or which have more than local signifi-

[58] Nicolson & Stewart, MacPhail's Sheriff Court Practice (2nd ed, 1998), paras 25–27.
[59] *Strathclyde Regional Council v. T*, 1984 S.L.T. (Sh.Ct) 18.
[60] SCSAR, App, Form 1 expressly provides for pleas-in-law.
[61] *Lanark County Council v. Frank Doonin Ltd*, 1974 S.L.T. (Sh.Ct) 13.
[62] Section 114(4).
[63] Sched. 20, para. 2.
[64] Paras 3 and 7.
[65] Para. 4.
[66] Para. 5.

cance, which raise significant local controversy, or give rise to difficult legal points or major, novel issues, will be the categories most likely to be recovered.

23.60 The options for determining an appeal under the Schedule are therefore to determine it on written representations only, to hold a hearing, or to hold a public local inquiry. Either party can require a hearing[67]; regardless of the wishes of the parties, the appointed person may hold a public local inquiry or other hearing, and must hold a local inquiry if the Secretary of State so directs.[68]

23.61 Where a local inquiry or hearing is held, an assessor may be appointed by the Secretary of State to sit with the inspector or reporter in an advisory capacity.[69] This is likely to be a useful procedure for technically complex appeals, which may raise highly specialist issues.

23.62 Under regulations 10–11 of the Regulations, the following procedures apply to appeals to the Secretary of State:

1. Before determining the appeal, the Secretary of State may if he thinks fit:
   (a) cause the appeal to take or continue in the form of a hearing (which may be held wholly or partly in private); or
   (b) cause a public local inquiry to be held.[70]
2. The Secretary of State shall act as mentioned in (a) or (b) above if a request is made to be heard by either party to the appeal.
3. The persons entitled to be heard are the appellant, the Environment Agency and any person on whom the appellant was required to serve a copy of the notice of appeal.[71] Others may be permitted to be heard by the inspector, and such permission shall not be unreasonably withheld.[72]
4. After the conclusion of a hearing, the person appointed to conduct it shall (unless appointed to determine the appeal) make a report in writing to the Secretary of State, which shall include his conclusions and his recommendations or his reasons for not making any recommendations.[73]
5. The Secretary of State must notify the appellant of his decision and provide a copy of the inspector's report.[74]
6. At the same time, the Secretary of State shall send a copy of the determination and report to the appropriate Agency and to any other person on whom the appellant served a copy of his notice of appeal.[75]

---

[67] Para. 4(1).
[68] Para. 4(2).
[69] Para. 4(4).
[70] Regulation 10(1).
[71] Regulation 10(2).
[72] Regulation 10(3).
[73] Regulation 10(4).
[74] Regulation 11(1).
[75] Regulation 11(2).

23.63 It will be appreciated that these rules provide only the bare bones of procedure. The three possible forms of determination—written representation, informal hearing and public local inquiry—correspond to those applicable in planning appeals, and it is reasonable to expect that the Secretary of State and planning inspectorate or the Scottish Office Reporters' Unit will at least to some extent seek to follow the principles of those tried procedures.[76] This certainly appears to be the case from the descriptions of the normal procedures for written representations, hearings and inquiries set out at paras 65–70 of Annex 4 to the Circular. In particular, parties should consider the *Code of Good Practice at Planning Inquiries* which appears at Annex 5 of Circular 15/96 as to the preparation and exchange of proofs of evidence and document management. The equivalent guidance in Scotland is the Scottish Office Development Department Circular 13/1997, *Tribunals and Inquiries Act 1992—Planning Inquiries and Hearings: Procedures and Good Practice.*

23.64 In the less formal context of informal hearings, it needs to be remembered that the rules of natural justice remain paramount, and that the use of such procedure places an onus on the inspector to adopt a fair, thorough and inquisitorial approach.[77] Indeed, given the disputed issues of fact and of expert evidence which may arise on Part IIA appeals, it is questionable whether any procedure other than formal cross-examination is likely to provide sufficient rigour, other than in relatively straightforward cases.

23.65 Similarly, if written representations are used, considerations of fairness and natural justice will be relevant,[78] though the courts are more likely to look to the substance of whether the relevant parties have been given a fair opportunity to consider the relevant material and to comment upon it.[79]

Alternative dispute resolution and appeals

23.66 Appeals under section 78L are likely in many cases to be lengthy, complex and expensive. The novelty of the legislation and the absence of full procedural rules for appeals means that there is potential for time being wasted in arriving at appropriate procedures, agreeing facts and identifying the key issues. Magistrates and planning inspectors alike will, to begin with, lack experience in determining some of the issues likely to arise; this may make the outcome of the appeals process unpredictable.

---

[76] See generally, S. Tromans and R. Turrall-Clarke, *Planning Law, Practice and Precedents* (London, Sweet & Maxwell), Chapter 6.

[77] *Dyason v. Secretary of State for the Environment* [1998] J.P.L. 778.

[78] See *Geha v. Secretary of State for the Environment* [1994] J.P.L. 717.

[79] *Stockton on Tees Borough Council v. Secretary of State for the Environment* [1988] J.P.L. 834; *Parkin v. Secretary of State for the Environment and East Devon District Council* [1993] J.P.L. 141.

23.67 All of these factors mean that the professional advisers to the parties to an appeal would do well to consider whether alternative dispute resolution (ADR) techniques may have a role to play in the appeal process. Whilst any attempt at ADR will have to be without prejudice to the formal appeal process, it may at the very least assist in clarifying the issues and the relative positions of the parties. Issues of what exactly is required by way of remediation and issues of apportionment of liability as between appropriate persons may be examples of matters which would be particularly susceptible to the use of ADR. However, it has to be remembered that one of the parties is a public body, acting in the public interest; the role that ADR can play will ultimately be limited accordingly.

23.68 It should also be remembered that a dispute on appeal will not necessarily simply lie between the appellant and the enforcing authority: the dispute may be between a number of potential appropriate persons as to which of them should bear responsibility, or in what proportions such responsibility should be shared. There may be advantages in those parties engaging in ADR in parallel with any formal appeal, and thus reaching their own solution rather than having one imposed by magistrates or the Secretary of State. The Guidance itself recognises the ability of appropriate persons to agree the allocation of costs between themselves.

Evidence generally

23.69 Normal principles of evidence will need to be observed in the Magistrates' Court and in the Sheriff Court. The issue of the burden of proof is one that will need to be considered carefully at various stages. It may shift; for example where the authority can show that X was the owner of land at a particular time, it may be a reasonable inference in the absence of any other information that he was the person in a position to control the land, the evidential burden then falling on him to show otherwise.[80]

23.70 The evidence of witnesses will generally require to be given on oath, and the originals of private documents (such as transfers of title, leases, contracts, etc.) will need to be produced. Documents which are records of a business or public authority (for example as to the period of operation of a particular site or facility) will be subject to section 9 of the Civil Evidence Act 1995 and may be received in evidence without further proof if the appropriate certificate signed by an officer of the business or authority is produced. As the appeal constitutes civil proceedings, the general rule will be that he who asserts must prove: thus, for example, the appellant who seeks to show that he is entitled to an exclusion from liability will have to prove that the requirements of the exclusion test are in his case satisfied.

23.71 If seeking to attack the remediation notice on grounds of a technical or scientific nature, the appellant will no doubt call one or more expert witnesses, since the matters at issue, such as risk assessment or remedia-

[80] *Nourish v. Adamson* [1998] J.P.L. 859, p. 863.

tion techniques, will be outside the experience of the court.[81] The normal principles of such evidence will apply, so that the views of the expert should be independent and will be subject to the duties owed by the expert to the court.[82] Evidence of published articles, etc., given by someone who is not an expert in the field may be regarded as inadmissible hearsay.[83] There is no absolute bar on an officer or employee of the authority or Agency giving their views as an expert witness, but the authority will need to be prepared to satisfy the tribunal that the individual is capable of giving expert evidence, and is familiar with the requirement of objectivity.[84]

23.72 The general principle is that justices may take into consideration matters which they know from their own knowledge, particularly with regard to the locality, but such knowledge must be properly applied within reasonable limits.[85] Magistrates should be aware that their knowledge may not be sufficiently precise to be sure as to exactly which company occupies land or conducts an undertaking.[86] Where a magistrate does happen to have specific knowledge of the history of a local site, it is possible that such specific knowledge (as opposed to general knowledge of the locality) can be brought into account: "There is no doubt that justices are entitled . . . to use their own local knowledge of what has been going on as to the use of a particular piece of land".[87] However if such knowledge is used it must be done with notice to the parties, so they have the opportunity to consider it.[88] Further, there is nothing to prevent magistrates appointing a suitable assessor to assist them in assimilating technical material. However, there is the important question as to who should bear the cost of appointing any assessor.

Evidence in Secretary of State appeals

23.73 Whereas an appeal to the magistrates' court or the Sheriff Court is part of the civil jurisdiction of those courts, and as such is subject to formal rules of evidence, appeals to the Secretary of State are administrative proceedings. Hearsay evidence and documentary evidence will therefore be admissible, though the issue will arise of what weight is to be attached to it.[89]

---

[81] See *Stones Justices Manual*, para. 2–343. Matters such as harm to health will fall into this category—see *London Borough of Southwark v. Simpson, New Law Publishing*, November 3, 1998; *O'Toole v. Knowsley Metropolitan Borough Council* [1999] E.H.L.R. 420.

[82] *The Ikarian Reefer* [1993] 2 Lloyd's Rep.68.

[83] *London Borough of Southwark v. Simpson* (above).

[84] *Field v. Leeds City Council, The Times,* January 18, 2000.

[85] *Stones Justices Manual*, para. 3034.

[86] *Norbrook Laboratories Ltd v. HSE* [1998] E.H.L.R. 207; *Millen v. Hackney London Borough Council* [1997] 2 All E.R. 906.

[87] *Bowman v. DPP* [1991] R.T.R. 263, 269 (Watkins L.J.).

[88] *ibid.*

[89] *Knights Motors v. Secretary of State for the Environment and Leicester City Council* [1984] J.P.L. 584.

23.74 The inspector, or reporter, will be entitled to bring to bear his experience and such technical knowledge as he possesses.[90] However, this must be based on evidence, and issues of natural justice will arise if the parties are not given the opportunity to comment on the relevant issues.[91] The issue of consultation with other Government departments or agencies will also be important, as their views may carry considerable weight.

23.75 The same principles govern the role of expert witnesses as in civil proceedings—that is to say, the role of the expert is to draw the attention of the inspector to material considerations, not to argue his own client's case.[92] Legal advisers should be mindful of this when working with the experts on their evidence.

Costs: magistrates court

23.76 By section 64 of the Magistrates' Courts Act 1980, on the hearing of a complaint, a magistrates' court has power in its discretion to make such order as to costs

(a) on making the order for which the complaint is made (*i.e.* quashing the remediation notice on appeal) for costs to be paid by the defendant/enforcing authority to the complainant/appellant;

(b) on dismissing the complaint/appeal, for costs to be paid by the complainant/appellant to the defendant/enforcing authority;

as it thinks just and reasonable. The amount of any such sum should be specified in the order.[93] The parties should have available the material upon which the court can make its determination on costs.[94]

23.77 Given the scale and complexity of appeals against remediation notices, particularly where a number of parties are involved, the costs may be high. The issue arose in *R. v. The Southend Stipendiary Magistrate, ex p. Rochford District Council*[95] in an appeal against a statutory nuisance abatement notice. There the hearing ran for eight days, with judgment being reserved. The appeal being successful. Costs of £75,000 plus VAT were awarded to the appellant (£94,655 plus VAT had been claimed). This award was challenged by judicial review. The court noted the wide discretion given by section 64 and held that there as no obligation to give reasons for the decision. The magistrates' decision would only be set aside if procedurally

---

[90] *Westminster Renslade Ltd v. Secretary of State for the Environment* [1983] J.P.L. 454; *Homebase Ltd v. Secretary of State for the Environment* [1993] J.P.L. B54.

[91] *Archer and Thompson v. Secretary of State for the Environment* [1991] J.P.L. 1027; *Relayshine Ltd v. Secretary of State for the Environment* [1994] E.G.C.S. 12. See also *Wordie Property Co. Ltd v. Secretary of State for Scotland* 1984 S.L.T. 345.

[92] *Buroughs Day v. Bristol City Council* [1996] N.P.C. 3.

[93] Section 64(2); see *R. v. Pwllheli Justices, ex p. Soane* [1948] 2 All E.R. 815.

[94] *R. v. West London Magistrates' Court, ex p. Kyprianou, The Times*, May 3, 1993.

[95] [1995] Env. L.R. 1.

flawed or perverse. It was correct procedurally to hear the arguments and give judgment in open court; formal taxation is not required, but in complex or large cases an adjournment to consider the issue may be appropriate.[96] Since the authority is under a public duty to serve a remediation notice, there may well be strong arguments that the authority should not be exposed to costs where it acted reasonably; however, the financial position of a successful appellant will also be relevant here.

Expenses: sheriff's court

23.78 As the Appeals Regulations make no provision for dealing with expenses, the sheriff must rely upon his inherent power to determine such questions.[97] In appeals which raise novel issues of difficulty or importance, the sheriff may decide not to exercise his inherent power and may make no award.[98] Expenses may be modified.[99]

Costs: Secretary of State

23.79 No provision is made for costs in the case of an appeal determined on the basis of written representations. By Schedule 20, para. 5(1) the provisions on costs of sections 250(2)–(5) of the Local Government Act 1972 apply to local inquiries or hearings. The same is true for Scotland: Schedule 20, para. 5(2) applies the provisions of sections 210(3)–(8) of the Local Government (Scotland) Act 1973. Sections 250(5) of the 1972 Act allows orders to be made as to the costs of the parties and as to the parties by whom such costs are to be paid. Subject to those provisions, the costs of any local inquiry are to be defrayed by the Secretary of State.

23.80 The same principles as apply to awards of costs on planning appeals are likely to apply to costs on appeals to the Secretary of State under Part IIA.[1] Cost awards will therefore be based on unreasonable behaviour by one or otherpartyandwillnotfollowtheevent.[2] A specific application for an award of costs will probably relate to the conduct of the appeal; for example, putting forward material at a late stage, lack of cooperation in the appeals process, refusal to discuss data in advance, or late withdrawal of an appeal or of specific grounds of appeal. The power to award costs

---

[96] *ibid.* pp. 5–6.

[97] *McQuater v. Ferguson*, 1911 S.C. 640; *Society of Accountants in Edinburgh v. Lord Advocate* [1924] S.L.T. 194; *Steele v. Lanarkshire Middle Ward District Committee* [1928] S.L.T. (Sh.Ct) 20.

[98] *Society of Accountants in Edinburgh v. Lord Advocate*, note 97 above.

[99] *Duncan v. Dundee Licensing Authority*, 1962 S.L.T. (Sh.Ct) 15.

[1] See Circular 8/93 *Costs Awards in Planning Appeals*. For Scotland, see the Scottish Office Circular 6/1990, *Awards of Expenses in Appeals and Other Planning Proceedings and in Compulsory Purchase Order Inquiries*. Generally see S. Tromans and R. Turrall-Clarke, *Planning Law, Practice and Precedents*, Chapter 6.

[2] Circular 8/93, Annex 1, para. 1; Circular 6/1990, para. 4.

relates to costs of and in connection with the inquiry or hearing into the appeal: thus the behaviour of the parties before submission of the appeal is not a matter to be taken into account.[3]

Reasons for appeal decisions

23.81 There is a general duty for a professional judge "to give reasons for a decision, though the amount of reasoning required will depend on the issue and how susceptible it is to detailed reasoning".[4] Explicit reasons can and should normally be given when deciding between conflicting expert evidence.[5] Failure to give reasons can be itself a ground of appeal.[6] Although it has been said that the requirement for reasons does not always or even usually apply in the magistrates' courts,[7] in the context of appeals against remediation notices there are strong arguments that reasons should be given both in order to secure fairness and natural justice, and also as a way of concentrating the mind of the decision-taker.[8] In particular, in the absence of explicit findings of fact and of rulings on law, the appellant may be left in doubt as to which appeal route to pursue.

Where the appeal is determined by an inspector it is to be expected that reasons will be given. While reasons need not cover in detail every argument raised, it is to be expected that they will be intelligible and will deal with the substantial issues, so as to enable the parties to know on what grounds the appeal has been decided. Paragraph 73 of Annex 4 of the Circular indicates that details of decision letters relating to special sites will be placed on the register and copies will be available from the Planning Inspectorate's Decision Library.

Appeals from decisions of magistrates' court

23.82 By regulation 13 an appeal against the decision of the magistrates on a section 78L appeal lies to the High Court at the instance of any party to the proceedings.

23.83 The procedure will be that of RSC Order 55 (Schedule 1 to the Civil Procedure Rules). The appeal is by way or rehearing and must be brought by notice of appeal, stating the grounds of appeal.[9] Bringing the appeal does not operate as a stay of the decision unless this relief is specifically

---

[3] [1983] J.P.L. 333 (Appeal Decision).
[4] See generally *R. v. Crown Court at Harrow, ex p. Dave* [1994] 1 All E.R. 315 and authorities referred to therein.
[5] *Eckersley v. Binnie* [1988] 18 Const. L.R. 1, 77.
[6] *Fiannery v. Halifax Estate Agencies Ltd, The Times*, March 4, 1999.
[7] *ibid.*
[8] *ibid.*
[9] Sched. 1, r. 55.3.

sought and ordered.[10] The notice of appeal must be entered within 28 days of the decision.[11] Further evidence of fact may be received by the High Court,[12] though this discretion may well be exercised sparingly.

23.84 Appeal from the High Court will lie to the Court of Appeal. Permission will be required, and because there has already been one level of appeal, a more restrictive approach will be taken, so that permission will only be granted if an important point of principle or practice arises.[13]

23.85 It should also be considered whether it may be possible to apply for a case to be stated by the magistrates under section 111 of the 1980 Act to the Divisional Court, though the more obvious route is by appeal to the High Court. The procedure for stating a case is contained in rules 76–81 of the Magistrates' Courts Rules 1981. It is possible, in civil proceedings only, to ask the magistrates' court to state a case at an interlocutory stage, though special circumstances may need to be shown.[14] This may be particularly useful if difficult points of interpretation arise early in an appeal.

23.86 Another possible though probably tenuous avenue of challenge to a magistrate's decision is judicial review, which can be used to consider whether the court has failed to exercise its jurisdiction correctly, or has come to some error of law which appears on the face of the record. Again, therefore, it is essentially an appeal in relation to the process rather than the substance of the decision. The tendency has been to require that statutory appeal rights be exhausted before granting leave for judicial review.[15] The existence of a statutory right of appeal does not necessarily preclude judicial review,[16] but the jurisdiction should be exercised cautiously and not simply in respect of immaterial or minor deviations from best practice.[17] Matters of law going to procedure or jurisdiction are most appropriately raised as a preliminary point in the first instance court: advocates should therefore not rely on the possible later availability of judicial review.

Appeals from decisions of sheriff

23.87 It remains to be seen what provision will be made for an appeal from decisions of the sheriff. Where there is no prescribed right of appeal, sections 27 and 28 of the Sheriff Courts (Scotland) Act 1907 provide that

---

[10] *ibid.*, r. 55.3(2).
[11] *ibid.*, r. 55.4(2).
[12] *ibid.*, r. 55.7(2).
[13] *Practice Direction (Court of Appeal, Civil decision): Leave to Appeal and Skeleton Arguments* [1999] 1 W.L.R. 2.
[14] *R. v. Chesterfield Justices, ex p. Kovacs* [1992] 2 All E.R. 325.
[15] *R. v. Bradford Justices, ex. p. Wilkinson* [1990] 1 W.L.R. 692.
[16] *R. v. Gloucester Crown Courts, ex p. Chester* [1998] C.O.D. 365.
[17] See *Hopson v. D.P.P.* [1997] C.O.D. 236, considering *R. v. Peterborough Magistrates' Court., ex p. Dowler* [1996] 2 Cr.App.R. 561.

there is the same scope for appeals as in sheriff court ordinary actions. This would mean that an appeal would be possible either to the Sheriff Principal, thence to the Inner House of the Court of Session and thence to the House of Lords, or directly to the Inner House of the Court of Session and thence to the House of Lords. However, section 28(2) of the 1907 Act also provides that sections 27–28 do not affect any right of appeal or exclusion of such right provided by any Act of Parliament. The effect of this is that in the absence of any express provisions relating to the appeal, it is necessary to examine the terms of Part IIA to ascertain whether the appeal rights provided by sections 27 and 28 are excluded by clear or necessary implication.[18]

23.88 Whether or not the rights of appeal in ordinary actions are available depends largely upon whether the sheriff is exercising an administrative or judicial function in dealing with applications under the Appeals Regulations.[19] If it is seen as an administrative function, then his decision will be final, whereas if it is judicial it will be subject to review.[20] However, even if the sheriff's decision is regarded as final, it would still be subject to judicial review. To ascertain whether or not the sheriff is exercising an administrative or judicial function, a variety of principles and tests may be employed.[21] For example, if the sheriff is exercising a new jurisdiction conferred by statute rather than an established one, his jurisdiction is more likely to be administrative and privative although this is not conclusive.[22] Furthermore, the proceedings may originate in a manner which indicates the sheriff is performing an administrative function. An appeal to the sheriff against a notice may indicate that he is performing an administrative function.[23] Where the appeal lies to the sheriff court, as in section 78L of Part IIA, rather than to the sheriff, that may indicate that the sheriff is to exercise his ordinary judicial jurisdiction.[24]

23.89 In the absence of clear authority in the Regulations, it cannot, therefore, be said with absolutely certainty that the normal appeal routes from decisions of the sheriff which are available in ordinary actions also apply in the case of appeals from decisions of the sheriff under section 78L of Part IIA. Given that appeals in relation to special sites go to an administrative

---

[18] *Arcari v. Dumbartonshire County Council*, 1948 S.C. 62; *Kave v. Hunter*, 1958 S.C. 208 *per* Lord President Clyde at p. 210; *Rodenhurst v. Chief Constable of Grampian Police*, 1992 S.C. 1.

[19] *Arcari v. Dumbartonshire County Council, supra* note 18, *per* Lord President Cooper at p. 66.

[20] *Kave v. Hunter, supra* note 18.

[21] See generally Nicolson & Stewart, *MacPhail's Sheriff Court Practice*, W. Green, 1998, paras 25-36–25-45.

[22] *Arcari v. Dumbartonshire County Council, supra* note 18, *per* Lord President Cooper at p. 66. See also *Portobello Magistrates v. Edinburgh Magistrates* (1882) 10 R. 130 *per* Lord Justice-Clerk Moncrieff at p. 137.

[23] *Arcari v. Dumbartonshire County Council, supra* note 18, *per* Lord President Cooper at p. 67. See also *Hopkin v. Ayr Local Taxation Officer*, 1964 S.L.T (Sh.Ct) 60 and *T. v. Secretary of State for Scotland*, 1987 S.C.L.R. 65.

[24] *Portobello Magistrates v. Edinburgh Magistrates, supra*, note 22.

body, namely the Scottish Ministers from whose decision only judicial review will lie, that might be an argument that the sheriff is in reality exercising an administrative jurisdiction. However, because section 78L provides that the appeal will lie to the sheriff court as noted above that might be determinative of the issue.

Appeals from decisions of Secretary of State

23.90 No provision is made in the Act or Regulations for appealing against decisions of the Secretary of State in relation to special sites. It is therefore a matter of challenge on ordinary principles of judicial review. Given that the rules applying to the Secretary of State require reasons to be given, challenges may be based on deficiencies in reasoning. However, the courts will most likely be wary of setting too high a standard for the reasoning required, or of setting aside decisions on the basis of unclear or defective reasoning where there is no real or substantial prejudice arising.[25] Ultimately, the question is whether the decision enables the parties to understand on what grounds the appeal was decided and is in sufficient detail to enable them to know what conclusions were reached on the main substantive issues.[26]

It remains unsatisfactory that the rules for further appeals are so different as between special sites and other sites.

Consequence of appeal decision

23.91 By section 78L(2) the outcome of the appeal will be that the notice is quashed, confirmed or modified. If confirmed, with or without modifications, then the notice takes effect from final determination under regulation 14(1). The Circular, Annex 2, paragraph 12.7, raises one potential practical difficulty, which is that if a number of appropriate persons are involved, and separate notices have been served on each, then if an appeal on one notice against the stated apportionment is successful, that will inevitably mean that the other apportionments are affected. If those other notices have not been appealed, there is no power to vary the apportionment, short of issuing fresh notices. The practical answer suggested in the Circular is to serve at the outset a single notice on all the persons involved, allowing the appellate authority to adjust the apportionment in the single notice. Annex 4, paragraph 78 also points out that in order to give effect to

---

[25] See especially *Save Britain's Heritage v. Secretary of State for the Environment* [1991] 2 All E.R. 10; *Bolton Metropolitan District Council v. Secretary of State for the Environment* (1995) 71 P & C.R. 309.

[26] *City of Edinburgh Council v. Secretary of State for Scotland* [1998] J.P.L. 224, approving *Hope v. Secretary of State for the Environment* [1975] 31 P & C.R. 120, at 123.

the appeal decision, it may be necessary for the enforcing authority to serve a new notice or notices on another person (for example, where there is a successful appeal that another person should have been served). Such notices will themselves be fully subject to the requirements of the Act, including rights of appeal.

Abandonment of appeals

23.92 The appellant can seek to abandon their appeal by making a request to magistrates' court or by notifying the Secretary of State under regulation 9(3). In both cases, the request may be refused if by that time the appellant has been notified under regulation 12 of unfavourable proposed modifications by the appellate authority. This effectively prevents the appellant authority "cutting his losses" in such circumstances: he will be stuck with the proposed modification, subject to this rights under regulation 12 to make representations or to be heard.

Chapter 24

Challenging Remediation Notices otherwise than by Appeal

Summary

The statutory process for appealing against remediation notices has been described at Chapter 23. Apart from this appeal system, there may (depending on the circumstances) be the possibility of challenging a remediation notice by judicial review proceedings, or by way of defence to a criminal prosecution for alleged non-compliance with the notice or as a defence to subsequent cost recovery proceedings by the authority. However, there are a number of potential difficulties in relation to all of these alternative courses, which will need to be considered carefully by those advising the recipient of the notice.

Judicial Review

24.01 A remediation notice is in principle capable of challenge in judicial review proceedings on the normal grounds—namely that the issue of the notice was *ultra vires*; or some procedural defect has affected the issue and service of the notice or the requisite prior procedures, such as consultation; or that service of the notice was motivated by irrelevant considerations or failed to take relevant considerations into account; or that no reasonable authority could have served such a notice. The applicant for judicial review in such circumstances will no doubt seek an order for certiorari to quash the notice. Any such proceedings will be subject to the normal rules of judicial review in terms of the requirement of leave, promptness and delay, and the grant of relief will be discretionary. The existence of a defect will therefore not necessarily result in the notice being quashed.

Issues for which judicial review not appropriate

24.02 The inherent limitations on the scope of judicial review mean that it will not be an appropriate means of challenging remediation notices on some of the most common grounds likely to occur in practice—except perhaps in the most extreme cases. Arguments of a technical nature that the land is not contaminated as a matter of fact, or that the requirements of the notice are unreasonably onerous, or legal arguments that it has not been served on the appropriate person, are much more suited to the statutory appeal process, involving as they do the detailed examination of issues of fact. Judicial review will be most effective where there has been a clear breach of the requirements of the legislation. An example might be where the notice has been served before the end of the moratorium period,[1] or where the authority has acted in some way which is patently contrary to the mandatory Guidance, where it has made a relevant mistake of law, or has acted *ultra vires* or unfairly.

Possible exclusion of judicial review

24.03 The provisions of Part IIA do not seek explicitly to exclude the remedy of judicial review. However, the argument may well arise that judicial review is not available where the applicant has the possibility of a statutory appeal against the notice. The courts are likely to be reluctant to find that the possibility of judicial review is ousted entirely.[2] A relevant issue here may be whether the grounds of appeal conferred by the Regulations include judicial review grounds.[3] The existence of an alternative remedy is however not in itself a ground for refusing judicial review relief. The question may in such cases be whether the existence of the statutory right of appeal is a reason for refusing to allow judicial review as a matter of discretion, judicial review being regarded in this respect as a "long stop" or remedy of last resort.[4] Where the appeal process is a legally more convenient or appropriate remedy, exceptional circumstances may be needed to convince the court to entertain judicial review.[5] Thus, the closer the grounds of judicial review are to the grounds for a statutory appeal, the more likely the courts are to decline jurisdiction.[6] The courts are also likely to be mindful, where the applicant for judicial review is seeking to raise what are effectively procedural defects in a notice, that such defects might well be capable of being remedied as part of the appeal process.[7] The adequacy, speed and convenience of the alternative remedy will also be relevant.[8]

---

[1] See Chapter 9.
[2] *Pyx Granite Co. Ltd v. Ministry of Housing and Local Government* [1960] A.C. 260 at 304, 286; *Leech v. Deputy Governor of Parkhurst Prison* [1988] A.C. 533, 580–581.
[3] *R. v. Dacorum District Council, ex p. Cannon*, *The Times*, July 9, 1996.
[4] *R. v. Panel on Take-Overs and Mergers, ex p. Guinness Plc* [1990] 1 Q.B. 146, 177–178; *R. v. Metropolitan Stipendiary Magistrate, ex p. London Waste Regulation Authority* [1993] 3 All E.R. 113, 120.
[5] *Harley Development Inc. v. Commissioner of Inland Revenue* [1996] 1 W.L.R. 727; *R. v. Inland Revenue Commissioners, ex p. Preston* [1985] A.C. 835, 852; *R. v. Secretary of State for the Home Department ex p. Sovati* [1986] 1 W.L.R. 477, 485.
[6] *R. v. I.R.C., ex p. Bishopp* [1999] C.O.D. 354.
[7] See Chapter 23.
[8] *Ex parte Waldron* [1986] Q.B. 824 at 852.

Where it appears that there are grounds for judicial review, the discretion to refuse relief will not be lightly exercised,[9] and it may be possible to argue that a statutory appeal to the magistrates, thence to the Divisional or High Court on a point of law, may be a less convenient way of dealing with the matter; as always much will depend on the exact facts and circumstances.

Collateral challenge in criminal proceedings

24.04 Another possibility is that the recipient of a remediation notice who believes the notice to be invalid may not comply with the notice and, if prosecuted for non-compliance, may seek to raise the invalidity by way of a defence in the criminal proceedings. The law on this issue is difficult and is largely derived from the cases on planning enforcement and on prosecution for breach of bye-laws discussed which are in the following paragraphs. However, this is an area where close regard must be paid to the specific statutory provisions and the overall scheme in which they operate, and it is dangerous to seek to derive general principles from the case-law under other legislation.

24.05 It is clear from the decision of the House of Lords in *R. v. Wicks*[10] that the correct approach in cases of collateral challenge is now to consider the interpretation of the provision under which the offence is charged. In *Wicks*, the prosecution was for failure to comply with an enforcement notice under section 179(1) of the Town and Country Planning Act 1990. The question according to the House of Lords was whether "enforcement notice" in this section meant:

(a) a notice which was not liable to be quashed on standard public law grounds (in that case it was alleged that the service of the notice was *ultra vires* because it had been motivated by bad faith and improper reasons); or
(b) a notice which complies with the formal requirements of the 1990 Act and which has not actually been quashed on appeal or judicial review.[11]

24.06 The correct approach was to be found by examining the scheme of the legislation, and having considered Part VII of the 1990 Act, dealing with enforcement, the conclusion was that "enforcement notice" meant a notice issued by the planning authority which was formally valid and which had not been quashed. It should not, however, be concluded from *Wicks* that the same interpretation should necessarily be applied to notices under

---

[9] See *R. v. Lincolnshire County Council and Wealden District Council, ex p. Atkinson, Wales and Stratford* (1996) 8 Admin. L.R. 529, 550.

[10] [1998] A.C. 92.

[11] *ibid.*, at p. 119.

analogous legislation. A different conclusion was drawn by the Divisional Court in *Dilieto v. Ealing London Borough Council*[12] in relation to a breach of condition notice served under a different section of the same Part of the 1990 Act. When prosecuted for failure to comply, the recipient argued (a) that the notice had been served out of time, and (b) that the relevant condition of the notice was so vague and imprecise as to be a nullity. It was held to be too simplistic an approach to follow *Wicks* simply because the relevant provisions were in the same Part of the same Act. In particular, whereas elaborate provision exists for challenging enforcement notice notices on appeal, no such provision is made in the case of breach of condition notices.[13] It was unlikely that Parliament intended an owner or occupier to have no means of challenging a breach of condition notice on the basis that it was out of time. Nor was it appropriate to require such a challenge to be made only by judicial review, as was argued by the prosecuting authority: the question would involve detailed examination of oral and written evidence—a function which was better fulfilled by magistrates in the context of a prosecution rather than by the High Court in judicial review proceedings.[14]

24.07 The general approach to the issue of statutory construction, according to the decision of the House of Lords in *Boddington v. British Transport Police*,[15] should begin from the premise that Parliament ought not to be taken to have removed the right of individuals to challenge the relevant administrative measures unless the intention to do so is clear. Such an intention may be more readily found where, as in *R. v. Wicks*, the statutory scheme in question is concerned with administrative acts specifically directed at the defendant, and where there is clear and ample opportunity under the legislation for the defendant to challenge the legality of those acts, before being charged with an offence.[16]

Nullity and invalidity

24.08 The distinction between a notice which is so defective as to be a nullity and one which may be affected by invalidity has often been drawn in the context of planning enforcement notices.[17] The traditional approach is that an enforcement notice will be a nullity where it is bad upon its face, so that

---

[12] [1998] 3 W.L.R. 1403.
[13] *ibid.*, at p. 1416.
[14] *ibid.*, at p. 1417. To like effect, see *R. v. Wicks* [1998] A.C. 92 at 106 and *R. v. Jenner* [1983] 1 W.L.R. 873 at 877, CA. Compare *R. v. Ettrick Trout Co. Ltd* [1994] Env. L.R. 165, CA, where it was regarded as inappropriate and an abuse of process to allow arguments as to the validity of a discharge consent condition to be raised before a jury by way of collateral challenge.
[15] [1999] 2 A.C. 143.
[16] *ibid.*, at p. 216.
[17] See for example S. Tromans and R. Turrall-Clarke, *Planning Law, Practice and Precedents*, Sweet & Maxwell, para. 9.25.

it is legally ineffective: in such circumstances no prosecution can lie for non-compliance with it, nor can any appeal against it be entertained.

24.09 Matters which can mean that a notice is a nullity include where it is "hopelessly ambiguous and uncertain",[18] for example by failing to state what is required to comply with it[19] or by inadequate identification of the land to which the notice relates.[20] Both of these fundamental defects clearly could occur in relation to remediation notices.[21] In one case reference was made to notices ". . . so utterly extravagant in their terms that anyone acquainted with the property or any other relevant fact would say unhesitatingly that there must be some mistake somewhere", in which case the recipient may, "if of sufficiently strong nerve, simply disregard them".[22] Issues of bad faith on the part of the authority in serving the notice might also raise the possibility of nullity. By contrast, a notice which is invalid (as opposed to a nullity) is effective until such time as it is set aside, and there may also be the possibility of the defect being rectified as part of the appeal process.[23]

24.10 It seems clear on the basis of *R. v. Wicks*[24] and *Boddington v. British Transport Police*.[25] that the distinction between nullity and invalidity is neither conclusive nor critical in the context of collateral challenge. In *Bugg v. DPP, DPP v. Percy*[26] Woolf L.J. had differentiated between byelaws or subordinate legislation affected by procedural defects (which were valid until such time as set aside by a court) and laws which were "substantially invalid":

> "No citizen is required to comply with a law which is bad on its face. If the citizen is satisfied that is the situation, he is entitled to ignore the law".[27]

24.11 However, that distinction was rejected by the House of Lords in *R. v. Wicks* for the purpose of deciding whether there was a defence to a criminal charge.[28] The Court of Appeal in *R. v. Wicks*[29] by contrast had based its judgment on that distinction; a result which can be heavily criticised for

---

[18] *Miller-Mead v. Ministry of Housing and Local Government* [1963] 2 Q.B. 196.

[19] *Dudley Bowers Amusements Enterprises Ltd v. Secretary of State for the Environment* [1986] J.P.L. 689.

[20] However, lack of clarity may not be fatal if in the circumstances the recipient was aware what was required or to which premises the notice related: *Coventry Scaffolding Company (London) Ltd. v. Parker* [1987] J.P.L. 127; *Pitman v. Secretary of State for the Environment* [1989] J.P.L. 831.

[21] See Chapter 12.

[22] *Graddage v. London Borough of Haringey* [1975] 1 All E.R. 224 at 231.

[23] See Chapter 23.

[24] [1998] A.C. 92.

[25] [1999] 2 A.C. 143.

[26] [1993] Q.B. 473.

[27] *ibid.*, at p. 500.

[28] [1998] A.C. 92 at 108–9, 116–7.

[29] [1996] J.P.L. 743.

perpetrating a flawed legal distinction and for denying the ability to raise matters of invalidity as a defence as of right, rather than as a matter of the court's discretion in judicial review proceedings.[30] In *Boddington*, the House of Lords went further, overruling *Bugg*, and rejecting any distinction between substantive (patent) and procedural (latent) invalidity.[31] The application of these principles to remediation notices is considered in the next paragraph.

Application of nullity and collateral challenge principles to remediation notices

24.12 How do the principles outlined in the preceding paragraphs apply to remediation notices? As required by *R. v. Wicks*, the starting point must be to consider the words of the offence provision at section 78M(1) of the 1990 Act, which states that a person on whom an enforcing authority serves a remediation notice who fails, without reasonable excuse, to comply with any of the requirements of the notice, shall be guilty of an offence. The question is therefore whether a "remediation notice" has been served on that person, and what is meant by a "remediation notice" in that context. Looking at the cases on the issue, especially *R. v. Wicks*,[32] *Dilieto v. Ealing London Borough Council*,[33] and *Boddington v. British Transport Police*[34] the following factors will be relevant:

1. *Boddington* suggests that the courts should be reluctant to prevent a defendant raising issues of validity as a defence to criminal proceedings: it is unjust to require magistrates to convict for an offence based on an invalid order or notice.[35]
2. However, Part IIA provides a statutory scheme for appealing on various grounds against remediation notices: this gives support to the view that those matters which may be grounds of appeal under the statutory system ought not to be allowed to be raised by way of defence in criminal proceedings. Parliament has chosen to confer the right to appeal subject to specific procedures, and different procedures would apply if the matter were raised in the context of a criminal prosecution.
3. It is also relevant to consider whether refusing to hear arguments based on collateral challenge would result in unfairness to the defendant. It is difficult to see how it would be unfair to require the defendant to follow the statutory appeal procedures where these are available.

---

[30] Barry Hough, *Collateral Challenge and Enforcement Notices* [1997] J.P.L. 111.
[31] [1999] 2 A.C. 143, 157G, 158E, 164H, 172E.
[32] [1998] A.C. 92.
[33] [1998] 3 W.L.R. 1403.
[34] [1999] 2 A.C. 143.
[35] *ibid.*, at 162E–G, 164D, 173B and G.

4. Where the defendant's challenge to the remediation notice is based on grounds other than those falling within the statutory appeal procedure, the issue is whether the matter should be challenged by judicial review rather than by raising a defence in the criminal courts. The courts here will have regard to the issue of fairness to the defendant (in the sense that he may be prejudiced by having to rely on the discretionary remedy of judicial review and by having to take lengthy and expensive judicial review proceedings in respect of a defective notice) and also the issue of good administration of justice (in that judicial review proceedings may be better or worse suited to dealing with different issues and may result in long delay before any prosecution can be brought).
5. On those principles, where the notice is quite obviously bad—for example, in failing to identify the land or to state what is required by way of remediation and who is responsible for it—there are grounds for arguing that the notice is not a "remediation notice" at all, and that the defendant is entitled to ignore it.[36] If the issue however is one of the procedures behind the notice—for example, defective consultation—then the courts may say that these are issues better suited for determination by judicial review. In that sense, the old and discredited distinction between patent and latent defects may be relevant, even though it is not conclusive. However, various dicta in *Boddington* suggest that the difficulties of dealing with complex legal issues in magistrates' courts should not be exaggerated.[37]
6. The offence under section 78(M)(1) is failure to comply with the notice "without reasonable excuse". It may be argued that an honest belief that the notice is void is a "reasonable excuse." Such an argument seems most unlikely to succeed however, as no *mens rea* is involved in the offence of not complying with a notice and so the state of mind of the defendant (and in particular, any mistake as to the legal effect of the notice) is not relevant.[38]

24.13 It is clearly the intention behind Part IIA that the issues arising under a remediation notice as to what should be done and who should do it, or contribute to its cost, will be dealt with under the statutory appeal process. A party who fails to protect their position by participating in that appeal process does so at their own risk. There may be substantial problems if that appeal process, which may have been lengthy, could be later "unpicked" by collateral challenge. Nevertheless, it is impossible to rule out entirely the possible success of such challenges, in particular in the

[36] As it was put by Lord Radcliffe in *Smith v. East Elloe Rural District Council* [1956] A.C. 736, 769–770, does the notice bear a "brand of invalidity upon its forehead".
[37] [1999] 2 A.C. 143 at 162E–G, 164D, 173B and G.
[38] See *DPP v. Morgan* [1976] A.C. 182; *R. v Bradish* [1990] 1 Q.B. 981.

case of defective notices; and particularly if that defect might have misled the recipient into not exercising their statutory right of appeal.

Challenge in cost recovery proceedings

24.14 As described above, the enforcing authority has the power, in specified circumstances, to carry out remediation itself.[39] In particular, the power can be used where it is considered necessary to avoid the occurrence of serious harm or pollution in cases of imminent danger, or where the recipient of an enforcement notice fails to comply with any of its requirements.[40] In either case, the authority has power to recover its reasonable costs from the appropriate person or persons,[41] and in some cases to serve a charging notice making the costs and accrued interest a charge on the premises.[42]

24.15 If a charging notice is served, then there is a statutory right of appeal against it within 21 days to the county court, which may confirm the notice, substitute a different amount, or order that the notice shall be of no effect.[43] However, no appeal procedure is provided in cases where no charging notice is served, and the appropriate person may wish to defend the civil proceedings for cost recovery by the authority on a number of possible bases:

(a) that the original remediation notice was a nullity;
(b) that the circumstances allowing the authority to carry out remediation itself were not in fact applicable;
(c) that the works carried out by the authority were not "appropriate by way of remediation"[44];
(d) that the costs were unreasonable[45];
(e) that the authority failed to have regard to hardship and to the statutory guidance on that issue[46]; or
(f) that the authority has apportioned too great a share of the cost to the defendant.[47]

24.16 The issue is whether the defendant will be allowed to raise these issues by way of defence in civil proceedings. This involves the question of procedural exclusivity; namely whether matter is such that can be raised as

---

[39] Section 78N. See Chapter 22.
[40] Sections 78N(3)(a) and (c).
[41] Section 78P(1).
[42] Section 78P(7).
[43] Sections 78P(8) and (9). See Chapter 22.
[44] Section 78N(1).
[45] Section 78P(1).
[46] Section 78P(2).
[47] Sections 78P(1) and 78F(7). See Chapter 18.

part of the civil proceedings, or whether judicial review is the only appropriate route of challenge to what is essentially an exercise of public law powers.[48]

24.17 In response to the exclusivity argument, if raised by the plaintiff enforcing authority, the defendant will no doubt seek to rely on the line of cases following *Wandsworth London Borough Council v. Winder*,[49] which emphasise the right of a defendant to put forward whatever defences are available. It could be said that, as in *Winder*, the defendant is simply defending the action on the ground that he is not liable for the whole sum claimed by the plaintiff.[50]

24.18 On the other hand, there have been cases where the Court has refused to allow challenges to action by public authorities by way of defence to civil proceedings, on the basis that it was not a true defence on the merits but rather a challenge on a public law basis to the decision of the authority to initiate the action or proceedings.[51] In such cases the appropriate course will be to consider granting a stay of the proceedings to allow leave for judicial review proceedings to be sought.

24.19 The courts' approach to this issue is inherently a flexible and somewhat unpredictable one, but is at its heart a question of whether the defence constitutes an abuse of the process of the court[52]; the onus resting with the plaintiff enforcing authority to show that the defence is such an abuse.

24.20 Some of the arguments indicated at (a)–(f) above by way of illustration would clearly be genuine issues to raise in the context of civil cost recovery proceedings—for example that the costs involved are unreasonable, or that the authority did not apply the criteria on hardship correctly. There would appear to be no abuse inherent in raising those matters in answer to the authority's claim. The position may however be different where the challenge is based on matters which could have been challenged at the time of the original remediation notice: for example, the contaminated condition of the land, the works required to be undertaken, or the share of cost apportioned to the defendant. All of these would be matters which the defendant could have raised at a much earlier stage, either by appealing against the remediation notice or by challenging it in judicial review proceedings—which in either case would have been subject to stringent time limits. Here the authority will have a strong argument that to allow the recipient of a notice to sit back without objection and wait until the

---

[48] *O'Reilly v. Mackman* [1983] 2 A.C. 237; *Cocks v. Thanet District Council* [1983] 2 A.C. 286.

[49] [1985] A.C. 461.

[50] See also *R. v. Inland Revenue Commissioners, ex p. T.C. Coombs & Co.* [1991] 2 A.C. 283 at 304; *British Steel Plc v. Commissioners of Customs and Excise* [1996] 1 All E.R. 1002 at 1013; *Credit Suisse v. Allendale Borough Council* [1996] 3 W.L.R. 984 at 926, 937; *Warwick District Council v. Freeman* (1995) 27 H.L.R. 616.

[51] *Waverley District Council v. Hilden* [1988] 1 W.L.R. 246; *Avon County Council v. Buscott* [1988] Q.B. 656.

[52] See *Mercury Communications Ltd v. Director General of Telecommunications* [1996] 1 W.L.R. 48 at 57 (Lord Slynn).

authority has expended considerable sums before raising any challenge would be to permit not only an abuse of the process of the court, but also would be contrary to the public interest in securing the clean-up of contaminated land. This argument would of course have less force in cases of urgency under section 78N(3)(a) where no remediation notice has been served and where there may have been no adequate opportunity to challenge or question the authority's chosen course of action (though even there it may have been possible to obtain competitive estimates for the work).

24.21 There may naturally be grey areas falling between the examples discussed above, but the overarching concept of abuse of process does appear to offer a more satisfactory framework for considering the issues, rather than seeking esoteric distinctions between public and private law. In general, the clearer the authority makes its intentions before exercising its section 78N powers, the more arguments it will have in its favour should its decision be attacked at the later cost-recovery stage.

24.22 What is the situation where a remediation notice is served, no appeal is made, but at the later stage of cost recovery the defendant asserts that the notice was a nullity? If the notice was a nullity then the authority will have acted *ultra vires* in acting under section 78N and consequently its costs will not be recoverable. Failure to appeal will not prevent the nullity argument being raised later in civil proceedings, though on the principles stated above, it probably will debar the defendant from objecting to matters which could have been the subject of a statutory appeal.[53]

[53] See *West Ham Corporation v. Charles Benabo and Sons* [1934] 2 K.B. 253 at 264; *Graddage v. London Borough of Haringey* [1975] 1 All E.R. 224 at 227, 231.

Chapter 25

Registers and Information

Summary

Enforcing authorities are required to maintain public registers containing various information relating to their functions under Part IIA. The duty is subject to restrictions on including information on two grounds: national security and commercial confidential information. Additionally, the Environment Agency/SEPA is required to publish reports on the state of contaminated land generally.

Registers

25.01 By section 78R(1), subject to exclusions relating to information affecting national security[1] and commercially confidential information,[2] every enforcing authority is required to maintain a register containing prescribed particulars of a number of matters as set out below. The form and descriptions of information contained in the register are to be prescribed by the Secretary of State.[3] The matters to be contained on the register and listed at section 78R(1)are:

(a) remediation notices served by the enforcing authority;
(b) appeals against remediation notices;
(c) remediation statements or remediation declarations under section 78H;
(d) appeals against charging notices;
(e) notices by the local authority effecting designation of land as a special site;
(f) notices of the Secretary of State effecting designation as a special site;
(g) notices terminating the designation of land as a special site;

[1] Section 78S.
[2] Section 78T.
[3] Section 78R(2).

(h) notifications of what has been done by way of remediation by a person served with a remediation notice or who is required to publish a remediation statement;
(j) notification given by owners or occupiers of what has been done on land by way of remediation;
(k) convictions for prescribed offences; and
(l) any other matters relating to contaminated land prescribed by the Secretary of State.

It will be appreciated from this list that the register is not intended to be a register of sites which are, or may be, contaminated as such. Rather, it is a register of what is effectively the enforcement history of a site once a remediation notice has been served (or in the case of special sites, their prior designation). As indicated below, those seeking information of a more special nature, such as the outcome of inspections of land carried out by local authorities, will need to consider other provisions. **25.02**

Where any particular is entered on a register by the appropriate Agency, the appropriate Agency shall send a copy of those particulars to the local authority in whose area the land in question is situated. Information received in this way is entered by the local authority on its own register.[4] **25.03**

Enforcing authorities must secure that the registers are available at all reasonable times for inspection by the public free of charge, and there must be facilities for members of the public to obtain copies of entries on payment of reasonable charges.[5] The Secretary of State may prescribe the places where such registers or facilities are to be made available, and the prescribed places are the principal office of the local authority or (if the enforcing authority is the Agency) its office for the area in which the contaminated land is situated.[6] The registers may be kept in any form.[7] In practice most authorities will probably maintain a computerised record. **25.04**

Other prescribed particulars

The particulars to be contained on the registers are prescribed pursuant to section 78R by the Contaminated Land (England) Regulations 2000, reg. 15(1) and (2) and Schedule 3. A number of matters, with prescribed particulars for each, are listed as follows. Schedule 3 to the Regulations refers to "full particulars" in this respect. This indicates that a summary or précis of the relevant information will not suffice. **25.05**

Remediation notices The required particulars in relation to remediation notices served are: **25.06**

(a) the name and address of the person on whom served;

[4] Sections 78R(4) and (6).
[5] Section 78R(8).
[6] The Contaminated Land (England) Regulations 2000, reg. 15(3).
[7] Section 78R(9).

(b) the location and extent of the contaminated land in question sufficient to enable it to be identified, whether by reference to a plan or otherwise;
(c) the relevant significant harm or pollution of controlled waters;
(d) the contaminating substances, and the location of any land from which they have escaped, if that is the case;
(e) the current use of the contaminated land;
(f) what each appropriate person is to do by way of remediation and within what period each of the things is to be done; and
(g) the date of the notice.

25.07 *Appeals* The particulars for appeals against remediation notices are:

(a) the appeal itself ("full particulars" would indicate that at least the date, the details of the appellant and the grounds of appeal would need to be given);
(b) any decision on the appeal.

25.08 *Remediation declarations* In relation to remediation declarations published by the authority under section 78H(6) by the authority where it cannot include particular actions in a remediation notice, the particulars are:

(a) the declaration itself (which must record the things the authority would have required in the notice, the reasons why they would have required them, and the grounds on which they are satisfied they cannot include them);
(b) the location and extent of the contaminated land, sufficient to enable it to be identified; and
(c) the particulars referred to at (c), (d) and (e) above in relation to remediation notices.

25.09 *Remediation statements* In relation to statements under section 78H(7) by the person responsible, or under section 78H(9) by the enforcing authority in default, the relevant matters are as for remediation declarations, namely:

(a) full particulars of the statement (which must contain the past, present or future remediation actions, the person carrying them out and the date by which they are expected to be carried out);
(b) the location and extent of the contaminated land; and
(c) the same particulars (c), (d) and (e) as remediation notices.

25.10 *Appeals against charging notices*

(a) any appeal against a charging notice served by the authority under section 78P; and
(b) any decision on the appeal.

25.11 *Special sites* The Environment Agency must register particulars of:

(a) any notice given by the local authority of the Secretary of State which has effect as the designation of a special site;
(b) the relevant provision of the Regulations by dint of which the land is a special site;
(c) any notice given by the Agency under section 78Q of its intention to adopt a remediation notice already served; and
(d) any notice given under section 78Q terminating the designation of land as a special site.

25.12 *Notification of claimed remediation* Full particulars of any notification given to the authority under section 78R(1)(h) or (j) of remediation claimed to have been carried out. This is discussed further below.

25.13 *Convictions* Any convictions for offences under section 78M for failure to comply with a remediation notice, including:

(a) the name of the offender;
(b) the date of conviction;
(c) the penalty imposed; and
(d) the name of the court.

25.14 *Agency guidance* In relation to site-specific guidance issued by the Agency to a local authority under section 78V(1), the date of the guidance must be placed on the register. It appears that the substance of the guidance, curiously, need not be.

25.15 *Other controls* This category covers the situations where the enforcing authority is precluded from serving a remediation notice because of the application of other statutory controls. The particulars in this case are:

(a) the location and extent of the contaminated land;
(b) the matters covered by particulars (c), (d) and (e) in the case of remediation notices;
(c) any steps of which the authority has knowledge, carried out under section 27 of the 1990 Act towards remedying the relevant harm or pollution;
(d) any steps of which the authority has knowledge, carried out under section 59 of the 1990 Act in relation to unlawfully deposited waste or the consequences of its deposit, including the name of the waste collection authority involved, if any; and
(e) any water discharge consent the existence of which precludes the enforcing authority from specifying a particular thing in a remediation notice.

Registers: notifications by owners, occupiers and other appropriate persons

25.16 Most of the matters listed as requiring to be included within the register are obvious and require no explanation. However, sections 78R(1)(h) and (j) are somewhat different. They require the enforcing authority to place on the register notifications by persons served with a remediation notice, or owners or occupiers, of what they claim has been done by way of remediation. This is the only statutory reference to such notifications contained in Part IIA.

25.17 The Regulations state that such notifications must contain[8]:

(a) the location and extent of the land sufficient to enable it to be identified;
(b) the name and address of the person who it is claimed has done each thing;
(c) a description of any thing it is claimed has been done; and
(d) the period within which it is claimed each such thing was done.

25.18 The statutory reference to such notifications was inserted by a Government amendment in response to an amendment proposed by the Law Society of England and Wales, and provides a means by which an appropriate person, owner or occupier can record in a public form what has been done to comply with a remediation notice or remediation statement. In this way some of the blight which might otherwise affect the land may be alleviated. However, it is important to note subsection (3), which negatives any implied representation by the authority that what is stated on the register as having been done has in fact been done, or has been done adequately. The authority's role is therefore simply one of a "postbox", recording notification which it receives; as it was put in debate, it is not the responsibility of the enforcing authority to indicate that the land has "a clean bill of health".[9] The owner of the land might well ask, in those circumstances, how the land is to get a "clean bill of health" and how the blighting effects of the service of a notice are to be removed where all that the notice required has been done.

Exclusions from the register: national security

25.19 By section 78S(1) no information may be included on the register if in the opinion of the Secretary of State its inclusion could be contrary to the interests of national security. It is for the Secretary of State to issue a direction under section 78S(2) relating to information which in his view

---

[8] Regulation 15(2).
[9] *Hansard*, H.L. Vol. 562, col. 1047.

affects national security. That direction may either be specific or general, that is related to information of a particular description. It may also require certain descriptions of information to be referred to the Secretary of State for his determination: such information must not be included on the register in the interim. The enforcing authority is required by section 78S(3) to notify the Secretary of State of any information which it excludes from the register in pursuance of such directions.

25.20 Section 78S(4) provides a procedure whereby any person who feels that information may be such as to fall within section 78S(1) can give a notice to the Secretary of State specifying the information and indicating its apparent nature. The enforcing authority must at the same time be notified and the relevant information may then not be included on the register until the Secretary of State determines that it should be included.

Commercially confidential information: procedures

25.21 By section 78T(1) no information relating to the affairs of an individual or business shall be included on the register without the consent of the individual or, unless pursuant to a direction from the Secretary of State, where that information is commercially confidential. Information is only commercially confidential for this purpose where it is determined to be so by the enforcing authority or by the Secretary of State on appeal, pursuant to the following procedures:

1. **25.22** Where it appears to the enforcing authority that information might be commercially confidential, it must give notice to the person or business to whom the information relates, effectively warning them that the information will be placed on the register unless excluded under the section.[10]
2. **25.23** The authority must give the person or business a reasonable opportunity of objecting to its inclusion and of making representations to justify that objection.[11]
3. **25.24** The authority must determine, taking into account any representations made, whether the information is or is not commercially confidential.[12]
4. **25.25** Where the determination is that the information is not commercially confidential, a period of 21 days from the date on which the decision is notified to the person concerned must be allowed to elapse before the information is entered on the register.[13]

---

[10] Section 78T(2)(a).
[11] Section 78T(2)(b).
[12] Section 78T(2).
[13] Section 78T(3)(a).

25.26 5. The person concerned may appeal to the Secretary of State within that period, and if an appeal is brought the information shall not be entered until seven days after the appeal is finally determined or is withdrawn.[14]

25.27 6. Appeal shall be by way of private hearing if either party requests it or if the Secretary of State so directs.[15]

25.28 7. Appeals are subject to section 15(10) of the 1990 Act[16] and to section 114 of the 1995 Act, which allows the Secretary of State to delegate determination of the appeal to an appointed person.[17] Schedule 20 of the 1995 Act has effect with regard to such delegated appeals, in respect of which the appointed person has the same powers as the Secretary of State.[18] No local inquiry may be held in relation to appeals under Section 78T.[19] The provisions of the Local Government Act 1972 sections 250(2)–(5) on evidence and costs will apply to hearings under Schedule 20.[20]

25.29 8. The Secretary of State may give directions to enforcing activities as to specified information, or descriptions of information, which are required in the public interest to be included in registers notwithstanding that the information may be commercially confidential.[21]

25.30 9. Information which is excluded on commercial confidentiality grounds ceases to be treated as confidential after four years from the date of the relevant determination.[22] The onus rests with the person who furnished it to apply to the authority for it to remain excluded. There is no obligation on the authority to give any warning or notice that this period is about to elapse. The same procedures for determination and appeals as above will then apply. There is an oddity in the drafting of subsection 78T(8), which refers to "the person who furnished the information". This may not by now be the same person as the person whose commercial interests are affected: indeed the two persons may never have been the same. The subsection should presumably have read: ". . . the person to whom or to whose business it then relates may apply, . . . etc."

---

[14] Section 78T(3).
[15] Section 78T(4).
[16] Section 78T(5).
[17] Section 78T(6).
[18] Sched. 20, para. 3.
[19] Sched. 20, para. 4(3).
[20] Sched. 20, para. 5. (In Scotland the Local Government (Scotland) Act 1973, sections 210(3)–(8).)
[21] Section 78T(7).
[22] Section 78T(8).

Meaning of "commercially confidential"

25.31 By section 78T(10) information is commercially confidential for the purpose of making a determination under the section if its inclusion on the register would prejudice to an unreasonable degree the commercial interests of the relevant person or business.

25.32 This wording follows that used in relation to commercial confidentiality in the context of IPC under Part I of the 1990 Act. The considerations involved are, however, likely to be somewhat different. In relation to the control of prescribed processes, the operator is likely to be concerned about information on the nature of the process, volumes, throughputs, fuel mix, and similar matters, which may give valuable information to trade competitors or suppliers. The main concern of an owner of contaminated land is likely to be the blighting effect on value, resulting from public knowledge that the land is contaminated or is subject to enforcement action. However, section 78T(11) expressly states that any prejudice relating to the value of the contaminated land in question, or otherwise to its occupation or ownership, is to be disregarded in making the determination. As the Circular makes clear, this means that information cannot be excluded from the register solely on the basis that its inclusion might affect its saleability or sale price.[23]

Reports by appropriate Agency on contaminated land in England, Wales and Scotland

25.33 Under Section 78U, so that the public may be reassured that action is being taken in respect of contaminated land, the appropriate Agency is required from time to time to prepare and publish a report on the state of contaminated land in England and Wales or in Scotland, as the case may be. Such reports are to be published "from time to time", or at any time requested by the Secretary of State.[24] For that purpose a local authority must respond to a written request to furnish the Agency with the relevant information, namely such information as the local authority may have, or may reasonably be expected to obtain, with respect to the condition of contaminated land in its area, acquired in the exercise of its functions under Part IIA.[25]

Handling information outside the statutory registers

25.34 As indicated above, the registers under section 78R provide only a partial picture. They are focused on the enforcement history of the site and do not relate to actions or information preceding the service of a remediation

---

[23] DETR Circular, *Contaminated Land,* Annex 2, para. 17.11.
[24] Section 78U(1).
[25] Sections 78U(2) and (3).

notice (or if the site is a special site, its designation as such). It is the process of inspecting land and determining whether it is contaminated which may in practice prove more controversial and which may have serious effects on value and marketability well before any notice is served. The question of how authorities handle such information is therefore important. Local authorities will need to consider the Environmental Information Regulations 1992, the provisions of the Local Government Act 1972 and general issues of confidentiality and data protection legislation.

The Environmental Information Regulations 1992

25.35 Information on the condition of land or groundwater, and the risks which they present, could clearly constitute "information relating to the environment" within regulation 2(1) of the Environmental Information Regulations 1992.[26] So equally, it appears could information on past activities resulting in contamination, or present activities to identify and deal with it.[27] The sources of such information could itself constitute information within the Regulations, in that it is relevant to the reliability and quality of the information.[28] As such, a local authority as a public body in possession of such information, which it may have gathered under its inspection functions, is obliged to make it available to any person who requests it,[29] unless it falls within the exceptions at regulation 4.

25.36 There are a number of exceptions which could be of relevance. By regulation 4(2) certain categories of information are capable of being treated as confidential should the authority so choose. These include:

> "information the disclosure of which—
> (a) . . .
> (b) would affect matters which . . . are the subject-matter of any investigation undertaken with a view to any [legal] proceedings or enquiry;
> (c) would affect the confidentiality of the deliberations of any relevant person [*i.e.* the authority];
> (d) would involve the supply of a document or other record which is still in the course of completion, or of any internal communication of a relevant person; or
> (e) would affect the confidentiality of matters to which any commercial or industrial confidentiality attaches . . ."

25.37 An authority could potentially argue that a number of these provisions apply, because it is obtaining the information with a view to legal

---

[26] S.I. 1992 No. 3240, reg. 2(1) and (2)(a).
[27] *ibid.*, reg. 2(2)(b) and (c).
[28] *R. v. British Coal Corporation, ex p. Ibstock Building Productions Ltd.* [1995] J.P.L. 836; [1995] Env. L.R. 277.
[29] Regulation 3.

proceedings (namely, issuing a remediation notice), or because the investigations are not finished, or because disclosure would affect the confidentiality of the determination process by the authority as to whether the land is contaminated, or because the information was provided on a confidential basis by a third party. In *Maile v. Wigan Metropolitan Borough Council*[30] it was held, in relation to a database identifying potentially contaminated sites commissioned by the local authority, that the exemptions in regulations 4(c) and (d) were applicable. The date and study were still unfinished, and in the view of Harrison J., disclosure of what was still preliminary material could cause unnecessary alarm to local residents and landowners. It will be a question in each case whether the facts are such as to fall within one of these exemptions, but it must be remembered that the authority still has to take a decision as to whether, as a matter of its discretion, the information should or should not be disclosed.

25.38 The classes of information which must be treated as confidential under the 1992 Regulations are limited by regulation 4(3) to confidential information the disclosure of which would contravene a statutory requirement or involve the breach of an agreement, personal information contained in records held on individuals, and information supplied voluntarily in circumstances where the authority is not entitled apart from the 1992 Regulations to disclose it. The name of a person supplying information to the authority would probably not consitute "personal information" relating to individual records.[31] However, it is conceivable that some information on contamination or on proposed remediation might constitute information provided voluntarily so that it cannot be disclosed.[32] This would depend on the circumstances, though it should be remembered that under section 108 of the Environment Act 1995 there is a wide power to compel the provision of information: where that power applies, it may be difficult to argue that the person supplying the information could not have been put under a legal obligation to provide it, as regulation 4(3)(c)(i) requires. Also, it would need to be considered whether the authority would have been entitled apart from the 1992 Regulations to disclose it (see regulation 4(3)(c)(ii)). This may be important where information is supplied, for example, in conjunction with a planning application; if it forms part of the application or any environmental statement then its disclosure is required under planning law.

25.39 As will be appreciated from this brief summary, difficult legal issues may well arise for local authorities, for those who have provided information, and for those seeking it, whether in the public interest or for commercial reasons of their own. Another factor which will have to be taken into account is the Human Rights Act 1998, and in particular Article 8 of the

---

[30] High Court, May 27, 1999, unreported, but noted at *Environmental Law Bulletin*, No. 58, September 1999, p. 20.

[31] See *R. v. British Coal Corporation, ex p. Ibstock Building Productions Ltd.* (above, n. 28).

[32] See *R. v. Secretary of State for the Environment, Transport and the Regions and Midlands Expressway Ltd.* [1999] Env. L.R. 447.

European Convention on Human Rights, dealing with respect for the home and family life. In *Guerra v. Italy*[33] it was held that failure to provide information to concerned local residents on emissions and risks relating to a chemical plant could constitute a breach of that Article. It is conceivable therefore that the same argument might be raised where an authority decides in its discretion not to provide information on the risks which might be presented by a contaminated site. The authority may have to show, for example, that it has weighed the harm arising from withholding that information against the blighting effect on the homes of other residents which might be caused by releasing the information prematurely.

Committee reports and papers

25.40 Information on potentially contaminated sites may well be contained in reports to committee and in background papers. If a report indicates that a site is contaminated, but the committee ultimately decides that it does not meet the statutory test for contaminated land, considerable blight could still be caused by the papers having been released. A decision will therefore need to be taken as to whether the press or public should be excluded from the relevant part of the meeting and whether the papers should be made available. This depends on whether the information in question is "exempt information" under Part VA of the Local Government Act 1972.

25.41 The categories of exempt information specified by section 100I of, and Schedule 12A to, the 1972 Act include:

(a) information relating to the business or financial affairs of a particular person (this might be relevant where issues of financial hardship fall to be discussed);
(b) certain types of expenditure under contracts proposed to be incurred by the authority (possibly relevant where the authority is contemplating undertaking remediation itself);
(c) information which would reveal that the authority proposes to give a statutory notice imposing requirements upon a person (this may be relevant in any case where a remediation notice is in contemplation); and
(d) instructions to and opinions of counsel, and advice received, information obtained or action to be taken in connection with legal proceedings or any other matter affecting the authority (this may be relevant where the authority seeks legal advice as to who is the appropriate person, or on enforcement action following service of a notice, or on the authority's own position where it is an appropriate person).

25.42 Further, by section 100A(2) nothing in Part VA authorises or requires the disclosure of confidential information in breach of confidence, though

[33] (1998) 26 E.H.R.R. 357.

the definition of "confidential information" given by section 100A(3) is relatively narrow.

25.43 The application of section 100A was considered in relation to a contaminated land database in *Maile v. Wigan Metropolitan Borough Council.*[34] On the evidence it appeared that the database had not in fact been used in preparing the relevant report, and hence was not required to be disclosed as background information. This illustrates the importance of handling the relevant information carefully and precisely.

---

[34] See n. 30 above.

Chapter 26

COMMERCIAL IMPLICATIONS

Summary

Liability for contaminated land is not a new risk. Increasingly, it has been recognised as a factor in commercial transactions during the 1990s. The new regime of Part IIA however heightens the risks and necessitates reconsideration of previous practices in terms of drafting contracts and other commercial instruments. This Chapter consider the risks of liability from the viewpoint of parties to various types of transactions and suggests some possible approaches to allocating risks.

***Note:* This Chapter is an expanded and updated version of the relevant part of Section 1 of the *First Supplement* to *Contaminated Land*, published in 1998, amended to take account of the final version of the statutory guidance.**

Commercial implications

26.01 What are the implications of Part IIA for those involved now or in the future with property or other commercial transactions? To some extent the principles remain much as they were prior to the introduction of Part IIA in terms of the general allocation of risk; however, the massively increased sophistication of the new contaminated land system requires reconsideration of previous practices. The implications are considered below from the following viewpoints:

(a) sellers of land;
(b) buyers of land;
(c) landlords;
(d) tenants;
(e) parties to options/development agreements;
(f) contractors and employers;
(g) sellers of company shares;
(h) buyers of company shares; and
(i) mortgagees.

It will be appreciated that contractual provisions designed to take advantage of the provisions for channelling liability under the contaminated land regime are strictly speaking relevant only to that regime. However, to the extent that a regulator (for example, the Environment Agency or SEPA under the new works notice legislation in relation to water pollution) has a discretion as to whom to target for enforcement action, and has sympathy with the policy rationale for the contaminated land exclusion and apportionment provisions, then the person whom such agreements are designed to protect might also seek to use their existence to influence that regulator's exercise of its discretion. **26.02**

Also, it is worth noting that the focus in what follows is on the parcelling out of potential liabilities between the contracting parties. The process might in some cases be eased by looking beyond this to ways, such as insurance, of reducing exposure to such liabilities. **26.03**

Sellers of land

So far as the seller of land is concerned, the basic position is as follows: **26.04**

1. If the seller himself caused the land to be contaminated, he will remain potentially liable as a Class A "causer" following completion.
2. If prior to the sale, the seller was aware of the presence of contamination, there is a risk that the seller may remain liable following completion on the basis that he had knowingly permitted the contamination to remain present during his period of ownership. This point is open to argument but there must at least be a risk of such responsibility.

Sellers of land: managing risk

The key commercial question for the seller is whether he wishes to avoid retaining the risks mentioned above following completion. The possible ways of seeking to achieve this are as follows. Reference should also be made to the detailed discussion of the Guidance on exclusion from liability in Chapter 17. **26.05**

Indemnities

A contractual indemnity from the purchaser in respect of such liability. The purchaser may well not be willing to give such an indemnity, or to give one only upon limited terms; and in any event the value of such indemnity will depend upon the worth of the party giving it. At best, therefore, an indemnity should be seen as a method of reducing rather than removing risk. However, it has the virtue of potentially covering costs and expenses **26.06** arising from all of the statutory sources of liability discussed above, and indeed also potential civil liabilities. In construing indemnities, one point which needs to be borne carefully in mind is the definition of "environmental law".

26.07 Given the lengthy period before the provisions giving rise to the risks of liability came into force, the recipient of the indemnity should have ensured, in order to obtain protection, that the definition was such as to include Part IIA in the period between enactment and entry into force, and was not confined to legislation in force at the time of the agreement. In some cases, the parties will have dealt with this by expressly referring to the relevant provisions contained in Part IIA of the Environmental Protection Act 1990 and the works notice provisions in sections 161A–161D of the Water Resources Act 1991 in the definition provisions.

Agreements as liabilities

26.08 The seller and the buyer may agree that the buyer will be responsible for the cost of any future remediation. Assuming (as, depending on the facts, arguably may become the case) that the buyer's state of knowledge and capacity to act will make it a "knowing permitter" at some point following the transaction, then according to paragraphs D.38 and D.39 of the Guidance the enforcing authority should make such determinations on exclusion and apportionment as are needed to give effect to any such agreement, and should not apply the remainder of the Guidance between the parties to the agreement.

26.09 However, only Part IIA (as opposed to the other potential sources of liability) contains a mechanism for such an agreement to determine on whom the regulator serves a notice. In any event, such an agreement will only apply as between the seller and the buyer, and would not affect the position of any subsequent purchaser or, indeed, any other appropriate person who is not a party to the agreement. Any such agreement would need to be notified in writing to the relevant authority. Another risk is that should the buyer cease to exist before any notice is served, then he would not be in the liability group at all, and accordingly the authority would not be able to give effect to the agreement.

26.10 The Guidance contains its own limitations as to the effect of such agreements. A copy of the agreement must be provided to the authority, and none of the parties must inform the authority that it challenges the application of the agreement. Secondly, the authority should disregard the agreement where the party whose share of costs would be increased under it would benefit from the "hardship" limitation on cost recovery. Accordingly, an adverse change in the financial circumstances of one party after the agreement may prejudice the other: see further the discussion in Chapter 17.

Payments to cover clean-up

26.11 If the extent of the contamination is known, and a clean-up method can be agreed and costed, then the seller may make a payment to the buyer to cover that remediation, either by way of a direct payment or by an express deduction from the purchase price. Provided that the payment is adequate, then in the event that the remediation is not carried out, or is not carried out effectively, the seller should be excluded by Test 2 at paragraphs D.51–D.56 from future liability under the contaminated land regime (although not necessarily under the other statutory sources of liability discussed above). In order to ensure the seller's protection, the position should be stated explicitly in the contracts.

26.12 To satisfy this test, the seller must not retain any further control over the condition of the land in question. The seller can, however, safely retain contractual rights to ensure that the work is carried out properly. As mentioned above, this test will not be suitable for all circumstances, and may well be confined in practice to cases where land is sold either for development or with a clear idea as to the remedial measures required, or where there is a specific and clearly identified contamination problem which is capable of being addressed by specific measures. In those circumstances Test 2 may be particularly useful.

"Sold with information"

26.13 The seller may seek to apply the "sold with information" test (Test 3, set out at paragraphs D.57–D.61 of the Guidance) in relation to liability under the contaminated land regime. This can be triggered either by the seller acquiring information as to the presence of contamination itself, and providing that information to the buyer, or alternatively by the seller allowing the buyer to carry out its own investigation, so that the buyer thereby generates the information. The information must be sufficient to allow the buyer to be aware of the relevant pollutant and the "broad measure of its presence". The information must be available before the sale becomes binding, and therefore either the investigation must be carried out before exchange of contracts, or exchange must be on a conditional basis to allow subsequent investigations to take place. The problem with the conditional contract approach is the difficulty of framing the conditions so as to achieve sufficient certainty as to the circumstances in which the contract does become binding. Also, the price may need to be left flexible (or some mechanism agreed for adjusting the price) following the completion of the investigations. Clearly, the preferable course would be to obtain the information prior to exchange.

26.14 Knowledge of contamination may be deemed to exist where the buyer is a "large commercial organisation" or a public body, and permission has been given to the buyer to carry out its own investigations. The Guidance, however, only suggests that this should "normally be taken as sufficient indication" that the purchaser had the necessary information. Obviously, it would be as well to make it clear in writing that the opportunity had been given (and the scope of that opportunity), either in correspondence or in the contract itself.

Restriction of future uses

26.15 The seller may seek to restrict the future uses or development of the land so as to minimise the risk that the contamination will migrate or otherwise be exacerbated, thereby triggering enforcement action. Buyers can in many cases, however, be expected to resist attempts to encroach in this way upon their autonomous use of the land.

Future access by the seller

26.16 Where there is a possibility (for example, under the contaminated land or works notice provisions) that the seller may require access to the land following the sale, in order to comply with statutory clean-up requirements, the seller could seek to secure the necessary rights of access in the sale contract. This would in particular prevent the buyer from claiming statutory compensation for the disturbance and other costs associated with compliance, pursuant to section 78G of the Environmental Protection Act 1990, section 161B of the Water Resources Act 1991 or (in Scotland) section 46B of the Control of Pollution Act 1974. The seller who will be seeking to rely on Test 2 or 3 (payments for remediation or sale with information) should be aware that the retention of such rights may prejudice his ability to rely on those tests as constituting retention of control under Test 2 or retention of rights to use and occupy under Test 3. This aspect is discussed further in Chapter 17.

26.17 In the light of this risk, rather than reserving rights of access, it may be preferable simply for the seller to seek the inclusion of an acknowledgement from the buyer that it will not apply for statutory compensation in relation to access by the seller (if necessary) to carry out such works.

Buyers of Land

26.18 So far as the contaminated land regime is concerned, the basic position for the buyer of land is as follows:

1. The buyer may fall within the first group of appropriate persons with regard to the pre-existing contamination on the land, if he become aware of its presence and may accordingly be said to have knowingly permitted it to remain following the purchase.
2. The buyer may become responsible as a second tier appropriate person in his capacity as owner, if it is the case that no person can be found who caused or knowingly permitted the contaminants to be present. Unlike the first kind of liability, this risk lasts only for so long as the buyer continues to own the land.

26.19 Two factual scenarios can be contemplated, discussed in the following paragraphs.

Buyer aware of contamination

26.20 The first scenario is where the buyer is aware at the time of purchase that the land is subject to contamination. The possibilities here are as follows.

26.21 The buyer may seek an indemnity from the seller with regard to liability for that contamination. The same difficulties and shortcomings here would apply as in relation to indemnities granted to the seller, namely the unwillingness to give unlimited indemnities and the possibility that the indemnifying party may not have the funds to satisfy the indemnity at some point in the future.

26.22 The parties may make an agreement as to who should be responsible. This solution suffers from the same potential problems as the equivalent method referred to above under the heading of "Sellers of land", namely that its statutory effect is limited to the contaminated land regime and, in any event, the seller may cease to exist, or the enforcing authority may take the view that it should not give effect to the agreement.

26.23 The price may be adjusted downwards to reflect the risk of future liability to the buyer. This is a different situation to where the payment is made to be used for remediation of land, and the buyer should ensure this is explicitly reflected in the agreement, to avoid any argument that the test of "payments made for remediation" should apply in the seller's favour.

26.24 The contamination may be cleaned up, thereby removing (hopefully) the risk of future liability. Either the contamination may be cleaned up by the seller before completion, or alternatively the price may be reduced by a sum appropriate to allow the buyer to carry out clean-up. The question here is partly one of who bears the risk that the remediation proves ineffective. If the remediation is to be carried out by the buyer of the land, then potentially the test of "payments made for remediation" may be applied to impose the risk solely on the buyer, and thus exclude the seller from further liability under the contaminated land regime. Where the remediation is carried out by the seller, he may well retain residual liability if remediation proves ineffective, but this will not necessarily release the buyer from also being liable. The risks to the buyer may be reduced by an indemnity given by the seller and/or by collateral warranties given by the engineers carrying out the remedial work. It is conceivable that the transaction could be structured in such a way that the buyer could be regarded as making a payment to the seller in order to carry out the remediation, but this would require careful structuring and drafting, and would not remove the risk that the seller might in future cease to exist, leaving the buyer as the sole member of the liability group.

Buyer does not know whether land contaminated or not

26.25 The other situation is where the buyer does not know for sure that the land is contaminated, but is aware of the risk that it might be. Awareness of this risk would not in itself be enough to trigger the "sold with information" test. However, if the buyer is a large commercial organisation or public body there is a risk that it might be deemed to have knowledge of the contamination by virtue of having had permission to carry out an investigation, but having failed to do so. Effectively, this puts the onus on sophisticated buyers of land to take professional advice as to the investigations which they should carry out, and to assess the risks accordingly. From the buyer's point of view, that risk may be reflected either in an indemnity from the seller (with the attendant difficulties previously described) or by a reduction in the purchase price to reflect that risk.

Landlords

26.26 From the landlord's point of view there are two key questions:

1. how best to avoid becoming liable for contamination caused by the tenant; and
2. whether it is appropriate or possible to transfer to the tenant the risk of liability for pre-existing contamination (which may or may not have been caused by the landlord).

Contamination caused by tenant

26.27 On the first issue, the tenant who causes contamination will clearly be potentially liable as a causer. The landlord however is also at risk of liability on the following bases:

1. If the landlord had sufficient knowledge of and control over the tenant's activities, it might be argued that the landlord knowingly permitted the contamination. However, it is clear from the first exclusion test that simply creating a tenancy would not of itself constitute such a level of control.
2. When the lease comes to an end, the landlord, having regained possession and having become aware of contamination, may be regarded as responsible for knowingly permitting its continued presence.
3. If the tenant ceases to exist so that it can no longer be "found", then the landlord may potentially be liable under the second tier of appropriate persons as the owner.

26.28 The approach to all these risks is essentially the same. The landlord should ensure that the tenant's covenants will allow the landlord to require the clean-up of contamination caused by the tenant during the lease, and prior to its termination. When accepting a surrender of the lease, the landlord should be aware of releasing the tenant from ongoing contractual liability in respect of contamination which the tenant has caused.

Pre-existing contamination

26.29 The second and more difficult issue relates to pre-existing contamination, in relation to which the landlord may be responsible as follows:

1. If the landlord himself caused the contamination he may be liable as a first tier "causer".
2. If the landlord was aware of the contamination prior to the grant of the lease he may have responsibility for having knowingly permitted its continued presence.
3. If the contamination was caused by some third party who has ceased to exist, then both the landlord and tenant may be potentially responsible as second tier parties in their capacity as owner and occupier. In cases where the tenant is paying a rack rent, the exclusion test under the Guidance at paragraph D.89 will result in the tenant being excluded from liability, leaving the landlord solely responsible.

26.30 The landlord's approach to these risks may well depend on how the contamination was caused. A well advised tenant would be unlikely to accept liability for contamination which the landlord had caused, or of which the landlord was aware prior to the grant of the lease but had failed to draw to the attention of the tenant. With regard to unknown contamination which has been caused by a third party, this essentially comes down to an issue of allocation of risk between landlord and tenant, and in some respects is akin to the risk of inherent defects in buildings. The distinction may be drawn between long leases (more than 21 years) where the situation can be regarded as akin to that of seller and buyer of land, as outlined above. With regard to shorter leases, the landlord should work on the assumption that it will retain responsibility (and indeed may be solely responsible) for pre-existing contamination unless the lease expressly provides otherwise. The issue therefore comes down ultimately to the drafting of the leasehold covenants, as discussed in the main text of *Contaminated Land*.

26.31 However, if there is a possibility that the landlord may find itself requiring access to the property to comply with a statutory notice served upon it, then the landlord might additionally consider ensuring that it does not become liable to the tenant for statutory compensation, as discussed above in relation to Sellers of land at paragraphs 26.12 and 26.17.

Tenants

26.32 A tenant under a long lease of 21 years or more should be regarded as broadly in the same position as a buyer of land. Tenants under shorter leases at a rack rent may have potential liabilities on the following bases:

1. For contamination which he himself causes.
2. For pre-existing contamination of which he becomes aware and which he may be said to have knowingly permitted to remain present.
3. For pre-existing contamination where no first tier causer or knowing permitter can be found, and where the tenant may be responsible in his capacity as occupier.

4. For contamination which it is the tenant's responsibility to rectify (or to bear the cost of rectifying) under the covenants in the lease.

26.33 Taking each of these in turn, with regard to the first category of liability, the tenant would presumably expect to be responsible for such contamination. In the case of the second and third categories, the tenant might well expect such liability to rest with the landlord, and this would indeed be the effect of applying the exclusion test to the third category. However, the tenant is under potential residual liability if it can be argued that he is a knowing permitter, in which case the exclusion tests relating to owners and occupiers would not be relevant. Therefore, the only way in which the tenant can completely guard against such liability is by providing expressly in the lease that the landlord is responsible for pre-existing contamination, a position which may or may not be acceptable to the landlord.

26.34 On the general principle of caveat lessee, the tenant will be deemed to take the property in its actual condition, defects and all.[1] This seems likely to include pre-existing contamination. The tenant should thus be aware that perhaps the main risk of liability lies under the fourth category, and this may result from inadvertently wide drafting of the "normal" leasehold covenants, for example those requiring compliance by the tenant with statutory requirements. Agreements and covenants in leases may be regarded as an agreement between the parties as to which of them shall be responsible, in which case the enforcing authority would be required under the Guidance to seek to give effect to that agreement in applying the exclusion and apportionment tests.

Parties to options or development agreements

26.35 Insofar as agreements for the development of land involve a transfer of title of the creation of leasehold interest, the basic principles are the same as for those types of transaction. However, there are a number of other issues which require separate consideration.

26.36 The first of these is potential risk to a party who is not the owner or occupier of land, and who is not directly involved in the actual development of the land. The party to an unexercised option will generally not be in occupation of the land. The involvement of such a party may be by a provision of financial assistance, or perhaps as a shareholder in a joint venture company used as the vehicle for the development. Such a party may be concerned that its involvement could result in responsibility for the contaminated condition of the site. Here the first exclusion test in the Guidance is helpful, in that it would exclude from liability anyone who might otherwise be liable by virtue only of having made a grant, loan or

---

[1] *Southwark London Borough Council v. Tanner; Baxter v. Camden London Borough Council (No. 2)* [1999] 3 W.L.R. 939.

other form of credit, or having invested in the undertaking of a company by acquiring share or loan capital (so long as they have not thereby acquired such control as the holding company has over a subsidiary as defined in section 736 of the Companies Act 1985).

26.37 The effect of this first exclusion test should be borne in mind by the parties to development agreements in deciding whether the arrangements reflect accurately the agreement reached between the parties as to the allocation of risk. If it is the intention that a party who might otherwise be excluded by this test from liability should in fact for commercial reasons bear a proportion of the risk, then this will need to be reflected in the documentation. The exclusion test will not (as mentioned above) protect a party who acquires a controlling interest in the company; nor will it protect someone who provides funding and who becomes actively involved in the control of activities which lead to contamination or which create conditions whereby contamination presents a risk of harm.

26.38 Another issue is how contamination may affect the value of the land and consequently the commercial structure of the transaction. It is possible that land may be identified as contaminated prior to the exercise of any option, or before or after the commencement of development. Indeed, it may be the development which generates the information that results in the land being identified as contaminated. Once land is formally identified as being contaminated, the works required for remediation—and consequently the cost—are no longer solely within the control of the owner and other parties to any development agreement. The local authority or Agency will also be involved. The parties to the agreement may have different priorities and ideas as to how, to what standard and within what timescale the problem is to be addressed. The agreement will need to cater for these eventualities.

Development: introduction of pathways or receptors

26.39 Another factor which needs to be considered in the context of development is Test 6, set out at paragraphs D.68–D.72 of the Guidance, and dealing with the introduction of pathways or receptors. The purpose behind this Test is to exclude from liability those who may have previously caused or knowingly permitted contamination, where the relevant harm or risk arises from subsequent actions of others in creating pathways for the transmission of pollution, or by introducing receptors which might be harmed. The Test requires consideration of whether any party has carried out a "relevant action" or made a "relevant omission". A relevant action is carrying out development on the land (development being defined by reference to the town and country planning terms of building, mining or other operations) or making a change in the use of the land such as to require a grant of planning permission. "Relevant omission" is defined as failure during the course of carrying out a relevant action, to take a step which would have ensured that a significant pollutant linkage was not brought into existence, or unreasonably failing to maintain or operate a system installed for the purpose of reducing or managing the relevant problems caused by the contamination.

26.40 In applying the test, the authority should consider whether there are members of the liability group who are liable by virtue of having carried out such acts or omissions which have resulted in the creation of a pathway or receptor, which means that the land is then regarded as contaminated within the statutory test. If there are also members of the liability group who are there solely because they carried out earlier actions, which would not on their own, apart from the later actions, have resulted in the existence of the pollution linkage, then those responsible for the earlier actions should be excluded from liability. Typical situations where this test might operate are therefore:

1. Party A causes or knowingly permits land to be contaminated. Party B then carries out building or engineering works which result in the movement of contamination by creating a new pathway. The test would exclude Party A from liability.
2. Party A causes or knowingly permits land to be contaminated. Party B changes the use of the land by constructing housing, offices or some other form of development, the occupants of which might be harmed by the contamination. Again, the test would exclude Party A from liability.
3. Party A develops land, in the process installing systems for the control of gas, leachate or some other form of contamination. Responsibility for maintaining those systems then passes to Party B, who unreasonably does not maintain them adequately, thereby creating a risk which was not previously present. The test would exclude Party A from liability.

26.41 As will be appreciated, the effect of the test is to channel liability to the person last involved in time in terms of carrying out the relevant acts but in many cases this will no doubt reflect the commercial intention of the parties involved. However, if it does not, then express provision will be needed to achieve the intended result.

Contractors and employers

26.42 It will be recalled that one of the limbs of the first exclusion Test relates to those performing contracts (paragraph D.48(k)). Essentially, a party may be excluded from liability where it falls within the liability group by virtue of providing a service or supplying goods under the terms of a contract with another member of the liability group in question. The Test applies irrespective of whether the contract in question is one for employment, provision of services or the provision of goods. Sub-contracts are to be treated on the same footing as contracts for the purpose of applying the test.

26.43 On its face, therefore, the test is a very broad one applying in favour of a contractor in a way which may or may not reflect the commercial intention of the parties concerned. However, the test states expressly that it will not apply where the action or omission by the contractor leading to him being regarded as an appropriate person was not in accordance with the terms of the contract. Also, it will not apply to protect directors or other senior officers providing services to their company, where consent, connivance or neglect is involved on their part.

26.44 The risk in this Test lies in the importance of acting in accordance with the terms of the contract. Activities not in accordance with the contract will expose the contractor to the risk of liability. This puts a premium on stating the terms of the contract precisely. Again, the question is fundamentally one of the commercial allocation of risk, and whether the consideration paid to the contractor adequately reflects the full risk which the contractor is assuming in carrying out the relevant activities. Those risks could be substantial in the context of the development of land. For example, if a contractor constructs buildings which result in the creation of targets for contamination, then if the contractor is a liable party, the effect of Test 6 (Introduction of Pathways and Receptors) as described above, could result in the contractor bearing sole responsibility for the entire cost of clean up. Those advising employers and contractors will therefore need to consider carefully the drafting of the contract in terms of the level of control actually assumed by the contractor, and whether the contractual terms reflect the true intention of the parties as to whether the contractor should in fact assume the risk of responsibility for contamination and, if so, in what circumstances. The way in which the Test is now worded means that the narrower the parameters within which the contractor operates, the greater the risk that he might be liable for activity outside their scope. The greater the contractual freedom the contractor has, the less likely he may be to fall outside the protection of Test 1. This may not necessarily coincide with the commercial intention of the parties involved.

Sellers of company shares

26.45 The sale of the company will obviously carry with it the full spectrum of potential responsibilities of that company for past, present or future contamination. Any residual risk would relate to whether the main party selling the shares could be said, through his past control over the company, to have caused or knowingly permitted the contamination. Under Exclusion Test 1 investing in a company, acquiring shares or loan capital should not of itself result in liability under the contaminated land regime, unless the level of control is equivalent to that of a holding company over its subsidiary under section 736 of the Companies Act 1985. Holding companies who are disposing of subsidiaries should therefore be aware that they may not be able to rely upon that exclusion test. However, it must be emphasised that being in a position of holding company will not necessarily be equated to having caused or knowingly permitted contamination. As discussed in Chapter 13, this would depend on the facts, and in particular the degree of knowledge and the level of control exercised by the holding company over the relevant acts of the subsidiary.

26.46 Whether an indemnity should be given by the vendor will be a matter for negotiation between the parties. In circumstances where the holding company feels that it may have exercised sufficient control to have a residual liability, this will also need to be considered carefully, since if it is the intention that the holding company should divest itself of all liability, it will need to reflect this in the agreement either by way of an indemnity from the buyer, or by express agreement to that effect which can be notified to the authority under the Guidance).

Buyers of company shares

26.47 So far as sites which are owned by the company being sold are concerned, the position, in terms of risk to the buyer of the company, is effectively the same as for a buyer of land. However, there is the further dimenson the company will also carry with it any liabilities which it may have by virtue of previous actions or omissions, and in particular these will include:

1. Liability for contamination which the company may have caused or knowingly permitted during its general business activities. The risk here will no doubt depend upon the nature of those activities. For example, in the case of a construction company which may have been involved in many contaminated sites over the years, the risk could be significant. Similarly, for a company which has been involved in the transport and distribution of contaminative substances such as oil, again the risk could be relatively widespread.
2. Liability relating to sites which the company has owned in the past, either where it caused contamination on its own site, or where it was aware of the presence of contamination on the site which it knowingly permitted to remain.

26.48 Both of these are difficult matters to deal with. A company may only have limited knowledge of contamination which it has caused in the past. There may only be incomplete or inadequate information as to sites which have been previously owned or occupied, and even where these are known investigation may be impossible. The matter can therefore either be dealt with by structuring the transaction in terms of an assets purchase in order to confine the risks to those assets currently owned by the company, or alternatively by way of indemnification. Issues relating to these aspects are considered in detail in *Contaminated Land*. Assets "hive down" arrangements also need some careful thought in the light of the Part IIA regime and are considered below.

Assets "hive down" and similar arrangements

26.49 A common practice is to package or "hive down" assets of a company into a new corporate vehicle ("Newco"), the shares of which are then sold. Even where a "Newco" is not involved, there may be some re-arrangement of assets within a group to be carried out before a company is sold. The possible operation of the regime of Part IIA will need thinking through carefully in the circumstances of such cases. Take, for example, the situation where the assets hived down from the seller to Newco include a site which is known to be contaminated. If the disposal to Newco was at arm's length, it might well be argued that Test 3 would apply so as to exclude the seller from future liability. However, if at the time the seller and Newco are related companies, then under paragraph D.46 of the Guidance, the seller will not be excluded from liability by Test 3. The subsequent sale of the shares of Newco to the buyer will not change that position.

26.50 Similar considerations could apply where a company to be sold, "Targetco", disposes of sites which are known to be contaminated prior to the sale, so that they are excluded from the sale to the buyer of Targetco. If Targetco disposes of those assets to a company within the same group, Test 3 will not operate to rid Targetco of any residual liability it may have in respect of those sites and, consequently, this will pass with Targetco to the buyer.

26.51 In both examples above, the end result may not necessarily be in accordance with the intention of the parties. In the first example, the seller might have thought it was selling the site in a known contaminated condition and that the buyer would be taking over all liabilities. In the second example, the buyer might have thought it was acquiring Targetco free of liability for the site which had previously been divested.

26.52 Test 3 is in fact qualified by paragraph D.59(c), which states that where a group of transactions or a wider agreement includes a sale of land, the sale of land should be taken to be at arm's length where the person seeking to be excluded can show that the net effect of the group of transactions or the agreement as a whole was a sale at arm's length. In the first example above, the seller might seek to argue that, looked at overall, the land was being sold to the buyer, via Newco, at arm's length, and that accordingly Test 3 should operate on the disposal to Newco. This is probably not an argument that would be open to the buyer in the second example. In any event, paragraph D.59(c) is not precisely worded, and it would be inadvisable for either side to make assumptions as to whether or not it would apply to their circumstances. The important thing is, therefore, to think through carefully how the Guidance might apply to the arrangements proposed (which may be driven by tax or other non-environmental considerations) and whether this accords with what is intended in terms of allocation of environmental risk.

Mortgagees

26.53 The potential risks to mortgagees of liability under the new contaminated land regime were the subject of debate in Parliament during the passage of the Environment Bill. The protection which mortgagees have lies in part in the definition of "owner" which makes it clear that liability should not extend to a mortgagee who is not in possession of the land. The mortgagee who wishes to enforce his security can therefore take an informed decision as to potential risks before doing so. The question of "involuntary possession"—for example, where the borrower simply returns the keys of the property to the lender—was considered during debate and the Government's response was that mortgagees would have to bear the risk in that particular situation. Leaving that issue aside, mortgagees can therefore control their own risks with regard to liability as owners.

26.54 The other question is then whether the mortgagee may be responsible for having caused or knowingly permitted contamination. Test 1 makes it clear that the act simply of providing a loan or other form of credit should not of itself result in liability. A more direct form of control would therefore be necessary, and it is submitted that it is extremely unlikely in practice that a mortgagee would in normal circumstances exercise such a level of control. The dilemma here for mortgagees perhaps lies mainly in the wish to take steps to protect the value of their security, without going so far as to exercise control over contamination prevention or remediation measures. This is the balance which needs to be struck in drafting the contractual arrangements, and in the practical measures put in place to ensure compliance. The documentation could, for example, make it clear that whilst the borrower must report to the lender any matters which may affect the value of the property, and must take steps to address those matters, the responsibility for the nature of the steps to be taken and their adequacy, rest with the borrower.

Lenders by investment

26.55 Lenders by way of investment in a company are in the same position as buyers of shares. Investment which falls short of acquiring the control of a holding company will not of itself give rise to liability. However, the actual degree of control exercised will be relevant and this will depend on matters such as whether a representative of the investor is appointed as a director of the company in question.

Trust companies

26.56 Commercial trustees, such as banks, are in a potentially exposed position as the owner of contaminated property, and do not benefit from any special protection or exclusion. They have a lien over the property of the trust in respect of expenses properly incurred, but this depends for its protection and comfort to the trustee on the adequacy of those assets. It may be that the best that can be said is that such companies should put proper risk management procedures into place before accepting trusteeships, and that they should explore the possibility of obtaining insurance against residual sites.

Appendix A

PRECEDENTS

1. Notice of identification of contaminated land (section 78B(3)).
2. Notice to other appropriate persons of identification of contaminated land (section 78B(4)).
3. Notice of decision that land is required to be designated a special site (section 78C(1)(b)).
4. Notice of decision that land is required to be designated as a special site in response to notice from appropriate Agency (section 78(5)).
5. Notice that decision that land is required to be designated as a special site has taken effect (section 78C(6)).
6. Notice of reference of decision as to special site status to the Secretary of State (section 78D(3)).
7. Notice requesting information.
8. Remediation notice (section 78E(1)).
9. Notice to accompany copies of remediation notice (section 78E(1)).
10. Remediation declaration (section 78H(6)).
11. Remediation statement (section 78H(7)).
12. Application for compensation for the grant of rights (section 78G(5)).
13. Complaint and notice of appeal against remediation notice (magistrates' court) (section 78L(1)).
14. Notice of appeal against remediation notice (Secretary of State) (section 78L(1)).
15. Prosecution for failure to comply with remediation notice (section 78M(1))
16. Notice before exercising remediation powers under section 78N.
17. Charging notice under section 78P.
18. Notice of enforcing authority as to possible commercial confidentiality (section 78T(2)).

PRECEDENT 1

NOTICE OF IDENTIFICATION OF CONTAMINATED LAND

Note: Section 78B(3) requires an authority which identifies contaminated land in its area to give notice of that fact to:

(a) the appropriate Agency;
(b) the owner;
(c) any person who appears to be in occupation; and
(d) each person who appears to be an appropriate person.

The notice must state by virtue of which paragraph (a)–(d) it is given. The service of notices under the 1990 Act is governed by section 160.

ENVIRONMENTAL PROTECTION ACT 1990, SECTION 78B(3)

To: [NAME AND ADDRESS OF RECIPIENT]:

1. Under section 78B(1) of the Environmental Protection Act 1990 [NAME OF AUTHORITY] has identified the land described in the Schedule to this Notice ("the Land"), situated within the Authority's area, as contaminated land.

2. Notice of that fact is given to you in the capacity indicated below and pursuant to the paragraph of section 78B(3) of the 1990 Act indicated below:

   (A) the appropriate Agency;
   (B) the owner of the Land;
   (C) a person appearing to the Enforcing Authority to be in occupation of [the whole] [part] of the Land;
   (D) a person appearing to the Enforcing Authority to be an appropriate person.
   [DELETE AS APPLICABLE]

3. Because this notice may potentially have financial consequences, you are advised to consult an appropriate independent professional adviser.

4. Should you or your adviser wish to make any representation in response to this notice, or to seek further information, please contact the following as soon as possible:

[NAME OF CONTACT]
at
[ADDRESS, TELEPHONE, FAX]

SCHEDULE
[DESCRIPTION OF THE LAND]

Dated:
[NAME OF AUTHORISED OFFICER]
[NAME AND ADDRESS OF ENFORCING AUTHORITY]

Precedent 2

NOTICE TO OTHER APPROPRIATE PERSONS OF IDENTIFICATION OF CONTAMINATED LAND

Note: Section 78B(4) provides that if after a notice has been given pursuant to section 78B(3)(d) (see Precedent 1) some other person appears to the authority to be an appropriate person, then the authority must give notice to that other person of that fact.

ENVIRONMENTAL PROTECTION ACT 1990, SECTION 78B(4)

To: [NAME AND ADDRESS OF RECIPIENT]:

1. Under section 78B(1) of the Environmental Protection Act 1990 [NAME OF AUTHORITY] has identified the land described in Schedule 1 to this Notice, situated within the Authority's area, as contaminated.

2. The Authority gave notice of that fact on [DATE] to those persons appearing to the Authority to be appropriate persons in relation to that land, more particularly identified in Schedule 2.

3. It now appears to the Enforcing Authority that you are also an appropriate person in relation to that land and this notice is accordingly given to you pursuant to section 78B(4) of the 1990 Act.

4. Because this notice may potentially have financial consequences, you are advised to consult an appropriate independent professional adviser.

5. Should you or your adviser wish to make any representation in response to this Notice, or to seek further information, please contact the following as soon as possible:

    [NAME OF CONTACT]
    at
    [ADDRESS, TELEPHONE, FAX]

SCHEDULE 1
[DESCRIPTION OF THE LAND]

SCHEDULE 2
[PERSONS APPEARING TO BE APPROPRIATE PERSONS]

Dated:
[NAME OF AUTHORISED OFFICER]
[NAME AND ADDRESS OF ENFORCING AUTHORITY]

PRECEDENT 3

NOTICE OF DECISION THAT LAND IS REQUIRED TO BE DESIGNATED AS A SPECIAL SITE

Note: If it appears to a local authority that contaminated land in its area is required to be designated as a special site, it must under section 78C(1)(b) give notice of that decision to the "relevant persons"; that is the appropriate Agency, owner of the land, any person who appears to the local authority to be in occupation of the whole or any part of the land; and each person who appears to the authority to be an appropriate person.

ENVIRONMENTAL PROTECTION ACT 1990, SECTION 78C(1)(b)

To: [NAME AND ADDRESS OF RECIPIENT]:

1. Under section 78C(1)(a) of the Environmental Protection Act 1990 [NAME OF AUTHORITY] has decided that the land described in the Schedule to this Notice being contaminated land situated in its area is land which is required to be designated as a special site.

2. Notice of that decision is given to you pursuant to section 78C(1)(b) as a relevant person as defined in section 78C(2), namely as:

   (A) the appropriate Agency;
   (B) the owner of the land;
   (C) a person appearing to the Authority to be in occupation of [the whole] [part] of the land;
   (D) a person appearing to the Authority to be an appropriate person.
   [DELETE AS APPLICABLE]

3. Should you wish to make any representation in response to this Notice, or to seek further information, please contact the following as soon as possible:

   [NAME OF CONTACT]
   at
   [ADDRESS, TELEPHONE, FAX]

SCHEDULE
[DESCRIPTION OF LAND]

Dated:
[NAME OF AUTHORISED OFFICER]
[NAME AND ADDRESS OF LOCAL AUTHORITY]

PRECEDENT 4

NOTICE OF DECISION AS TO WHETHER LAND IS REQUIRED TO BE DESIGNATED AS A SPECIAL SITE IN RESPONSE TO NOTICE FROM APPROPRIATE AGENCY

Note: By section 78C(4) the appropriate Agency may serve notice on the local authority if it considers that land is required to be designated as a special site. The local authority must then decide whether that is the case, and by section 78C(5) must give notice of its decision to the "relevant persons".

ENVIRONMENTAL PROTECTION ACT 1990, SECTION 7C(5)

To: [NAME AND ADDRESS OF RECIPIENT]:

1. In relation to the land situated in its area and described in the Schedule to this Notice ("the Land") [NAME OF LOCAL AUTHORITY] has received a notice from [NAME OF APPROPRIATE AGENCY] that the Agency considers the Land to be contaminated land which is required to be designated as a special site.
2. The Local Authority has decided that the Land [is land which is required to be designated as a special site] [is not land which is required to be designated as a special site]
   [DELETE AS APPLICABLE]
3. This Notice of that decision is given to you pursuant to section 78C(5) as a relevant person as defined in section 78C(2), namely as:

   (A) the appropriate Agency;
   (B) the owner of the Land;
   (C) a person appearing to the Local Authority to be in occupation of [the whole] [part] of the Land;
   (D) a person appearing to the Local Authority to be an appropriate person.

   [DELETE AS APPLICABLE]
4. Should you wish to make any representation in response to this Notice, or to seek further information, please contact the following as soon as possible:

   [NAME OF CONTACT]
   at
   [ADDRESS, TELEPHONE, FAX]

SCHEDULE
[DESCRIPTION OF THE LAND]

Dated:
[NAME OF AUTHORISED OFFICER]
[NAME AND ADDRESS OF LOCAL AUTHORITY]

Precedent 5

NOTICE THAT DECISION THAT LAND IS REQUIRED TO BE DESIGNATED AS A SPECIAL SITE HAS TAKEN EFFECT

Note: Where a Local Authority makes a decision that land is required to be designated as a special site it must give notice of that fact to (inter alia) the appropriate Agency under either section 78C(1)(b) or section 78C(5) of the Act. The Local Authority's decision then takes effect under section 78C(6) either after the expiration of 21 days from such notice, or sooner if the appropriate Agency notifies its agreement to the decision.

By section 78C(6) the Local Authority is required to give a further notice to the "relevant persons" that the decision has taken effect.

ENVIRONMENTAL PROTECTION ACT 1990, SECTION 78C(6)

To: [NAME AND ADDRESS OF RECIPIENT]:

1. In relation to the land situated in its areas and described in the Schedule to this Notice ("the Land") [NAME OF AUTHORITY] decided under [section 78C(1)(b)] [section 78C(5)(a)] of the 1990 Act that the Land is contaminated land which is required to be designated as a special site.
2. Notice of that decision was given to the appropriate Agency on [DATE].
3. In accordance with Section 78C(6) that decision took effect on [DATE].
4. This Notice that the decision has taken effect is given to you pursuant to section 78C(6) as a relevant person as defined in section 78C(2), namely as:

   (A) the appropriate Agency;
   (B) the owner of the Land;
   (C) a person appearing to the Local Authority to be in occupation of [the whole] [part] of the Land;
   (D) a person appearing to the Local Authority to be an appropriate person.

   [DELETE AS APPLICABLE]

5. Should you wish to make any representation in response to this Notice, or to seek further information, please contact the following as soon as possible:

[NAME OF CONTACT]
at
[ADDRESS, TELEPHONE, FAX]

SCHEDULE
[DESCRIPTION OF THE LAND]

Dated:
[NAME OF AUTHORISED OFFICER]
[NAME AND ADDRESS OF LOCAL AUTHORITY]

Precedent 6

NOTICE OF REFERENCE OF DECISION AS TO SPECIAL SITE STATUS TO THE SECRETARY OF STATE

Note: In the event of disagreement between the Local Authority and the appropriate Agency as to whether land should be notified as a special site, the decision is referred by the Local Authority to the Secretary of State under section 78D(1). The Local Authority is required by section 78D(3) to give notice of that fact to the "relevant persons". Responsibility for notifying the relevant persons of the ultimate decision lies with the Secretary of State under section 78(4)(b).

ENVIRONMENTAL PROTECTION ACT 1990, SECTION 78D(3)

To: [NAME AND ADDRESS OF RECIPIENT]:

1. In relation to the land situated in its area and described in the Schedule to this Notice ("the Land") [NAME OF AUTHORITY] gave notice of its decision as to whether the Land is required to be designated as a special site ("the Decision") to [NAME OF AGENCY] pursuant to [section 78C(1)(b)][section 78C(5)(b)] of the 1990 Act.

2. The Agency has given notice to the Local Authority that it disagrees with the Decision.

3. Pursuant to section 78D(1) of the 1990 Act, the Local Authority has referred the Decision to the Secretary of State for determination under section 78D(4) of the 1990 Act.

4. This Notice that the Decision has been referred to the Secretary of State is given to you pursuant to section 78D(3) as a relevant person as defined by section 78C(2) and 78D(7), namely as:

   (A) the appropriate Agency;
   (B) the owner of the land;
   (C) a person appearing to the Authority to be in occupation of [the whole] [part] of the land;
   (D) a person appearing to the Authority to be an appropriate person.
   [DELETE AS APPLICABLE]

5. Should you wish to make any representation in response to this Notice, or to seek further information, please contact the following as soon as possible:

[NAME OF CONTACT]
at
[ADDRESS, TELEPHONE, FAX]

SCHEDULE
[DESCRIPTION OF THE LAND]

Dated:
[NAME OF AUTHORISED OFFICER]
[NAME AND ADDRESS OF LOCAL AUTHORITY]

PRECEDENT 7

NOTICE REQUESTING INFORMATION

Note: In practice, before an enforcing authority serves a remediation notice it will be vital to obtain as much information as possible on the relevant circumstances which will affect decisions as to who are the appropriate persons and what should be their apportioned shares of liability. Failure to obtain and assimilate such information may result in a later successful appeal, with wasted expense. The enforcing authority may wish, initially at least, to seek to obtain the information on a voluntary basis, which is the basis on which this notice is drafted. There are however, reserve powers to obtain information on a mandatory basis under section 108 of the Environment Act 1995 or under section 16 of the Local Government (Miscellaneous Provisions) Act 1976. The authority may wish to serve the notice either before or after it has formally identified the land as contaminated.

ENVIRONMENTAL PROTECTION ACT 1990, SECTIONS 78B, E AND F

To: [NAME AND ADDRESS OF RECIPIENT/REGISTERED OFFICE IF COMPANY]

EITHER

[1. Under section 78B(1) of the Environmental Protection Act 1990 [NAME OF ENFORCING AUTHORITY] has identified the land described in the Schedule to this Notice ("the Land"), as contaminated land.]

OR

[1. Pursuant to section 78B(1) of the Environmental Protection Act 1990 [NAME OF ENFORCING AUTHORITY] is considering whether the land described in the Schedule to this Notice ("the Land"), should be identified as contaminated land.]

2. The Enforcing Authority believes that you may have information relevant to the exercise of its duties in relation to the Land.

3. Accordingly, you are requested to provide the following information:

(A)	Are you, or any company in which you have been a shareholder, director, company secretary or manager, the owner of all or part of the Land? If you are the owner of part of the Land please identify the extent of your ownership [on the attached plan].	Yes/No
(B)	Are you, or any company in which you have been a shareholder, director, company secretary or manager, the occupier of all or part of the Land, whether as a tenant or otherwise? If you are the occupier or tenant of part of the Land please identify the extent of your occupation or tenancy [on the attached plan].	Yes/No
(C)	Have you, or any company in which you have been a shareholder, director, company secretary or manager, been an owner of all or part of the Land at any time in the past? If "yes" please give the dates when you were the owner.	Yes/No
(D)	Have you, or any company in which you have been a shareholder, director, company secretary or manager, been an occupier of all or part of the Land at any time in the past, whether as a tenant or otherwise? If "yes"please give the dates when you were in occupation or were a tenant.	Yes/No
(E)	Are there any other persons who are or have been joint owners or joint occupiers with you now or in the past, whether as a tenant or otherwise? If "yes" please provide details.	Yes/No
(F)	Do you know the identity of the present or any former owners of all or part of the Land? If "yes", please provide details.	Yes/No
(G)	Do you know the identity of the present or any former occupiers of all or part of the Land, whether as a tenant or otherwise? If "yes" please provide details.	Yes/No

(H)	Are you aware of the presence of any contaminating substances in, on or under the Land? If "yes", please provide details.	Yes/No
(I)	Are you aware of the identity of any person or persons who may have caused or permitted the Land to be contaminated, for example, by spilling or dumping substances? If "yes" please give details.	Yes/No
(J)	Are you aware of any investigations already carried out as to whether the Land is or may be contaminated? If "yes", please give details and indicate whether any reports or results of such investigations are in your possession or that of any other person.	Yes/No
(K)	What is the nature of the activities which to your knowledge have been carried out on the land?	Yes/No
(L)	Are you aware of any current or previous licence, consent, permission or authorisation held by you or any company or other person involving the introduction on to the land of any substances (such as waste, solvents, oils or chemicals) capable of contaminating the land? If "yes", please give details.	
(M)	Are you aware of any contractual agreements relating to responsibility for contamination present on the Land? If "yes" please give details.	Yes/No
(N)	Are there any other matters you wish to draw to the attention of the Enforcing Authority? Please provide details on a separate sheet if necessary.	Yes/No

Please make your response on the enclosed copy of this Notice, sign and date it and return it within [21] days to the address below. If you are completing the response on behalf of a Company, please indicate in what capacity you are responding and whether you are authorised to do so.

SCHEDULE 1
[DESCRIPTION OF THE LAND]

Dated:
[NAME OF AUTHORISED OFFICER]
[NAME AND ADDRESS OF ENFORCING AUTHORITY]

PRECEDENT 8

REMEDIATION NOTICE

Note: This is a suggested form for a remediation notice under section 78E(1). See generally Chapter 12. Copies of the notice must be sent to each person specified in regulation 5(1) of the Contaminated Land (England) Regulations 2000. There is no prescribed form of notice, though the intention of the DETR and Environment Agency is to draw up a model form.

ENVIRONMENTAL PROTECTION ACT 1990, SECTION 78E(1) THE CONTAMINATED LAND (ENGLAND) REGULATIONS 2000

To: [NAME AND ADDRESS OF PERSON OR PERSONS ON WHOM SERVED]:

This Notice is served on you by [NAME OF AUTHORITY] pursuant to section 78E of the Environmental Protection Act 1990 in relation to contaminated land identified as such by the Authority under section 78B(1) [and designated as a special site pursuant to [section 78C] [section 78D]] of the 1990 Act.

This Notice specifies in the Schedule below what you are to do by way of remediation and the periods within which you are to do each of the specified things.

[Where two or more persons are appropriate persons in relation to any particular thing to be done by way of remediation, the Schedule also states the proportion determined under section 78F(7) of the cost of doing those things which each person is liable to bear.]

The further matters required to be stated by the Contaminated Land (England) Regulations 2000 are set out below as Particulars to this Notice.

SCHEDULE

Things required by way of remediation (referring, if necessary, to Annexed specifications)	Period within which each thing is to be done	[Proportion of cost for which each appropriate person is liable]

PARTICULARS

(A) Name and address of the person or persons on whom the notice is served:

(B) Location and extent of the relevant land:
[DESCRIBE AND REFER TO PLAN]

(C) Date of the notice given under Section 78B identifying the land as contaminated:

(D) Reason why the person on whom the notice is served is considered to be an appropriate person by the Authority:

(i) the person caused or knowingly permitted the substances, or any of the substances, by reason of which the land is contaminated land, to be present in, on or under the relevant land;
(ii) the person is the owner of the relevant land;
(iii) the person is the occupier of the relevant land.
[DELETE AS APPLICABLE]

(E) Particulars of the significant harm or pollution of controlled waters by reason of which the land is contaminated land:

(F) The substances by reason of which the land is contaminated land [and location of the land from which they have escaped (if applicable)]:

(G) The reasons of the Authority for its decision as to the remediation action required in the Schedule and how the Secretary of State's guidance has been applied:

(H) Whether two or more persons are appropriate persons in relation to the contaminated land in question and if so, the name and address of each such person and the thing by way of remediation for which each bears responsibility:
[Yes] [No]
[IF YES, THE NAME AND ADDRESS OF EACH]

[THE REMEDIATION ACTION FOR WHICH EACH BEARS RESPONSIBILITY]

(I) Whether two or more persons would, apart from section 78F(6), be appropriate persons in relation to a particular remediation action and, if so, the Authority's reasons for determining which of them is to be treated as being an appropriate person, showing how the Secretary of State's guidance has been applied:
[IF YES, THE REASONS]

(J) If proportions of cost are stated in the Schedule, the reasons for the apportionment, showing how the Secretary of State's guidance has been applied:

(K) Whether the Authority knows the name and address of:

(a) the owner of the relevant land
[Yes] [No]
[IF YES, NAMES AND ADDRESSES]

(b) any person who appears to be in occupation of the whole or part of the relevant land
[Yes] [No]
[IF YES, NAME AND ADDRESSES]

(L) Whether the Authority knows the name and address of any person whose consent is required under section 78G of the Act before any thing required by this notice can be done and if so, the names and addresses:
[Yes] [No]
[IF YES, NAMES AND ADDRESSES]

(M) Whether it appears to the Authority that the land is in such a condition by reason of substances in, on or under the land that there is imminent danger of serious harm or serious pollution of controlled waters being caused:
[Yes] [No]

(N) A person on whom a remediation Notice is served may be guilty of an offence for failure, without reasonable excuse, to comply with any of its requirements;

(O) On conviction of an offence for failure to comply with a remediation notice, the offender shall be liable, on summary conviction, to a fine not exceeding £5,000 and to a further fine of £500 for each day on which the failure to comply continues after conviction of the offence and before the Authority begins to exercise its powers to carry out remediation under section 78N. Where a person commits an offence in a case where the contaminated land to which the notice relates is industrial, trade or business premises, the maximum fine on summary conviction is £20,000 and the further daily fine is £2,000;

(P) Name and address of the Authority serving the Notice:

(Q) Date of the Notice:

RIGHT OF APPEAL

1. A person on whom a remediation Notice is served may appeal against the Notice under Section 78L of the 1990 Act. The appeal must be made within a period of 21 days beginning with the day on which the Notice is served.

2. The appeal should be made:

   [to the Magistrates Court for the area in which the relevant land is situated]
   [to the Secretary of State]
   [DELETE AS APPLICABLE]

3. The grounds on which an appeal may be made are set out in regulation 7 of the Contaminated Land (England) Regulations 2000 and are as follows:

   [SET OUT GROUNDS (a)–(s) FROM THE REGULATIONS]

4. By regulation 14(1) of the Contaminated Land (England) Regulations 2000 the effect of a duly made appeal is to suspend the Notice so that it is of no effect pending the final determination or abandonment of the appeal.

[SIGNATURE OF AUTHORISED ISSUING OFFICER IF DESIRED.[1]]
[NAME OF AUTHORISED OFFICER]
[NAME OF ENFORCING AUTHORITY]

[1] See further Chapter 12.

PRECEDENT 9

NOTICE TO ACCOMPANY COPIES OF REMEDIATION NOTICE

Note: By Regulation 5 of the Contaminated Land (England) Regulations 2000 when serving a remediation notice the Enforcing Authority must also send a copy of it to the persons specified in the regulation.

ENVIRONMENTAL PROTECTION ACT 1990, SECTION 78E(1)
THE CONTAMINATED LAND (ENGLAND) REGULATIONS 2000

To:[NAME AND ADDRESS OF RECIPIENT]:

1. [NAME OF ENFORCING AUTHORITY] has pursuant to section 78E(1) of the Environmental Protection Act 1990 served [a remediation notice] [remediation notices] in relation to contaminated land within its area.

2. The Enforcing Authority is required to send you a copy of the [notice] [notices] by regulation 5(1) of the Contaminated Land (England) Regulations 2000 because:

   (A) you are known to be a person whose consent is required under section 78G before any thing required by the notice can be done;
   (B) you are a person who was required to be consulted under section 78H before the notice was served;
   (C) you are the Environment Agency;
   (D) you are the local authority in whose area the relevant land is situated.

   [DELETE AS APPLICABLE]

3. [A copy] [copies] of the remediation [notice is] [notices are] attached.

4. Should you wish to make any representation in response to this Notice, or to seek further information, please contact the following person as soon as possible:

   [NAME OF CONTACT]
   at
   [ADDRESS, TELEPHONE, FAX]

Dated:
[SIGNATURE OF AUTHORISED OFFICER, IF DESIRED]
[NAME OF AUTHORISED OFFICER]
[NAME AND ADDRESS OF ENFORCING AUTHORITY]

PRECEDENT 10

REMEDIATION DECLARATION

Note: Under section 78H(6) where the enforcing authority is precluded by sections 78E(4) or (5) from specifying a particular thing in a remediation notice (because it is unreasonable in terms of cost or seriousness of harm), the enforcing authority must prepare and publish a remediation declaration referring to the reasons why that thing would otherwise have been specified, and the reason why the authority is precluded from its inclusion in the notice.

ENVIRONMENTAL PROTECTION ACT 1990, SECTION 78H(6)—REMEDIATION DECLARATION

1. [NAME OF ENFORCING AUTHORITY] has identified the land specified in the Schedule to this Declaration, situated within the Enforcing Authority's area, as contaminated land under section 78B of the 1990 Act.

2. There are a number of things which the Enforcing Authority would have included in a remediation notice in relation to the land, were it not precluded from including those things by [section 78E(4)] [section 78E(5)] and [section 78H(5)(a)] of the 1990 Act.

3. Accordingly, the Enforcing Authority is required to prepare and publish this Declaration under section 78H(6) of the 1990 Act.

4. The things which the Enforcing Authority would have included in a remediation notice, and the reasons why it would have specified those things are:

Things that would have been specified	Reasons why they would have been specified

5. The grounds on which the Enforcing Authority is satisfied that it has precluded from specifying each such thing in a remediation notice are:

Things that would have been specified	Grounds on which precluded from specifying each thing

SCHEDULE
(DESCRIPTION OF LAND)

Dated:
[NAME OF AUTHORISED OFFICER]
[NAME AND ADDRESS OF ENFORCING AUTHORITY]

PRECEDENT 11

REMEDIATION STATEMENT

Note: An enforcing authority may be precluded from serving a remediation notice under section 78H(5)(b)–(d) because:

- *(b) appropriate things are being done or will be done without service of a notice; or*
- *(c) the person on whom the notice would be served is the authority itself; or*
- *(d) the powers conferred on the authority by section 78N to do what is appropriate by way of remediation are exercisable.*

In such cases, the "responsible person" (as defined by section 78H(8)) must under section 78H(7) prepare and publish a remediation statement recording the specified matters.

ENVIRONMENTAL PROTECTION ACT 1990, SECTION 78H(7)—REMEDIATION STATEMENT

1. This Statement is prepared by [NAME] as the "responsible person" under section 78H(8) of the 1990 Act in respect of the land specified in the Schedule to this Statement which has been identified by [NAME OF AUTHORITY] as contaminated land under section 78B of the 1990 Act.
2. The Authority has been precluded from serving a remediation notice in respect of the land by section [78H(5)(b)] [78H(5)(c)] [78H(5)(d)] of the 1990 Act and accordingly this Statement is required to be made under section 78H(7) of the 1990 Act.
3. The things which [are being] [have been] [are expected to be] done by way of remediation and the periods within which they are being or are expected to be done, are as follows:

Things by way of remediation	Whether completed, or the period within which each thing is being done or is expected to be done

4. The name and address of the person [who is doing] [who has done] [who is expected to do] each of those things is:

   [NAME AND ADDRESS]

SCHEDULE
(DESCRIPTION OF LAND)

Dated:
[NAME AND ADDRESS OF PERSON MAKING STATEMENT]

PRECEDENT 12

APPLICATION FOR COMPENSATION FOR THE GRANT OF RIGHTS UNDER SECTION 78G

Note: Section 78G(2) requires any person whose consent is required in order to comply with a remediation notice to grant, or joint in granting such rights as are necessary. By section 78G(5) such a person may make an application in the prescribed form and within a prescribed period, seeking compensation for the grant of such rights and related matters. The Contaminated Land (England) Regulations 2000 prescribe a period of 12 months from the date of the grant of the right, or six months from the date on which the rights were first exercised for that purpose.

ENVIRONMENTAL PROTECTION ACT 1990, SECTION 78G(5) CONTAMINATED LAND (ENGLAND) REGULATIONS 2000—APPLICATION FOR COMPENSATION

To: [NAME AND ADDRESS OF APPROPRIATE PERSON]

1. On [DATE] rights were granted pursuant to section 78G(2) of the 1990 Act to enable requirements of a remediation notice served under section 78E of the 1990 Act to be complied with [and on [DATE] the rights were first exercised.]
2. In accordance with section 78G(5) of the Act, I/we apply for compensation as a person who granted, or joined in granting, those rights.
3. A copy of the grant [and of the plans attached to it] is annexed to this Application.
4. The exact nature of the interest in land in respect of which compensation is applied for is: [. . .]
5. The amount of compensation applied for is [TOTAL AMOUNT] calculated in relation to paragraph 4 of Schedule 2 to the 2000 Regulations as follows:

    (A) Depreciation in the value of any relevant interest resulting from the grant of the rights [AMOUNT AND CALCULATIONS];
    (B) Depreciation in the value of any other interest in land to which I am entitled, resulting from the exercise of the rights [AMOUNT AND CALCULATIONS];
    (C) Loss or damage in relation to any relevant interest to which I am entitled attributable to the grant of the rights and falling within paragraph 4(c) of Schedule 2 to the Regulations [AMOUNT AND CALCULATIONS];

(D) Damage to, or injurious affection of, any interest in land to which I am entitled which is not a relevant interest, resulting from the grant of the rights or the exercise of them [AMOUNT AND CALCULATIONS];

(E) Loss in respect of work carried out by me or on my behalf which is rendered abortive by the grant of the rights or exercise of them [AMOUNT AND CALCULATIONS];

(F) Reasonable valuation or legal expenses [AMOUNT AND CALCULATIONS].

Dated:
[NAME AND ADDRESS OF APPLICANT]

PRECEDENT 13

COMPLAINT AND NOTICE OF APPEAL INITIATING APPEAL AGAINST REMEDIATION NOTICE (MAGISTRATES COURT)

Note: Section 78L of the 1990 Act provides for an appeal against a remediation notice served by a local authority to be made to the magistrates' court. By regulation 8(1) of the Contaminated Land (England) Regulations 2000 such an appeal is by way of complaint for an order. Regulation 8(2) requires the appellant to file with the court at the same time as making the complaint a notice of appeal, and regulation 8(2) also requires a copy of the remediation notice to be filed. Copies of the notice of appeal must be served on the persons specified at regulation 8(2)(a) and a statement of the names and addresses of those persons must be filed with the court. The form for a complaint is provided by Form 98 (see the Magistrates' Courts Act 1980, sections 51 and 52, and the Magistrates Courts Rules 1981, rule 4).

COMPLAINT FOR ORDER

[ ] MAGISTRATES' COURT (Code)

Date:

Defendant: [LOCAL AUTHORITY]

Address:

Matter of Complaint: Environmental Protection Act 1990, section 78L(1) Contaminated Land (England) Regulations 2000, regulation 8(1).

The Complainant appeals against the remediation notice served by the Defendant dated [DATE OF NOTICE] on the grounds specified in the Notice of Appeal which is filed with this Complaint.

The Complaint of: [NAME OF COMPLAINANT]

Address:

Telephone Number:

Who states that the Defendant was responsible for the matter of complaint of which particulars are given above.

Taken before me [JUSTICE/JUSTICE'S CLERK]

APPENDIX A

NOTICE OF APPEAL

Environmental Protection Act 1990, section 78L(1)
Contaminated Land (England) Regulations 2000, Regulation 8(2)

Name of Appellant:

Address:

1. The Appellant appeals against the Remediation Notice dated [DATE] served by [NAME OF LOCAL AUTHORITY] a copy of which is filed with this Notice of Appeal.
2. The grounds on which the Appellant appeals against the Remediation Notice are as follows:
   [LIST GROUNDS OF APPEAL]
3. A statement of the names and addresses of persons on whom a copy of this Notice of Appeal have been served pursuant to regulation 8(2)(a)(ii), (iii) or (iv) is also filed with this Notice of Appeal.

Dated:
[NAME OF APPELLANT/AGENT]

Precedent 14

NOTICE OF APPEAL AGAINST REMEDIATION NOTICE (SECRETARY OF STATE)

Note: Section 78L of the 1990 Act provides for an appeal against a remediation notice served by the Environment Agency or SEPA to be made to the Secretary of State. By regulation 9 of the Contaminated Land (England) Regulations 2000 such an appeal is by notice, copies of which are to be served on the persons specified at regulation 9(2).

NOTICE OF APPEAL—ENVIRONMENTAL PROTECTION ACT 1990, SECTION 78L(1) CONTAMINATED LAND (ENGLAND) REGULATIONS 2000, REGULATION 9

To: [SECRETARY OF STATE]

Name of Appellant:

Address:

1. The Appellant appeals against the Remediation Notice dated [DATE] served by [NAME OF AGENCY] a copy of which is lodged with this Notice of Appeal.
2. The grounds on which the Appellant appeals against the Remediation Notice are as follows:
   [LIST GROUNDS OF APPEAL]
3. The Appellant wishes the appeal
   [to be in the form of a Hearing]
   [to disposed of on the basis of written representations]
4. The Appellant has served copies of this Notice and of the Remediation Notice on the following persons as required by regulaton 9(2)(a)(ii), (iii) or (iv):
   [LIST NAMES AND ADDRESSES]

   Dated:
   [NAME OF APPELLANT/AGENT]

PRECEDENT 15

PROSECUTION FOR FAILURE TO COMPLY WITH REMEDIATION NOTICE

Note: By section 78M(1) it is an offence for a person on whom a remediation notice is served to fail, without reasonable excuse, to comply with any of the requirements of the notice.

ENVIRONMENTAL PROTECTION ACT 1990, SECTION 78M(1)

[ ] Magistrates' court (Code)

INFORMATION

Informant: [NAME OF AUTHORITY]

Address:

Defendant:

Address:

Date of offence:

Place of offence:

Statute under which offence charged: section 78M(1) of the Environmental Protection Act 1990.

1. The Defendant is an appropriate person on whom the Authority served a Remediation Notice under section 78E(1) of the 1990 Act dated [DATE] requiring the Defendant to do the things specified in the notice by way of remediation in respect of the contaminated land identified in the notice.

[2. The Defendant appealed against the Remediation Notice which on appeal was upheld on [DATE] subject to the following variations: [VARIATIONS]]

[2. The Defendant appealed against the Remediation Notice but abandoned the appeal on [DATE]]
[USE/DELETE AS APPLICABLE]

3. The period for compliance with the Remediation Notice expired on [DATE] and since that date the Defendant has failed without reasonable excuse to comply with [any] [the following] requirements of the Remediation Notice:
[DETAILS OF NON COMPLIANCE]

Statement of Offence

That the said [DEFENDANT] at [PLACE] on [DATE] being a person on whom [NAME OF AUTHORITY] had served a Remediation Notice, failed without reasonable excuse to comply with any of the requirements of the Notice, contrary to section 78M(1) of the Environmental Protection Act 1990.

Precedent 16

NOTICE BY ENFORCING AUTHORITY BEFORE EXERCISING POWERS UNDER SECTION 78N

Note: Section 78N gives power to the enforcing authority, in a range of situations, to do itself what is appropriate by way of remediation to the relevant land or waters. The Act does not require notice to be given by the authority before exercising those powers, but it will be good practice to do so, in particular where the appropriate person has not complied with the remediation notice.

ENVIRONMENTAL PROTECTION ACT 1990, SECTION 78N

To: [NAME OF APPROPRIATE PERSON]
[ADDRESS]

1. [NAME OF ENFORCING AUTHORITY] served a Remediation Notice on you, dated [DATE] requiring you to do those things specified in the Notice in relation to the relevant land specified in the Notice.
2. You have failed to comply with the Remediation Notice in the following respects [LIST REQUIREMENTS OF NOTICE NOT COMPLIED WITH].
3. Accordingly, the Enforcing Authority has power under section 78N(1) and (3)(c) of the 1990 Act to do those things itself.
4. The Enforcing Authority intends to do those things listed in the Schedule to this Notice, the cost of which is indicated in the Schedule. The Enforcing Authority will be entitled under section 78P(1) to recover from you the reasonable cost of doing those things, or such part of the cost as is reasonable having regard to section 78P(2). [In such circumstances the Enforcing Authority may also charge interest on the sums recoverable and may impose a charge on your interest in any premises consisting of or including the contaminated land in question for such sums and accrued interest][2]
5. The Enforcing Authority will not commence the relevant works before [DATE]. Should you wish to make representations in respect of this notice before that date you are advised to contact:
   [NAME OF CONTACT]
   [ADDRESS, TELEPHONE]

---

[2] Include this sentence only if the appropriate person is a causer and knowing permitter and is an owner of the land.

SCHEDULE
[DETAILS OF PROPOSED WORKS, COST, ESTIMATES]

[NAME OF AUTHORISED OFFICER]
[ADDRESS]

PRECEDENT 17

CHARGING NOTICE UNDER SECTION 78P

Note: Where the enforcing authority exercises its powers under section 78N to carry out remediation itself, it is entitled, in four circumstances (sections 78N(3)(a), (c) (e) and (f)) to recover its reasonable costs incurred from the appropriate person or persons. Under sections 78P(4)–(7) the authority may serve a charging notice on the appropriate person or persons where they are an owner of premises consisting of or including the contaminated land as well as being a causer or knowing permitter. The effect of such Notice is to make the costs carry interest at such reasonable rate as the authority determines, and to make the cost and accrued interest a charge on the owner's interest. A copy of the notice must be served on the same date on every other person who, to the knowledge of the authority, has an interest in the land capable of being affected by the charge.

CHARGING NOTICE—ENVIRONMENTAL PROTECTION ACT 1990, SECTION 78P(4)–(7)

To: [NAME OF OWNER]
[ADDRESS]

1. [NAME OF ENFORCING AUTHORITY] has in exercise of its powers under section 78N(1) of the 1990 Act carried out works by way of remediation ("the Works") to the land identified in Schedule 1 to this Notice. The Works are listed in Schedule 2 to this Notice.

[2. You were informed of the Enforcing Authority's intention to carry out the Works in a Notice dated [DATE] and served on [DATE]].

3. The costs incurred by the Authority in carrying out those works are specified in Schedule 2 to this Notice.

4. The Enforcing Authority has decided pursuant to section 78P(1) and (2) of the 1990 Act that it should recover from you a sum in respect of those costs [being the proportion determined pursuant to section 78F(7) of the 1990 Act, there being more than one appropriate person.]

5. This Notice is served on you pursuant to section 78P(3) as the owner of premises consisting of or including the contaminated land in question, who caused or knowingly permitted the substances by reason of which the land is contaminated to be in, on or under the land.

6. The amount of the cost claimed by the Authority as recoverable is [AMOUNT].

7. The effect of this Notice is as follows:

(A) the cost will carry interest at [RATE OF INTEREST] from the date of service of this Notice until the whole amount is paid; and
(B) subject to your right to appeal against this notice, the cost and accrued interest shall be a charge on the premises referred to at paragraph 5 above.

[8. The Enforcing Authority has served copies of this Notice on the following persons who, to the knowledge of the Authority, have an interest in those premises capable of being affected by the charge: [NAMES AND ADDRESSES]]

9. You [or such persons] may appeal against this notice to a county court within the period of twenty-one days beginning with the date of service.

SCHEDULE 1
(DESCRIPTION OF THE CONTAMINATED LAND)

SCHEDULE 2
[WORKS CARRIED OUT BY THE ENFORCING AUTHORITY AND COSTS]

Date:
[NAME OF AUTHORISED OFFICER]
[NAME OF ENFORCING AUTHORITY]
[ADDRESS]
[CONTACT NAMES AND
TELEPHONE NUMBER]

Precedent 18

NOTICE BY ENFORCING AUTHORITY AS TO POSSIBLE COMMERCIAL CONFIDENTIALITY

Note: Section 78T(1) provides that no information relating to the affairs of any individual or business shall be included in the public registers without their consent, if and so long as the information is commercially confidential. Section 78T(2) requires an authority who feels that information it has obtained might be commercially confidential to give notice to the person or business concerned.

ENVIRONMENTAL PROTECTION ACT 1990, SECTION 78T(2)

To: [NAME OF RECIPIENT]
[ADDRESS]

1. [NAME OF AUTHORITY] holds the information, in relation to its functions on contaminated land under Part IIA of the Environmental Protection Act 1990, the general nature of which is given in the Schedule to this Notice. The information is required to be included in the register relating to contaminated land maintained by the Authority under section 78R(1) of the 1990 Act unless the information is excluded under section 78T (exclusion from registers of certain confidential information).

2. It appears to the Authority that the information might be commercially confidential as relating to your affairs or to the affairs of your business. *Note:* Section 78T(1) of the 1990 Act states that information is commercially confidential in relation to any individual or person if its being contained in the register would prejudice to an unreasonable degree the commercial interests of that person. Section 78T(11) of the Act requires prejudice to be disregarded insofar as it relates only to the value of the contaminated land in question or otherwise to the ownership or occupation of that land.

3. You are entitled to object to the inclusion of the information on the register on the grounds that it is commercially confidential and to make representations to the Authority for the purpose of justifying such objection.

4. Objection or representations should be made to:

[CONTACT NAME]
[ADDRESS]
[TELEPHONE]
within [PERIOD][3] beginning with the date of service of this notice.

SCHEDULE
[THE INFORMATION][4]

Date:
[NAME OF AUTHORISED OFFICER]
[NAME OF ENFORCING AUTHORITY]
[ADDRESS]

---

[3] No period for such representations is prescribed. The authority must give a reasonable opportunity for objections. It is suggested that 21 days should be a reasonable opportunity for this purpose.

[4] Because the object of the notice is the possible confidentiality of information, and the Notice itself may reach third parties, the general nature of the information should be stated so as to avoid disclosing its precise substance.

Appendix B

Environmental Protection Act 1990 (c. 43)

Part IIA: Contaminated Land

Preliminary

78A—(1) The following provisions have effect for the interpretation of this Part.

(2) "Contaminated land" is any land which appears to the local authority in whose area it is situated to be in such a condition, by reason of substances in, or under the land, that—

(a) significant harm is being caused or there is a significant possibility of such harm being caused; or
(b) pollution of controlled waters is being, or is likely to be, caused;

and, in determining whether any land appears to be such land, a local authority shall, subject to subsection (5) below, act in accordance with guidance issued by the Secretary of State in accordance with section 78YA below with respect to the manner in which that determination is to be made.

(3) A "special site" is any contaminated land—

(a) which has been designated as such a site by virtue of section 78C(7) or 78D(6) below; and
(b) whose designation as such has not been terminated by the appropriate Agency under section 78Q(4) below.

(4) "Harm" means harm to the health of living organisms or other interference with the ecological systems of which they form part and, in the case of man, includes harm to his property.

(5) The questions—

(a) what harm is to be regarded as "significant",
(b) whether the possibility of significant harm being caused is "significant",
(c) whether pollution of controlled waters is being, or is likely to be caused,

shall be determined in accordance with guidance issued for the purpose by the Secretary of State in accordance with section 78YA below.

(6) Without prejudice to the guidance that may be issued under subsection (5) above, guidance under paragraph (a) of that subsection may make provision for different degrees of importance to be assigned to, or for the disregard of,—

(a) different descriptions of living organisms or ecological systems;
(b) different descriptions of places; or
(c) different descriptions of harm to health or property, or other interference;

and guidance under paragraph (b) of that subsection may make provision for different degrees of possibility to be regarded as "significant" (or as not being "signficant") in relation to diffrerent descriptions of significant harm.

(7) "Remediation" means—

(a) the doing of anything for the purpose of assessing the condition of—
  (i) the contaminated land in question;
  (ii) any controlled waters affected by that land; or
  (iii) any land adjoining or adjacent to that land;

(b) the doing of any works, the carrying out of any operations or the taking of any steps in relation to any such land or waters for the purpose—
  (i) of preventing or minimising, or remedying or mitigating the effects of, any significant harm, or any pollution of controlled waters, by reason of which the contaminated land is such land; or
  (ii) of restoring the land or waters to their former state; or

(c) the making of subsequent inspections from time to time for the purpose of keeping under review the condition of the land or waters;

and

cognate expressions shall be construed accordingly;

(8) Controlled waters are "affected by" contaminated land if (and only if) it appears to the enforcing authority that the contaminated land in question is, for the purposes of subsection (2) above, in such a condition, by reason of substances in, on or under the land, that pollution of those waters is being, or is likely to be caused.

(9) The following expressions have the meaning respectively assigned to them:

"the appropriate Agency" means—

(a) in relation to England and Wales, the Environment Agency;
(b) in relation to Scotland, the Scottish Environment Protection Agency;

"appropriate person" means any person who is an appropriate person, determined in accordance with section 78F below, to bear responsibility for any thing which is to be done by way of remediation in any particular case;
"charging notice" has the meaning given by section 78P(3)(b) below;
"controlled waters"—

(a) in relation to England and Wales, has the same meaning as in Part II of the Water Resources Act 1991; and
(b) in relation to Scotland, has the same meaning as in section 30A of the Control of Pollution Act 1974;

"creditor" has the same meaning as in the Conveyancing and Feudal Reform (Scotland) Act 1970;
"enforcing authority" means—

(a) in relation to a special site, the appropriate Agency;
(b) in relation to contaminated land other than a special site, the local authority in whose area the land is situated;

"heritable security" has the same meaning as in the Conveyancing and Feudal Reform (Scotland) Act 1970;
"local authority" in relation to England and Wales means—

(a) any unitary authority;
(b) any district council, so far as it is not a unitary authority;
(c) the Common Council of the City of London and, as respects the Temples, the Sub-Treasurer of the Inner Temple and the Under-Treasurer of the Middle Temple respectively;

and in relation to Scotland means a council for an area constituted under section 2 of the Local Government, etc. (Scotland) Act 1994;

"notice" means notice in writing;
"notification" means notification in writing;
"owner", in relation to any land in England and Wales, means a person (other than a mortgagee not in possession) who, whether in his own right or as trustee for any other person, is entitled to receive the rack rent of the land, or, where the land is not let at a rack rent, would be so entitled if it were so let;
"owner", in relation to any land in Scotland, means a person (other than a creditor in a heritable security not in possession of the security subjects) for the time being entitled to receive or who would, if the land were let, be entitled to receive, the rents of the land in connection with which the word is used and includes a trustee, factor, guardian or curator and in the case of public or municipal land includes the persons to whom the management of the land is entrusted;
"pollution of controlled waters" means the entry into controlled waters of any poisonous, noxious or polluting matter or any solid waste matter;
"prescribed" means prescribed by regulations;
"regulations" means regulations made by the Secretary of State;
"remediation declaration" has the meaning given by section 78H(6) below;
"remediation notice" has the meaning given by section 78E(1) below;
"remediation statement" has the meaning given by section 78H(7) below;
"required to be designated as a special site" shall be construed in accordance with section 78C(8) below;
"substance" means any natural or artificial substance, whether in solid or liquid form or in the form of a gas or vapour;
"unitary authority" means—

(a) the council of a county, so far as it is the council of an area for which there are no district councils;
(b) the council of any district comprised in an area for which there is no county council;
(c) the council of a London borough;
(d) the council of a county borough in Wales.

Identification of contaminated land

78B—(1) Every local authority shall cause its area to be inspected from time to time for the purpose—

(a) of identifying contaminated land; and
(b) of enabling the authority to decide whether any such land is land which is required to be designated as a special site.

(2) In performing its functions under subsection (1) above a local authority shall act in accordance with any guidance issued for the purpose by the Secretary of State in accordance with section 78YA below.

(3) If a local authority identifies any contaminated land in its area, it shall give notice of that fact to—

(a) the appropriate Agency;
(b) the owner of the land;
(c) any person who appears to the authority to be in occupation of the whole or any part of the land; and
(d) each person who appears to the authority to be an appropriate person;

and any notice given under this subsection shall state by virtue of which of paragraphs (a) to (d) above it is given.

(4) If, at any time after a local authority has given any person a notice pursuant to subsection (3)(d) above in respect of any land, it appears to the enforcing authority that another person is an appropriate person, the enforcing authority shall give notice to that other person—

(a) of the fact that the local authority has identified the land in question as contaminated land; and
(b) that he appears to the enforcing authority to be an appropriate person.

Identification and designation of special sites

78C—(1) If at any time it appears to a local authority that any contaminated land in its area might be land which is required to be designated as a special site, the authority—

(a) shall decide whether or not the land is land which is required to be so designated; and
(b) if the authority decides that the land is land which is required to be so designated, shall give notice of that decision to the relevant persons.

(2) For the purposes of this section, "the relevant persons" at any time in the case of any land are the persons who at that time fall within paragraphs (a) to (d) below, that is to say—

(a) the appropriate Agency;
(b) the owner of the land;
(c) any person who appears to the local authority concerned to be in occupation of the whole or any part of the land; and
(d) each person who appears to that authority to be an appropriate person.

(3) Before making a decision under paragraph (a) of subsection (1) above in any particular case, a local authority shall request the advice of the appropriate Agency, and in making its decision shall have regard to any advice given by that Agency in response to the request.

(4) If at any time the appropriate Agency considers that any contaminated land is land which is required to be designated as a special site, that Agency may give notice of that fact to the local authority in whose area the land is situated.

(5) Where notice under subsection (4) above is given to a local authority, the authority shall decide whether the land in question—

(a) is land which is required to be designated as a special site, or
(b) is not land which is required to be so designated,

and shall give notice of that decision to the relevant persons.

(6) Where a local authority makes a decision falling within subsection (1)(b) or (5)(a) above, the decision shall, subject to section 78D below, take effect on the day after whichever of the following events first occurs, that is to say—

(a) the expiration of the period of twenty-one days beginning with the day on which the notice required by virtue of subsection (1)(b) or, as the case may be, (5)(a) above is given to the appropriate Agency; or
(b) if the appropriate Agency gives notification to the local authority in question that it agrees with the decision, the giving of that notification;

and where a decision takes effect by virtue of this subsection, the local authority shall give notice of that fact to the relevant persons.

(7) Where a decision that any land is land which is required to be designated as a special site takes effect in accordance with subsection (6) above, the notice given under subsection (1)(b) or, as the case may be, (5)(a) above shall have effect, as from the time when the decision takes effect, as the designation of that land as such a site.

(8) For the purposes of this Part, land is required to be designated as a special site if, and only if, it is land of a description prescribed for the purposes of this subsection.

(9) Regulations under subsection (8) above may make different provision for different cases or circumstances or different areas or localities and may, in particular, describe land by reference to the area or locality in which it is situated.

(10) Without prejudice to the generality of his power to prescribe any description of land for the purposes of subsection (8) above, the Secretary of State, in deciding whether to prescribe a particular description of contaminated land for those purposes, may, in particular, have regard to—

(a) whether land of the description in question appears to him to be land which is likely to be in such a condition, by reason of substances in, on or under the land that—
   (i) serious harm would or might be caused, or
   (ii) serious pollution of controlled waters would be, or would be likely to be, caused; or

(b) whether the appropriate Agency is likely to have expertise in dealing with the kind of significant harm, or pollution of controlled waters, by reason of which land of the description in question is contaminated land.

Referral of special site decisions of the Secretary of State

78D—(1) In any case where—

(a) a local authority gives notice of a decision to the appropriate Agency pursuant to subsection (1)(b) or (5)(b) of section 78C above, but
(b) before the expiration of the period of twenty-one days beginning with the day on which that notice is so given, that Agency gives the local authority notice that it disagrees with the decision, together with a statement of its reasons for disagreeing,

the authority shall refer the decision to the Secretary of State and shall send to him a statement of its reasons for reaching the decision.

(2) Where the appropriate Agency gives notice to a local authority under paragraph (b) of subsection (1) above, it shall also send to the Secretary of State a copy of the notice and of the statement given under that paragraph.

(3) Where a local authority refers a decision to the Secretary of State under subsection (1) above, it shall give notice of that fact to the relevant persons.

(4) Where a decision of a local authority is referred to the Secretary of State under subsection (1) above, he—

(a) may confirm or reverse the decision with respect to the whole or any part of the land to which it relates; and
(b) shall give notice of his decision on the referral—
(i) to the relevant persons; and
(ii) to the local authority.

(5) Where a decision of a local authority is referred to the Secretary of State under subsection (1) above, the decision shall not take effect until the day after that on which the Secretary of State gives the notice required by subsection (4) above to the persons there mentioned and shall then take effect as confirmed or reversed by him.

(6) Where a decision which takes effect in accordance with subsection (5) above is to the effect that at least some land is land which is required to be designated as a special site, the notice given under subsection (4)(b) above shall have effect, as from the time when the decision takes effect, as the designation of that land as such a site.

(7) In this section "the relevant persons" has the same meaning as in section 78C above.

Duty of enforcing authority to require remediation of contaminated land, etc.

78E—(1) In any case where—

(a) any land has been designated as a special site by virtue of section 78C(7) or 78D(6) above, or
(b) a local authority has identified any contaminated land (other than a special site) in its area,

the enforcing authority shall, in accordance with such procedure as may be prescribed and subject to the following provisions of this Part, serve on each person who is an appropriate person a notice (in this Part referred to as a "remediation notice") specifying what that person is to do by way of remediation and the periods within which he is required to do each of the thigs so specified.

(2) Different remediation notices requiring the doing of different things by way of remediation may be served on different persons in consequence of the presence of different substances in, on or under any land or waters.

(3) Where two or more persons are appropriate persons in relation to any particular thing which is to be done by way of remediation, the remediation notice served on each of them shall state the proportion, determined under section 78F(7) below, of the cost of doing that thing which each of them respectively is liable to bear.

(4) The only things by way of remediation which the enforcing authority may do, or require to be done, under or by virtue of this Part are things which it considers reasonable, having regard to—

(a) the cost which is likely to be involved; and
(b) the seriousness of the harm, or pollution of controlled waters, in question.

(5) In determining for any purpose of this Part—

(a) what is to be done (whether by an appropriate person, the enforcing authority or any other person) by way of remediation in any particular case,
(b) the standard to which any land is, or waters are, to be remediated pursuant to the notice, or
(c) what is, or is not, to be regarded as reasonable for the purposes of subsection (4) above,

the enforcing authority shall have regard to any guidance issued for the purpose by the Secretary of State.

(6) Regulations may make provision for or in connection with—

(a) the form or content of remediation notices; or
(b) any steps of a procedural nature which are to be taken in connection with, or in consequence of the service of a remediation notice.

Determination of the appropriate person to bear responsibility for remediation

78F—(1) This section has effect for the purpose of determining who is the appropriate person to bear responsibility for any particular thing which the enforcing authority determines is to be done by way of remediation in any particular case.

(2) Subject to the following provisions of this section, any person, or any of the persons, who caused or knowingly permitted the substances, or any of the substances, by reason of which the contaminated land in question is such land to be in, on or under that land is an appropriate person.

(3) A person shall only be an appropriate person by virtue of subsection (2) above in relation to things which are to be done by way of remediation which are to any extent referable to substances which he caused or knowingly permitted to be present in, on or under the contaminated land in question.

(4) If no person has, after reasonable inquiry, been found who is by virtue of subsection (2) above an appropriate person to bear responsibility for the things which are to be done by way of remediation, the owner or occupier for the time being of the contaminated land in question is an appropriate person.

(5) If, in consequence of subsection (3) above, there are things which are to be done by way of remediation in relation to which no person has, after reasonable inquiry, been found who is an appropriate person by virtue of subsection (2) above, the owner or occupier for the time being of the contaminated land in question is an appropriate person in relation to those things.

(6) Where two or more persons would, apart from this subsection. be appropriate persons in relation to any particular thing which is to be done by way of remediation, the enforcing authority shall determine in accordance with guidance issued for the purpose by the Secretary of State whether any, and if so which, of them is to be treated as not being an appropriate person in relation to that thing.

(7) Where two or more persons are appropriate persons in relation to any particular thing which is to be done by way of remediation, they shall be liable to bear the cost of doing that thing in proportions determined by the enforcing authority in accordance with guidance issued for the purpose by the Secretary of State.

(8) Any guidance issued for the purposes of subsection (6) or (7) above shall be issued in accordance with section 78YA below.

(9) A person who has caused or knowingly permitted any substance ("substance A") to be in, on or under any land shall also be taken for the purposes of this section to have caused or knowingly permitted there to be in, on or under that land any substance which is there as a result of a chemical reaction or biological process affecting substance A.

(10) A thing which is to be done by way of remediation may be regarded for the purposes of this Part as referable to the presence of any substance notwithstanding that the thing in question would not have to be done—

(a) in consequence only of the presence of that substance in any quantity; or
(b) in consequence only of the quantity of that substance which any particular person caused, or knowingly permitted to be present.

Grant of, and compensation for, rights of entry, etc.

78G—(1) A remediation notice may require an appropriate person to do things by way of remediation, notwithstanding that he is not entitled to do those things.

(2) Any person whose consent is required before any thing required by a remediation notice may be done shall grant, or join in granting, such rights in relation to any of the relevant land or waters as will enable the appropriate person to comply with any requirements imposed by the remediation notice.

(3) Before serving a remediation notice, the enforcing authority shall reasonably endeavour to consult every person who appears to the authority—

(a) to be the owner or occupier of any of the relevant land or waters, and
(b) to be a person who might be required by subsection (2) above to grant, or join in granting, any rights,

concerning the rights which that person may be so required to grant.

(4) Subsection (3) above shall not preclude the service of a remediation notice in any case where it appears to the enforcing authority that the contaminated land in question is in such a condition, by reason of substances in, on or under the land, that there is imminent danger of serious harm, or serious pollution of controlled waters, being caused.

(5) A person who grants, or joins in granting, any rights pursuant to subsection (2) above shall be entitled on making an application within such period as may be prescribed and in such manner as may be prescribed to such person as may be prescribed, to be paid by the appropriate person compensation of such amount as may be determined in such manner as may be prescribed.

(6) Without prejudice to the generality of the regulations that may be made by virtue of subsection (5) above, regulations by virtue of that subsection may make such provision in relation to compensation under this section as may be made by regulations by virtue of subsection (4) of section 35A above in relation to compensation under that section.

(7) In this section, "relevant land or waters" means—

(a) the contaminated land in question;
(b) any controlled waters affected by that land; or
(c) any land adjoining or adjacent to that land or those waters.

Restrictions and prohibitions on serving remediation notices

78H—(1) Before serving a remediation notice, the enforcing authority shall reasonably endeavour to consult—

(a) the person on whom the notice is to be served,
(b) the owner of any land to which the notice relates,
(c) any person who appears to that authority to be in occupation of the whole or any part of the land, and
(d) any person of such other description as may be prescribed, concerning what is to be done by way of remediation.

(2) Regulations may make provision for, or in connection with, steps to be taken for the purposes of subsection (1) above.

(3) No remediation notice shall be served on any person by reference to any contaminated land during any of the following periods, that is to say—

(a) the period:
(i) beginning with the identification of the contaminated land in question pursuant to section 78B(1) above, and
(ii) ending with the expiration of the period of three months beginning with the day on which the notice required by subsection (3)(d) or, as the case may be, (4) of section 78B above is given to that person in respect of that land;

(b) if a decision falling within paragraph (b) of section 78C(1) above is made in relation to the contaminated land in question, the period beginning with the making of the decision and ending with the expiration of the period of three months beginning with—
(i) in a case where the decision is not referred to the Secretary of State under section 78D above, the day on which the notice required by section 78C(6) above is given, or
(ii) in a case where the decision is referred to the Secretary of State under section 78D above, the day on which he gives the notice required by subsection (4)(b) of that section;

(c) if the appropriate Agency gives a notice under subsection (4) of section 78C above to a local authority in relation to the contaminated land in question, the period beginning with the day on which that notice is given and ending with the expiration of the period of three months beginning with—
(i) in a case where notice is given under subsection (6) of that section, the day on which that notice is given;
(ii) in a case where the authority makes a decision falling within subsection (5)(b) of that section and the appropriate Agency fails to give notice under paragraph (b) of section 78D(1) above, the day following the expiration of the period of twenty-one days mentioned in that paragraph; or
(iii) in a case where the authority makes a decision falling within section 78C(5)(b) above which is referred to the Secretary of State under section 78D above, the day on which the Secretary of State gives the notice required by subsection (4)(b) of that section.

(4) Neither subsection (1) nor subsection (3) above shall preclude the service of a remediation notice in any case where it appears to the enforcing authority that the land in question is in such a condition, by reason of substances in, on or under the land, that there is imminent danger of serious harm, or serious pollution of controlled waters, being caused.

(5) The enforcing authority shall not serve a remediation notice on a person if and so long as any one or more of the following conditions is for the time being satisfied in the particular case, that is to say—

(a) the authority is satisfied, in consequence of section 78E(4) and (5) above, that there is nothing by way of remediation which could be specified in a remediation notice served on that person;

(b) the authority is satisfied that appropriate things are being, or will be, done by way of remediation without the service of a remediation notice on that person;

(c) it appears to the authority that the person on whom the notice would be served is the authority itself; or

(d) the authority is satisfied that the powers conferred on it by section 78N below to do what is appropriate by way of remediation are exercisable.

(6) Where the enforcing authority is precluded by virtue of section 78E(4) or (5) above from specifying in a remediation notice any particular thing by way of remediation which it would otherwise have specified in such a notice, the authority shall prepare and publish a document (in this Part referred to as a "remediation declaration") which shall record—

(a) the reasons why the authority would have specified that thing; and
(b) the grounds on which the authority is satisfied that it is precluded from specifying that thing in such a notice.

(7) In any case where the enforcing authority is precluded, by virtue of paragraph (b), (c) or (d) of subsection (5) above, from serving a remediation notice, the responsible person shall prepare and publish a document (in this Part referred to as a "remediation statement") which shall record—

(a) the things which are being, have been, or are expected to be, done by way of remediation in the particular case;
(b) the name and address of the person who is doing, has done, or is expected to do, each of those things; and
(c) the periods within which each of those things is being, or is expected to be, done.

(8) For the purposes of subsection (7) above, the "responsible person" is—

(a) in a case where the condition in paragraph (b) of subsection (5) above is satisfied, the person who is doing or has done, or who the enforcing authority is satisfied will do, the things there mentioned; or
(b) in a case where the condition in paragraph (c) or (d) of that subsection is satisfied, the enforcing authority.

(9) If a person who is required by virtue of subsection (8)(a) above to prepare and publish a remediation statement fails to do so within a reasonable time after the date on which a remediation notice specifying the things there mentioned could, apart from subsection (5) above, have been served, the enforcing authority may itself prepare and publish the statement and may recover its reasonable costs of doing so from that person.

(10) Where the enforcing authority has been precluded by virtue only of subsection (5) above from serving a remediation notice on an appropriate person but—

(a) none of the conditions in that subsection is for the time being satisfied in the particular case, and
(b) the authority is not precluded by any other provision of this Part from serving a remediation notice on that appropriate person,

the authority shall serve a remediation notice on that person; and any such notice may be so served without any further endeavours by the authority to consult persons pursuant to subsection (1) above, if and to the extent that that person has been consulted pursuant to that subsection concerning the things which will be specified in the notice.

Restrictions on liability relating to the pollution of controlled waters

78J—(1) This section applies where any land is contaminated land by virtue of paragraph (b) of subsection (2) of section 78A above (whether or not the land is also contaminated land by virtue of paragraph (a) of that subsection).

(2) Where this section applies, no remediation notice given in consequence of the land in question being contaminated land shall require a person who is an appropriate person by virtue of section 78F(4) or (5) above to do anything by way of remediation to that or any other land, or any waters, which he could not have been required to do by such a notice had paragraph (b) of section 78A(2) above (and all other references to pollution of controlled waters) been omitted from this Part.

(3) If, in a case where this section applies, a person permits, has permitted, or might permit. water from an abandoned mine or part of a mine—

(a) to enter any controlled waters, or
(b) to reach a place from which it is or, as the case may be, was likely, in the opinion of the enforcing authority, to enter such waters,

no remediation notice shall require him in consequence to do anything by way of remediation (whether to the contaminated land in question or to any other land or waters) which he could not have been required to do by such a notice had paragraph (b) of section 78A(2) above (and all other references to pollution of controlled waters) been omitted from this Part.

(4) Subsection (3) above shall not apply to the owner or former operator of any mine or part of a mine if the mine or part in question became abandoned after 31st December 1999.

(5) In determining for the purposes of subsection (4) above whether a mine or part of a mine became abandoned before, on or after 31st December 1999 in a case where the mine or part has become abandoned on two or more occasions, of which—

(a) at least one falls on or before that date, and
(b) at least one falls after that date,

the mine or part shall be regarded as becoming abandoned after that date (but without prejudice to the operation of subsection (3) above in relation to that mine or part at, or in relation to, any time before the first of those occasions which falls after that date).

(6) Where, immediately before a part of a mine becomes abandoned, that part is the only part of the mine not falling to be regarded as abandoned for the time being, the abandonment of that part shall not be regarded for the purposes of subsection (4) or (5) above as constituting the abandonment of the mine, but only of that part of it.

(7) Nothing in subsection (2) or (3) above prevents the enforcing authority from doing anything by way of remediation under section 78N below which it could have done apart from that subsection, but the authority shall not be entitled under section 78P below to recover from any person any part of the cost incurred by the authority in doing by way of remediation anything which it is precluded by subsection (2) or (3) above from requiring that person to do.

(8) In this section "mine" has the same meaning as in the Mines and Quarries Act 1954.

Liability in respect of contaminating substances which escape to other land

78K—(1) A person who has caused or knowingly permitted any substances to be in, on or under any land shall also be taken for the purposes of this Part to have caused or, as the case may be, knowingly permitted those substances to be in, on or under any other land to which they appear to have escaped.

(2) Subsections (3) and (4) below apply in any case where it appears that any substances are or have been in, on or under any land (in this section referred to as "land A") as a result of their escape, whether directly or indirectly, from other land in, on or under which a person caused or knowingly permitted them to be.

(3) Where this subsection applies, no remediation notice shall require a person—

(a) who is the owner or occupier of land A, and
(b) who has not caused or knowingly permitted the substances in question to be in, on or under that land,

to do anything by way of remediation to any land or waters (other than land or waters of which he is the owner or occupier) in consequence of land A appearing to be in such a condition, by reason of the presence of those substances in, on or under it, that significant harm is being caused, or there is a significant possibility of such harm being caused, or that pollution of controlled waters is being, or is likely to be caused.

(4) Where this subsection applies, no remediation notice shall require a person—

(a) who is the owner or occupier of land A, and
(b) who has not caused or knowingly permitted the substances in question to be in, on or under that land,

to do anything by way of remediation in consequence of any further land in, on or under which those substances or any of them appear to be or to have been present as a result of their escape from land A ("land B") appearing to be in such a condition, by reason of the presence of those substances in, on or under it, that significant harm is being caused, or there is a significant possibility of such harm being caused, or that pollution of controlled waters is being, or is likely to be caused, unless he is also the owner or occupier of land B.

(5) In any case where—

(a) a person ("person A") has caused or knowingly permitted any substances to be in, on, or under any land,
(b) another person ("person B") who has not caused or knowingly permitted those substances to be in, on or under that land becomes the owner or occupier of that land, and
(c) the substances, or any of the substances, mentioned in paragraph (a) above appear to have escaped to other land,

no remediation notice shall require person B to do anything by way of remediation to that other land in consequence of the apparent acts or omissions of person A, except to the extent that person B caused or knowingly permitted the escape.

(6) Nothing in subsection (3), (4) or (5) above prevents the enforcing authority from doing anything by way of remediation under section 78N below which it could have done apart from that subsection, but the authority shall not be entitled under

section 78P below to recover from any person any part of the cost incurred by the authority in doing by way of remediation anything which it is precluded by subsection (3), (4) or (5) above from requiring that person to do.

(7) In this section, "appear" means appear to the enforcing authority, and cognate expressions shall be construed accordingly.

Appeals against remediation notices

78L—(1) A person on whom a remediation notice is served may, within the period of twenty-one days beginning with the day on which the notice is served, appeal against the notice—

(a) if it was served by a local authority, to a magistrates' court or, in Scotland, to the sheriff by way of summary application; or
(b) if it was served by the appropriate Agency, to the Secretary of State;

and in the following provisions of this section "the appellate authority" means the magistrates' court, the sheriff or the Secretary of State, as the case may be.

(2) On any appeal under subsection (1) above the appellate authority—

(a) shall quash the notice, if it is satisfied that there is a material defect in the notice; but
(b) subject to that, may confirm the remediation notice, with or without modification, or quash it.

(3) Where an appellate authority confirms a remediation notice, with or without modification, it may extend the period specified in the notice for doing what the notice requires to be done.

(4) Regulations may make provision with respect to—

(a) the grounds on which appeals under subsection (1) above may he made;
(b) the cases in which, grounds on which, court or tribunal to which, or person at whose instance, an appeal against a decision of a magistrates' court or sheriff court in pursuance of an appeal under subsection (1) above shall lie; or
(c) the procedure on an appeal under subsection (1) above or on an appeal by virtue of paragraph (b) above.

(5) Regulations under subsection (4) above may (among other things)—

(a) include provisions comparable to those in section 290 of the Public Health Act 1936 (appeals against notices requiring the execution of works);
(b) prescribe the cases in which a remediation notice is, or is not, to be suspended until the appeal is decided, or until some other stage in the proceedings;
(c) prescribe the cases in which the decision on an appeal may in some respects be less favourable to the appellant than the remediation notice against which he is appealing;
(d) prescribe the cases in which the appellant may claim that a remediation notice should have been served on some other person and prescribe the procedure to be followed in those cases;
(e) make provision as respects:

(i) the particulars to be included in the notice of appeal;
(ii) the persons on whom notice of appeal is to be served and the particulars, if any, which are to accompany the notice; and
(iii) the abandonment of an appeal;

(f) make different provision for different cases or classes of case.

(6) This section, so far as relating to appeals to the Secretary of State, is subject to section 114 of the Environment Act 1995 (delegation or reference of appeals etc).

Offences of not complying with a remediation notice

78M—(1) If a person on whom an enforcing authority serves a remediation notice fails, without reasonable excuse, to comply with any of the requirements of the notice, he shall be guilty of an offence.

(2) Where the remediation notice in question is one which was required by section 78E(3) above to state, in relation to the requirement which has not been complied with, the proportion of the cost involved which the person charged with the offence is liable to bear, it shall be a defence for that person to prove that the only reason why he has not complied with the requirement is that one or more of the other persons who are liable to bear a proportion of that cost refused, or was not able, to comply with the requirement.

(3) Except in a case falling within subsection (4) below, a person who commits an offence under subsection (1) above shall be liable, on summary conviction, to a fine not exceeding level 5 on the standard scale and to a further fine of an amount equal to one-tenth of level 5 on the standard scale for each day on which the failure continues after conviction of the offence and before the enforcing authority has begun to exercise its powers by virtue of section 78N(3)(c) below.

(4) A person who commits an offence under subsection (1) above in a case where the contaminated land to which the remediation notice relates is industrial, trade or business premises shall he liable on summary conviction to a fine not exceeding £20,000 or such greater sum as the Secretary of State may from time to time by order substitute and to a further fine of an amount equal to one-tenth of that sum for each day on which the failure continues after conviction of the offence and before the enforcing authority has begun to exercise its powers by virtue of section 78N(3)(c) below.

(5) If the enforcing authority is of the opinion that proceedings for an offence under this section would afford an ineffectual remedy against a person who has failed to comply with any of the requirements of a remediation notice which that authority has served on him, that authority may take proceedings in the High Court or, in Scotland, in any court of competent jurisdiction, for the purpose of securing compliance with the remediation notice.

(6) In this section, "industrial, trade or business premises" means premises used for any industrial, trade or business purposes or premises not so used on which matter is burnt in connection with any industrial, trade or business process, and premises are used for industrial purposes where they are used for the purposes of any treatment or process as well as where they are used for the purpose of manufacturing.

(7) No order shall be made under subsection (4) above unless a draft of the order has been laid before, and approved by a resolution of, each House of Parliament.

Powers of the enforcing authority to carry out remediation

78N—(1) Where this section applies, the enforcing authority shall itself have power, in a case falling within paragraph (a) or (b) of section 78E(1) above to do what is appropriate by way of remediation to the relevant land or waters.

(2) Subsection (1) above shall not confer power on the enforcing authority to do anything by way of remediation if the authority would, in the particular case. be

precluded by section 78YB below from serving a remediation notice requiring that thing to be done.

(3) This section applies in each of the following cases, that is to say—

(a) where the enforcing authority considers it necessary to do anything itself by way of remediation for the purpose of preventing the occurrence of any serious harm, or serious pollution of controlled waters, of which there is imminent danger;

(b) where an appropriate person has entered into a written agreement with the enforcing authority for that authority to do, at the cost of that person, that which he would otherwise be required to do under this Part by way of remediation;

(c) where a person on whom the enforcing authority serves a remediation notice fails to comply with any of the requirements of the notice;

(d) where the enforcing authority is precluded by section 78J or 78K above from including something by way of remediation in a remediation notice;

(e) where the enforcing authority considers that, were it to do some particular thing by way of remediation, it would decide, by virtue of subsection (2) of section 78P below or any guidance issued under that subsection—

(i) not to seek to recover under subsection (1) of that section any of the reasonable cost incurred by it in doing that thing; or

(ii) to seek so to recover only a portion of that cost;

(f) where no person has, after reasonable inquiry, been found who is an appropriate person in relation to any particular thing.

(4) Subject to section 78E(4) and (5) above, for the purposes of this section, the things which it is appropriate for the enforcing authority to do by way of remediation are—

(a) in a case falling within paragraph (a) of subsection (3) above, anything by way of remediation which the enforcing authority considers necessary for the purpose mentioned in that paragraph;

(b) in a case falling within paragraph (b) of that subsection, anything specified in, or determined under, the agreement mentioned in that paragraph;

(c) in a case falling within paragraph (c) of that subsection, anything which the person mentioned in that paragraph was required to do by virtue of the remediation notice;

(d) in a case falling within paragraph (d) of that subsection, anything by way of remediation which the enforcing authority is precluded by section 78J or 78K above from including in a remediation notice;

(e) in a case falling within paragraph (e) or (f) of that subsection, the particular thing mentioned in the paragraph in question.

(5) In this section "the relevant land or waters" means—

(a) the contaminated land in question;

(b) any controlled waters affected by that land; or

(c) any land adjoining or adjacent to that land or those waters.

Recovery of, and security for, the cost of remediation by the enforcing authority

78P—(1) Where, by virtue of section 78N(3)(a), (c), (e) or (f) above, the enforcing authority does any particular thing by way of remediation, it shall be entitled, subject to sections 78J(7) and 78K(6) above, to recover the reasonable cost incurred in doing it from the appropriate person or, if there are two or more appropriate persons in relation to the thing in question, from those persons in proportions determined pursuant to section 78F(7) above.

(2) In deciding whether to recover the cost, and, if so, how much of the cost, which it is entitled to recover under subsection (1) above, the enforcing authority shall have regard—

(a) to any hardship which the recovery may cause to the person from whom the cost is recoverable; and
(b) to any guidance issued by the Secretary of State for the purposes of this subsection.

(3) Subsection (4) below shall apply in any case where—

(a) any cost is recoverable under subsection (1) above from a person—

(i) who is the owner of any premises which consist of or include the contaminated land in question; and
(ii) who caused or knowingly permitted the substances, or any of the substances, by reason of which the land is contaminated land to be in, on or under the land; and

(b) the enforcing authority serves a notice under this subsection (in this Part referred to as a "charging notice") on that person.

(4) Where this subsection applies—

(a) the cost shall carry interest, at such reasonable rate as the enforcing authority may determine, from the date of service of the notice until the whole amount is paid; and
(b) subject to the following provisions of this section, the cost and accrued interest shall be a charge on the premises mentioned in subsection (3)(a)(i) above.

(5) A charging notice shall—

(a) specify the amount of the cost which the enforcing authority claims is recoverable;
(b) state the effect of subsection (4) above and the rate of interest determined by the authority under that subsection; and
(c) state the effect of subsections (7) and (8) below.

(6) On the date on which an enforcing authority serves a charging notice on a person, the authority shall also serve a copy of the notice on every other person who, to the knowledge of the authority, has an interest in the premises capable of being affected by the charge.

(7) Subject to any order under subsection (9)(b) or (c) below, the amount of any cost specified in a charging notice and the accrued interest shall be a charge on the premises—

(a) as from the end of the period of twenty-one days beginning with the service of the charging notice, or
(b) where an appeal is brought under subsection (8) below, as from the final determination or (as the case may be) the withdrawal, of the appeal.

until the cost and interest are recovered.

(8) A person served with a charging notice or a copy of a charging notice may appeal against the notice to a county court within the period of twenty-one days beginning with the date of service.

(9) On an appeal under subsection (8) above, the court may—

(a) confirm the notice without modification:
(b) order that the notice is to have effect with the substitution of a different amount for the amount originally specified in it: or
(c) order that the notice is to be of no effect.

(10) Regulations may make provision with respect to—

(a) the grounds on which appeals under this section may be made; or
(b) the procedure on any such appeal.

(11) An enforcing authority shall, for the purpose of enforcing a charge under this section, have all the same powers and remedies under the Law of Property Act 1925, and otherwise, as if it were a mortgagee by deed having powers of sale and lease, of accepting surrenders of leases and of appointing a receiver.

(12) Where any cost is a charge on premises under this section, the enforcing authority may by order declare the cost to be payable with interest by instalments within the specified period until the whole amount is paid.

(13) In subsection (12) above—

"interest" means interest at the rate determined by the enforcing authority under subsection (4) above; and
"the specified period" means such period of thirty years or less from the date of service of the charging notice as is specified in the order.

(14) Subsections (3) to (13) above do not extend to Scotland.

Special sites

78Q—(1) If, in a case where a local authority has served a remediation notice, the contaminated land in question becomes a special site, the appropriate Agency may adopt the remediation notice and, if it does so,—

(a) it shall give notice of its decision to adopt the remediation notice to the appropriate person and to the local authodty:
(b) the remediation notice shall have effect, as from the time at which the appropriate Agency decides to adopt it, as a remediation notice given by that Agency; and
(c) the validity of the remediation notice shall not be affected by—
 (i) the contaminated land having become a special site;
 (ii) the adoption of the remediation notice by the appropriate Agency; or
 (iii) anything in paragraph (b) above.

(2) Where a local authority has, by virtue of section 78N above, begun to do any thing, or any series of things, by way of remediation—

(a) the authority may continue doing that thing, or that series of things, by virtue of that section, notwithstanding that the contaminated land in question becomes a special site, and
(b) section 78P above shall apply in relation to the reasonable cost incurred by the authority in doing that thing or those things as if that authority were the enforcing authority.

(3) If and so long as any land is a special site, the appropriate Agency may from time to time inspect that land for the purpose of keeping its condition under review.

(4) If it appears to the appropriate Agency that a special site is no longer land which is required to be designated as such a site, the appropriate Agency may give notice—

(a) to the Secretary of State, and
(b) to the local authority in whose area the site is situated,

terminating the designation of the land in question as a special site as from such date as may be specified in the notice.

(5) A notice under subsection (4) above shall not prevent the land, or any of the land, to which the notice relates being designated as a special site on a subsequent occasion.

(6) In exercising its functions under subsection (3) or (4) above, the appropriate Agency shall act in accordance with any guidance given for the purpose by the Secretary of State.

Registers

78R—(1) Every enforcing authority shall maintain a register containing prescribed particulars of or relating to—

(a) remediation notices served by that authority;
(b) appeals against any such remediation notices;
(c) remediation statements or remediation declarations prepared and published under section 78H above;
(d) in relation to an enforcing authority in England and Wales, appeals against charging notices served by that authority;
(e) notices under subsection (1)(b) or (5)(a) of section 78C above which have effect by virtue of subsection (7) of that section as the designation of any land as a special site;
(f) notices under subsection (4)(b) of section 78D above which have effect by virtue of subsection (6) of that section as the designation of any land as a special site;
(g) notices given by or to the enforcing authority under section 78Q(4) above terminating the designation of any land as a special site;
(h) notifications given to that authority by persons—
   (i) on whom a remediation notice has been served, or
   (ii) who are or were required by virtue of section 78H(8)(a) above to prepare and publish a remediation statement,

   of what they claim has been done by them by way of remediation;
(j) notifications given to that authority by owners or occupiers of land—
   (i) in respect of which a remediation notice has been served, or
   (ii) in respect of which a remediation statement has been prepared and published,

   of what they claim has been done on the land in question by way of remediation;
(k) convictions for such offences under section 78M above as may be prescribed;
(l) such other matters relating to contaminated land as may be prescribed;

but that duty is subject to sections 78S and 78T below.

(2) The form of, and the descriptions of information to be contained in, notifications for the purposes of subsection (1)(h) or (j) above may be prescribed by the Secretary of State.

(3) No entry made in a register by virtue of subsection (1)(h) or (j) above constitutes a representation by the body maintaining the register or, in a case

where the entry is made by virtue of subsection (6) below, the authority which sent the copy of the particulars in question pursuant to subsection (4) or (5) below—

(a) that what is stated in the entry to have been done has in fact been done; or
(b) as to the manner in which it has been done.

(4) Where any particulars are entered on a register maintained under this section by the appropriate Agency, the appropriate Agency shall send a copy of those particulars to the local authority in whose area is situated the land to which the particulars relate.

(5) In any case where—

(a) any land is treated by virtue of section 78X(2) below as situated in the area of a local authority other than the local authority in whose area it is in fact situated, and
(b) any particulars relating to that land are entered on the register maintained under this section by the local authority in whose area the land is so treated as situated,

that authority shall send a copy of those particulars to the local authority in whose area the land is in fact situated.

(6) Where a local authority receives a copy of any particulars sent to it pursuant to subsection (4) or (5) above, it shall enter those particulars on the register maintained by it under this section.

(7) Where information of any description is excluded by virtue of section 78T below from any register maintained under this section, a statement shall be entered in the register indicating the existence of information of that description.

(8) It shall be the duty of each enforcing authority—

(a) to secure that the registers maintained by it under this section are available, at all reasonable times, for inspection by the public free of charge; and
(b) to afford to members of the public facilities for obtaining copies of entries, on payment of reasonable charges;

and, for the purposes of this subsection, places may be prescribed by the Secretary of State at which any such registers or facilities as are mentioned in paragraph (a) or (b) above are to be available or afforded to the public in pursuance of the paragraph in question.

(9) Registers under this section may be kept in any form.

Exclusion from registers of information affecting national security

78S—(1) No information shall be included in a register maintained under section 78R above if and so long as, in the opinion of the Secretary of State, the inclusion in the register of that information, or information of that description, would be contrary to the interests of national security.

(2) The Secretary of State may, for the purpose of securing the exclusion from registers of information to which subsection (1) above applies, give to enforcing authorities directions—

(a) specifying information, or descriptions of information, to be excluded from their registers; or
(b) specifying descriptions of information to be referred to the Secretary of State for his determination;

and no information referred to the Secretary of State in pursuance of paragraph (b) above shall be included in any such register until the Secretary of State determines that it should be so included.

(3) The enforcing authority shall notify the Secretary of State of any information which it excludes from the register in pursuance of directions under subsection (2) above.

(4) A person may, as respects any information which appears to him to be information to which subsection (1) above may apply, give a notice to the Secretary of State specifying the information and indicating its apparent nature; and, if he does so—

(a) he shall notify the enforcing authority that he has done so; and
(b) no information so notified to the Secretary of State shall be included in any such register until the Secretary of State has determined that it should be so included.

Exclusion from registers of certain confidential information

78T—(1) No information relating to the affairs of any individual or business shall be included in a register maintained under section 78R above, without the consent of that individual or the person for the time being carrying on that business, if and so long as the information—

(a) is, in relation to him, commercially confidential; and
(b) is not required to be included in the register in pursuance of directions under subsection (7) below;

but information is not commercially confidential for the purposes of this section unless it is determined under this section to be so by the enforcing authority or, on appeal, by the Secretary of State.

(2) Where it appears to an enforcing authority that any information which has been obtained by the authority under or by virtue of any provision of this Part might be commercially confidential, the authority shall—

(a) give to the person to whom or whose business it relates notice that that information is required to be included in the register unless excluded under this section; and
(b) give him a reasonable opportunity—
 (i) of objecting to the inclusion of the information on the ground that it is commercially confidential; and
 (ii) of making representations to the authority for the purpose of justifying any such objection;

and, if any representations are made, the enforcing authority shall, having taken the representations into account, determine whether the information is or is not commercially confidential.

(3) Where, under subsection (2) above, an authority determines that information is not commercially confidential—

(a) the information shall not be entered in the register until the end of the period of twenty-one days beginning with the date on which the determination is notified to the person concerned;
(b) that person may appeal to the Secretary of State against the decision;

and, where an appeal is brought in respect of any information, the information shall not be entered in the register until the end of the period of seven days following the day on which the appeal is finally determined or withdrawn.

(4) An appeal under subsection (3) above shall, if either party to the appeal so requests or the Secretary of State so decides, take or continue in the form of a hearing (which must be held in private).

(5) Subsection (10) of section 15 above shall apply in relation to an appeal under subsection (3) above as it applies in relation to an appeal under that section.

(6) Subsection (3) above is subject to section 114 of the Environment Act 1995 (delegation or reference of appeals, etc.).

(7) The Secretary of State may give to the enforcing authorities directions as to specified information, or descriptions of information, which the public interest requires to be included in registers maintained under section 78R above notwithstanding that the information may be commercially confidential.

(8) Information excluded from a register shall be treated as ceasing to be commercially confidential for the purposes of this section at the expiry of the period of four years beginning with the date of the determination by virtue of which it was excluded; but the person who furnished it may apply to the authority for the information to remain excluded from the register on the ground that it is still commercially confidential and the authority shall determine whether or not that is the case.

(9) Subsections (3) to (6) above shall apply in relation to a determination under subsection (8) above as they apply in relation to a determination under subsection (2) above.

(10) Information is, for the purposes of any determination under this section, commercially confidential, in relation to any individual or person, if its being contained in the register would prejudice to an unreasonable degree the commercial interests of that individual or person.

(11) For the purposes of subsection (10) above, there shall be disregarded any prejudice to the commercial interests of any individual or person so far as relating only to the value of the contaminated land in question or otherwise to the ownership or occupation of that land.

Reports by the appropriate Agency on the state of contaminated land

78U—(1) The appropriate Agency shall—

(a) from time to time, or
(b) if the Secretary of State at any time so requests,

prepare and publish a report on the state of contaminated land in England and Wales or in Scotland, as the case may be.

(2) A local authority shall, at the written request of the appropriate Agency, furnish the appropriate Agency with such information to which this subsection applies as the appropriate Agency may require for the purpose of enabling it to perform its functions under subsection (1) above.

(3) The information to which subsection (2) above applies is such information as the local authority may have, or may reasonably be expected to obtain, with respect to the condition of contaminated land in its area, being information which the authority has acquired or may acquire in the exercise of its functions under this Part.

Site specific guidance by the appropriate Agency concerning contaminated land

78V—(1) The appropiate Agency may issue guidance to any local authority with respect to exercise or performance of the authority's powers or duties under this

Part in relation to any particular contaminated land; and in exercising or performing those powers or duties in relation to that land the authority shall have regard to any such guidance so issued.

(2) If and to the extent that any guidance issued under subsection (1) above to a local authority is inconsistent with any guidance issued under this Part by the Secretary of State, the local authority shall disregard the guidance under that subsection.

(3) A local authority shall, at the written request of the appropriate Agency, furnish the appropriate Agency with such information to which this subsection applies as the appropriate Agency may require for the purpose of enabling it to issue guidance for the purposes of subsection (1) above.

(4) The information to which subsection (3) above applies is such information as the local authority may have, or may reasonably be expected to obtain, with respect to any contaminated land in its area, being information which the authority has acquired, or may acquire, in the exercise of its functions under this Part.

The appropriate Agency to have regard to guidance given by the Secretary of State

78W—(1) The Secretary of State may issue guidance to the appropriate Agency with respect to the exercise or performance of that Agency's powers or duties under this Part; and in exercising or performing those powers or duties the appropriate Agency shall have regard to any such guidance so issued.

(2) The duty imposed on the appropriate Agency by subsection (1) above is without prejudice to any duty imposed by any other provision of this Part on that Agency to act in accordance with guidance issued by the Secretary of State.

Supplementary provisions

78X—(1) Where it appears to a local authority that two or more different sites, when considered together, are in such a condition, by reason of substances in, on or under the land, that—

(a) significant harm is being caused or there is a significant possibility of such harm being caused, or
(b) pollution of controlled waters is being, or is likely to be, caused,

this Part shall apply in relation to each of those sites, whether or not the condition of the land at any of them, when considered alone, appears to the authority to be such that significant harm is being caused, or there is a significant possibility of such harm being caused, or that pollution of controlled waters is being or is likely to be caused.

(2) Where it appears to a local authority that any land outside, but adjoining or adjacent to, its area is in such a condition, by reason of substances in, on or under the land, that significant harm is being caused, or there is a significant possibility of such harm being caused, or that pollution of controlled waters is being, or is likely to be, caused within its area—

(a) the authority may, in exercising its functions under this Part, treat that land as if it were land situated within its area; and
(b) except in this subsection, any reference—
    (i) to land within the area of a local authority, or
    (ii) to the local authority in whose area any land is situated,

shall be construed accordingly;

but this subsection is without prejudice to the functions of the local authority in whose area the land is in fact situated.

(3) A person acting in a relevant capacity—

(a) shall not thereby be personally liable, under this Part, to bear the whole or any part of the cost of doing any thing by way of remediation, unless that thing is to any extent referable to substances whose presence in, on or under the contaminated land in question is a result of any act done or omission made by him which it was unreasonable for a person acting in that capacity to do or make; and
(b) shall not thereby be guilty of an offence under or by virtue of section 78M above unless the requirement which has not been complied with is a requirement to do some particular thing for which he is personally liable to bear the whole or any part of the cost.

(4) In subsection (3) above, "person acting in a relevant capacity" means—

(a) a person acting as an insolvency practitioner, within the meaning of section 388 of the Insolvency Act 1986 (including that section as it applies in relation to an insolvent partnership by virtue of any order made under section 421 of that Act);
(b) the official receiver acting in a capacity in which he would be regarded as acting as an insolvency practitioner within the meaning of section 388 of the Insolvency Act 1986 if subsection (5) of that section were disregarded;
(c) the official receiver acting as receiver or manager;
(d) a person acting as a special manager under section 177 or 370 of the Insolvency Act 1986;
(e) the Accountant in Bankruptcy acting as permanent or interim trustee in a sequestration (within the meaning of the Bankruptcy (Scotland) Act 1985);
(f) a person acting as a receiver or receiver and manager—
    (i) under or by virtue of any enactment; or
    (ii) by virtue of his appointment as such by an order of a court or by any other instrument.

(5) Regulations may make different provision for different cases or circumstances.

Application to the Isles of Scilly

78Y—(1) Subject to the provisions of any order under this section, this part shall not apply in relation to the Isles of Scilly.

(2) The Secretary of State may, after consultation with the Council of the Isles of Scilly, by order provide for the application of any provisions of this Part to the Isles of Scilly; and any such order may provide for the application of those provisions to those Isles with such modifications as may be specified in the order.

(3) An order under this section may—

(a) make different provision for different cases, including different provision in relation to different persons, circumstances or localities; and
(b) contain such supplemental, consequential and transitional provision as the Secretary of State considers appropriate, including provision saving provision repealed by or under any enactment.

Supplementary provisions with respect to guidance by the Secretary of State

78YA—(1) Any power of the Secretary of State to issue guidance under this Part shall only be exercisable after consultation with the appropriate Agency and such other bodies or persons as he may consider it appropriate to consult in relation to the guidance in question.

(2) A draft of any guidance proposed to be issued under section 78A(2) or (5), 78B(2) or 78F(6) or (7) above shall be laid before each House of Parliament and the guidance shall not be issued until after the period of 40 days beginning with the day on which the draft was so laid or, if the draft is laid on different days, the later of the two days.

(3) If, within the period mentioned in subsection (2) above, either House resolves that the guidance, the draft of which was laid before it, should not be issued, the Secretary of State shall not issue that guidance.

(4) In reckoning any period of 40 days for the purposes of subsection (2) or (3) above, no account shall be taken of any time during which Parliament is dissolved or prorogued or during which both Houses are adjourned for more than four days.

(5) The Secretary of State shall arrange for any guidance issued by him under this Part to be published in such manner as he considers appropriate.

Interaction of this Part with other enactments

78YB—(1) A remediation notice shall not be served if and to the extent that it appears to the enforcing authority that the powers of the appropriate Agency under section 27 above may be exercised in relation to—

(a) the significant harm (if any), and
(b) the pollution of controlled waters (if any).

by reason of which the contaminated land in question is such land.

(2) Nothing in this Part shall apply in relation to any land in respect of which there is for the time being in force a site licence under Part II above, except to the extent that any significant harm, or pollution of controlled waters, by reason of which that land would otherwise fall to be regarded as contaminated land is attributable to causes other than—

(a) breach of the conditions of the licence; or
(b) the carrying on, in accordance with the conditions of the licence, of any activity authorised by the licence.

(3) If, in a case falling within subsection (1) or (7) of section 59 above, the land in question is contaminated land, or becomes such land by reason of the deposit of the controlled waste in question, a remediation notice shall not be served in respect of that land by reason of that waste or any consequences of its deposit, if and to the extent that it appears to the enforcing authority that the powers of a waste regulation authority or waste collection authority under that section may be exercised in relation to that waste or the consequences of its deposit.

(4) No remediation notice shall require a person to do anything the effect of which would be to impede or prevent the making of a discharge in pursuance of a consent given under Chapter II of Part III of the Water Resources Act 1991 (pollution offences) or, in relation to Scotland, in pursuance of a consent given under Part II of the Control of Pollution Act 1974.

This Part and radioactivity

78YC—Except as provided by regulations. nothing in this Part applies in relation to harm, or pollution of controlled waters, so far as attributable, to any radioactivity possessed by any substance; but regulations may—

(a) provide for prescribed provisions of this Part to have effect with such modifications as the Secretary of State considers appropriate for the purpose of dealing with harm, or pollution of controlled waters, so far as attributable to any radioactivity possessed by any substances; or
(b) make such modifications of the Radioactive Substances Act 1993 or any other Act as the Secretary of State considers appropriate.

Appendix C

Contaminated Land (England) Regulations 2000 (S.I. 2000 No. 227)

Made	*2nd February 2000*
Laid before Parliament	*9th February 2000*
Coming into force	*1st April 2000*

The Secretary of State for the Environment, Transport and the Regions in exercise of the powers conferred on him by sections 78C(8) to (10), 78E(6), 78G(5) and (6), 78L(4) and (5) and 78R(1), (2) and (8) of the Environmental Protection Act 1990, and all other powers enabling him in that behalf, hereby makes the following Regulations:

Citation, commencement, extent and interpretation

1.—(1) These Regulations may be cited as the Contaminated Land (England) Regulations 2000 and shall come into force on 1st April 2000.
(2) These Regulations extend to England only.
(3) In these Regulations, unless otherwise indicated, any reference to a numbered section is to the section of the Environmental Protection Act 1990 which bears that number.

Land required to be designated as a special site

2.—(1) Contaminated land of the following descriptions is prescribed for the purposes of section 78C(8) as land required to be designated as a special site–

(a) land to which regulation 3 applies;
(b) land which is contaminated land by reason of waste acid tars in, on or under the land;
(c) land on which any of the following activities have been carried on at any time–

  (i) the purification (including refining) of crude petroleum or of oil extracted from petroleum, shale or any other bituminous substance except coal; or
  (ii) the manufacture or processing of explosives;

(d) land on which a prescribed process designated for central control has been or is being carried on under an authorisation where the process does not comprise solely things being done which are required by way of remediation;

(e) land within a nuclear site;
(f) land owned or occupied by or on behalf of–

(i) the Secretary of State for Defence;
(ii) the Defence Council;
(iii) an international headquarters or defence organisation; or
(iv) the service authority of a visiting force,

being land used for naval, military or air force purposes;

(g) land on which the manufacture, production or disposal of–

(i) chemical weapons;
(ii) any biological agent or toxin which falls within section 1(1)(a) of the Biological Weapons Act 1974 (restriction on development of biological agents and toxins); or
(iii) any weapon, equipment or means of delivery which falls within section 1(1)(b) of that Act (restriction on development of biological weapons),

has been carried on at any time;

(h) land comprising premises which are or were designated by the Secretary of State by an order made under section 1(1) of the Atomic Weapons Establishment Act 1991 (arrangements for development etc of nuclear devices);
(i) land to which section 30 of the Armed Forces Act 1996 (land held for the benefit of Greenwich Hospital) applies; and
(j) land which–

(i) is adjoining or adjacent to land of a description specified in sub-paragraphs (b) to (i) above; and
(ii) is contaminated land by virtue of substances which appear to have escaped from land of such a description.

(2) For the purposes of paragraph (1)(b) above, "waste acid tars" are tars which–

(a) contain sulphuric acid;
(b) were produced as a result of the refining of benzole, used lubricants or petroleum; and
(c) are or were stored on land used as a retention basin for the disposal of such tars.

(3) In paragraph (1)(d) above, "authorisation" and "prescribed process " have the same meaning as in Part I of the Environmental Protection Act 1990 (integrated pollution control and air pollution control by local authorities) and the reference to designation for central control is a reference to designation under section 2(4) (which provides for processes to be designated for central or local control).

(4) In paragraph (1)(e) above, "nuclear site" means–

(a) any site in respect of which, or part of which, a nuclear site licence is for the time being in force; or
(b) any site in respect of which, or part of which, after the revocation or surrender of a nuclear site licence, the period of responsibility of the licensee has not come to an end;

and "nuclear site licence", "licensee" and "period of responsibility " have the meaning given by the Nuclear Installations Act 1965.

(5) For the purposes of paragraph (1)(f) above, land used for residential purposes or by the Navy, Army and Air Force Institutes shall be treated as land used for naval, military or air force purposes only if the land forms part of a base occupied for naval, military or air force purposes.

(6) In paragraph (1)(f) above–

"international headquarters" and "defence organisation" mean, respectively, any international headquarters or defence organisation designated for the purposes of the International Headquarters and Defence Organisations Act 1964;

"service authority" and "visiting force" have the same meaning as in Part I of the Visiting Forces Act 1952.

(7) In paragraph (1)(g) above, "chemical weapon" has the same meaning as in subsection (1) of section 1 of the Chemical Weapons Act 1996 disregarding subsection (2) of that section.

Pollution of controlled waters

3. For the purposes of regulation 2(1)(a), this regulation applies to land where–

(a) controlled waters which are, or are intended to be, used for the supply of drinking water for human consumption are being affected by the land and, as a result, require a treatment process or a change in such a process to be applied to those waters before use, so as to be regarded as wholesome within the meaning of Part III of the Water Industry Act 1991 (water supply);

(b) controlled waters are being affected by the land and, as a result, those waters do not meet or are not likely to meet the criterion for classification applying to the relevant description of waters specified in regulations made under section 82 of the Water Resources Act 1991 (classification of quality of waters); or

(c) controlled waters are being affected by the land and–

(i) any of the substances by reason of which the pollution of the waters is being or is likely to be caused falls within any of the families or groups of substances listed in paragraph 1 of Schedule 1 to these Regulations; and

(ii) the waters, or any part of the waters, are contained within underground strata which comprise wholly or partly any of the formations of rocks listed in paragraph 2 of Schedule 1 to these Regulations.

Content of remediation notices

4.—(1) A remediation notice shall state (in addition to the matters required by section 78E(1) and (3))–

(a) the name and address of the person on whom the notice is served;

(b) the location and extent of the contaminated land to which the notice relates (in this regulation referred to as the "contaminated land in question"), sufficient to enable it to be identified whether by reference to a plan or otherwise;

(c) the date of any notice which was given under section 78B to the person on whom the remediation notice is served identifying the contaminated land in question as contaminated land;

(d) whether the enforcing authority considers the person on whom the notice is served is an appropriate person by reason of–

(i) having caused or knowingly permitted the substances, or any of the substances, by reason of which the contaminated land in question is contaminated land, to be in, on or under that land;

(ii) being the owner of the contaminated land in question; or

(iii) being the occupier of the contaminated land in question;

(e) particulars of the significant harm or pollution of controlled waters by reason of which the contaminated land in question is contaminated land;

(f) the substances by reason of which the contaminated land in question is contaminated land and, if any of the substances have escaped from other land, the location of that other land;
(g) the enforcing authority's reasons for its decisions as to the things by way of remediation that the appropriate person is required to do, which shall show how any guidance issued by the Secretary of State under section 78E(5) has been applied;
(h) where two or more persons are appropriate persons in relation to the contaminated land in question–
   (i) that this is the case;
   (ii) the name and address of each such person; and
   (iii) the thing by way of remediation for which each such person bears responsibility;
(i) where two or more persons would, apart from section 78F(6), be appropriate persons in relation to any particular thing which is to be done by way of remediation, the enforcing authority's reasons for its determination as to whether any, and if so which, of them is to be treated as not being an appropriate person in relation to that thing, which shall show how any guidance issued by the Secretary of State under section 78F(6) has been applied;
(j) where the remediation notice is required by section 78E(3) to state the proportion of the cost of a thing which is to be done by way of remediation which each of the appropriate persons in relation to that thing is liable to bear, the enforcing authority's reasons for the proportion which it has determined, which shall show how any guidance issued by the Secretary of State under section 78F(7) has been applied;
(k) where known to the enforcing authority, the name and address of–
   (i) the owner of the contaminated land in question; and
   (ii) any person who appears to the enforcing authority to be in occupation of the whole or any part of the contaminated land in question;
(l) where known to the enforcing authority, the name and address of any person whose consent is required under section 78G(2) before any thing required by the remediation notice may be done;
(m) where the notice is to be served in reliance on section 78H(4), that it appears to the enforcing authority that the contaminated land in question is in such a condition, by reason of substances in, on or under the land, that there is imminent danger of serious harm, or serious pollution of controlled waters, being caused;
(n) that a person on whom a remediation notice is served may be guilty of an offence for failure, without reasonable excuse, to comply with any of the requirements of the notice;
(o) the penalties which may be applied on conviction for such an offence;
(p) the name and address of the enforcing authority serving the notice; and
(q) the date of the notice.

(2) A remediation notice shall explain–
(a) that a person on whom it is served has a right of appeal against the notice under section 78L;
(b) how, within what period and on what grounds an appeal may be made; and
(c) that a notice is suspended, where an appeal is duly made, until the final determination or abandonment of the appeal.

Service of copies of remediation notices

5.—(1) Subject to paragraph (2) below, the enforcing authority shall, at the same time as it serves a remediation notice, send a copy of it to each of the following persons, not being a person on whom the notice is to be served–

(a) any person who was required to be consulted under section 78G(3) before service of the notice;
(b) any person who was required to be consulted under section 78H(1) before service of the notice;
(c) where the local authority is the enforcing authority, the Environment Agency; and
(d) where the Environment Agency is the enforcing authority, the local authority in whose area the contaminated land in question is situated.

(2) Where it appears to the enforcing authority that the contaminated land in question is in such a condition by reason of substances in, on or under it that there is imminent danger of serious harm, or serious pollution of controlled waters, being caused, the enforcing authority shall send any copies of the notice pursuant to paragraph (1) above as soon as practicable after service of the notice.

Compensation for rights of entry etc.

6. Schedule 2 to these Regulations shall have effect–
(a) for prescribing the period within which a person who grants, or joins in granting, any rights pursuant to section 78G(2) may apply for compensation for the grant of those rights;
(b) for prescribing the manner in which, and the person to whom, such an application may be made; and
(c) for prescribing the manner in which the amount of such compensation shall be determined and for making further provision relating to such compensation.

Grounds of appeal against a remediation notice

7.—(1) The grounds of appeal against a remediation notice under section 78L(1) are any of the following–
(a) that, in determining whether any land to which the notice relates appears to be contaminated land, the local authority–
  (i) failed to act in accordance with guidance issued by the Secretary of State under section 78A(2), (5) or (6); or
  (ii) whether by reason of such a failure or otherwise, unreasonably identified all or any of the land to which the notice relates as contaminated land;
(b) that, in determining a requirement of the notice, the enforcing authority–
  (i) failed to have regard to guidance issued by the Secretary of State under section 78E(5); or
  (ii) whether by reason of such a failure or otherwise, unreasonably required the appellant to do any thing by way of remediation;
(c) that the enforcing authority unreasonably determined the appellant to be the appropriate person who is to bear responsibility for any thing required by the notice to be done by way of remediation;
(d) subject to paragraph (2) below, that the enforcing authority unreasonably failed to determine that some person in addition to the appellant is an appropriate person in relation to any thing required by the notice to be done by way of remediation;
(e) that, in respect of any thing required by the notice to be done by way of remediation, the enforcing authority failed to act in accordance with guidance issued by the Secretary of State under section 78F(6);
(f) that, where two or more persons are appropriate persons in relation to any thing required by the notice to be done by way of remediation, the enforcing authority–

(i) failed to determine the proportion of the cost stated in the notice to be the liability of the appellant in accordance with guidance issued by the Secretary of State under section 78F(7); or

(ii) whether, by reason of such a failure or otherwise, unreasonably determined the proportion of the cost that the appellant is to bear;

(g) that service of the notice contravened a provision of subsection (1) or (3) of section 78H (restrictions and prohibitions on serving remediation notices) other than in circumstances where section 78H(4) applies;

(h) that, where the notice was served in reliance on section 78H(4) without compliance with section 78H(1) or (3), the enforcing authority could not reasonably have taken the view that the contaminated land in question was in such a condition by reason of substances in, on or under the land, that there was imminent danger of serious harm, or serious pollution of controlled waters, being caused;

(i) that the enforcing authority has unreasonably failed to be satisfied, in accordance with section 78H(5)(b), that appropriate things are being, or will be, done by way of remediation without service of a notice;

(j) that any thing required by the notice to be done by way of remediation was required in contravention of a provision of section 78J (restrictions on liability relating to the pollution of controlled waters);

(k) that any thing required by the notice to be done by way of remediation was required in contravention of a provision of section 78K (liability in respect of contaminating substances which escape to other land);

(l) that the enforcing authority itself has power, in a case falling within section 78N(3)(b), to do what is appropriate by way of remediation;

(m) that the enforcing authority itself has power, in a case falling within section 78N(3)(e), to do what is appropriate by way of remediation;

(n) that the enforcing authority, in considering for the purposes of section 78N(3)(e), whether it would seek to recover all or a portion of the cost incurred by it in doing some particular thing by way of remediation–

(i) failed to have regard to any hardship which the recovery may cause to the person from whom the cost is recoverable or to any guidance issued by the Secretary of State for the purposes of section 78P(2); or

(ii) whether by reason of such a failure or otherwise, unreasonably determined that it would decide to seek to recover all of the cost;

(o) that, in determining a requirement of the notice, the enforcing authority failed to have regard to guidance issued by the Environment Agency under section 78V(1);

(p) that a period specified in the notice within which the appellant is required to do anything is not reasonably sufficient for the purpose;

(q) that the notice provides for a person acting in a relevant capacity to be personally liable to bear the whole or part of the cost of doing any thing by way of remediation, contrary to the provisions of section 78X(3)(a);

(r) that service of the notice contravened a provision of section 78YB (interaction of Part IIA of the Environmental Protection Act 1990 with other enactments), and–

(i) in a case where subsection (1) of that section is relied on, that it ought reasonably to have appeared to the enforcing authority that the powers of the Environment Agency under section 27 might be exercised;

(ii) in a case where subsection (3) of section 78YB is relied on, that it ought reasonably to have appeared to the enforcing authority that the powers of a waste regulation authority or waste collection authority under section 59 might be exercised; or

(s) that there has been some informality, defect or error in, or in connection with, the notice, in respect of which there is no right of appeal under the grounds set out in sub-paragraphs (a) to (r) above.

(2) A person may only appeal on the ground specified in paragraph (1)(d) above in a case where–

(a) the enforcing authority has determined that he is an appropriate person by virtue of subsection (2) of section 78F and he claims to have found some other person who is an appropriate person by virtue of that subsection;
(b) the notice is served on him as the owner or occupier for the time being of the contaminated land in question and he claims to have found some other person who is an appropriate person by virtue of that subsection; or
(c) the notice is served on him as the owner or occupier for the time being of the contaminated land in question, and he claims that some other person is also an owner or occupier for the time being of the whole or part of that land.

(3) If and in so far as an appeal against a remediation notice is based on the ground of some informality, defect or error in, or in connection with, the notice, the appellate authority shall dismiss the appeal if it is satisfied that the informality, defect or error was not a material one.

Appeals to a magistrates' court

8.—(1) An appeal under section 78L(1) to a magistrates' court against a remediation notice shall be by way of complaint for an order and, subject to section 78L(2) and (3) and regulations 7(3), 12 and 13, the Magistrates' Courts Act 1980 shall apply to the proceedings.
(2) An appellant shall, at the same time as he makes a complaint,–

(a) file a notice ("notice of appeal") and serve a copy of it on–

(i) the enforcing authority;
(ii) any person named in the remediation notice as an appropriate person;
(iii) any person named in the notice of appeal as an appropriate person; and
(iv) any person named in the remediation notice as the owner or occupier of the whole or any part of the land to which the notice relates;

(b) file a copy of the remediation notice to which the appeal relates and serve a copy of it on any person named in the notice of appeal as an appropriate person who was not so named in the remediation notice; and
(c) file a statement of the names and addresses of any persons falling within paragraph (ii), (iii) or (iv) of sub-paragraph (a) above.

(3) The notice of appeal shall state the appellant's name and address and the grounds on which the appeal is made.
(4) On an appeal under section 78L(1) to a magistrates' court–

(a) the justices' clerk or the court may give, vary or revoke directions for the conduct of proceedings, including–

(i) the timetable for the proceedings;
(ii) the service of documents;
(iii) the submission of evidence; and
(iv) the order of speeches;

(b) any person falling within paragraph (2)(a)(ii), (iii) or (iv) above shall be given notice of, and an opportunity to be heard at, the hearing of the complaint and any hearing for directions, in addition to the appellant and the enforcing authority; and

(c) the court may refuse to grant a request by the appellant to abandon his appeal against a remediation notice, where the request is made after the court has notified the appellant in accordance with regulation 12(1) of a proposed modification of that notice.

(5) Rule 15 of the Family Proceedings Courts (Matrimonial Proceedings etc.) Rules 1991 (delegation by justices' clerk) shall apply for the purposes of an appeal under section 78L(1) to a magistrates' court as it applies for the purposes of Part II of those Rules.

(6) In this regulation, "file" means deposit with the justices' clerk.

Appeals to the Secretary of State

9.—(1) An appeal to the Secretary of State against a remediation notice shall be made to him by a notice ("notice of appeal") which shall state–

(a) the name and address of the appellant;
(b) the grounds on which the appeal is made; and
(c) whether the appellant wishes the appeal to be in the form of a hearing or to be disposed of on the basis of written representations.

(2) The appellant shall, at the same time as he serves a notice of appeal on the Secretary of State,–

(a) serve a copy of it on–
  (i) the Environment Agency;
  (ii) any person named in the remediation notice as an appropriate person;
  (iii) any person named in the notice of appeal as an appropriate person; and
  (iv) any person named in the remediation notice as the owner or occupier of the whole or any part of the land to which the notice relates;

  and serve on the Secretary of State a statement of the names and addresses of any persons falling within paragraph (ii), (iii) or (iv) above; and

(b) serve a copy of the remediation notice to which the appeal relates on the Secretary of State and on any person named in the notice of appeal as an appropriate person who is not so named in the remediation notice.

(3) Subject to paragraph (5) below, if the appellant wishes to abandon an appeal, he shall do so by notifying the Secretary of State in writing and the appeal shall be treated as abandoned on the date the Secretary of State receives that notification.

(4) The Secretary of State may refuse to permit an appellant to abandon his appeal against a remediation notice where the notification by the appellant in accordance with paragraph (3) above is received by the Secretary of State at any time after the Secretary of State has notified the appellant in accordance with regulation 12(1) of a proposed modification of that notice.

(5) Where an appeal is abandoned, the Secretary of State shall give notice of the abandonment to any person on whom the appellant was required to serve a copy of the notice of appeal.

Hearings and local inquiries

10.—(1) Before determining an appeal, the Secretary of State may, if he thinks fit–

(a) cause the appeal to take or continue in the form of a hearing (which may, if the person hearing the appeal so decides, be held, or held to any extent, in private); or

(b) cause a local inquiry to be held,

and the Secretary of State shall act as mentioned in sub-paragraph (a) or (b) above if a request is made by either the appellant or the Environment Agency to be heard with respect to the appeal.

(2) The persons entitled to be heard at a hearing are–

(a) the appellant;
(b) the Environment Agency; and
(c) any person (other than the Agency) on whom the appellant was required to serve a copy of the notice of appeal.

(3) Nothing in paragraph (2) above shall prevent the person appointed to conduct the hearing of the appeal from permitting any other person to be heard at the hearing and such permission shall not be unreasonably withheld.

(4) After the conclusion of a hearing, the person appointed to conduct the hearing shall, unless he has been appointed under section 114(1)(a) of the Environment Act 1995 (power of Secretary of State to delegate his functions of determining appeals) to determine the appeal, make a report in writing to the Secretary of State which shall include his conclusions and his recommendations or his reasons for not making any recommendations.

Notification of Secretary of State's decision on an appeal

11.—(1) The Secretary of State shall notify the appellant in writing of his decision on an appeal and shall provide him with a copy of any report mentioned in regulation 10(4).

(2) The Secretary of State shall, at the same time as he notifies the appellant, send a copy of the documents mentioned in paragraph (1) above to the Environment Agency and to any other person on whom the appellant was required to serve a copy of the notice of appeal.

Modification of a remediation notice

12.—(1) Before modifying a remediation notice under section 78L(2)(b) in any respect which would be less favourable to the appellant or any other person on whom the notice was served, the appellate authority shall–

(a) notify the appellant and any persons on whom the appellant was required to serve a copy of the notice of appeal of the proposed modification;
(b) permit any persons so notified to make representations in relation to the proposed modification; and
(c) permit the appellant or any other person on whom the remediation notice was served to be heard if any such person so requests.

(2) Where, in accordance with paragraph (1) above, the appellant or any other person is heard, the enforcing authority shall also be entitled to be heard.

Appeals to the High Court

13. An appeal against any decision of a magistrates' court in pursuance of an appeal under section 78L(1) shall lie to the High Court at the instance of any party to the proceedings in which the decision was given (including any person who exercised his entitlement under regulation 8(4)(b) to be heard at the hearing of the complaint).

Suspension of a remediation notice

14.—(1) Where an appeal is duly made against a remediation notice, the notice shall be of no effect pending the final determination or abandonment of the appeal.

(2) An appeal against a remediation notice is duly made for the purposes of this regulation if it is made within the period specified in section 78L(1) and the requirements of regulation 8(2) and (3) (in the case of an appeal to a magistrates' court) or regulation 9(1) and (2) (in the case of an appeal to the Secretary of State) have been complied with.

Registers

15.—(1) Schedule 3 to these Regulations shall have effect for prescribing–

(a) for the purposes of subsection (1) of section 78R, the particulars of or relating to the matters to be contained in a register maintained under that section; and

(b) other matters in respect of which such a register shall contain prescribed particulars pursuant to section 78R(1)(l).

(2) The following descriptions of information are prescribed for the purposes of section 78R(2) as information to be contained in notifications for the purposes of section 78R(1)(h) and (j)–

(a) the location and extent of the land sufficient to enable it to be identified;

(b) the name and address of the person who it is claimed has done each of the things by way of remediation;

(c) a description of any thing which it is claimed has been done by way of remediation; and

(d) the period within which it is claimed each such thing was done.

(3) The following places are prescribed for the purposes of subsection (8) of section 78R as places at which any registers or facilities for obtaining copies shall be available or afforded to the public in pursuance of paragraph (a) or (b) of that subsection–

(a) where the enforcing authority is the local authority, its principal office; and

(b) where the enforcing authority is the Environment Agency, its office for the area in which the contaminated land in question is situated.

Schedule 1

Regulation 3(c)

Special Sites

1. The following families and groups of substances are listed for the purposes of regulation 3(c)(i)–

organohalogen compounds and substances which may form such compounds in the aquatic environment;
organophosphorus compounds;
organotin compounds;
substances which possess carcinogenic, mutagenic or teratogenic properties in or via the aquatic environment;
mercury and its compounds;
cadmium and its compounds;
mineral oil and other hydrocarbons;
cyanides.

2. The following formations of rocks are listed for the purposes of regulation 3(c)(ii)–

Pleistocene Norwich Crag;
Upper Cretaceous Chalk;

Lower Cretaceous Sandstones;
Upper Jurassic Corallian;
Middle Jurassic Limestones;
Lower Jurassic Cotteswold Sands;
Permo-Triassic Sherwood Sandstone Group;
Upper Permian Magnesian Limestone;
Lower Permian Penrith Sandstone;
Lower Permian Collyhurst Sandstone;
Lower Permian Basal Breccias, Conglomerates and Sandstones;
Lower Carboniferous Limestones.

Schedule 2

Regulation 6

Compensation for Rights of Entry Etc.

Interpretation

1. In this Schedule–
"the 1961 Act" means the Land Compensation Act 1961;
"grantor" means a person who has granted, or joined in the granting of, any rights pursuant to section 78G(2);
"relevant interest" means an interest in land out of which rights have been granted pursuant to section 78G(2).

Period for Making an Application

2. An application for compensation shall be made within the period beginning with the date of the grant of the rights in respect of which compensation is claimed and ending on whichever is the latest of the following dates–

(a) twelve months after the date of the grant of those rights;
(b) where an appeal is made against a remediation notice in respect of which the rights in question have been granted, and the notice is of no effect by virtue of regulation 14, twelve months after the date of the final determination or abandonment of the appeal; or
(c) six months after the date on which the rights were first exercised.

Manner of Making an Application

3. —(1) An application shall be made in writing and delivered at or sent by pre-paid post to the last known address for correspondence of the appropriate person to whom the rights were granted.

(2) The application shall contain, or be accompanied by–

(a) a copy of the grant of rights in respect of which the grantor is applying for compensation, and of any plans attached to that grant;
(b) a description of the exact nature of any interest in land in respect of which compensation is applied for; and
(c) a statement of the amount of compensation applied for, distinguishing the amounts applied for under each of sub-paragraphs (a) to (e) of paragraph 4 below, and showing how the amount applied for under each sub-paragraph has been calculated.

Loss and Damage for which Compensation Payable

4. Subject to paragraph 5(3) and (5)(b) below, compensation is payable under section 78G for loss and damage of the following descriptions–

(a) depreciation in the value of any relevant interest to which the grantor is entitled which results from the grant of the rights;
(b) depreciation in the value of any other interest in land to which the grantor is entitled which results from the exercise of the rights;
(c) loss or damage, in relation to any relevant interest to which the grantor is entitled, which–

(i) is attributable to the grant of the rights or the exercise of them;
(ii) does not consist of depreciation in the value of that interest; and
(iii) is loss or damage for which he would have been entitled to compensation by way of compensation for disturbance, if that interest had been acquired compulsorily under the Acquisition of Land Act 1981 in pursuance of a notice to treat served on the date on which the rights were granted;

(d) damage to, or injurious affection of, any interest in land to which the grantor is entitled which is not a relevant interest, and which results from the grant of the rights or the exercise of them; and
(e) loss in respect of work carried out by or on behalf of the grantor which is rendered abortive by the grant of the rights or the exercise of them.

Basis on which Compensation Assessed

5.—(1) The following provisions shall have effect for the purpose of assessing the amount to be paid by way of compensation under section 78G.

(2) The rules set out in section 5 of the 1961 Act (rules for assessing compensation) shall, so far as applicable and subject to any necessary modifications, have effect for the purpose of assessing any such compensation as they have effect for the purpose of assessing compensation for the compulsory acquisition of an interest in land.

(3) No account shall be taken of any enhancement of the value of any interest in land, by reason of any building erected, work done or improvement or alteration made on any land in which the grantor is, or was at the time of erection, doing or making, directly or indirectly concerned, if the Lands Tribunal is satisfied that the erection of the building, the doing of the work, the making of the improvement or the alteration was not reasonably necessary and was undertaken with a view to obtaining compensation or increased compensation.

(4) In calculating the amount of any loss under paragraph 4(e) above, expenditure incurred in the preparation of plans or on other similar preparatory matters shall be taken into account.

(5) Where the interest in respect of which compensation is to be assessed is subject to a mortgage–

(a) the compensation shall be assessed as if the interest were not subject to the mortgage; and
(b) no compensation shall be payable in respect of the interest of the mortgagee (as distinct from the interest which is subject to the mortgage).

(6) Compensation under section 78G shall include an amount equal to the grantor's reasonable valuation and legal expenses.

Payment of Compensation and Determination of Disputes

6.—(1) Compensation payable under section 78G in respect of an interest which is subject to a mortgage shall be paid to the mortgagee or, if there is more than one mortgagee, to the first mortgagee and shall, in either case, be applied by him as if it were proceeds of sale.

(2) Amounts of compensation determined under this Schedule shall be payable–

(a) where the appropriate person and the grantor or mortgagee agree that a single payment is to be made on a specified date, on that date;
(b) where the appropriate person and the grantor or mortgagee agree that payment is to be made in instalments at different dates, on the date agreed as regards each instalment; and
(c) in any other case, subject to any direction of the Lands Tribunal or the court, as soon as reasonably practicable after the amount of the compensation has been finally determined.

(3) Any question of the application of paragraph 5(3) above or of disputed compensation shall be referred to and determined by the Lands Tribunal.

(4) In relation to the determination of any such question, sections 2 and 4 of the 1961 Act (procedure on reference to the Lands Tribunal and costs) shall apply as if–

(a) the reference in section 2(1) of that Act to section 1 of that Act were a reference to sub-paragraph (3) of this paragraph; and
(b) references in section 4 of that Act to the acquiring authority were references to the appropriate person.

Schedule 3

Regulation 15

Registers

A register maintained by an enforcing authority under section 78R shall contain full particulars of the following matters–

Remediation Notices

1. In relation to a remediation notice served by the authority–

(a) the name and address of the person on whom the notice is served;
(b) the location and extent of the contaminated land to which the notice relates (in this paragraph referred to as the "contaminated land in question"), sufficient to enable it to be identified whether by reference to a plan or otherwise;
(c) the significant harm or pollution of controlled waters by reason of which the contaminated land in question is contaminated land;
(d) the substances by reason of which the contaminated land in question is contaminated land and, if any of the substances have escaped from other land, the location of that other land;
(e) the current use of the contaminated land in question;
(f) what each appropriate person is to do by way of remediation and the periods within which they are required to do each of the things; and
(g) the date of the notice.

Appeals against Remediation Notices

2. Any appeal against a remediation notice served by the authority.

3. Any decision on such an appeal.

Remediation Declarations

4. Any remediation declaration prepared and published by the enforcing authority under section 78H(6).

5. In relation to any such remediation declaration–

(a) the location and extent of the contaminated land in question, sufficient to enable it to be identified whether by reference to a plan or otherwise; and
(b) the matters referred to in sub-paragraphs (c), (d) and (e) of paragraph 1 above.

Remediation Statements

6. Any remediation statement prepared and published by the responsible person under section 78H(7) or by the enforcing authority under section 78H(9).

7. In relation to any such remediation statement–

(a) the location and extent of the contaminated land in question, sufficient to enable it to be identified whether by reference to a plan or otherwise; and
(b) the matters referred to in sub-paragraphs (c), (d) and (e) of paragraph 1 above.

Appeals against Charging Notices

8. In the case of an enforcing authority, any appeal under section 78P(8) against a charging notice served by the authority.

9. Any decision on such an appeal.

Designation of special sites

10. In the case of the Environment Agency, as respects any land in relation to which it is the enforcing authority, and in the case of a local authority, as respects any land in its area–

(a) any notice given by a local authority under subsection (1)(b) or (5)(a) of section 78C, or by the Secretary of State under section 78D(4)(b), which, by virtue of section 78C(7) or section 78D(6) respectively, has effect as the designation of any land as a special site;
(b) the provisions of regulation 2 or 3 by virtue of which the land is required to be designated as a special site;
(c) any notice given by the Environment Agency under section 78Q(1)(a) of its decision to adopt a remediation notice; and
(d) any notice given by or to the enforcing authority under section 78Q(4) terminating the designation of any land as a special site.

Notification of Claimed Remediation

11. Any notification given to the authority for the purposes of section 78R(1)(h) or (j).

Convictions for Offences under Section 78M

12. Any conviction of a person for any offence under section 78M in relation to a remediation notice served by the authority, including the name of the offender, the date of conviction, the penalty imposed and the name of the Court.

Guidance Issued under Section 78V(1)

13. In the case of the Environment Agency, the date of any guidance issued by it under subsection (1) of section 78V and, in the case of a local authority, the date of any guidance issued by the Agency to it under that subsection.

Other Environmental Controls

14. Where the authority is precluded by virtue of section 78YB(1) from serving a remediation notice–

(a) the location and extent of the contaminated land in question, sufficient to enable it to be identified whether by reference to a plan or otherwise;
(b) the matters referred to in sub-paragraphs (c), (d) and (e) of paragraph 1 above; and
(c) any steps of which the authority has knowledge, carried out under section 27, towards remedying any significant harm or pollution of controlled waters by reason of which the land in question is contaminated land.

15. Where the authority is precluded by virtue of section 78YB(3) from serving a remediation notice in respect of land which is contaminated land by reason of the deposit of controlled waste or any consequences of its deposit–

(a) the location and extent of the contaminated land in question, sufficient to enable it to be identified whether by reference to a plan or otherwise;
(b) the matters referred to in sub-paragraphs (c), (d) and (e) of paragraph 1 above; and
(c) any steps of which the authority has knowledge, carried out under section 59, in relation to that waste or the consequences of its deposit, including in a case where a waste collection authority (within the meaning of section 30(3)) took those steps or required the steps to be taken, the name of that authority.

16. Where, as a result of a consent given under Chapter II of Part III of the Water Resources Act 1991 (pollution offences), the authority is precluded by virtue of section 78YB(4) from specifying in a remediation notice any particular thing by way of remediation which it would otherwise have specified in such a notice,–

(a) the consent;
(b) the location and extent of the contaminated land in question, sufficient to enable it to be identified whether by reference to a plan or otherwise; and
(c) the matters referred to in sub-paragraphs (c), (d) and (e) of paragraph 1 above.

Appendix D

DETR Circular 02/2000 Environmental Protection Act 1990:

Part IIA: Contaminated Land

1 I am directed by the Secretary of State for the Environment, Transport and the Regions to draw your attention to the entry into force of the new statutory regime for the identification and remediation of contaminated land with effect from 1 April 2000.

2 For this purpose, the Secretary of State has made the Environment Act 1995 (Commencement No.16 and Saving Provision) (England) Order 2000 (S.I. 2000/340(C.8)), which brings into force Part IIA of the Environmental Protection Act 1990 (the "1990 Act"). Part IIA was inserted into the 1990 Act by section 57 of the Environment Act 1995. He has also made the Contaminated Land (England) Regulations 2000 (SI 2000/227), which have been made under sections 78C, 78E, 78G, 78L 78R and 78X.

Purpose of this circular

3 This circular has two functions: first it promulgates the statutory guidance which is an essential part of the new regime; secondly, it sets out the way in which the new regime is expected to work, by providing a summary of Government policy in this field, a description of the new regime, a guide to the Regulations and a note on the saving provision in the Commencement Order.

4 The new regime as described in this circular does not apply to any radioactive contamination of land. Part IIA makes provision for the regime to be applied to such contamination with such modifications as the Secretary of State considers appropriate. Consultation started in February 1998 on how such an application might be made.

5 This circular applies only to England. Responsibility for implementing Part IIA in Scotland and Wales rests with the Scottish Executive and the National Assembly for Wales, respectively.

Statutory guidance

6 I am therefore further directed by the Secretary of State for the Environment, Transport and the Regions to say that he hereby issues the statutory guidance in Annex 3 to this circular. This guidance is issued under the following powers:

(a) *The Definition of Contaminated Land* - Chapter A of Annex 3 to this circular sets out guidance issued under section 78A(2) and (5);

(b) *The Identification of Contaminated Land* - Chapter B of Annex 3 to this circular sets out guidance issued under section 78B(2);
(c) *The Remediation of Contaminated Land* - Chapter C of Annex 3 to this circular sets out guidance issued under section 78E(5);
(d) *Exclusion from, and Apportionment of, Liability for Remediation* - Chapter D of Annex 3 to this circular sets out guidance issued under section 78F(6) and (7); and
(e) *The Recovery of the Costs of Remediation* - Chapter E of Annex 3 to this Circular sets out guidance issued under section 78P(2).

7 Section 78YA states that before the Secretary of State can issue any guidance under Part IIA, he must consult the Environment Agency and such other persons as he considers it appropriate to consult. Drafts of all the guidance were published for consultation in September 1996, October 1998 and October 1999. The guidance contained in Annex 3 to this Circular has been prepared in the light of responses to those consultation exercises.

8 In addition, section 78YA requires the Secretary of State to lay a draft of any guidance he proposes to issue under sections 78A(2) or (5), 78B(2) or 78F(5) or (6) before each House of Parliament for approval under the negative resolution procedure. The guidance now issued in Chapters A, B and D of Annex 3 to this Circular was laid in draft before both Houses on 7 February 2000.

Financial and manpower implications

9 The Explanatory and Financial Memorandum to the Environment Bill stated that the creation of the Part IIA regime would have neither any financial nor any manpower implications, as it largely restated existing functions of local authorities and the Environment Agency. However, in the light of responses received to the consultation on the draft statutory guidance, published in September 1996, the Government decided that the successful operation of the new regime would necessitate the provision of additional resources for local authorities and the Environment Agency.

10 Accordingly, as part of the outcome of the Comprehensive Spending Review announced in July 1998, and after consulting the Local Government Association, the Government announced that an additional £50 million would be provided over the next three years to help local authorities develop inspection strategies, carry out site investigations and take forward enforcement action. This funding would be in addition to £45 million already planned to be spent over the same period through the Contaminated Land Supplementary Credit Approval (SCA) programme, which provides support for capital costs incurred by local authorities in inspecting and remediating land. Of the new funding, £12 million has been added to the figure for Total Standard Spending for local authorities in each of the three years, with the remainder being added to the SCA programme.

11 The additional cost burdens placed on the Environment Agency were taken into account in setting the level of grant-in-aid to be paid to the Agency.

Regulatory Impact Assessment

12 A Regulatory Impact Assessment (RIA) on the implementation of the Part IIA regime has been prepared. A draft of the earlier style of Compliance Cost Assessment was published for consultation in November 1996. A further draft was included with the draft Circular published for

consultation in October 1999. Comments received in response to these consultations have been taken into account in the final RIA, which was published on 7 February 2000.

13 Copies of the RIA are available from the address shown in paragraph 15 below.

Enquiries

14 Enquiries about particular sites and how they may be affected by the new regime should be directed, in the first instance, to the local authority in whose area they are situated.

15 Enquiries about this Circular should be addressed to:

Land Quality Team
Marine, Land and Liability Division
DETR
3/B4 Ashdown House
123 Victoria Street
London SW1E 6DE

Phone: (020) 7944 5287
Fax: (020) 7944 5279
Email: landquality—enquiries@DETR.GOV.UK

Contents

Annex 1 A statement of government policy

Sustainable development

1 In his foreword to *A better quality of life: A strategy for sustainable development for the UK* the Prime Minister, the Rt Hon Tony Blair MP, said:

"The last hundred years have seen a massive increase in the wealth of this country and the well-being of its people. But focusing solely on economic growth risks ignoring the impact — both good and bad — on people and on the environment. Had we taken account of these links in our decision making, we might have reduced or avoided costs such as contaminated land or social exclusion."

Preventing new contamination

2 Contaminated land is an archetypal example of our failure in the past to move towards sustainable development. We must learn from that failure. The first priority for the Government's policy on land contamination is therefore to prevent the creation of new contamination. We have, or are creating, a range of regimes aimed at achieving this. Of these, the most significant are:

(a) *Integrated Pollution Control (IPC)* Part I of the Environmental Protection Act 1990 ("the 1990 Act") places a requirement on operators of prescribed industrial processes to operate within the terms of permits issued by the Environment Agency to control harmful environmental discharges;

(b) *Pollution Prevention and Control (PPC)* — A new regime will shortly be introduced to replace IPC, and to implement the European Union's Integrated Pollution Prevention and Control directive; that includes the specific requirement that permits for industrial plants and installations must include conditions to prevent the pollution of soil; and

(c) *Waste Management Licensing* — Part II of the 1990 Act places controls over the handling, treatment and disposal of wastes; in the past, much land contamination has been the result of unregulated, or badly-managed, waste disposal activities.

3 Although the prevention of new contamination is of critical importance, the focus of this Circular is on land which has been contaminated in the past.

Our inherited legacy of contaminated land

4 As well as acting to prevent new contamination, we have also to deal with a substantial legacy of land which is already contaminated, for example by past industrial, mining and waste disposal activities. It is not known, in detail, how much land is contaminated. This can be found out only through wide-ranging and detailed site investigation and risk assessment. The answer will be critically dependent on the definition used to establish what land counts as being "contaminated".

5 Various estimates have been made of the extent of the problem. In its report *Contaminated Land*, published in 1993, the Parliamentary Office of Science and Technology referred to expert estimates of between 50,000 and 100,000 potentially contaminated sites across the UK, with estimates of the extent of land ranging between 100,000 and 200,000 hectares.

However, the report did note that international experience suggested that only a small proportion of potentially contaminated sites posed an immediate threat to human health and the environment. More recently, the Environment Agency has estimated that that there may be some 300,000 hectares of land in the UK affected to some extent by industrial or natural contamination.

6 The existence of contamination presents its own threats to sustainable development:

(a) it impedes social progress, depriving local people of a clean and healthy environment;
(b) it threatens wider damage to the environment and to wildlife;
(c) it inhibits the prudent use of our land and soil resources, particularly by obstructing the recycling of previously-developed land and increasing development pressures on greenfield areas; and
(d) the cost of remediation represents a high burden on individual companies, home- and other land-owners, and the economy as a whole.

7 In this context, the Government's objectives with respect to contaminated land are:

(a) to identify and remove unacceptable risks to human health and the environment;
(b) to seek to bring damaged land back into beneficial use; and
(c) to seek to ensure that the cost burdens faced by individuals, companies and society as a whole are proportionate, manageable and economically sustainable.

8 These three objectives underlie the "suitable for use" approach to the remediation of contaminated land, which the Government considers is the most appropriate approach to achieving sustainable development in this field.

The "suitable for use" approach

9 The "suitable for use" approach focuses on the risks caused by land contamination. The approach recognises that the risks presented by any given level of contamination will vary greatly according to the use of the land and a wide range of other factors, such as the underlying geology of the site. Risks therefore need to be assessed on a site-by-site basis.

10 The "suitable for use" approach then consists of three elements:

(a) **ensuring that land is suitable for its current use** – in other words, identifying any land where contamination is causing unacceptable risks to human health and the environment, assessed on the basis of the current use and circumstances of the land, and returning such land to a condition where such risks no longer arise ("remediating" the land); the new contaminated land regime provides general machinery to achieve this;
(b) **ensuring that land is made suitable for any new use, as planning permission is given for that new use** – in other words, assessing the potential risks from contamination, on the basis of the proposed future use and circumstances, before official permission is given for the development and, where necessary to avoid unacceptable risks to human health and the environment, remediating the land before the new use commences; this is the role of the town and country planning and building control regimes; and
(c) **limiting requirements for remediation to the work necessary to prevent unacceptable risks to human health or the environ-**

ment in relation to the current use or future use of the land for which planning permission is being sought – in other words, recognising that the risks from contaminated land can be satisfactorily assessed only in the context of specific uses of the land (whether current or proposed), and that any attempt to guess what might be needed at some time in the future for other uses is likely to result either in premature work (thereby risking distorting social, economic and environmental priorities) or in unnecessary work (thereby wasting resources).

11 Within this framework, it is important to recognise both that the use of any particular area of land may cover several different activities and that some potential risks arising from contamination (particularly impacts on water and the wider environment) may arise independently of the use of the land. In practical terms, the current use of any land should be taken to be any use which:

(a) is currently being made of the land, or is likely to be made of it; and
(b) is consistent with any existing planning permission or is otherwise lawful under town and country planning legislation.

(This approach is explained in more detail in paragraph *A.26 of Annex 3* to this Circular.)

12 Regulatory action may be needed to make sure that necessary remediation is carried out. However, limiting remediation costs to what is needed to avoid unacceptable risks will mean that we will be able to recycle more previously-developed land than would otherwise be the case, increasing our ability to make beneficial use of the land. This helps to increase the social, economic and environmental benefits from regeneration projects and to reduce unnecessary development pressures on greenfield sites.

13 The "suitable for use" approach provides the best means of reconciling our various environmental, social and economic needs in relation to contaminated land. Taken together with tough action to prevent new contamination, and wider initiatives to promote the reclamation of previously-developed land, it will also help to bring about progressive improvements in the condition of the land which we pass on to future generations.

14 Within the "suitable for use" approach, it is always open to the person responsible for a site to do more than can be enforced through regulatory action. For example, a site owner may plan to introduce at a future date some new use for the land which would require more stringent remediation, and may conclude that, in these circumstances, it is more economic to anticipate those remediation requirements. However, this is a judgement which only the person responsible for the site is in a position to make.

15 The one exception to the "suitable for use" approach to regulatory action applies where contamination has resulted from a specific breach of an environmental licence or permit. In such circumstances, the Government considers that it is generally appropriate that the polluter is required, under the relevant regulatory regime, to remove the contamination completely. To do otherwise would be to undermine the regulatory regimes aimed at preventing new contamination.

Action to deal with contamination

Voluntary remediation action

16 The Government aims to maintain the quality of the land in this country and to improve it progressively where it has been degraded in the past.

Redeveloping areas where previous development has reached the end of its useful life not only contributes to social and economic regeneration of the local communities but is also an important driver in achieving this progressive environmental improvement.

17 The Government is determined to limit the unnecessary development of greenfield areas, and has in particular set a target for 60% of new housing to be built on previously-developed land. Various initiatives aimed at achieving the objective of increasing the recycling of land are outlined in *Planning for Communities of the Future*. Further proposals for action have been made by the Urban Task Force, chaired by Lord Rogers of Riverside, in its report *Towards an Urban Renaissance*. The Government will be responding to these recommendations in an Urban White Paper.

18 It is, of course, necessary to ensure that when previously-developed land is redeveloped any potential risks associated with contamination are properly identified and remediated. The planning and building control systems, described at paragraphs 45 to 49 below, provide the means of achieving this.

19 There are very few cases where land cannot be restored to some beneficial use. However, the actual or potential existence of contamination on a site can inhibit the willingness or ability of a developer to do so. The Government is acting in three specific ways to overcome the potential obstacles to the redevelopment of land affected by contamination:

(a) *by providing public subsidy* – substantial funding is made available through the Single Regeneration Budget, English Partnerships and the regional development agencies to support site redevelopment costs for projects aimed at particular social and economic regeneration objectives;

(b) *by promoting research and development* – the programmes of the science research councils, the Environment Agency, the DETR and the DTI aim to increase scientific understanding and the availability and take-up of improved methods of risk assessment and remediation; and

(c) *by providing an appropriate policy and legal framework* – the "suitable for use" approach ensures that remediation requirements are reasonable and tailored to the needs of individual sites; a significant objective underlying the new contaminated land regime is to improve the clarity and certainty of potential regulatory action on contamination, thereby assisting developers to make informed investment appraisals.

Regulatory Action

20 The regeneration process is already dealing with much of our inherited legacy of contaminated land. But there will be circumstances where contamination is causing unacceptable risks on land which is either not suitable or not scheduled for redevelopment. For example, there may be contamination on sites now regarded as greenbelt or rural land, or contamination may be affecting the health of occupants of existing buildings on the land or prejudicing wildlife on the site or in its surroundings. We therefore need systems in place both to identify problem sites of this kind and, more significantly, to ensure that the problems are dealt with and the contamination remediated.

21 A range of specific clean-up powers exists to deal with cases where contamination is the result of offences against, or breaches of, pollution prevention regimes. The main examples of these are described in paragraphs 51 to 58 below.

22 Part IIA of the Environmental Protection Act 1990 creates a new framework for the identification and remediation of contaminated land in

circumstances where there has not been any identifiable breach of a pollution prevention regime.

23 Although Part IIA itself is new, it largely replaces existing regulatory powers and duties. Borough and district councils have long-standing duties to identify particular environmental problems, including those resulting from land contamination, and to require their abatement. The origins of these powers is found in the mid-19th century legislation which created the concept of the statutory nuisance. They were codified in the Public Health Act 1936 and have most recently been set out in Part III of the Environmental Protection Act 1990, which modernised the statutory nuisance regime.

24 In addition, the Environment Agency has powers under Part VII of the Water Resources Act 1991 to take action to prevent or remedy the pollution of controlled waters, including circumstances where the pollution arises from contamination in the land.

The new contaminated land regime

Objectives for the new regime

25 The main objective underlying the introduction of the Part IIA Contaminated Land regime is to provide an improved system for the identification and remediation of land where contamination is causing unacceptable risks to human health or the wider environment, assessed in the context of the current use and circumstances of the land.

26 As stated in paragraph 23 above, the new regime broadly reflects the approaches already in place under the statutory nuisance regime and Part VII of the Water Resources Act 1991. The Government's primary objectives for introducing the new regime are:

(a) to improve the focus and transparency of the controls, ensuring authorities take a strategic approach to problems of land contamination;

(b) to enable all problems resulting from contamination to be handled as part of the same process; previously separate regulatory action was needed to protect human health and to protect the water environment;

(c) to increase the consistency of approach taken by different authorities; and

(d) to provide a more tailored regulatory mechanism, including liability rules, better able to reflect the complexity and range of circumstances found on individual sites.

27 In addition to providing a more secure basis for direct regulatory action, the Government considers that the improved clarity and consistency of the new regime, in comparison with its predecessors, is also likely to encourage voluntary remediation. This forms an important secondary objective for implementation of the Part IIA regime.

28 Companies who may be responsible for contamination, for example on land they currently own or on former production sites, will be able to assess the likely requirements of regulators acting under Part IIA. They will then be able to plan their own investment programmes to carry out remediation in advance of actual regulatory intervention.

29 Similarly, the Part IIA regime will assist in the recycling of previously-developed land. The new regime cannot be used directly to require the redevelopment of land, only its remediation. However, the Government considers that implementation of the regime will assist developers by reducing uncertainties about so-called "residual liabilities", in particular the perceived risk of further regulatory intervention. In particular it will:

(a) reinforce the "suitable for use" approach, enabling developers to design and implement appropriate and cost-effective remediation schemes as part of their redevelopment projects;
(b) clarify the circumstances in which future regulatory intervention might be necessary (for example, if the initial remediation scheme proved not to be effective in the long term); and
(c) set out the framework for statutory liabilities to pay for any further remediation, should that be necessary.

Outline of Part IIA and associated documents

30 The primary legislation in Part IIA contains the structure and main provisions of the new regime. It consists of sections 78A to 78YC. An explanation of how the regime will operate is set out in the *Guide to the New Regime*, at Annex 2 to this Circular.

31 Within the structure of the Part IIA legislation, the statutory guidance set out in Annex 3 to this Circular provides the detailed framework for the following key elements of the new regime:

(a) the definition of contaminated land (Chapter A);
(b) the identification of contaminated land (Chapter B);
(c) the remediation of contaminated land (Chapter C);
(d) exclusion from, and apportionment of, liability for remediation (Chapter D); and
(e) the recovery of the costs of remediation and the relief from hardship (Chapter E).

32 Regulations made under Part IIA deal with:

(a) the descriptions of land which are required to be designated as "special sites";
(b) the contents of, and arrangements for serving, remediation notices;
(c) compensation to third parties for granting rights of entry etc. to land;
(d) grounds of appeal against a remediation notice, and procedures relating to any such appeal; and
(e) particulars to be contained in registers compiled by enforcing authorities, and the locations at which such registers must be available for public inspection.

33 Annex 4 to this Circular provides a detailed description of the Contaminated Land (England) Regulations 2000.

Main features of the new regime

34 The primary regulatory role under Part IIA rests with local authorities:

(a) in Greater London, this means the London borough councils, the City of London and the Temples; and
(b) elsewhere it means the borough or district councils or, where appropriate, the unitary authority.

35 This reflects their existing functions under the statutory nuisance regime, and will also complement their roles as planning authorities. In outline, the role of these authorities under Part IIA will be:

(a) to cause their areas to be inspected to identify contaminated land;
(b) to determine whether any particular site is contaminated land;
(c) to act as enforcing authority for all contaminated land which is not designated as a "special site" (the Environment Agency will be the enforcing authority for special sites).

36 The enforcing authorities will have four main tasks:

(a) to establish who should bear responsibility for the remediation of the land (the "appropriate person" or persons);
(b) to decide, after consultation, what remediation is required in any individual case and to ensure that such remediation takes place, either through agreement with the appropriate person, or by serving a remediation notice on the appropriate person if agreement is not possible or, in certain circumstances, through carrying out the work themselves;
(c) where a remediation notice is served, or the authority itself carries out the work, to determine who should bear what proportion of the liability for meeting the costs of the work; and
(d) to record certain prescribed information about their regulatory actions on a public register.

37 Contaminated land is land which appears to the local authority to be in such a condition, by reason of substances in, on or under the land, that significant harm is being caused, or there is a significant possibility of such harm being caused, or that pollution of controlled waters is being, or is likely to be, caused. This definition is to be applied in accordance with other definitions in Part IIA and statutory guidance set out in this Circular. These definitions and the guidance are based on the assessment of risks to human health and the environment. The regime thus reflects the "suitable for use" approach.

38 Under the provisions concerning liabilities, responsibility for paying for remediation will, where feasible, follow the "polluter pays" principle. In the first instance, any persons who caused or knowingly permitted the contaminating substances to be in, on or under the land will be the appropriate person(s) to undertake the remediation and meet its costs. However, if it is not possible to find any such person, responsibility will pass to the current owner or occupier of the land. (This latter step does not apply where the problem caused by the contamination is solely one of water pollution: this reflects the potential liabilities for water pollution as they existed prior to the introduction of Part IIA.) Responsibility will also be subject to limitations, for example where hardship might be caused; these limitations are set out in Part IIA and in the statutory guidance in this Circular.

39 The Environment Agency will have four principal roles with respect to contaminated land under Part IIA. It will:

(a) assist local authorities in identifying contaminated land, particularly in cases where water pollution is involved;
(b) provide site-specific guidance to local authorities on contaminated land;
(c) act as the "enforcing authority" for any land designated as a "special site" (the descriptions of land which are required to be designated in this way are prescribed in the Regulations); and
(d) publish periodic reports on contaminated land.

40 In addition, the Agency has inherited the contaminated land research programme previously run by the then Department of the Environment. The Agency will continue to carry out technical research and, in conjunction with DETR, publish scientific and technical advice.

Measuring Progress

41 DETR will be developing performance indicators to assess overall progress in the task of identifying and remediating our inherited legacy of

contaminated land. This will rely on information gathered by the Environment Agency as part of its role in preparing periodic reports on contaminated land.

42 The indicators will, potentially, include both:

(a) measures of the scale of regulatory activities carried out by local authorities and the Environment Agency under Part IIA; and
(b) indicators of overall progress in the task of identifying and remediating contaminated land, whether this is the result of voluntaryaction or a response to regulatory action under Part IIA.

43 It is the Government's intention in due course to establish targets for overall progress. However, at this stage, it is not possible to set meaningful targets as too little is known about the true extent of contaminated land. This will change once local authorities have worked up their inspection strategies and started carrying out the detailed inspection of individual sites.

Published technical advice

44 DETR, the Environment Agency and other bodies have published a range of technical advice documents relating to contaminated land. A bibliography is on the DETR website at www.detr.gov.uk/. This will be kept up to date as further documents are produced.

Interaction with other regimes

Planning and development control

45 Land contamination, or the possibility of it, is a material consideration for the purposes of town and country planning. This means that a planning authority has to consider the potential implications of contamination both when it is developing structure or local plans (or unitary development plans) and when it is considering individual applications for planning permission.

46 The planning authority should satisfy itself that the potential for contamination is properly assessed, and the development incorporates any necessary remediation. Where necessary, any planning permission should include appropriate site investigation and remediation conditions. Under the "suitable for use" approach, risks should be assessed, and remediation requirements set, on the basis of both the current use and circumstances of the land and its proposed new use. (This is in contrast to the approach under Part IIA, where only the current use and circumstances are considered.)

47 Guidance to planning authorities is set out in *Planning Policy Guidance: Planning and Pollution Control (PPG 23)*, published in 1994, and DOE Circular 11/95 *The Use of Conditions in Planning Permissions*. DETR is currently preparing further planning guidance on land contamination, which will amplify the guidance in PPG 23, explain the interface with the Part IIA regime from a planning perspective, and provide planning authorities with technical and practical advice on land contamination issues. In the meantime, the guidance contained in PPG23 remains valid, although planning authorities should note that references to the term "contaminated land" in that document should be interpreted in the general sense rather than according to the particular definition used for the purposes of the Part IIA regime.

48 In some cases, the carrying out of remediation activities may itself constitute "development" within the meaning given at section 55 of the

Town and Country Planning Act 1990, and therefore require planning permission.

49 In addition to the planning system, the Building Regulations 1991 (made under the Building Act 1984) may require measures to be taken to protect the fabric of new buildings, and their future occupants, from the effects of contamination. *Approved Document Part C (Site Preparation and Resistance to Moisture)* gives guidance on these requirements.

50 In any case where new development is taking place, it will be the responsibility of the developer to carry out the necessary remediation. In most cases, the enforcement of any remediation requirements will be through planning conditions and building control, rather than through a remediation notice issued under Part IIA.

Integrated Pollution Control (IPC) and Pollution Prevention and Control (PPC)

51 Section 27 of the Environmental Protection Act 1990 gives the Environment Agency the power to take action to remedy harm caused by a breach of IPC controls under section 23(1)(a) or (c) of the Act. This could apply to cases of land contamination arising from such causes.

52 In any case where an enforcing authority acting under Part IIA considers that the section 27 power is exercisable, it is precluded by section 78YB(1) from serving a remediation notice to remedy the same harm.

53 In some cases, remediation activities may themselves constitute processes which cannot be carried out without a permit issued under the IPC regime.

54 The Government is currently developing a new regime of Pollution Prevention and Control (PPC). This will replace the current IPC regime, and will transpose into national law the requirements of the EC Integrated Pollution Prevention and Control Directive (96/61/EC). The regime will include a new system of enforcement notices, which will enable the Environment Agency to require the operator of permitted plants or installations to remedy the effects of any breaches of their permits. The PPC regime will have the same relationship to Part IIA as has the IPC regime.

Waste Management Licensing

55 There are three areas of potential interaction between the Part IIA regime and the waste management licensing system under Part II of the Environmental Protection Act 1990.

56 Firstly, there may be significant harm or pollution of controlled waters arising from land for which a site licence is in force under Part II. Where this is the case, under section 78YB(2), the Part IIA regime does not normally apply; that is, the land cannot formally be identified as "contaminated land" and no remediation notice can be served. If action is needed to deal with a pollution problem in such a case, this would normally be enforced through a "condition" attached to the site licence. However, Part IIA does apply if the harm or pollution on a licensed site is attributable to a cause other than a breach of the site licence, or the carrying on of an activity authorised by the licence in accordance with its terms and conditions.

57 Secondly, under section 78YB(3), an enforcing authority acting under Part IIA cannot serve a remediation notice in any case where the contamination results from an illegal deposit of controlled waste. In these circumstances, the Environment Agency and the waste disposal authority have powers under section 59 of the 1990 Act to remove the waste, and to deal with the consequences of its having been present.

58 Thirdly, remediation activities on contaminated land may themselves fall within the definitions of "waste disposal operations" or "waste recovery operations", and be subject to the licensing requirements under the Part II system. Guidance on the meaning of the relevant definitions and the operation of the licensing system is provided in DOE Circular 11/94.

Statutory nuisance

59 Until the implementation of the Part IIA contaminated land regime, the statutory nuisance system under Part III of the 1990 Act was the main regulatory mechanism for enforcing the remediation of contaminated land.

60 Parliament has considered that the Part IIA regime, as explained in the statutory guidance, sets out the right level of protection for human health and the environment from the effects of land contamination. It has therefore judged it inappropriate to leave in place the possibility of using another, less precisely defined, system which could lead to the imposition of regulatory requirements on a different basis.

61 From the entry into force of the new contaminated land regime, most land contamination issues are therefore removed from the scope of the Statutory Nuisance regime. This is the effect of an amendment to the definition of a statutory nuisance in section 79 of the 1990 Act, consisting of the insertion of sections 78(1A) and (1B); this amendment was made by paragraph 89 of Schedule 22 of the Environment Act 1995. Any matter which would otherwise have been a statutory nuisance will no longer be treated as such, to the extent that it consists of, or is caused by, land "being in a contaminated state". The definition of land which is "in a contaminated state", and where the statutory nuisance regime is therefore excluded, covers all land where there are substances in, on or under the land which are causing harm or where there is a possibility of harm being caused.

62 However the statutory nuisance regime will continue to apply for land contamination issues in any case where an abatement notice under section 80(1), or an order of the court under section 82(2)(a), has already been issued and is still in force. This will ensure that any enforcement action taken under the statutory nuisance regime can continue, and will not be interrupted by the implementation of the Part IIA regime.

63 It should also be noted that the statutory nuisance regime will continue to apply to the effects of deposits of substances on land which give rise to such offence to human senses (such as stenches) as to constitute a nuisance, since the exclusion of the statutory nuisance regime applies only to harm (as defined in section 78A(4)) and the pollution of controlled waters.

Water Resources Act 1991

64 Sections 161 to 161D of the Water Resources Act 1991 give the Environment Agency powers to take action to prevent or remedy the pollution of controlled waters. The normal enforcement mechanism under these powers is a "works notice" served under section 161A, which specifies what actions have to be taken and in what time periods. This is served on any person who has "caused or knowingly permitted" the potential pollutant to be in the place from which it is likely to enter controlled waters, or to have caused or knowingly permitted a pollutant to enter controlled waters. Where it is not appropriate to serve such a notice, because of the need for urgent action or where no liable person can be found, the Agency has the power to carry out the works itself.

65 There is an obvious potential for overlap between these powers and the Part IIA regime in circumstances where substances in, on or under land are likely to enter controlled waters. The decision as to which regime is used in any case may have important implications, as there are differences between the two enforcement mechanisms.

66 The Environment Agency has published a policy statement, *Environment Agency Policy and Guidance on the Use of Anti-Pollution Works Notices*. This sets out how the Agency intends to use the works notice powers, particularly in cases where there is an overlap with the Part IIA regime. The statement has been agreed with the DETR. In summary, the effect of the policy, taken together with the legislation, is that:

(a) the local authority, acting under Part IIA, should consult the Environment Agency before determining that land is contaminated land in respect of pollution of controlled waters;

(b) in any case where a local authority has identified contaminated land which is potentially affecting controlled waters, the statutory guidance set out in this Circular requires that authority to consult the Environment Agency, and to take into account any comments the Agency makes with respect to remediation requirements;

(c) where the Agency identifies any case where actual or potential water pollution is arising from land affected by contamination, the Agency will notify the relevant local authority, thus enabling that authority formally to identify the land as "contaminated land" for the purposes of the Part IIA regime; and

(d) in any case where land has been identified as "contaminated land" under the Part IIA regime, the Part IIA enforcement mechanisms would normally be used, rather than the works notice system. This is because Part IIA imposes a duty to serve a remediation notice, whereas the Agency is given only a power to serve a works notice.

67 The Water Resources Act powers may be particularly useful in cases where there is historic pollution of groundwater, but where the Part IIA regime does not apply. This may occur, for example, where the pollutants are entirely contained within the relevant body of groundwater or where the "source" site cannot be identified.

68 No remediation notice can require action to be carried out which would have the effect of impeding or preventing a discharge into controlled waters for which a "discharge consent" has been issued under Chapter II of Part III of the Water Resources Act 1991.

Radioactivity

69 Under section 78YC of the 1990 Act, the normal Part IIA regime does not apply with respect to harm, or water pollution, which is attributable to any radioactivity possessed by any substance.

70 However, this section does give powers to the Secretary of State to make regulations applying the Part IIA regime, with any necessary modifications, to problems of radioactive contamination. The DETR published a consultation paper in February 1998 outlining a possible approach to applying the Part IIA regime. More detailed proposals are currently being developed in the light of responses to that consultation.

Other regimes

71 Other regimes which may have implications for land contamination, or which may overlap with Part IIA, include the following:

(a) *Food Safety* – Part I of the Food and Environment Protection Act 1985 gives ministers emergency powers to issue orders for the purpose of prohibiting specified agricultural activities in a designated area, in order to protect consumers from exposure to contaminated food. The 1985 Act provides for ministers to designate authorities for the enforcement of emergency control orders. Following the coming into force of the Food Standards Act 1999, which establishes the new Food Standards Agency, the above powers are exercisable by the Secretary of State for Health, acting in the light of advice from the Food Standards Agency. The Minister of Agriculture, Fisheries and Food may, however, be designated as an enforcement authority under the 1999 Act, since in some cases his officials will be best placed to monitor compliance in the field. The Food Standards Agency may, in addition, exercise certain functions in relation to emergency control orders issued by the Secretary of State. This includes for example the power to issue consents in relation to certain activities and the power to give directions on compliance with the provisions of an order. Enforcing authorities under Part IIA should liaise with the Food Standards Agency about any possible use of the powers in Part I of the 1985 Act.

(b) *Health and Safety* – The Health and Safety at Work etc Act 1974, the Construction (Design and Management) Regulations 1994 (S.I. 1994/3140) and their associated controls are concerned with risks to the public or employees at business and other premises; risks of these kinds could arise as a result of land contamination. Liaison between Part IIA enforcing authorities and the Health and Safety Executive will help to ensure that unnecessary duplication of controls is avoided, and that the most appropriate regime is used to deal with any problems.

(c) *Landfill Tax* – The Finance Act 1996 introduced a tax on the disposal of wastes, including those arising from the remediation and reclamation of land. However, an exemption from this tax can be obtained where material is being removed from contaminated land in order to prevent harm, or to facilitate the development of the land for particular purposes. An exemption certificate has to be specifically applied for, through HM Customs and Excise, in each case where it might apply. No exemption certificate will be granted where the material is being removed in order to comply with the requirements of a remediation notice served under section 78E of the 1990 Act. This provides a fiscal incentive for those responsible for carrying out remediation under Part IIA to do so by agreement, rather than waiting for the service of a remediation notice.

(d) *Major Accident Hazards* – The Control of Major Accident Hazards Regulations 1999 (SI 1999/743) (COMAH) require operators of establishments handling prescribed dangerous substances to prepare on-site emergency plans, and the local authorities to prepare off-site emergency plans. The objectives of these emergency plans include providing for the restoration and clean-up of the environment following a major accident. The Health and Safety Executive are responsible for overseeing the COMAH Regulations.

ANNEX 2 A DESCRIPTION OF THE NEW REGIME FOR CONTAMINATED LAND

1 Introduction

1.1 Part IIA of the Environmental Protection Act 1990—which was inserted into that Act by section 57 of the Environment Act 1995—provides a new regulatory regime for the identification and remediation of contaminated land. In addition to the requirements contained in the primary legislation, operation of the regime is subject to regulations and statutory guidance.

1.2 This annex to the circular describes, in general terms, the operation of the regime, setting out the procedural steps the enforcing authority takes, and some of the factors which may underlie its decisions at each stage. Where appropriate it refers to the primary legislation, regulations or statutory guidance. However, the material in this part of the Circular does not form a part of that statutory guidance, and it should not be taken to qualify or contradict any requirements in the guidance, or to provide any additional guidance. It represents the Department's views and interpretations of the legislation, regulations and guidance. Readers should seek their own legal advice where necessary.

Definitions

1.3 Throughout the text, various terms are used which have specific meanings under the primary legislation, or in the regulations or the statutory guidance. Where this is the case, the terms are printed in SMALL CAPITALS. The Glossary of Terms at Annex 6 to the circular either repeats these definitions or shows where they can be found.

1.4 Unless the contrary is shown, references in this document to "sections" are to sections of the Environmental Protection Act 1990 (as amended) and references to "regulations" are references to the Contaminated Land (England) Regulations 2000. References to the statutory guidance include the relevant Chapter in Annex 3 to this Circular and the specific paragraph (so that, for example, a reference to paragraph 13 of Chapter B is shown as "paragraph B.13".) Such references are to the most relevant paragraph(s): those paragraph(s) must, of course, be read in the context of the relevant guidance as a whole.

2 The definition of contaminated land

The Definition in Part IIA

2.1 Section 78A(2) defines CONTAMINATED LAND for the purposes of Part IIA as:
"any land which appears to the LOCAL AUTHORITY in whose area it is situated to be in such a condition, by reason of substances in, on or under the land, that—

(a) SIGNIFICANT HARM is being caused or there is a SIGNIFICANT POSSIBILITY of such harm being caused; or
(b) POLLUTION OF CONTROLLED WATERS is being, or is likely to be, caused".

2.2 This definition reflects the intended role of the Part IIA regime, which is to enable the identification and remediation of land on which contamination is causing unacceptable risks to human health or the wider environment. It does not necessarily include all land where contamination is present, even though such contamination may be relevant in the context of other regimes. For example, contamination which might cause risks in the context of a new development of land could be a "material planning consideration" under the Town and Country Planning Act 1990.

2.3 The definition does not cover any HARM or POLLUTION OF CONTROLLED WATERS which is attributable to any radioactivity possessed by any substance (*section 78YC*). However, the Secretary of State has powers to make regulations applying some or all of the Part IIA regime—with modifications where appropriate—to cases of radioactive contamination (*section 78YC(a)*). Consultations began in February 1998 on the content of such regulations. Those regulations will deal with the procedure for sites where both radioactive and non-radioactive contamination is present. For the time being, any non-radioactive contamination on such sites may be addressed under the Part IIA regime as described here.

Significant harm

2.4 The definition of CONTAMINATED LAND includes the notion of "SIGNIFICANT HARM" and the "SIGNIFICANT POSSIBILITY" of such HARM being caused. The LOCAL AUTHORITY is required to act in accordance with statutory guidance issued by the SECRETARY OF STATE in determining what is "significant" in either context (*section 78A(2) & (5)*). This statutory guidance is set out at Chapter A of Annex 3 to this circular.

2.5 The statutory guidance uses the concept of a "POLLUTANT LINKAGE"—that is, a linkage between a CONTAMINANT and a RECEPTOR, by means of a PATHWAY. The statutory guidance then explains:

(a) the types of RECEPTOR to which SIGNIFICANT HARM can be caused (HARM to any other type of RECEPTOR can never be regarded as SIGNIFICANT HARM);
(b) the degree or nature of HARM to each of these RECEPTORS which constitutes SIGNIFICANT HARM *(Chapter A, Table A*); and
(c) for each RECEPTOR, the degree of possibility of the SIGNIFICANT HARM being caused which will amount to a SIGNIFICANT POSSIBILITY *(Chapter A, Table B, & paragraphs A.27 to A.33)*.

2.6 Before the LOCAL AUTHORITY can make the judgement that any land appears to be CONTAMINATED LAND on the basis that SIGNIFICANT HARM is being caused, or that there is a SIGNIFICANT POSSIBILITY of such harm being caused, the authority must therefore identify a SIGNIFICANT POLLUTANT LINKAGE. This means that each of the following has to be identified:

(a) a CONTAMINANT;
(b) a relevant RECEPTOR; and
(c) a PATHWAY by means of which either:
   (i) that CONTAMINANT is causing SIGNIFICANT HARM to that RECEPTOR, or
   (ii) there is a SIGNIFICANT POSSIBILITY of such harm being caused by that CONTAMINANT to that RECEPTOR *(paragraphs A.11 and A.19)*.

Pollution of controlled waters

2.7 The LOCAL AUTHORITY is also required to act in accordance with statutory guidance issued by the SECRETARY OF STATE in determining whether

POLLUTION OF CONTROLLED WATERS is being, or is likely to be, caused *(section 78A(5))*. This guidance is also set out at Chapter A of Annex 3 to this circular.

2.8 Before the LOCAL AUTHORITY can make the judgement that any land appears to be CONTAMINATED LAND on the basis that the POLLUTION OF CONTROLLED WATERS is being caused or is likely to be caused, the authority must identify a SIGNIFICANT POLLUTANT LINKAGE, where a body of CONTROLLED WATERS forms the RECEPTOR *(paragraphs A.11 and A.19)*.

2.9 There is no power to issue guidance on what constitutes the POLLUTION OF CONTROLLED WATERS. This term is defined in section 78A(9), in terms which are close to those used in Part III of the Water Resources Act 1991. However, when considering cases where it is thought that very small quantities of a CONTAMINANT might satisfy that definition, it is necessary also to consider the guidance on what remediation it is reasonable to require (see paragraphs 6.30 to 6.32 below).

2.10 Such cases may well give rise to some problems. The Government has indicated its intention of reviewing the wording of the legislation on this aspect and of seeking amendments to the primary legislation.

3 Identification of contaminated land

Inspection of a local authority's area

3.1 Each LOCAL AUTHORITY has a duty to cause its area to be inspected from time to time for the purpose of identifying CONTAMINATED LAND (*section 78B(1)*). In doing so, it has to act in accordance with statutory guidance issued by the SECRETARY OF STATE. This statutory guidance is set out at Chapter B of Annex 3 to this circular.

Strategy for inspection

3.2 The LOCAL AUTHORITY needs to take a strategic approach to the inspection of its area *(paragraph B.9)*. It is to set out this approach as a written strategy, which it is to publish within 15 months of the issuing of the statutory guidance, that is by July 2001 *(paragraph B.12)*.

3.3 Taking a strategic approach enables the LOCAL AUTHORITY to identify, in a rational, ordered and efficient manner, the land which merits detailed individual inspection, identifying the most pressing and serious problems first and concentrating resources on the areas where CONTAMINATED LAND is most likely to be found.

3.4 The strategy is also to contain procedures for liaison with other regulatory bodies, which may have information about land contamination problems, and for responding to information and complaints from members of the public, businesses and voluntary organisations *(paragraphs B.15 and B.16)*.

Inspecting land

3.5 The LOCAL AUTHORITY may identify a particular area of land where it is possible that a POLLUTANT LINKAGE exists. The authority could do so as a result of:

- (a) its own gathering of information as part of its strategy;
- (b) receiving information from another regulatory body, such as the ENVIRONMENT AGENCY; or

(c) receiving information or a complaint from a member of the public, a business or a voluntary organisation.

3.6 Where this is the case, the LOCAL AUTHORITY needs to consider whether to carry out a detailed inspection to determine whether or not the land actually appears to be CONTAMINATED LAND. Normally, the LOCAL AUTHORITY will be interested only in land which is in its area. But if it considers SIGNIFICANT HARM or the POLLUTION OF CONTROLLED WATERS might be caused within its area as a result of contamination on land outside its area, it may also inspect that other land (*section 78X(2)*).

3.7 The LOCAL AUTHORITY may already have detailed information concerning the condition of the land. This may have been provided, for example, by the ENVIRONMENT AGENCY or by a person such as the owner of the land. Alternatively, such a person may offer to provide such information within a reasonable and specified time. It may therefore be helpful for the authority to consult the owner of the land and other persons, in order to find out whether information already exists, or could be made available to the authority.

3.8 Where information is already available, or will become available, the LOCAL AUTHORITY needs to consider whether the information provides, or would provide, a sufficient basis on which it can determine whether or not the land appears to be CONTAMINATED LAND. If the information meets this test, the authority does not need to carry out any further investigation of the land (*paragraph B.23*) and will proceed to make a determination on that basis (see paragraph 3.33 below).

3.9 Where the LOCAL AUTHORITY does not have sufficient information, it needs to consider whether to make an inspection of the land. For this purpose it needs to consider whether:

(a) there is a reasonable possibility that a POLLUTANT LINKAGE exists on the land (*paragraph B.22(a)*); and

(b) if the land were eventually determined to be CONTAMINATED LAND, whether it would fall to be designated a SPECIAL SITE (see paragraphs 3.12 to 3.16 below).

3.10 If the answer to the first of these questions is "yes", and the second is "no", the LOCAL AUTHORITY needs to authorise an inspection of the land. It has specific powers under section 108 of the Environment Act 1995 to authorise suitable persons to carry out any such investigation. This can involve entering premises, taking samples or carrying out related activities for the purpose of enabling the authority to determine whether any land is CONTAMINATED LAND. In some circumstances, the authorised person can also ask other persons questions, which they are obliged to answer, and make copies of written or electronic records.

3.11 If there is to be an inspection of the land, the LOCAL AUTHORITY needs to consider whether it needs to authorise an intrusive investigation (for example, exploratory excavations) into the land. Under the statutory guidance, the authority should authorise an intrusive investigation only where it considers that it is likely (rather than only "reasonably possible") that a CONTAMINANT is actually present and that, given the current use of the land (as defined at paragraph A.26) a RECEPTOR is present or is likely to be present (*paragraph B.22(b)*).

Potential special sites

3.12 Part IIA creates a particular category of CONTAMINATED LAND called "SPECIAL SITES". For any SPECIAL SITE, the ENVIRONMENT AGENCY, rather than the LOCAL AUTHORITY, is the ENFORCING AUTHORITY for the purposes of the Part IIA regime.

3.13 The descriptions of the types of land which are required to be designated as SPECIAL SITES are set out in the Regulations (*regulations 2 & 3*; see also Annex 4 to this Circular). The procedure for the designation of a SPECIAL SITE is described at paragraphs 18.1 to 18.34 below, along with other procedural issues relating to SPECIAL SITES.

3.14 The actual designation of a SPECIAL SITE cannot take place until the land in question has been formally identified as CONTAMINATED LAND by the LOCAL AUTHORITY. However, the Government considers it appropriate for detailed investigation of any potential SPECIAL SITE to be carried out by the ENVIRONMENT AGENCY, acting on behalf of the LOCAL AUTHORITY.

3.15 To answer the second of the questions in paragraph 3.9 above, the LOCAL AUTHORITY needs to consider, for any land where the answer to the first question is "yes", whether either:

(a) the land or site is of a type such that it would inevitably be designated a SPECIAL SITE were it identified as CONTAMINATED LAND (for example, because the land has been used at some time for the manufacture or processing of explosives *(regulation 2(1)(c)(ii))*; or

(b) the particular POLLUTANT LINKAGE which is being investigated is of a kind which would require the land to be designated a SPECIAL SITE were it found to be a SIGNIFICANT POLLUTANT LINKAGE (for example, where POLLUTION OF CONTROLLED WATERS might stop water for human consumption being regarded as wholesome *(regulation 3(a)))*.

3.16 Where either of these circumstances applies, the statutory guidance states that the LOCAL AUTHORITY should always seek to arrange with the ENVIRONMENT AGENCY for that Agency to carry out the detailed investigation of the land *(paragraphs B.28 and B.29)*. Where necessary, the LOCAL AUTHORITY will authorise a person nominated by the agency to use the powers of entry conferred by section 108 of the Environment Act 1995 *(paragraph B.30)*.

Inspection using statutory powers of entry

3.17 If the premises to be inspected are used for residential purposes, or if the inspection will necessitate taking heavy equipment onto the premises, the authorised person needs to give the occupier of the premises at least seven days notice of his proposed entry onto the premises. The authorised person can then enter the premises if he obtains either the consent of the occupier or, if this is not forthcoming, a warrant issued by a magistrate *(section 108(6) and Schedule 18, Environment Act 1995)*.

3.18 In other cases, consultation with the occupier prior to entry onto the premises may still be helpful, particularly so that any necessary health and safety precautions can be identified and then incorporated into the inspection. In some instances, specific consents or regulatory permissions may be needed for access to, or work on, the site.

3.19 In an EMERGENCY, these powers of entry can be exercised forthwith if this is necessary *(section 108(6))*. For these purposes, a case is an EMERGENCY if it appears to the authorised person–

"(a) that there is an immediate risk of serious pollution of the environment or serious harm to human health, or

(b) that circumstances exist which are likely to endanger life or health "and that immediate entry to any premises is necessary to verify the existence of that risk or those circumstances or to ascertain the cause of that risk or those circumstances or to effect a remedy" *(section 108(15), Environment Act 1995)*.

3.20 Compensation may be payable by the LOCAL AUTHORITY for any disturbance caused by an INSPECTION USING STATUTORY POWERS OF ENTRY *(paragraph 6 of Schedule 18 of the Environment Act 1995)*

Objectives for the inspection of land

3.21 The primary objective in inspecting land is to enable the LOCAL AUTHORITY to obtain the information needed to decide whether or not the land appears to be CONTAMINATED LAND.

3.22 It is not always necessary for the LOCAL AUTHORITY to produce a complete characterisation of the nature and extent of CONTAMINANTS, PATHWAYS and RECEPTORS on the land, or of other matters relating to the condition of the land. The authority may be able to identify, in accordance with the statutory guidance set out at Chapters A and B, one or more SIGNIFICANT POLLUTANT LINKAGES, basing its decision on less than a complete characterisation. Once any land has been identified as CONTAMINATED LAND, fuller investigation and characterisation of identified SIGNIFICANT POLLUTANT LINKAGES can, if necessary, form part of an ASSESSMENT ACTION required under a REMEDIATION NOTICE or described in a REMEDIATION STATEMENT *(paragraphs C.65 & C.66)*. The identification of any further SIGNIFICANT POLLUTANT LINKAGES will remain the responsibility of the LOCAL AUTHORITY.

3.23 In some cases, the information obtained from an inspection may lead the LOCAL AUTHORITY to the conclusion that, whilst the land does not appear to be CONTAMINATED LAND on the basis of that information assessed on the balance of probabilities, it is still possible that the land is CONTAMINATED LAND. This might occur, for example, where the mean concentration of a CONTAMINANT in soil samples lies just below an appropriate guideline value for that CONTAMINANT. In cases of this kind, the LOCAL AUTHORITY will need to consider whether to carry out further inspections or pursue other lines of enquiry to enable it either to discount the possibility that the land is CONTAMINATED LAND, or to conclude that the land does appear to be CONTAMINATED LAND. In the absence of any such further inspection or enquiry, the local authority will need to proceed to make its determination on the basis that it cannot be satisfied, on the balance of probabilities, that the land falls within the statutory definition of CONTAMINATED LAND.

3.24 In other cases, an inspection may yield insufficient information to enable the LOCAL AUTHORITY to determine, in the manner described at paragraphs 3.26 to 3.35 below, whether or not the land appears to be CONTAMINATED LAND. In such cases, the LOCAL AUTHORITY will need to consider whether carrying out further inspections (for example, taking more samples) or pursing other lines of enquiry (for example, carrying out or commissioning more detailed scientific analysis of a substance or its properties) would be likely to provide the necessary information. If it is not possible to obtain the necessary information, the LOCAL AUTHORITY will need to proceed to make its determination on the basis that it cannot be satisfied, on the balance of probabilities, that the land falls within the statutory definition of CONTAMINATED LAND. The LOCAL AUTHORITY may, nevertheless, decide that the question should be reopened at some future date, or when further information becomes available.

3.25 A secondary objective in inspecting land is to enable the LOCAL AUTHORITY to identify any CONTAMINATED LAND which is required to be designated as a SPECIAL SITE.

Determining whether land is contaminated land

3.26 Any determination by the LOCAL AUTHORITY that particular land appears to be CONTAMINATED LAND is made on one or more of the following bases, namely that:

(a) SIGNIFICANT HARM is being caused;

(b) there is a SIGNIFICANT POSSIBILITY of such harm being caused;
(c) POLLUTION OF CONTROLLED WATERS is being caused; or
(d) POLLUTION OF CONTROLLED WATERS is likely to be caused *(paragraph B.38)*.

Consistency with other regulatory bodies

3.27 If the LOCAL AUTHORITY is considering whether the land might be CONTAMINATED LAND by virtue of an ECOLOGICAL SYSTEM EFFECT *(Chapter A, Table A)*, the authority needs to consult English Nature *(paragraph B.42)*.

3.28 Similarly, if the LOCAL AUTHORITY is considering whether land might be CONTAMINATED LAND by virtue of any POLLUTION OF CONTROLLED WATERS, the authority needs to consult the ENVIRONMENT AGENCY *(paragraph B.43)*.

3.29 In either case, this is to ensure that the LOCAL AUTHORITY adopts an approach which is consistent with that adopted by the other regulatory bodies, and benefits from the experience and expertise available within that other body.

3.30 If the land is covered by a waste management site licence, the LOCAL AUTHORITY needs to consider, taking into account any information provided by the ENVIRONMENT AGENCY in its role as the waste regulation authority, whether all of the SIGNIFICANT HARM or POLLUTION OF CONTROLLED WATERS by reason of which the land might be CONTAMINATED LAND is the result of either:

(a) a breach of the conditions of the site licence; or
(b) activities authorised by, and carried on in accordance with the conditions of, the licence.

3.31 If all of the SIGNIFICANT HARM or POLLUTION OF CONTROLLED WATERS falls into either of these categories, the land cannot be identified as CONTAMINATED LAND for the purposes of Part IIA *(section 78YB(2))*. Any regulatory action on the land is the responsibility of the ENVIRONMENT AGENCY, acting as the waste regulation authority in the context of the waste management licensing regime in Part II of the Environmental Protection Act 1990.

3.32 Under other provisions in section 78YB, the land may be identified as CONTAMINATED LAND, but REMEDIATION may be enforced under other regimes rather than under Part IIA (see paragraphs 7.2 to 7.11 below).

Making the determination

3.33 The LOCAL AUTHORITY needs to carry out an appropriate, scientific and technical assessment of the circumstances of the land, using all of the relevant and available evidence. The authority then determines whether any of the land appears to it to meet the definition of CONTAMINATED LAND set out in section 78A(2). Where the authority has received information or advice given by other regulatory bodies referred to in paragraphs 3.27 to 3.31 above, it must have regard to that information or advice *(paragraphs B.42 and B.43)*. Chapter B provides statutory guidance on the manner in which the LOCAL AUTHORITY makes this determination *(Chapter B, Part 4)*. This includes guidance on the physical extent of the land which should be covered by any single determination *(paragraphs B.32 to B.36)*.

3.34 There may be cases where the presence of one or more contaminants is discovered on land which is undergoing, or is about to undergo, development. Where this occurs, the LOCAL AUTHORITY will need to consider what action is appropriate under both Part IIA and town and country planning legislation (see Annex 1, paragraphs 45 to 50). Where the LOCAL AUTHORITY is not the local planning authority, the two authorities will need to consult.

3.35 The LOCAL AUTHORITY needs to prepare a written record of any determination that land is CONTAMINATED LAND, providing a summary of the basis on which the land has been identified as such land *(paragraph B.52)*. This will include information on the specific SIGNIFICANT POLLUTANT LINKAGE, or linkages, found.

Information arising from the inspection of land

3.36 As the LOCAL AUTHORITY inspects its area, it will generate a substantial body of information about the condition of different sites in its area.

3.37 Where land has been identified as being CONTAMINATED LAND, and consequent action taken, the LOCAL AUTHORITY has to include specified details about the condition of the land, and the REMEDIATION ACTIONS carried out on it, in its REGISTER (*section 78R;* see section 17 of this Annex and Annex 4, paragraphs 79 to 100). Having this information on the REGISTER makes it readily available to the public and to those with an interest in the land.

3.38 But the LOCAL AUTHORITY may also be asked, for example as part of a "local search" for a property purchase, to provide information about other areas of land which have not been identified as CONTAMINATED LAND. This might include, for example, information on whether the authority had inspected the land and, if so, details of any site investigation reports prepared.

3.39 The Environmental Information Regulations 1992 (SI 1992/3240 as amended) may apply to any information about land contamination. This means that, depending on the circumstances and the particular information requested, the authority may be obliged to provide the information when requested to do so. However, this is subject to the requirements in the 1992 Regulations relating to commercial confidentiality, national defence and public security.

3.40 Even where land has not been identified as CONTAMINATED LAND, information collected under Part IIA may also be useful for the wider purpose of the LOCAL AUTHORITY and other regulatory bodies, including:

(a) planning and building control functions; and
(b) other relevant statutory pollution control regimes (for example, powers to require the removal of illegally-deposited controlled wastes).

4 Identifying and notifying those who may need to take action

Notification of the identification of contaminated land

Identification of interested persons

4.1 For any piece of land identified as being CONTAMINATED LAND, the LOCAL AUTHORITY needs to establish:

(a) who is the OWNER of the land (*defined in section 78A(9)*);
(b) who appears to be in occupation of all or part of the land; and
(c) who appears to be an APPROPRIATE PERSON to bear responsibility for any REMEDIATION ACTION which might be necessary (*defined in section 78F;* see paragraphs 9.3 to 9.20 below).

4.2 At this early stage, the LOCAL AUTHORITY may not be able to establish with certainty who falls into each of these categories, particularly the last of them. As it obtains further information, the authority needs to reconsider these questions. It needs to act, however, on the basis of the best information available to it at any particular time.

The notification

4.3 The LOCAL AUTHORITY needs to notify, in writing, the persons set out in paragraph 4.1 above, as well as the ENVIRONMENT AGENCY, of the fact that the land has been identified as being CONTAMINATED LAND (*section 78B(3)*). The notice given to any of these persons will inform them of the capacity—for example, OWNER or APPROPRIATE PERSON—in which they have been sent it.

4.4 The LOCAL AUTHORITY (or, in the case of a SPECIAL SITE, the ENVIRONMENT AGENCY) may, at any subsequent time, identify some other person who appears to be an APPROPRIATE PERSON, either as well as or instead of those previously identified. Where this happens, the relevant authority needs to notify that person that he appears to be an APPROPRIATE PERSON with respect to land which has been identified as CONTAMINATED LAND (*section 78B(4)*).

4.5 The issuing of a notice under either of these headings has the effect of starting the process of consultation on what REMEDIATION might be appropriate. The LOCAL AUTHORITY (or the ENVIRONMENT AGENCY) may therefore wish to consider whether to provide any additional information to the recipients of the notification, in order to facilitate this consultation. The following categories of information may be useful for these purposes:

(a) a copy of the written record of the determination made by the authority that the land appears to be CONTAMINATED LAND (*paragraph B.52*);
(b) information on the availability of site investigation reports, with copies of the full reports being available on request;
(c) an indication of the reason why particular persons appear to the authority to be APPROPRIATE PERSONS; and
(d) the names and addresses of other persons notified at the same time or previously, indicating the capacity in which they were notified (eg as OWNER or as APPROPRIATE PERSON).

4.6 The authority will also need to inform each APPROPRIATE PERSON about the tests for EXCLUSION from, and APPORTIONMENT of, liabilities set out in the statutory guidance in Chapter D (*paragraph D.33*). This will enable those persons to know what information they might wish to provide the authority, in order to make a case for their EXCLUSION from liability, or for a particular APPORTIONMENT of liability.

4.7 The notification to the ENVIRONMENT AGENCY enables the Agency to decide whether:

(a) it considers that the land should be designated a SPECIAL SITE, on the basis that it falls within one or more of the relevant descriptions (*regulations 2 and 3;* see also paragraphs 7 to 15 of Annex 4);
(b) it wishes to provide site-specific guidance to the LOCAL AUTHORITY, for example on what REMEDIATION might be required (see paragraphs 6.8 to 6.9 below); or
(c) it requires further information from the LOCAL AUTHORITY about the land, in order for the ENVIRONMENT AGENCY to prepare its national report (*section 78U*).

4.8 If the ENVIRONMENT AGENCY requires any further information from the LOCAL AUTHORITY, it should request this in writing. The LOCAL AUTHORITY should provide such information as it has, or can "reasonably be expected to obtain" (*sections 78U(3) & 78V(3)*).

Identifying possible special sites

4.9 Having identified any CONTAMINATED LAND, the LOCAL AUTHORITY needs to consider whether the land also meets any of the descriptions which would require it to be designated as a SPECIAL SITE. These descriptions are prescribed in the Contaminated Land (England) Regulations 2000 (*regulations 2 & 3;* see also paragraphs 7 to 15 of Annex 4). If the LOCAL AUTHORITY concludes that it should designate any land, it will need to notify the ENVIRONMENT AGENCY.

4.10 The authority needs to reconsider this question whenever it obtains further relevant information about the land, for example after the carrying out of any ASSESSMENT ACTION under the terms of a REMEDIATION NOTICE.

4.11 A description of the procedures for the designation of a SPECIAL SITE, and the implications of any such designation, are set out in paragraphs 18.1 to 18.34 below.

Role of the enforcing authority

4.12 After the LOCAL AUTHORITY has identified any SIGNIFICANT POLLUTANT LINKAGE, thus determined that the land is CONTAMINATED LAND and then carried out the necessary notifications, it is for the ENFORCING AUTHORITY (that is, the ENVIRONMENT AGENCY for any SPECIAL SITE and the LOCAL AUTHORITY for any other site) to take further action.

5 Urgent remediation action

5.1 Where it appears to the ENFORCING AUTHORITY that there is an imminent danger of serious HARM or serious POLLUTION OF CONTROLLED WATERS being caused as a result of a SIGNIFICANT POLLUTANT LINKAGE that has been identified, that authority may need to ensure that urgent REMEDIATION is carried out.

5.2 The ENFORCING AUTHORITY needs to keep this question under review as it receives further information about the condition of the CONTAMINATED LAND. It may decide that urgent REMEDIATION is needed at any stage in the procedures set out below. It is likely that any REMEDIATION ACTION carried out on an urgent basis will be only a part of the total REMEDIATION SCHEME for the RELEVANT LAND OR WATERS, as not all of the REMEDIATION ACTIONS will need to be carried out urgently.

5.3 The terms "imminent" and "serious" are not defined in Part IIA. The ENFORCING AUTHORITY needs to judge each case on the normal meaning of the words and the facts of that case. However, the statutory guidance in Part 5 of Chapter C sets out a number of considerations relating to the assessment of the seriousness of any HARM or POLLUTION OF CONTROLLED WATERS which may be relevant.

5.4 Where the ENFORCING AUTHORITY is satisfied that there is a need for urgent REMEDIATION, two requirements which normally apply to the service of REMEDIATION NOTICES are disapplied (*sections 78G(4) & 78H(4)*). These are the requirements for:

- (a) prior consultation (*section 78H(1)*; see paragraphs 6.10 to 6.17 below); and
- (b) a three month interval between:
  - (i) the notification to the APPROPRIATE PERSON that the land has been identified as CONTAMINATED LAND or the land's designation as a SPECIAL SITE, and

(ii) the service of the remediation notice (*section 78H(3)*; see paragraphs 12.4 and 12.5 below).

5.5 However, other requirements in the primary legislation and in the statutory guidance continue to apply, in particular with respect to:

(a) the standard of REMEDIATION and what REMEDIATION ACTIONS may be required *(section 78E(4) and Chapter C;* see paragraphs 6.18 to 6.28 below); and

(b) the identification of the APPROPRIATE PERSON and any EXCLUSIONS from, or APPORTIONMENTS of, responsibility to bear the cost of REMEDIATION (*section 78F and Chapter D*; see paragraphs 9.3 to 9.49 below).

5.6 In general where there is a need for urgent REMEDIATION ACTION, the ENFORCING AUTHORITY will act by serving a REMEDIATION NOTICE on an urgent basis (that is, without necessarily consulting or waiting for the end of the three month period referred to in paragraph 5.4(b) above). However, if the ENFORCING AUTHORITY considers that serving a REMEDIATION NOTICE in this way would not result in the REMEDIATION happening soon enough, it may decide to carry out the REMEDIATION itself. The authority has the power to do this only where it considers that:

(a) there is an imminent danger of serious HARM or serious POLLUTION OF CONTROLLED WATERS, being caused; and

(b) it is necessary for the authority to carry out REMEDIATION itself to prevent that harm or pollution (*section 78N(3)(a)*).

5.7 These circumstances may apply, in particular, if the ENFORCING AUTHORITY cannot readily identify any APPROPRIATE PERSON on whom it could serve a REMEDIATION NOTICE. There may also be cases where the ENFORCING AUTHORITY considers that urgent REMEDIATION is needed and has already specified the necessary REMEDIATION ACTIONS in a REMEDIATION NOTICE, but the requirements of that notice have been suspended pending the decision in an appeal against the notice (see paragraphs 13.5 to 13.7 below).

5.8 If the ENFORCING AUTHORITY carries out any urgent REMEDIATION itself, it needs to prepare and publish a REMEDIATION STATEMENT describing the REMEDIATION ACTIONS it has carried out (*section 78H(7)*). It needs also to consider whether to seek to recover, from the appropriate person, the reasonable costs the authority has incurred in carrying out the REMEDIATION (*section 78P(1) and Chapter E*; see paragraphs 16.1 to 16.11 below).

6 Identifying appropriate remediation requirements

Introduction

6.1 Where any land has been identified as being CONTAMINATED LAND, the ENFORCING AUTHORITY has a duty to require appropriate REMEDIATION. The statutory guidance in Chapter C of Annex 3 to this circular sets out the standard to which any land or waters should be remediated.

6.2 For the purposes of Part IIA, the term REMEDIATION has a wider meaning than it has under its common usage (*section 78A(7)*). It includes ASSESSMENT ACTION, REMEDIAL TREATMENT ACTION and MONITORING ACTION *(paragraphs C.7 and C.8)*. Part 7 of the statutory guidance at Chapter C of Annex 3 identifies circumstances in which action falling within each of these three categories may be appropriate.

6.3 In relation to any particular piece of CONTAMINATED LAND, it may be necessary to carry out more than one thing by way of REMEDIATION. To describe the various things which may need to be done, the statutory guidance uses the following terms:

(a) a "REMEDIATION ACTION" is any individual thing which is being, or is to be done, by way of REMEDIATION;
(b) a "REMEDIATION PACKAGE" is all the REMEDIATION ACTIONS, within a REMEDIATION SCHEME, which are referable to a particular SIGNIFICANT POLLUTANT LINKAGE; and
(c) a "REMEDIATION SCHEME" is the complete set or sequence of REMEDIATION ACTIONS (referable to one or more SIGNIFICANT POLLUTANT LINKAGES) to be carried out with respect to the RELEVANT LAND OR WATERS.

Phased remediation

6.4 The overall process of REMEDIATION may well be phased, with different REMEDIATION ACTIONS being required at different times. For example, ASSESSMENT ACTION may be needed in order to establish what REMEDIAL TREATMENT ACTION would be effective. Once the results of that ASSESSMENT ACTION are known, the REMEDIAL TREATMENT ACTION itself might then be carried out, with MONITORING ACTIONS being needed to ensure that it has been effective. In another case, there might be a need for different REMEDIAL TREATMENT ACTIONS to be carried out in sequence.

6.5 Wherever the complete REMEDIATION SCHEME cannot be specified in a single REMEDIATION NOTICE or REMEDIATION STATEMENT, and needs to be phased, the process of consulting and determining what particular REMEDIATION ACTIONS are required need to be repeated for each such notice or statement.

Agreed Remediation

6.6 It is the Government's intention that, wherever practicable, REMEDIATION should proceed by agreement rather than by formal action by the ENFORCING AUTHORITY. In this context, the authority and the person who will carry out the REMEDIATION may identify by mutual agreement the particular REMEDIATION ACTIONS which would achieve REMEDIATION to the necessary standard (see paragraphs 6.33 and 6.34 below). The REMEDIATION may be carried out without a REMEDIATION NOTICE being served, but with the agreed REMEDIATION ACTIONS being described in a published REMEDIATION STATEMENT (see paragraphs 8.1 to 8.28 below).

6.7 However, where appropriate REMEDIATION is not being carried out, or where agreement cannot be reached on the REMEDIATION ACTIONS required, the authority has a duty to serve a REMEDIATION NOTICE. Any such notice must specify particular REMEDIATION ACTIONS to be carried out and the times within which they must be carried out (*section 78E(1)*).

Site-specific guidance from the Environment Agency

6.8 The ENVIRONMENT AGENCY has the power to provide site-specific guidance to the LOCAL AUTHORITY, where that LOCAL AUTHORITY is the ENFORCING AUTHORITY for any CONTAMINATED LAND (*section 78V(1)*). It may choose to do so, in particular, where either:

(a) it has particular technical expertise available, for example derived from its other pollution control functions; or

(b) the manner in which the REMEDIATION might be carried out could affect its responsibilities for protecting the water environment.

6.9 In any case where such guidance is given, the LOCAL AUTHORITY has to have regard to it when deciding what REMEDIATION is required (*section 78V(1)*).

Consultation

Remediation requirements

6.10 Before the ENFORCING AUTHORITY serves any REMEDIATION NOTICE it will, in general, need to make reasonable endeavours to consult the following persons with an interest in the CONTAMINATED LAND, or in the REMEDIATION (*section 78H(1)*):

(a) the person on whom the notice is to be served (ie the APPROPRIATE PERSON);
(b) the OWNER of the land to which the notice would relate; and
(c) any other person who appears to the authority to be in occupation of the whole, or any part of, the land.

6.11 This means that any recipient of a REMEDIATION NOTICE is consulted before the notice is served, at a minimum about the details of what he is being required to do, and the time within which he must do it. However, consultation is not a requirement in cases of urgency (see paragraph 5.4 above).

6.12 In addition to the consultation directly required by section 78H(1), the ENFORCING AUTHORITY is likely to find a wider process of discussion and consultation useful. This could cover, for example:

(a) whether the land should, in fact, be identified as CONTAMINATED LAND; this question might be re-visited, for example, in cases where the land OWNER, or the APPROPRIATE PERSON, had additional sampling information;
(b) what would need to be achieved by the REMEDIATION, in terms of the reduction of the possibility of SIGNIFICANT HARM being caused, or of the likelihood of the POLLUTION OF CONTROLLED WATERS, and in terms of the remedying of any effects of that harm or pollution; and
(c) what particular REMEDIATION ACTIONS would achieve that REMEDIATION.

6.13 This wider process of discussion may also help:

(a) to identify opportunities for agreed REMEDIATION which can be carried out without the service of a REMEDIATION NOTICE; and
(b) where a REMEDIATION NOTICE is served, to resolve as many disagreements as possible before the service of the notice, thus limiting the scope of any appeal against the notice under section 78L.

Granting of rights

6.14 The ENFORCING AUTHORITY also needs to consult on the rights which may need to be granted to the recipient of any REMEDIATION NOTICE to entitle him to carry out the REMEDIATION. For example, where the APPROPRIATE PERSON does not own the CONTAMINATED LAND, he may need the consent of the OWNER of the land to enter it. Under section 78G(2), any person whose consent is required has to grant, or join in granting, the necessary rights. He is then entitled to compensation *(section 78G & regulation 6; see paragraphs 21 to 38 of Annex 4)*.

6.15 Except in cases of urgency (see paragraph 5.4 above), the ENFORCING AUTHORITY needs to consult:

(a) the owner or occupier of any of the RELEVANT LAND OR WATERS; and
(b) any other person who might have to grant, or join in granting, any rights to the recipient of a REMEDIATION NOTICE (section 78G(3)).

Liabilities

6.16 If there are two or more APPROPRIATE PERSONS, the ENFORCING AUTHORITY should make reasonable endeavours to consult each of those persons on any EXCLUSION from, or APPORTIONMENT of, liability (*paragraph D.36*). This allows anyone who might be affected to provide the information on which an EXCLUSION or APPORTIONMENT can be based. In addition to information provided by the APPROPRIATE PERSONS, the authority needs to seek its own information, where this is reasonable (*paragraph D.36*).

6.17 The ENFORCING AUTHORITY may also find it useful to discuss wider questions relating to liabilities with those whom it has identified as being APPROPRIATE PERSONS. For example, they may be able to identify other persons who ought to be identified as APPROPRIATE PERSONS, either in addition or instead.

Identifying an appropriate remediation scheme

6.18 The ENFORCING AUTHORITY'S objective is to identify the appropriate REMEDIATION SCHEME, which will include the REMEDIAL TREATMENT ACTION or actions which, taken together, will ensure that the RELEVANT LAND OR WATERS are remediated to the necessary standard (*Chapter C, Part 3*). In some cases, the particular REMEDIATION ACTIONS to be carried out may be identified by mutual agreement between the authority and the persons who will carry them out. In other cases, that authority has to identify the particular actions itself.

6.19 Where the authority is identifying the actions itself, it is specifically required to ensure that they are "reasonable", having regard to the cost which is likely to be involved and the seriousness of the HARM or of the POLLUTION OF CONTROLLED WATERS in question (*section 78E(4)*). The authority needs to assess, in particular, the costs involved as against the benefits arising from the REMEDIATION (*paragraph C.30;* but see also paragraph 6.31 below).

6.20 It may be necessary for ASSESSMENT ACTIONS to be carried out before the appropriate REMEDIAL TREATMENT ACTION or actions can be identified (*paragraph C.65*). Where this is the case, the first step will be to identify the appropriate ASSESSMENT ACTION or actions. Once that ASSESSMENT ACTION has been carried out, it will be necessary to complete the identification of the remaining stages of the REMEDIATION SCHEME, identifying appropriate REMEDIAL TREATMENT ACTIONS in the light of the information obtained. This may require a sequence of REMEDIATION STATEMENTS or REMEDIATION NOTICES.

6.21 Throughout the process of identifying the appropriate REMEDIATION SCHEME, the ENFORCING AUTHORITY needs to keep under review whether there is a need for urgent REMEDIATION to be carried out (see section 5 of this Annex).

A single significant pollutant linkage

6.22 Where only a single SIGNIFICANT POLLUTANT LINKAGE has been identified on the CONTAMINATED LAND, the ENFORCING AUTHORITY, in conjunction with those it is consulting, needs to consider what is needed, with respect to that linkage, to:

(a) prevent, or reduce the likelihood of, the occurrence of any SIGNIFICANT HARM or POLLUTION OF CONTROLLED WATERS; and
(b) remedy, or mitigate, the effect of any such harm or water pollution which has been, or might be, caused.

6.23 The ENFORCING AUTHORITY then needs to identify the REMEDIATION PACKAGE which would represent the BEST PRACTICABLE TECHNIQUES of REMEDIATION for that SIGNIFICANT POLLUTANT LINKAGE. Such techniques will include appropriate measures to provide quality assurance and to verify what has been done.

6.24 The assessment of what represents such BEST PRACTICABLE TECHNIQUES is made in terms of:

(a) the extent to which the REMEDIATION PACKAGE would achieve the objectives identified in paragraph 6.22 above *(Part 4 of Chapter C)*;
(b) whether the package, and the individual REMEDIATION ACTIONS concerned, would be reasonable, having regard to their cost and to the seriousness of the HARM or of the POLLUTION OF CONTROLLED WATERS to which they relate *(Part 5 of Chapter C)*; and
(c) whether the package represents the best combination of practicability, effectiveness and durability *(Part 6 of Chapter C)*.

6.25 Any such REMEDIATION PACKAGE needs to include measures to achieve quality assurance and verification. Where appropriate, such measures may take the form of MONITORING ACTIONS *(paragraphs C.68 and C.69)*.

More than one significant pollutant linkage

6.26 If more than one SIGNIFICANT POLLUTANT LINKAGE has been identified, the REMEDIATION will have to deal with the SIGNIFICANT HARM or the POLLUTION OF CONTROLLED WATERS resulting from, or threatened by, each of those linkages. However, it may be neither practicable nor efficient simply to consider the REMEDIATION needed with respect to each linkage separately. There may, for example, be cost savings which can be achieved by carrying out particular REMEDIATION ACTIONS which deal with more than one SIGNIFICANT POLLUTANT LINKAGE. In other cases, if the separate REMEDIATION PACKAGES for each of the SIGNIFICANT POLLUTANT LINKAGES were carried out independently, the individual REMEDIATION ACTIONS might conflict or overlap.

6.27 The ENFORCING AUTHORITY therefore needs to try to identify a REMEDIATION SCHEME which deals with the RELEVANT LAND OR WATERS as a whole, avoids conflict or overlap between the REMEDIATION needed for the various SIGNIFICANT POLLUTANT LINKAGES, and does not involve unnecessary expense *(paragraph C.27)*. This may result in a REMEDIATION ACTION which replaces, or subsumes, what would otherwise be several separate REMEDIATION ACTIONS in different REMEDIATION PACKAGES.

6.28 The first step in this process is for the ENFORCING AUTHORITY to assess the standard of REMEDIATION to be achieved by the REMEDIATION SCHEME with respect to each SIGNIFICANT POLLUTANT LINKAGE.

6.29 In doing this, the ENFORCING AUTHORITY needs to identify, for each SIGNIFICANT POLLUTANT LINKAGE, the extent to which the relevant SIGNIFICANT HARM or POLLUTION OF CONTROLLED WATERS should be reduced, and its effects mitigated. The standard for this reduction or mitigation is set by reference to what would be achieved by the BEST PRACTICABLE TECHNIQUES of REMEDIATION for that linkage, if it were the only linkage required to be remediated *(paragraphs C.18 and C.26)*. In making this assessment, however, the authority works on the basis of REMEDIATION which could actually be carried out, given the wider circumstances of the land or waters, including the presence of other POLLUTANTS. In other words, in considering what might be achieved in relation to any particular SIGNIFICANT POLLUTANT LINKAGE, the ENFORCING AUTHORITY cannot ignore practical limitations on what might be done that are imposed by other problems on the same site.

Very slight levels of water pollution

6.30 As stated above (see paragraph 2.9 above), the definition of "POLLUTION OF CONTROLLED WATERS" is simply the "entry into CONTROLLED WATERS of any poisonous, noxious or polluting matter or any solid waste matter". Some commentators have suggested that the entry of very small amounts of matter into CONTROLLED WATERS might satisfy this definition, and thus lead to the identification of land as CONTAMINATED LAND. As has been said above, the Government is proposing to review the wording of the legislation on this aspect and to seek amendments to the primary legislation.

6.31 However, even if land is identified as CONTAMINATED LAND in this way—on the basis of the actual or likely entry of only a very small amount of a POLLUTANT into CONTROLLED WATERS—this should not lead to the imposition of major liabilities: there are other balances elsewhere in the regime to prevent this. In particular, any REMEDIATION that can be required must be "reasonable", having regard to the cost which is likely to be involved and the seriousness of the POLLUTION OF CONTROLLED WATERS involved *(section 78E(4) and Chapter C, Part 4)*. If there is only a very low degree of contamination on any land which gives, or is likely to give, rise to POLLUTION OF CONTROLLED WATERS which is minor in terms of its seriousness, it will be reasonable to incur only a correspondingly low level of expenditure in attempting to remediate that land.

6.32 Nevertheless, the simple fact of land being identified as CONTAMINATED LAND in this way may cause its own problems—for example, for landowners. It is therefore important that the circumstances of such cases are clearly entered on the REGISTER kept by the ENFORCING AUTHORITY. If REMEDIATION is not carried out because it would not be reasonable, a REMEDIATION DECLARATION needs to be published by the ENFORCING AUTHORITY (*section 78H(6)*) and entered on its REGISTER (*section 78R(1)(c)*). In this way, a public record is created explaining that no REMEDIATION is required under Part IIA, even though the land has been formally identified as CONTAMINATED LAND.

Assessing remediation schemes proposed by others

6.33 In general, the ENFORCING AUTHORITY needs to adopt a similar approach when it is assessing a REMEDIATION SCHEME proposed by the APPROPRIATE PERSON, the land OWNER or any other person to that which it adopts when itself identifying an appropriate REMEDIATION SCHEME (*paragraph C.3(b)*). In deciding whether it is satisfied that such a scheme would be appropriate and sufficient, it needs to consider whether that scheme would achieve a standard of REMEDIATION equivalent to that which would be achieved by the use of the BEST PRACTICABLE TECHNIQUES of REMEDIATION for each SIGNIFICANT POLLUTANT LINKAGE (*paragraph C.28*)

6.34 However, the ENFORCING AUTHORITY does not always need to consider whether the proposed scheme would, of itself, be "reasonable" in the sense required by section 78E(4) (ie. having regard to the cost likely to be involved and the seriousness of the particular harm or water pollution).

This is because the person proposing the scheme may wish to carry out REMEDIATION on a wider basis than could be required under the terms of a REMEDIATION NOTICE. For example, the proposed scheme may include works to deal with matters which do not form SIGNIFICANT POLLUTANT LINKAGES, or may involve a more expensive approach to REMEDIATION.

6.35 Where an acceptable REMEDIATION SCHEME is proposed by others, and that scheme is likely to proceed without the service of a REMEDIATION NOTICE, no such notice needs to be served. In such cases, the procedure set out section 8 of this Annex will apply.

7 Limitations on remediation notices

7.1 In addition to circumstances where REMEDIATION takes place without the service of a REMEDIATION NOTICE (see section 8 of this Annex), there are a number of restrictions on the service or contents of a REMEDIATION NOTICE.

Interactions with other provisions in the 1990 act

7.2 REMEDIATION cannot be required under Part IIA where the SIGNIFICANT HARM or the POLLUTION OF CONTROLLED WATERS in question results from an offence under the integrated pollution control regime or the waste management licensing regime, and powers are available under the relevant regime to deal with that HARM or POLLUTION OF CONTROLLED WATERS.

7.3 Nevertheless, even in such cases, the ENFORCING AUTHORITY needs to consider whether additional REMEDIATION is required on the RELEVANT LAND OR WATERS under Part IIA, to deal with matters which cannot be dealt with under those other powers.

7.4 If no such additional REMEDIATION is necessary, the ENFORCING AUTHORITY takes no further action, under Part IIA, with respect to the CONTAMINATED LAND in question. However, it then needs to include information about the exercise of these powers on its REGISTER (*Schedule 3, Contaminated Land (England) Regulations 2000;* see also Annex 4, paragraph 91).

Integrated pollution control

7.5 If the SIGNIFICANT HARM or POLLUTION OF CONTROLLED WATERS in question results from the carrying out of a process covered by the Integrated Pollution Control (IPC) regime or the Local Air Pollution Control (LAPC) regime, the ENVIRONMENT AGENCY may have powers under section 27 of the Environmental Protection Act 1990 to remedy that HARM or POLLUTION OF CONTROLLED WATERS.

7.6 Section 27 gives the Agency the power to carry out remedial steps where:

(a) the process has been carried out either without the necessary authorisation, or in contravention of an enforcement or prohibition notice;
(b) harm has been caused and it is possible to remedy that harm;
(c) the Secretary of State gives his written approval to the exercise of the powers; and
(d) the occupier of any affected land, other than the land on which the process is being carried out, gives his permission.

7.7 If a LOCAL AUTHORITY is the ENFORCING AUTHORITY and it considers that this might apply, it needs to consult the ENVIRONMENT AGENCY to find out whether the powers under section 27 are available to the Agency. In any case where the powers under section 27 may be exercised by the ENVIRONMENT AGENCY, a REMEDIATION NOTICE cannot include a REMEDIATION ACTION which would be carried out in order to achieve a purpose which could be achieved by the exercise of those powers (*section 78YB(1)*).

7.8 The SECRETARY OF STATE will be making regulations under the Pollution Prevention and Control Act 1999 to transpose the requirements of the Integrated Pollution Prevention and Control Directive (96/61/EC) into UK law. The new Pollution Prevention and Control (PPC) regime will replace the current regimes under Part I of the 1990 Act (ie IPC and LAPC). The regulations may therefore include provisions amending section 78YB(1) to refer both to section 27 of the 1990 Act and to any equivalent clean-up provision in the new PPC regime.

Waste management licensing

7.9 The ENVIRONMENT AGENCY (in its capacity as the "waste regulation authority"), and the waste collection authority for the area, have powers under section 59 of the Environmental Protection Act 1990 to deal with illegally-deposited controlled waste. These powers may permit the Agency or authority to remove, or require the removal of the waste, and to take other steps to eliminate or reduce the consequences of the deposit of the waste.

7.10 Section 59 applies where controlled waste has been deposited:

(a) without a waste management licence being in force authorising the deposit (except where regulations provide an exemption from licensing); or

(b) in a manner which is not in accordance with a waste management licence.

7.11 If a LOCAL AUTHORITY is the ENFORCING AUTHORITY and it considers that these circumstances might apply, it needs to consult the ENVIRONMENT AGENCY and to consider its position where it is the waste collection authority. If the powers under section 59 may be exercised, any REMEDIATION NOTICE cannot include a REMEDIATION ACTION which would be carried out in order to achieve a purpose which could be achieved by the exercise of those powers (*section 78YB(3)*).

Other precluded remediation actions

Actions which would be unreasonable

7.12 In identifying an appropriate REMEDIATION SCHEME, the ENFORCING AUTHORITY may have been precluded from specifying particular REMEDIATION ACTIONS on the grounds that they would not be reasonable, having regard to their likely cost and the seriousness of the HARM or the POLLUTION OF CONTROLLED WATERS to which they relate. In some cases, such restrictions may lead to a situation in which no REMEDIATION ACTION may be required (see, for one example, paragraph 6.31 above). Alternatively, the preclusion of a particular REMEDIATION ACTION or actions may lead to the adoption of an alternative REMEDIATION SCHEME.

7.13 Where particular REMEDIATION ACTIONS have been precluded because they would not be reasonable, the ENFORCING AUTHORITY needs to prepare and publish a REMEDIATION DECLARATION which records:

(a) the reasons why the authority would have specified the REMEDIATION ACTIONS in a REMEDIATION NOTICE; and

(b) the grounds on which it is satisfied that it is precluded from including them in any such notice—that is, why it considers that they are unreasonable (*section 78H(6)*).

7.14 The ENFORCING AUTHORITY also needs to enter details of the REMEDIATION DECLARATION on its REGISTER (*section 78R(1)(c)*; see paragraphs 17.1 to 17.19 below and Annex 4, paragraph 88).

Actions which would be contrary to the statutory guidance

7.15 In rare circumstances, there may also be a particular REMEDIATION ACTION which the ENFORCING AUTHORITY would include in a REMEDIATION NOTICE, but it cannot do so because that action is not consistent with the statutory guidance in Chapter C. In any such case, the authority needs to proceed in the same way as if that REMEDIATION ACTION had been precluded on the ground that it was unreasonable (*sections 78E(5) and 78H(6)*).

Discharges into controlled waters

7.16 The ENFORCING AUTHORITY also needs to consider whether any REMEDIATION ACTION in the REMEDIATION SCHEME would have the effect of impeding or preventing any discharge into CONTROLLED WATERS for which a consent has been given under Part III of the Water Resources Act 1991.

7.17 If this is the case, the ENFORCING AUTHORITY is precluded from specifying the REMEDIATION ACTION in question in any REMEDIATION NOTICE (*section 78YB(4)*). However, it will be good practice for the ENFORCING AUTHORITY to consider in such circumstances whether there is a REMEDIATION ACTION which could address the problems posed by the SIGNIFICANT POLLUTANT LINKAGE without impeding or preventing the discharge.

7.18 However, if a REMEDIATION ACTION cannot be specified because of the restriction in section 78YB(4), the ENFORCING AUTHORITY needs to include information about the circumstances on its REGISTER (*Schedule 3, Contaminated Land (England) Regulations 2000*; see also Annex 4, paragraph 92).

8 Remediation taking place without the service of a remediation notice

8.1 Having identified the appropriate REMEDIATION SCHEME for the RELEVANT LAND OR WATERS, the ENFORCING AUTHORITY needs to consider whether that REMEDIATION is being, or will be, carried out without any REMEDIATION NOTICE being served.

8.2 This might be the case, in particular, where:

(a) the APPROPRIATE PERSON, or some other person, already plans, or undertakes during the consultation process, to carry out particular REMEDIATION ACTIONS (see paragraphs 8.3 to 8.8 below); or

(b) REMEDIATION with an equivalent effect is taking, or will take, place as a result of enforcement action under other powers (see paragraphs 8.9 to 8.17 below).

Volunteered remediation

8.3 The ENFORCING AUTHORITY may be informed, before or during the course of consultation on REMEDIATION requirements, that the APPROPRIATE PERSON or some other person already intends, or now intends, to carry out particular REMEDIATION ACTIONS on a voluntary basis.

8.4 This may apply, in particular, where:

(a) the OWNER of the land has a programme for carrying out REMEDIATION on a number of different areas of land for which he is responsible which aims to tackle those cases in order of environmental priority;
(b) the land is already subject to development proposals;
(c) the APPROPRIATE PERSON brings forward proposals to develop the land in order to fund necessary REMEDIATION; or
(d) the APPROPRIATE PERSON wishes to avoid being served with a REMEDIATION NOTICE.

8.5 Where a development of CONTAMINATED LAND is proposed, an ENFORCING AUTHORITY which is the local planning authority will need to consider what steps it needs to take under town and country planning legislation to ensure that appropriate REMEDIATION ACTIONS are included in the development proposals and that these will ensure that contamination is properly dealt with. (Where the enforcing authority is not the local planning authority, the two authorities will need to consult.)

8.6 In all cases, the ENFORCING AUTHORITY needs to consider the standard of REMEDIATION which would be achieved by the proposed REMEDIATION ACTIONS. If it is satisfied that they would achieve an appropriate standard of REMEDIATION:

(a) it is precluded from serving any REMEDIATION NOTICE (*section 78H(5)(b)*); and
(b) the person who is carrying out, or will carry out, the REMEDIATION is required to prepare and publish a REMEDIATION STATEMENT (*sections 78H(7) & 78H(8)(a)*; see paragraphs 8.18 to 8.22 below).

8.7 Even if the ENFORCING AUTHORITY is not satisfied that an appropriate standard of REMEDIATION would be achieved by the REMEDIATION ACTIONS originally proposed, it may be able to persuade the person who made the proposals to bring forward a revised and satisfactory REMEDIATION SCHEME.

8.8 If this is not possible, the ENFORCING AUTHORITY's duty to serve a REMEDIATION NOTICE may apply (*section 78E(1)*; see paragraphs 12.1 to 12.9 below).

Enforcement action under other powers

8.9 Enforcement action under other regulatory powers may already be underway, or could be taken, which would bring about the REMEDIATION of the RELEVANT LAND OR WATERS.

8.10 REMEDIATION under Part IIA cannot overlap with enforcement action under section 27 (Integrated Pollution Control and Local Air Pollution Control) or section 59 (waste management licensing); see paragraphs 7.2 to 7.11 above. However, there may be potential overlaps with the applicability of other regimes.

8.11 The ENFORCING AUTHORITY needs to consider whether enforcement could be taken under any other powers, and liaise with the relevant regulatory bodies to find out if it is already in progress or is planned.

8.12 If such enforcement action is in progress, or is planned, the ENFORCING AUTHORITY needs to consider the standard of REMEDIATION which would be achieved as a result of that enforcement action.

8.13 If the ENFORCING AUTHORITY is satisfied that the enforcement action would result in the achievement of an appropriate standard of REMEDIATION:

(a) it is precluded from serving any REMEDIATION NOTICE (*section 78H(5)(b)*); and
(b) the person who is carrying out, or will carry out, the action is required to prepare and publish a REMEDIATION STATEMENT (*sections 78H(7) & 78H(8)(a)*; see paragraphs 8.18 to 8.22 below).

8.14 If the authority considers that enforcement action could be taken under other powers, but it is not in progress, the authority should liaise with the relevant regulatory body, seeking to ensure that the most appropriate regulatory powers are used.

8.15 The ENFORCING AUTHORITY is required to enter details of the use of these other regulatory powers onto its REGISTER (*Schedule 3, paragraphs 14 & 15, Contaminated Land (England) Regulations 2000*; see Annex 4, paragraphs 91 and 92).

8.16 The authority's duty to serve a REMEDIATION NOTICE (*section 78E(1)*; see paragraphs 12.1 to 12.9 below) may apply where either:

(a) enforcement action is not being taken under other powers, and none is intended; or
(b) the enforcement action under those other powers would not achieve an appropriate standard of REMEDIATION for all of the SIGNIFICANT POLLUTANT LINKAGES identified.

8.17 There is a potential for overlap between Part IIA and the works notice powers of the ENVIRONMENT AGENCY (*section 161A of the Water Resources Act 1991 and the Anti-Pollution Works Regulations 1999*). Where an incidence of actual, or potential, water pollution does fall within the remit of both regimes, ENFORCING AUTHORITIES acting under Part IIA will be under a duty to serve a REMEDIATION NOTICE, whereas the ENVIRONMENT AGENCY is merely granted a power to act under section 161A of the 1991 Act. As set out in the Agency's policy statement, *Environment Agency Policy and Guidance on the Use of Anti-Pollution Works Notices*, which was agreed with DETR, this means that enforcement action will generally take place under Part IIA (see Annex 1, paragraphs 64 to 67).

Remediation statements

8.18 In any case where no REMEDIATION NOTICE may be served because appropriate REMEDIATION is taking place, or will take place without any such notice being served, the person responsible for the remediation is required to prepare and publish a REMEDIATION STATEMENT (*sections 78H(7) & 78H(8)(a)*). This does not apply in the cases described at paragraphs 8.9 to 8.17 above.

8.19 Section 78H(7) requires the following information to be recorded in a REMEDIATION STATEMENT:

"(a) the things which are being, have been, or are expected to be, done by way of REMEDIATION in the particular case;
"(b) the name and address of the person who is doing, has done, or is expected to do, each of those things; and
"(c) the periods within which each of those things is being, or is expected to be done".

8.20 The ENFORCING AUTHORITY is required to enter details of the REMEDIATION STATEMENT onto its REGISTER (*section 78R(1)(c)*; see paragraphs 17.1 to 17.19 below and Annex 4, paragraph 88).

8.21 If the person who is required to prepare and publish the REMEDIATION STATEMENT fails to do so, the ENFORCING AUTHORITY has powers to do so itself. This applies after a reasonable time has elapsed since the date on which the authority could have served a REMEDIATION NOTICE, but for the fact that appropriate REMEDIATION was taking place, or was like to place, without the service of a notice (*section 78H(9)*).

8.22 In any case of this kind, the ENFORCING AUTHORITY needs to consider whether it should prepare and publish a REMEDIATION STATEMENT itself for inclusion on its REGISTER. If it does so, it is entitled to recover any reasonable costs it incurs from the person who should have prepared and published the statement (*section 78H(9)*).

Reviewing circumstances

8.23 The ENFORCING AUTHORITY needs to keep under review the REMEDIATION which is actually carried out on the RELEVANT LAND OR WATERS, as well as the question of whether any additional REMEDIATION is necessary. If, at any time, it ceases to be satisfied that appropriate REMEDIATION has been, is being, or will be, carried out it may need to serve a REMEDIATION NOTICE.

8.24 The authority may cease to be satisfied if, in particular:

(a) there has been, or is likely to be, a failure to carry out the REMEDIATION ACTIONS described in the REMEDIATION STATEMENT, or a failure to do so within the times specified; or
(b) further REMEDIATION ACTIONS now appear necessary in order to achieve the appropriate standard of REMEDIATION for the RELEVANT LAND OR WATERS.

8.25 If any of the REMEDIATION ACTIONS described in the REMEDIATION STATEMENT are not being carried out, the ENFORCING AUTHORITY needs to consider whether:

(a) the REMEDIATION ACTIONS in question still appear to be necessary in order to achieve an appropriate standard of REMEDIATION; and
(b) they are still "reasonable" for the purposes of section 78E(4).

8.26 If both of these apply, and the ENFORCING AUTHORITY is not precluded from serving a REMEDIATION NOTICE for any other reason, the authority will be under a duty to serve a REMEDIATION NOTICE, specifying the REMEDIATION ACTIONS in question. It may do this without any additional consultation, if the person on whom the notice would be served has already been consulted about those actions (*section 78H(10)*).

8.27 Even if the REMEDIATION ACTIONS described in the REMEDIATION STATEMENT are being carried out as planned, the ENFORCING AUTHORITY may consider that additional REMEDIATION is necessary. This may apply, in particular, where:

(a) the REMEDIATION was intended to be phased, and further REMEDIATION ACTIONS can now be identified as being necessary; or
(b) further SIGNIFICANT POLLUTANT LINKAGES are identified, or linkages which have already been identified are discovered to be more serious than previously thought.

8.28 Where it identifies further REMEDIATION as necessary, the ENFORCING AUTHORITY needs to consider how to ensure that the necessary REMEDIATION ACTIONS are carried out. This involves repeating the procedures set out above relating to consultation, and considering whether the additional REMEDIATION will be carried out without a REMEDIATION NOTICE being served. The authority cannot, for example, serve a REMEDIATION NOTICE specifying any additional REMEDIATION ACTIONS unless the person receiving the notice has been consulted on its contents (except in cases of urgency; see paragraphs 5.1 to 5.8 above).

9 Determining liability

9.1 If the ENFORCING AUTHORITY is not satisfied, at this stage, that appropriate REMEDIATION is being, or will be, carried out without a REMEDIATION

NOTICE being served, it needs to consider who might be served with such a notice. This section of this Annex deals with the questions of who appears to be an APPROPRIATE PERSON and, if there is more than one such person, whether any of these should be EXCLUDED from liability and, where necessary, of how the liability for carrying out any REMEDIATION ACTION should be APPORTIONED between the APPROPRIATE PERSONS who remain. Further questions, covered in section 10 of this Annex, need to be considered before the ENFORCING AUTHORITY can decide whether a REMEDIATION NOTICE should be served on anyone.

9.2 Where the ENFORCING AUTHORITY is precluded from serving a REMEDIATION NOTICE by virtue of section 78H(5)(d), because it has the power to carry out the REMEDIATION itself, the authority needs to follow the same processes for determining liabilities, including any EXCLUSIONS and APPORTIONMENTS, in order to determine from whom it can recover its reasonable costs incurred in doing the work (see also paragraphs 16.1 to 16.11 below).

The definition of the "appropriate person"

9.3 Part IIA defines two different categories of APPROPRIATE PERSON, and sets out the circumstances in which persons in these categories might be liable for REMEDIATION.

9.4 The first category is created by section 78F(2), which states that:

"... any person, or any of the persons, who caused or knowingly permitted the substances, or any of the substances, by reason of which the CONTAMINATED LAND in question is such land to be in, on or under that land is an APPROPRIATE PERSON."

9.5 Such a person (referred to in the statutory guidance as a CLASS A PERSON) will be the APPROPRIATE PERSON only in respect of any REMEDIATION which is referable to the particular substances which he caused or knowingly permitted to be in, on or under the land (*section 78F(3)*). This means that the question of liability has to be considered separately for each SIGNIFICANT POLLUTANT LINKAGE identified on the land.

9.6 The second category arises in cases where it is not be possible to find a CLASS A PERSON, either for all of the SIGNIFICANT POLLUTANT LINKAGES identified on the land, or for a particular SIGNIFICANT POLLUTANT LINKAGE. These circumstances are addressed in section 78F(4) and (5), which provide that:

"(4) If no person has, after reasonable inquiry, been found who is by virtue of subsection (2) above an appropriate person to bear responsibility for the things which are to be done by way of REMEDIATION, the OWNER or occupier for the time being of the land in question is an APPROPRIATE PERSON.

(5) If, in consequence of subsection (3) above, there are things which are to be done by way of REMEDIATION in relation to which no person has, after reasonable inquiry, been found who is an APPROPRIATE PERSON by virtue of subsection (2) above, the OWNER or occupier for the time being of the CONTAMINATED LAND in question is an APPROPRIATE PERSON in relation to those things."

9.7 A person who is an APPROPRIATE PERSON under sections 78F(4) or (5) is referred to in the statutory guidance as a CLASS B PERSON.

The meaning of "caused or knowingly permitted"

9.8 The test of "causing or knowingly permitting" has been used as a basis for establishing liability in environmental legislation for more than 100 years. In the context of Part IIA, what is "caused or knowingly permitted" is the presence of a POLLUTANT in, on or under the land.

9.9 In the Government's view, the test of "causing" will require that the person concerned was involved in some active operation, or series of operations, to which the presence of the pollutant is attributable. Such involvement may also take the form of a failure to act in certain circumstances.

9.10 The meaning of the term "knowingly permit" was considered during the debate on Lords' Consideration of Commons' Amendments to the then Environment Bill on 11 July 1995. The then Minister for the Environment, the Earl Ferrers, stated on behalf of the Government that:
"The test of "knowingly permitting" would require both knowledge that the substances in question were in, on or under the land and the possession of the power to prevent such a substance being there." *(House of Lords Hansard [11 July 1995], col 1497)*

9.11 Some commentators have questioned the extent to which this test might apply with respect to banks or other lenders, where their clients have themselves caused or knowingly permitted the presence of pollutants. With respect to that question, Earl Ferrers said:
"I am advised that there is no judicial decision which supports the contention that a lender, by virtue of the act of lending the money only, could be said to have "knowingly permitted" the substances to be in, on or under the land such that it is contaminated land. This would be the case if for no other reason than the lender, irrespective of any covenants it may have required from the polluter as to its environmental behaviour, would have no permissive rights over the land in question to prevent contamination occurring or continuing." *(House of Lords Hansard [11 July 1995], col 1497)*

9.12 It is also relevant to consider the stage at which a person who is informed of the presence of a pollutant might be considered to have knowingly permitted that presence, where he had not done so previously. In the Government's view, the test would be met only where the person had the ability to take steps to prevent or remove that presence and had a reasonable opportunity to do so.

9.13 Some commentators have, in particular, questioned the position of a person who, in his capacity as OWNER or occupier of land, is notified by the LOCAL AUTHORITY about the identification of that land as being CONTAMINATED LAND under section 78B(3). They have asked whether the resulting "knowledge" would trigger the "knowingly permit" test. In the Government's view, it would not. The legislation clearly distinguishes between those who cause or knowingly permit the presence of pollutants and those who are simply owners or occupiers of the land. In particular, this is evident in sections 78F, 78J and 78K which all relate to the different potential liabilities of OWNERS or occupiers as opposed to persons who have "caused or knowingly permitted" the presence of the POLLUTANTS.

9.14 Similarly, section 78H(1) requires consultation with OWNERS and occupiers for the specific purpose of determining "what shall be done by way of REMEDIATION" and not for the purpose of determining liability. In the Government's view, this implies that a person who merely owns or occupies the land in question cannot be held to have "knowingly permitted" as a consequence of that consultation alone.

9.15 It is ultimately for the courts to decide the meaning of "caused" and "knowingly permitted" as these terms apply to the Part IIA regime, and whether these tests are met in any particular case. However, indications of how the test should be construed can be obtained from case law under other legislation where the same or similar terms are used.

The potential liabilities of owners and occupiers of land

9.16 Only where no CLASS A PERSON can be found who is responsible for any particular REMEDIATION ACTION will the OWNER or occupier be liable for REMEDIATION by virtue solely of that ownership or occupation. OWNERS and occupiers may, of course, be CLASS A PERSONS because of their own past actions or omissions.

9.17 It is ultimately for the courts to decide whether, in any case, it can be said that no CLASS A PERSON has been found. In the Government's view, the context in which the word is used in Part IIA implies that a person must be in existence in order to be found. Section 78F(4) provides that the OWNER or occupier shall bear responsibility only "if no person has, after reasonable inquiry, been found who is an APPROPRIATE PERSON to bear responsibility for the things which are to be done by way of REMEDIATION". A person who is no longer in existence cannot meet that description. Under section 78E(1), the responsibility of an APPROPRIATE PERSON for REMEDIATION is established by the service of a REMEDIATION NOTICE. Service implies the existence of the person on whom the notice is served. In general, therefore, this means that a natural person would have to be alive and a legal person such as a company must not have been dissolved. However, it may be possible in some circumstances for the authority to act against the estate of a deceased person or to apply to a court for an order to annul the dissolution of a company.

9.18 Similarly, it is ultimately for the courts to determine what would constitute "reasonable inquiry" for the purposes of trying to find a CLASS A PERSON.

9.19 Section 78A(9) defines the term OWNER as follows:
"in relation to any land in England and Wales, means a person (other than a mortgagee not in possession) who, whether in his own right or as trustee for any other person, is entitled to receive the rack rent of the land, or, where the land is not let at a rack rent, would be so entitled if it were so let".

9.20 The term "occupier" is not defined in Part IIA and it will therefore carry its ordinary meaning. In the Government's view, it would normally mean the person in occupation and in many cases that will be the tenant or licensee of the premises.

The procedure for determining liabilities

9.21 Part 3 of the statutory guidance set out at Chapter D of Annex 3 provides a procedure for the ENFORCING AUTHORITY to follow to determine which of the APPROPRIATE PERSONS in any case should bear what liability for REMEDIATION. That procedure consists of the five distinct stages set out below.

9.22 Not all of these stages will be relevant to all cases. Most sites are likely to involve only one SIGNIFICANT POLLUTANT LINKAGE and thus have only one LIABILITY GROUP. In many cases, such a LIABILITY GROUP will consist of only one APPROPRIATE PERSON. However, more complicated situations will arise, requiring the application of all five stages. These steps may appear complex, but they are needed to fulfil the aims of the legislation in implementing the "polluter pays" principle while trying to avoid making APPROPRIATE PERSONS bear more than their fair share of the cost.

First stage—identifying potential appropriate persons and liability groups

9.23 The ENFORCING AUTHORITY will have already identified, on a preliminary basis, those persons who appear to it to be APPROPRIATE PERSONS in order to notify them of the identification of the CONTAMINATED LAND (see paragraph 4.1 above).

9.24 At this stage, the authority needs to reconsider this question, and identify all of the persons who appear to be APPROPRIATE PERSONS to bear responsibility for REMEDIATION. Depending on the information it has obtained, it may consider that:

(a) some or all of those who previously appeared to be APPROPRIATE PERSONS still appear to be such persons;
(b) some or all of those persons no longer appear to be APPROPRIATE PERSONS; or
(c) some other persons appear to be APPROPRIATE PERSONS, either in addition to those previously identified, or instead of them.

9.25 An example of circumstances in which the identity of those who appear to be APPROPRIATE PERSONS might change is if the authority had not previously found a person who had caused or knowingly permitted the POLLUTANT to be present (a CLASS A PERSON), but could now do so. At the time it identified the CONTAMINATED LAND, the authority would have identified the OWNER and the occupier of the land as being APPROPRIATE PERSONS. However, these persons would no longer appear to be APPROPRIATE PERSONS, unless they were also CLASS A PERSONS.

9.26 If, as a result of this process of reconsideration, the ENFORCING AUTHORITY identifies new persons who appear to be APPROPRIATE PERSONS, it needs to notify them of the fact that they have been identified as such (*section 78B(4)*, see paragraphs 4.3 to 4.6 above).

9.27 The ENFORCING AUTHORITY will have identified one or more SIGNIFICANT POLLUTANTS on the land and the SIGNIFICANT POLLUTANT LINKAGES of which they form part.

A single significant pollutant

9.28 Where there is a single SIGNIFICANT POLLUTANT, and a single SIGNIFICANT POLLUTANT LINKAGE, the ENFORCING AUTHORITY needs to make reasonable enquiries to find all those who have caused or knowingly permitted the SIGNIFICANT POLLUTANT in question to be in, on or under the land (*section 78F(2)*). Any such persons are then "CLASS A PERSONS" and together constitute a "CLASS A LIABILITY GROUP" for the SIGNIFICANT POLLUTANT LINKAGE.

9.29 If no such CLASS A PERSONS can be found, the ENFORCING AUTHORITY needs to consider whether the SIGNIFICANT POLLUTANT LINKAGE of which it forms part relates solely to the POLLUTION OF CONTROLLED WATERS, rather than to any SIGNIFICANT HARM. If this is the case, there will be no LIABILITY GROUP for that SIGNIFICANT POLLUTANT LINKAGE (*section 78J(2)*), and it should be treated as an ORPHAN LINKAGE (see paragraph 11.3 below).

9.30 In any other case where no CLASS A PERSONS can be found for a SIGNIFICANT POLLUTANT, the ENFORCING AUTHORITY needs to identify all of the OWNERS or occupiers of the CONTAMINATED LAND in question. These persons are then "CLASS B PERSONS" and together constitute a "CLASS B LIABILITY GROUP" for the SIGNIFICANT POLLUTANT LINKAGE.

9.31 If the ENFORCING AUTHORITY cannot find any CLASS A PERSONS or any CLASS B PERSONS in respect of a SIGNIFICANT POLLUTANT LINKAGE, there will be no LIABILITY GROUP for that linkage and it should be treated as an ORPHAN LINKAGE (see paragraph 11.3 below).

Two or more significant pollutants

9.32 Where there are several SIGNIFICANT POLLUTANTS, and therefore two or more SIGNIFICANT POLLUTANT LINKAGES, the ENFORCING AUTHORITY should consider each linkage in turn, carrying out the steps set out in paragraphs

9.28 to 9.31 above, in order to identify the LIABILITY GROUP (if one exists) for each of the linkages.

In all cases

9.33 Having identified one or more LIABILITY GROUPS, the ENFORCING AUTHORITY should consider whether any of the members of those groups are exempted from liability under the provisions in Part IIA. This could apply where:

(a) a person who would otherwise be a CLASS A PERSON is exempted from liability arising with respect to water pollution from an abandoned mine (*section 78J(3)*);
(b) a CLASS B PERSON is exempted from liability arising from the escape of a pollutant from one piece of land to other land (*section 78K*); or
(c) a person is exempted from liability by virtue of his being a person "ACTING IN A RELEVANT CAPACITY" (such as acting as an insolvency practitioner) (*section 78X(4)*).

9.34 If all of the members of a LIABILITY GROUP benefit from one or more of these exemptions, the ENFORCING AUTHORITY should treat the SIGNIFICANT POLLUTANT LINKAGE in question as an ORPHAN LINKAGE (see paragraph 11.3 below).

9.35 Individual persons may be members of more than one LIABILITY GROUP. This might apply, for example, if they had caused or knowingly permitted the presence of more than one SIGNIFICANT POLLUTANT.

9.36 Where the membership of all of the LIABILITY GROUPS is the same, there may be opportunities for the ENFORCING AUTHORITY to abbreviate the remaining stages of the procedure for determining liabilities. However, the tests for EXCLUSION and APPORTIONMENT may produce different results for different SIGNIFICANT POLLUTANT LINKAGES, and so the ENFORCING AUTHORITY will need to exercise caution before trying to simplify the procedure in any case.

Second stage—characterising remediation actions

9.37 Each REMEDIATION ACTION will be carried out to achieve a particular purpose with respect to one or more identified SIGNIFICANT POLLUTANT LINKAGES. Where there is only a single SIGNIFICANT POLLUTANT LINKAGE on the CONTAMINATED LAND in question, all the REMEDIATION ACTIONS will be referable to that linkage, and the ENFORCING AUTHORITY will not need to consider how the different REMEDIATION ACTIONS relate to different linkages. Therefore the authority will not need to carry out this stage and the third stage of the procedure where there is only a single SIGNIFICANT POLLUTANT LINKAGE.

9.38 However, where there are two or more SIGNIFICANT POLLUTANT LINKAGES on the CONTAMINATED LAND, the ENFORCING AUTHORITY needs to establish, for each REMEDIATION ACTION, whether it is:

(a) referable solely to the SIGNIFICANT POLLUTANT in a single SIGNIFICANT POLLUTANT LINKAGE (a SINGLE-LINKAGE ACTION); or
(b) referable to the SIGNIFICANT POLLUTANTS in more than one SIGNIFICANT POLLUTANT LINKAGE (a SHARED ACTION).

9.39 Where a REMEDIATION ACTION is a SHARED ACTION, there are two possible relationships between it and the SIGNIFICANT POLLUTANT LINKAGES to which it is referable. The ENFORCING AUTHORITY needs to establish whether the SHARED ACTION is:

(a) a COMMON ACTION—that is, an action which addresses together all of the SIGNIFICANT POLLUTANT LINKAGES to which it is referable, and which would have been part of the REMEDIATION PACKAGE for each of those linkages if each of them had been addressed separately;

(b) a COLLECTIVE ACTION—that is, an action which addresses together all of the SIGNIFICANT POLLUTANT LINKAGES to which it is referable, but which would not have been part of the REMEDIATION PACKAGE for every one of those linkages if each of them had been addressed separately, because:

(i) the action would not have been appropriate in that form for one or more of the linkages (since some different solution would have been more appropriate);

(ii) the action would not have been needed to the same extent for one or more of the linkages (since a less far-reaching version of that type of action would have sufficed); or

(iii) the action represents a more economic way of addressing the linkages together which would not be possible if they were addressed separately.

A COLLECTIVE ACTION replaces actions that would have been appropriate for the individual SIGNIFICANT POLLUTANT LINKAGES if they had been addressed separately, as it achieves the purposes which those other actions would have achieved.

Third stage—attributing responsibility to liability groups

9.40 This stage of the procedure does not apply in the simpler cases. Where there is only a single SIGNIFICANT POLLUTANT LINKAGE, the LIABILITY GROUP for that linkage bears the full cost of carrying out any REMEDIATION ACTION. Where the linkage is an ORPHAN LINKAGE, the ENFORCING AUTHORITY has the power to carry out the REMEDIATION itself, at its own cost (see paragraph 11.3 below).

9.41 Similarly, for any SINGLE-LINKAGE ACTION, the LIABILITY GROUP for the SIGNIFICANT POLLUTANT LINKAGE in question bears the full cost of carrying out that action.

9.42 However, for each SHARED ACTION the ENFORCING AUTHORITY needs to apply the statutory guidance set out in Part 9 of Chapter D, in order to attribute to each of the different LIABILITY GROUPS their share of responsibility for that action.

9.43 After that statutory guidance has been applied to all SHARED ACTIONS, it may be the case that a CLASS B LIABILITY GROUP which has been identified does not have to bear the costs for any REMEDIATION ACTIONS, since the full cost of the REMEDIATION ACTIONS required will have been borne by others. Where this is the case, the ENFORCING AUTHORITY does not need to carry out any of the rest of this procedure with respect to that LIABILITY GROUP.

Fourth stage—excluding members of a liability group

9.44 The ENFORCING AUTHORITY then needs to consider, for each LIABILITY GROUP which has two or more members, whether any of those members should be EXCLUDED from liability:

(a) for each CLASS A LIABILITY GROUP with two or more members, the authority applies the statutory guidance on EXCLUSION set out in Part 5 of Chapter D; and

(b) for each CLASS B LIABILITY GROUP with two or more members, the authority applies the statutory guidance on EXCLUSION set out in Part 7 of Chapter D.

Fifth stage—apportioning liability between members of a liability group

9.45 The ENFORCING AUTHORITY next needs to determine how any costs attributed to each LIABILITY GROUP should be apportioned between the members of that group who remain after any EXCLUSIONS have been made.

9.46 For any LIABILITY GROUP which has only a single remaining member, that person bears all of the costs falling to that LIABILITY GROUP. This means that he bears the cost of any SINGLE-LINKAGE ACTION referable to the SIGNIFICANT POLLUTANT LINKAGE, and the share of the cost of any SHARED ACTION attributed to the group as a result of the ATTRIBUTION process set out in Part 9 of Chapter D.

9.47 For any LIABILITY GROUP which has two or more remaining members, the ENFORCING AUTHORITY applies the relevant statutory guidance on APPORTIONMENT between those members. Each of the remaining members of the group will then bear the proportion determined under that guidance of the total costs falling to the group. The relevant APPORTIONMENT guidance is:

(a) for any CLASS A LIABILITY GROUP, the statutory guidance set out in Part 6 of Chapter D; and
(b) for any CLASS B LIABILITY GROUP, the statutory guidance set out in Part 8 of Chapter D.

Agreements on liabilities

9.48 The statutory guidance set out in Part 3 of Chapter D provides the procedure which the ENFORCING AUTHORITY should normally follow. However, two or more APPROPRIATE PERSONS may agree between themselves the basis on which they think costs should be borne, or apportioned between themselves, for any REMEDIATION for which they are responsible. If the ENFORCING AUTHORITY is provided a copy of such an agreement and none of the parties to the agreement has informed the authority that it challenges the application of the agreement, the authority needs to allocate liabilities between the parties to the agreement so as to reflect the terms of the agreement, rather than necessarily reflecting the outcome which would otherwise result from the normal processes of EXCLUSION and APPORTIONMENT (*paragraph D.38*).

9.49 However, the ENFORCING AUTHORITY should not do this if the effect of following the agreement would be to increase the costs to be borne by the public purse. In these circumstances, it should disregard the agreement and follow the five stage process outlined above (*paragraph D.39*).

10 Limits on costs to be borne by the appropriate person

10.1 When the ENFORCING AUTHORITY has APPORTIONED the costs of each REMEDIATION ACTION between the various APPROPRIATE PERSONS, and before proceeding to serve any REMEDIATION NOTICE on that basis, the authority must consider whether there are reasons why any of the APPROPRIATE PERSONS on whom that notice would be served should not be required to meet in full the share of the cost of carrying out the REMEDIATION ACTIONS which has been APPORTIONED to him. The importance of this question is that it may preclude the ENFORCING AUTHORITY from serving a REMEDIATION NOTICE in respect of those actions on any of the APPROPRIATE PERSONS at all (see paragraph 10.4 below).

10.2 To decide this question, the ENFORCING AUTHORITY needs to consider the hypothetical circumstances which would apply if the authority had carried out itself the REMEDIATION ACTION or actions for which each APPROPRIATE PERSON is liable. Specifically, the authority needs to consider whether, in these hypothetical circumstances, it would seek to recover from each APPROPRIATE PERSON all of the share of the costs which has been APPORTIONED to that person.

10.3 In making its decision, the authority must have regard to:

(a) any hardship which may be caused to the person in question (see paragraphs 10.8 to 10.10 below); and (b) the statutory guidance in Chapter E of Annex 3 (*section 78P(2)*).

10.4 If the ENFORCING AUTHORITY decides that, in these hypothetical circumstances, it would seek to recover from each APPROPRIATE PERSON all of the share of its reasonable costs APPORTIONED to that person, the authority can proceed to serve the necessary REMEDIATION NOTICES on the basis of its apportionment.

10.5 However, if the ENFORCING AUTHORITY decides, with respect to any REMEDIATION ACTION, that it would seek to recover from any APPROPRIATE PERSON none, or only a part, of that person's apportioned share of the authority's reasonable costs:

(a) it is precluded from serving a REMEDIATION NOTICE specifying that action both on the APPROPRIATE PERSON in question and on anyone else who is an APPROPRIATE PERSON in respect of that action (*section 78H(5)(d)*); and
(b) the authority has the power to carry out the REMEDIATION ACTION in question itself (*section 78N(3(e)*; see also paragraphs 11.7 to 11.11 below).

10.6 Where, in a case of this kind, the ENFORCING AUTHORITY does then decide to exercise its powers and carry out particular REMEDIATION ACTIONS, the authority will be entitled to recover its reasonable costs of doing so when it has completed the work. In deciding how much of those costs it will seek to recover, the authority will need to work on the basis of circumstances as they exist at that point. In practice, however, the decision that the authority has taken on the hypothetical basis described in paragraph 10.2 above will normally settle the questions of limits on the actual recovery of costs. Nevertheless, if there is evidence that the circumstances of the APPROPRIATE PERSON have changed in some relevant respect after the ENFORCING AUTHORITY has made its initial decision on this question, it will need to reconsider its decision as to how much of its reasonable costs it will seek to recover.

10.7 Further details about actual cost recovery are given in section 16 of this Annex.

The meaning of the term "hardship"

10.8 The term "hardship" is not defined in Part IIA, and therefore carries its ordinary meaning—hardness of fate or circumstance, severe suffering or privation.

10.9 The term has been widely used in other legislation, and there is a substantial body of case law about its meaning under that other legislation. For example, it has been held appropriate to take account of injustice to the person claiming hardship, in addition to severe financial detriment. Although the case law may give a useful indication of the way in which the term has been interpreted by the courts, the meaning ascribed to the term in individual cases is specific to the particular facts of those cases and the legislation under which they were brought.

10.10 In deciding whether there would be hardship, and its extent, the matters considered in Chapter E may well be relevant.

11 Remediation action by the enforcing authority

11.1 Before serving any REMEDIATION NOTICE, the ENFORCING AUTHORITY needs to consider whether it has the power to carry out any of the REMEDIATION ACTIONS itself. Where this applies, the authority is precluded from serving a REMEDIATION NOTICE requiring anyone else to carry out that REMEDIATION ACTION (*section 78H(5)*).

The power to carry out remediation

11.2 In general terms, the ENFORCING AUTHORITY has the power to carry out a REMEDIATION ACTION itself in cases where:

(a) the ENFORCING AUTHORITY considers it necessary to take urgent action itself (*section 78N(3)(a)*; see paragraphs 5.1 to 5.8 above);
(b) there is no APPROPRIATE PERSON to bear responsibility for the action (*section 78N(3)(f)*; see paragraph 11.3 below);
(c) the ENFORCING AUTHORITY is precluded from requiring one or more persons, who would otherwise be APPROPRIATE PERSONS, to carry out the action (*sections 78N(3)(d) & (e)*; see paragraph 11.4 below);
(d) the ENFORCING AUTHORITY has agreed with the APPROPRIATE PERSON that the authority should carry out the REMEDIATION ACTION (*section 78N(3)(b)*; see paragraphs 11.5 to 11.6 below); or
(e) the REMEDIATION ACTION has been specified in a REMEDIATION NOTICE, which has not been complied with (*section 78N(3)(c)*; see paragraph 15.15 below).

There is no appropriate person

11.3 The ENFORCING AUTHORITY has the power to carry out a REMEDIATION ACTION if, after reasonable enquiry, it has been unable to find an APPROPRIATE PERSON for that action (*section 78N(3)(f)*).

The Appropriate person cannot be required to carry out a remediation action

11.4 The ENFORCING AUTHORITY needs to consider whether it has the power to carry out a REMEDIATION ACTION on the basis that the APPROPRIATE PERSON cannot be required to carry it out. This applies where:

(a) the ENFORCING AUTHORITY considers that if it carried out the REMEDIATION ACTION itself, it would not seek to recover fully from that APPROPRIATE PERSON the proportion of the costs which that person would otherwise have to bear if the action were included in a REMEDIATION NOTICE (*sections 78N(3)(e) & 78P(2)*; see also paragraphs 10.1 to 10.10 above);
(b) the REMEDIATION ACTION is referable solely to one or more SIGNIFICANT POLLUTANT LINKAGES which relate to the POLLUTION OF CONTROLLED WATERS (and not to any SIGNIFICANT HARM), and either:

(i) the APPROPRIATE PERSON is a CLASS B PERSON (*section 78J(2)*), or
(ii) the APPROPRIATE PERSON is a CLASS A PERSON solely by virtue of his having permitted the discharge of water from a mine which was abandoned before the end of 1999 (*section 78J(3)*);

(c) the SIGNIFICANT POLLUTANT LINKAGE to which the REMEDIATION ACTION is referable is the result of the escape of the POLLUTANT from other land onto the CONTAMINATED LAND in question, and both:

(i) the APPROPRIATE PERSON is a CLASS B PERSON, and
(ii) the REMEDIATION ACTION is intended to deal with SIGNIFICANT HARM or the POLLUTION OF CONTROLLED WATERS on land other than the CONTAMINATED LAND in question, to which the POLLUTANT has escaped (*section 78K*); or

(d) requiring the APPROPRIATE PERSON to carry out the REMEDIATION ACTION would have the effect of making him personally liable to bear the costs, and:

(i) he is a "PERSON ACTING IN A RELEVANT CAPACITY" such as an insolvency practitioner (*section 78X(4)*), and
(ii) the REMEDIATION ACTION is not to any extent referable to any POLLUTANT which is present as a result of any act or omission which it was unreasonable for a person acting in that capacity to do or make (*section 78X(3)(a)*).

Written agreement

11.5 Even if none of the grounds set out in paragraph 11.4 above applies, the ENFORCING AUTHORITY may wish to consider whether it would, nonetheless, be appropriate for the authority to carry out a REMEDIATION ACTION itself on behalf of the APPROPRIATE PERSON. This might be appropriate, in particular, in the case of home-owners identified as APPROPRIATE PERSONS.

11.6 If the ENFORCING AUTHORITY considers that it wishes do this, it needs to seek the written agreement of the APPROPRIATE PERSON for:

(a) the ENFORCING AUTHORITY to carry out the REMEDIATION ACTION itself, on behalf of the APPROPRIATE PERSON; and
(b) the APPROPRIATE PERSON to reimburse the authority for any costs which he would otherwise have had to bear for the REMEDIATION (*section 78N(3)(b)*).

Action by the enforcing authority

11.7 The ENFORCING AUTHORITY'S powers to carry out REMEDIATION under section 78N may be triggered with respect to all of the APPROPRIATE PERSONS for a particular REMEDIATION ACTION, or only with respect to some of them. Whichever is the case, the authority is precluded from including the REMEDIATION ACTION in question in a REMEDIATION NOTICE served on anyone (*section 78H(5)*).

11.8 However, where the ENFORCING AUTHORITY carries out a REMEDIATION ACTION using its powers with respect to urgent action (*section 78N(3)(a)*) or limitations on costs (*section 78N(3)(e)*; see paragraphs 10.1 to 10.6 above), it is entitled to recover its reasonable costs from all of the APPROPRIATE PERSONS for that REMEDIATION ACTION (*section 78P(1)*). In deciding how much of those costs to recover from any particular APPROPRIATE PERSON, the authority must have regard to hardship which may be caused to that person and to the statutory guidance set out in Chapter E of Annex 3 (*section 78P(2)*).

11.9 For example, there may be two APPROPRIATE PERSONS (persons "1" and "2") for a particular REMEDIATION ACTION. The ENFORCING AUTHORITY may consider that the cost which "person 1" would have to bear would cause him hardship. On this basis, the authority has a power to carry out the REMEDIATION ACTION itself, and cannot include that action in a notice served on either of the APPROPRIATE PERSONS (see paragraph 10.5 above). Once the authority has carried out the action, it can recover from "person 2" the same proportion of its costs as a REMEDIATION NOTICE served on him would have specified, and from "person 1" as much of the remainder as would not cause hardship or be inconsistent with the statutory guidance in Chapter E.

11.10 Where the ENFORCING AUTHORITY is precluded from serving a REMEDIATION NOTICE because it has powers under section 78N to carry out the REMEDIATION itself, it will be under a duty to prepare and publish a REMEDIATION STATEMENT recording:

"(a) the things which are being, have been, or are expected to be, done by way of REMEDIATION in the particular case;

"(b) the name and address of the person who is doing, has done, or is expected to do, each of those things; and

"(c) the periods within which each of those things is being, or is expected to be done" (*section 78H(7)*).

11.11 The ENFORCING AUTHORITY must then include details of the REMEDIATION STATEMENT on its REGISTER (*section 78R(1)(c) and regulation 15*; see paragraphs 17.1 to 17.19 below and Annex 4, paragraph 88).

12 Serving a remediation notice

12.1 The basis for serving a REMEDIATION NOTICE is that the ENFORCING AUTHORITY considers that there are REMEDIATION ACTIONS, identified as part of the REMEDIATION SCHEME, which:

(a) have not been, are not being, and will not be carried out without the service of a REMEDIATION NOTICE; and

(b) in respect of which the authority has no power under section 78N to carry out itself and for which it is not, itself, the APPROPRIATE PERSON.

12.2 Before serving a REMEDIATION NOTICE, the ENFORCING AUTHORITY needs to decide whether it has made reasonable endeavours to consult the APPROPRIATE PERSON and the other relevant persons (described in paragraph 6.10 to 6.17 above) on the nature of the REMEDIATION which is to be carried out (*section 78H(1)*).

12.3 When the authority is satisfied that it has consulted sufficiently, and subject to the timing requirements outlined in paragraphs 12.4 and 12.5 below, the authority will be under a duty to serve a REMEDIATION NOTICE on each APPROPRIATE PERSON requiring the relevant REMEDIATION ACTION to be carried out (*section 78E(1)*).

Timing of the service of a remediation notice

12.4 THE ENFORCING AUTHORITY will have notified each APPROPRIATE PERSON that he appears to be such a person (*section 78B(3) & (4)*; see paragraphs 4.1 to 4.6 above). The date of this notification to any person determines the earliest date on which the ENFORCING AUTHORITY can serve a REMEDIATION NOTICE on that person. Except in a case of urgency (see paragraphs 5.1 to 5.8 above), at least three months must elapse between the date of the notification to the person concerned and the service of a REMEDIATION NOTICE on that person (*section 78H(3)(a)*).

12.5 However, later dates apply if the LOCAL AUTHORITY has given notice of a decision that the land is required to be designated a SPECIAL SITE, or if the ENVIRONMENT AGENCY has given an equivalent notice to the LOCAL AUTHORITY (see paragraphs 18.7 and 18.13 below). Once such a notice has been given, the ENFORCING AUTHORITY cannot serve a REMEDIATION NOTICE (except in cases of urgency) until three months have elapsed since:

(a) notice was given by the LOCAL AUTHORITY that the designation of the land as a SPECIAL SITE is to take effect; or
(b) notice was given by the SECRETARY OF STATE that the designation of the land as a SPECIAL SITE is, or is not, to take effect (*sections 78H(3)(b) & (c)*; see also section 18 of this Annex).

The remediation notice

12.6 The ENFORCING AUTHORITY must include in any REMEDIATION NOTICE particular information about the CONTAMINATED LAND, the REMEDIATION, the APPROPRIATE PERSON and rights of appeal against the notice. The requirements for the contents of a REMEDIATION NOTICE are formally set out in sections 78E(1) and (3), and regulation 4 of the Contaminated Land (England) Regulations 2000 (see Annex 4, paragraphs 16 to 20).

12.7 In any case where there are two or more APPROPRIATE PERSONS for any REMEDIATION ACTION, the ENFORCING AUTHORITY may serve a single REMEDIATION NOTICE on all of those persons. (Acting in this way will make the process of readjusting the APPORTIONMENT of costs after a successful appeal considerably simpler, as the APPELLATE AUTHORITY will be able to amend the single REMEDIATION NOTICE and the way it affects each of the APPROPRIATE PERSONS; if separate notices are served, this would not be possible, and new notices would have to be served.)

12.8 As well as serving the REMEDIATION NOTICE on the APPROPRIATE PERSONS, the ENFORCING AUTHORITY must send a copy:

(a) to any person who they have consulted under section 78G(3) about the granting of rights over the land or waters to the APPROPRIATE PERSON;
(b) to any person who was consulted under section 78H(1); and
(c) if the ENFORCING AUTHORITY is the LOCAL AUTHORITY, to the ENVIRONMENT AGENCY, and if the ENFORCING AUTHORITY is the ENVIRONMENT AGENCY, to the LOCAL AUTHORITY (*regulation 5(1)*).

12.9 The ENFORCING AUTHORITY is under a duty to include prescribed details of the REMEDIATION NOTICE on its REGISTER (*section 78R(1)(a) and regulation 15*; see paragraphs 17.1 to 17.19 below and Annex 4, paragraph 85).

13 Appeals against a remediation notice

13.1 Any person who receives a REMEDIATION NOTICE has twenty-one days within which he can appeal against the notice (*section 78L(1)*). Any appeal is made:

(a) to a magistrates' court, if the notice was served by a LOCAL AUTHORITY; or
(b) to the SECRETARY OF STATE, if the notice was served by the ENVIRONMENT AGENCY.

13.2 The grounds for any such appeal are prescribed in regulation 7. Regulations 8-14 prescribe the procedures for any appeal. These regulations are described in Annex 4 to this circular.

13.3 If an appeal is made, the REMEDIATION NOTICE is suspended until final determination or abandonment of the appeal (*regulation 14*).

13.4 If any appeal is made against a REMEDIATION NOTICE, the ENFORCING AUTHORITY must enter prescribed particulars of the appeal, and the decision reached on the appeal, on its REGISTER (*section 78R(1)(b) and regulation 15*).

Action during a suspension of a notice

13.5 Where the requirement to carry out particular REMEDIATION ACTIONS is suspended during an appeal, the ENFORCING AUTHORITY needs to consider whether this makes it necessary for the authority itself to carry out urgent REMEDIATION (*section 78N(3)(a)*; see paragraphs 5.1 to 5.8 above).

13.6 If the ENFORCING AUTHORITY does carry out urgent REMEDIATION itself in these circumstances, it does not need to prepare and publish a REMEDIATION STATEMENT, unless the REMEDIATION has not already been described in the original REMEDIATION NOTICE.

13.7 Having carried out any REMEDIATION ACTION, the ENFORCING AUTHORITY needs to consider whether to seek to recover its reasonable costs (*section 78P(1)*). Its ability to do so may, however, be affected by the decision in the appeal against the REMEDIATION NOTICE. For example, it would not be able to recover its costs from the recipient of a notice who successfully appealed on the grounds that he was not the APPROPRIATE PERSON.

14 Variations in remediation requirements

14.1 It may become apparent, whilst REMEDIATION ACTIONS are being carried out, that the overall REMEDIATION SCHEME for the RELEVANT LAND OR WATERS is no longer appropriate.
For example:

(a) further SIGNIFICANT POLLUTANT LINKAGES may be identified, requiring further REMEDIATION ACTIONS to be carried out;

(b) a REMEDIATION ACTION which is being carried out may be discovered to be:

(i) ineffective, given the circumstances of the RELEVANT LAND OR WATERS,

(ii) unsafe, in terms of pollution or health and safety risks, given the circumstances of the RELEVANT LAND OR WATERS, or

(iii) unnecessary, in the light of new information about the condition of the land; or

(c) a further REMEDIATION ACTION may be identified which would be reasonable and would achieve a purpose which could not previously be achieved by any reasonable REMEDIATION ACTION.

14.2 If other REMEDIATION ACTIONS are identified as being appropriate, this may require the preparation and publication of a new REMEDIATION STATEMENT or the serving of a new REMEDIATION NOTICE.

15 Follow-up action

15.1 The ENFORCING AUTHORITY needs to consider whether the REMEDIATION ACTIONS described in the REMEDIATION STATEMENT or specified in the REMEDIATION NOTICE have been carried out and, if so, whether they have been carried out adequately and satisfactorily. In many cases, the authority will do so on the basis of information generated by the quality assurance and verification procedures included within the REMEDIATION ACTIONS (*paragraphs C.25 and C.67*).

15.2 Whatever it decides, the ENFORCING AUTHORITY also needs to consider whether any further REMEDIATION is appropriate. This applies particularly in circumstances where the completed REMEDIATION ACTIONS form only a single phase of the overall process of REMEDIATION for the RELEVANT LAND OR WATERS. If it decides that further REMEDIATION is appropriate, the authority repeats the procedures set out above for consultation, identifying appropriate REMEDIATION ACTIONS and requiring that REMEDIATION to be carried out by service of a REMEDIATION NOTICE.

Remediation action has been carried out

Notifications of "claimed remediation"

15.3 Any person who has carried out any REMEDIATION which was required by a REMEDIATION NOTICE or described in a REMEDIATION STATEMENT can notify the ENFORCING AUTHORITY, providing particular details of the REMEDIATION he claims to have carried out (*regulation 15(2)*). The OWNER or occupier of the CONTAMINATED LAND is also entitled to notify the authority.

15.4 If the ENFORCING AUTHORITY receives any notification of this kind, it will be under a duty to include on its REGISTER prescribed details of the REMEDIATION which it is claimed has been carried out (*sections 78R(1)(h) & (j) and regulation 15*; see paragraphs 17.1 to 17.19 below and Annex 4, paragraph 89).

15.5 Part IIA provides that the inclusion of an entry of this kind on the REGISTER is not to be taken as a representation by the authority maintaining the REGISTER that the entry is accurate with respect to what is claimed to have been done, or the manner in which it may have been done (*section 78R(3)*).

"Signing off"

15.6 Although Part IIA does not include any formal "signing off" procedure, the ENFORCING AUTHORITY may wish to consider writing to the APPROPRIATE PERSON, confirming the position with respect to any further enforcement action. In a case where a REMEDIATION NOTICE has been served and appears to have been complied with, this could confirm that the authority currently sees no grounds, on the basis of available information, for further enforcement action. In other cases—where a REMEDIATION NOTICE has not been served—the ENFORCING AUTHORITY might confirm that it does not consider that it needs to serve a REMEDIATION NOTICE, which it would need to do if appropriate REMEDIATION had not been carried out.

Remediation has not been carried out

If a remediation statement has not been followed

15.7 If a REMEDIATION ACTION described in a REMEDIATION STATEMENT is not carried out in the manner and within the time period described, the ENFORCING AUTHORITY needs to consider whether it is necessary for a REMEDIATION NOTICE to be served requiring that REMEDIATION ACTION to be carried out.

15.8 The ENFORCING AUTHORITY has a duty to serve such a REMEDIATION NOTICE if:

(a) it considers that appropriate REMEDIATION is not being carried out and it is not satisfied that it will be carried out without the service of a notice; and
(b) it is not precluded for any other reason from serving a notice on the APPROPRIATE PERSON (*section 78H(10)*).

15.9 In these circumstances, the ENFORCING AUTHORITY can serve the REMEDIATION NOTICE without making any further efforts to consult, provided that the REMEDIATION ACTIONS specified in the notice have previously been the subject of consultation with the person in question (*section 78H(10)*).

If a remediation notice is not complied with

15.10 If a REMEDIATION ACTION specified in a REMEDIATION NOTICE is not carried out within the time required, the ENFORCING AUTHORITY needs to consider whether to prosecute the APPROPRIATE PERSON who has failed to comply with the REMEDIATION NOTICE. It will normally be desirable for the authority to inform the APPROPRIATE PERSON that it is considering bringing such a prosecution before it actually does so. This may give that person an opportunity to avoid prosecution by carrying out the requirements of the REMEDIATION NOTICE.

15.11 Part IIA makes it an offence for any person to fail to comply with a REMEDIATION NOTICE "without reasonable excuse" (*section 78M(1)*). The question of whether a person had a "reasonable excuse" in any case is a matter of fact to be decided on the basis of the particular circumstances of that case.

15.12 One defence is specified in Part IIA. This applies where:

(a) the APPROPRIATE PERSON was required by the REMEDIATION NOTICE to bear only a proportion of the cost of the REMEDIATION ACTION which has not been carried out; and

(b) that person can show that the only reason why he did not comply with the REMEDIATION NOTICE was that one or more of the other APPROPRIATE PERSONS who should have borne other shares of the cost refused, or were not able, to do so (*section 78M(2)*).

15.13 In general, a person convicted of the offence of non-compliance with a REMEDIATION NOTICE is liable to a fine not exceeding level 5 on the standard scale; at the date of this circular, that is £5,000. Until either he complies with the REMEDIATION NOTICE, or the ENFORCING AUTHORITY uses its powers to act in default (see paragraph 15.15 below), he is also liable for additional daily fines up of up to one tenth of level 5; that is, at the date of this circular, L500 (*section 78M(3)*).

15.14 However, where the CONTAMINATED LAND to which the notice relates is INDUSTRIAL, TRADE OR BUSINESS PREMISES, the limit on the fine is higher: the fine may be up to £20,000, with daily fines of up to £2,000 (*section 78M(4)*). Part IIA provides a power to increase those limits by order: the Government's intention is to use that power where necessary to maintain the differential with level 5 on the standard scale.

15.15 In addition, the authority needs to consider whether to carry out the REMEDIATION ACTION itself (*section 78N(3)(c)*). It can decide to do so whether or not it decides to prosecute the APPROPRIATE PERSON. If it does carry out the REMEDIATION, it is entitled to recover its reasonable costs from the APPROPRIATE PERSON (*sections 78P(1)*).

16 Recovering the costs of carrying out remediation

16.1 In general, where the ENFORCING AUTHORITY has carried out REMEDIATION itself, it is entitled to recover the reasonable costs it has incurred in doing so (*section 78P(1)*). The ENFORCING AUTHORITY has no power to recover any costs it incurred in inspecting the land to determine whether it was CONTAMINATED LAND.

16.2 In deciding whether to recover its costs and, if so, how much of its costs, the ENFORCING AUTHORITY must have regard to:

(a) any hardship which the recovery might cause to the APPROPRIATE PERSON (see paragraphs 10.8 to 10.10 above) and
(b) the statutory guidance set out in Chapter E of Annex 3 (*section 78P(2)*; see also paragraphs 10.8 to 10.10 above).

16.3 However, the ENFORCING AUTHORITY has no power under section 78P to recover its costs where:

(a) the authority itself was the APPROPRIATE PERSON;
(b) the person who would otherwise have been an APPROPRIATE PERSON for a REMEDIATION ACTION could not have beenrequired to carry out that action under the terms of a REMEDIATION NOTICE, because it related to the POLLUTION OF CONTROLLED WATERS or to the escape of the POLLUTANT from other land (*section 78N(3)(d)*); or
(c) the authority carried out the REMEDIATION with the written agreement of the APPROPRIATE PERSON (*section 78N(3)(b)*).

16.4 In the first two of these cases, the ENFORCING AUTHORITY has itself to bear the cost of carrying out the REMEDIATION (see paragraphs 16.12 to 16.14 below).

16.5 If the ENFORCING AUTHORITY carries out the REMEDIATION with the written agreement of the APPROPRIATE PERSON (section *78N(3)(b)*), reimbursement by the APPROPRIATE PERSON will be under the terms of the written agreement.

16.6 If the ENFORCING AUTHORITY decides to recover all or a part of its costs, it needs to consider whether to do so immediately (which will involve an action in the county court or High Court, if payment is not made) or to postpone recovery and, where this is possible, safeguard its right to cost recovery by imposing a charge on the land in question. A CHARGING NOTICE may also be served to safeguard the authority's interests where immediate recovery is intended.

Charging notices

16.7 If the ENFORCING AUTHORITY decides to safeguard its rights to cost recovery by imposing a charge on the land in question, it does so by serving a CHARGING NOTICE (*section 78P(3)*). The authority is entitled to serve a CHARGING NOTICE if the APPROPRIATE PERSON from whom it is recovering its costs is both:

(a) a CLASS A PERSON; and
(b) the OWNER of all or part of the CONTAMINATED LAND (*section 78P(3)*).

16.8 On the same day as the ENFORCING AUTHORITY serves any CHARGING NOTICE, it must send a copy of the notice to every other person who, to the knowledge of the authority, has an interest in the premises capable of being affected by the charge (*section 78P(6)*).

16.9 Any person served with a CHARGING NOTICE, or who receives a copy of it, can appeal against it to a county court (*section 78P(8)*). If any such appeal is made, the ENFORCING AUTHORITY must include prescribed particulars of that appeal on its REGISTER (*section 78R(1)(d)*; see paragraphs 17.1 to 17.19 below and Annex 4, paragraph 97). The CHARGING NOTICE itself will not appear on the REGISTER. The power to make regulations on the grounds of appeal against a CHARGING NOTICE and the related procedure has not been exercised. It is therefore for the county court to determine what grounds of appeal it will accept; the ordinary county court procedures for appeals will apply.

16.10 A CHARGING NOTICE can declare the cost to be payable with interest by instalments, within a specified period, until the whole amount is repaid (*section 78P(12)*).

16.11 If the ENFORCING AUTHORITY needs to enforce the charge, it has the same powers and remedies under the Law of Property Act 1925 as if the authority were a mortgagee by deed having powers of sale and lease, of accepting surrenders of leases and of appointing a receiver (*section 78P(11)*).

Central government support to local authorities

16.12 The Department of the Environment, Transport and the Regions runs a programme of Supplementary Credit Approvals (SCAs) for capital costs incurred by local authorities in dealing with land contamination where they:

(a) own the land;
(b) are responsible for its contamination; or
(c) have other statutory responsibilities for carrying out remediation, including the use of powers to carry out REMEDIATION under section 78N.

16.13 Support under this programme is not available for work needed solely to facilitate the development, redevelopment or sale of the land. Financial support for remediation in connection with the development or redevelopment of land may be available through the single regeneration budget or under the programmes of English Partnerships and the regional development agencies.

16.14 All local authorities which are entitled to receive SCAs are invited annually to bid for support from this programme for particular schemes. Schemes are assessed against environmental criteria and prioritised. Where a bid is successful, the authority is issued an SCA which permits it to raise money to finance the remediation. The revenue implications of servicing this borrowing are then taken into account in the Revenue Support Grant calculations for subsequent years.

16.15 DETR also provides financial support to the ENVIRONMENT AGENCY.

17 Registers

17.1 Each ENFORCING AUTHORITY has a duty to maintain a REGISTER (*section 78R(1)*). The register will include details of REMEDIATION NOTICES which have been served and certain other documents in relation to each area of CONTAMINATED LAND for which the authority is responsible. The REGISTER will also include information about the condition of the land in question. For a LOCAL AUTHORITY, the REGISTER must be kept at its principal office. For the ENVIRONMENT AGENCY, the REGISTER must be kept at the area office for the area in which the land is situated.

17.2 The particular details to be included in each REGISTER are prescribed in regulation 15 of, and Schedule 3 to, the Contaminated Land (England) Regulations 2000 (see Annex 4). Neither these Regulations, nor the primary legislation in Part IIA, state when details should be added to the REGISTER. In the Government's view, this implies that they should be added as soon as reasonably practicable after the information they contain is generated; so, for example, the prescribed details of a REMEDIATION NOTICE should be added as soon as reasonably practicable after the service of that notice.

17.3 Before including any information on its REGISTER, the ENFORCING AUTHORITY needs to consider whether that information should be excluded on the basis that:

(a) its inclusion would be against the interests of national security (see paragraphs 17.8 to 17.9 below); or
(b) the information is commercially confidential (see paragraphs 17.10 to 17.19 below).

Copying entries between authorities

17.4 For most areas of CONTAMINATED LAND, the LOCAL AUTHORITY for that area will be the ENFORCING AUTHORITY. However, for particular areas of CONTAMINATED LAND this may not be the case. This applies if:

(a) the CONTAMINATED LAND has been designated a SPECIAL SITE, in which case the ENVIRONMENT AGENCY is the ENFORCING AUTHORITY; or
(b) the land has been identified as CONTAMINATED LAND by the LOCAL AUTHORITY for an adjoining or adjacent area, as a result of SIGNIFICANT HARM or the POLLUTION OF CONTROLLED WATERS which might be caused in that LOCAL AUTHORITY'S own area (*section 78X(2)*).

17.5 Where this is the case, the ENFORCING AUTHORITY needs to copy all entries it makes into its own REGISTER for the land in question, to the LOCAL AUTHORITY in whose area the land is actually situated (*section 78R(4) & (5)*).

17.6 The LOCAL AUTHORITY which receives these copied entries needs to include them on its own REGISTER (*section 78R(6)*). This means that the REGISTER maintained by any LOCAL AUTHORITY provides a comprehensive set of information about all of the CONTAMINATED LAND identified in its area, whichever authority is the ENFORCING AUTHORITY.

Public access to registers

17.7 Each ENFORCING AUTHORITY is under a duty to keep its REGISTER available for free inspection by the public at all reasonable times (*section 78R(8)(a)*). In addition, it will be under a duty to provide facilities for members of the public to obtain copies of REGISTER entries. It can make reasonable charges for this (*section 78R(8)(b)*).

Exclusion on the grounds of national security

17.8 The ENFORCING AUTHORITY must not include any information on its REGISTER if, in the opinion of the SECRETARY OF STATE, its inclusion would be against the interests of national security (*section 78S(1)*). The SECRETARY OF STATE can give directions to ENFORCING AUTHORITIES specifying information, or descriptions of information, which are to be excluded from any REGISTER or referred to the SECRETARY OF STATE for his determination (*section 78S(2)*). At the date of this circular, no such directions have been given.

17.9 Any person who considers that the inclusion of particular information on a REGISTER would be against the interests of national security can notify the SECRETARY OF STATE and the ENFORCING AUTHORITY of this. The SECRETARY OF STATE will then consider whether, in his opinion, the information should be included or excluded. The ENFORCING AUTHORITY must not include on its REGISTER any information covered by this kind of notification unless and until the SECRETARY OF STATE determines that it can be included (*section 78S(4)*).

Exclusion on the grounds of commercial confidentiality

17.10 The ENFORCING AUTHORITY must not, without the relevant person's permission, include any information on its REGISTER which:

(a) relates to the affairs of any individual or business; and

(b) is commercially confidential to that individual or the person carrying on that business (*section 78T(1)*).

17.11 For these purposes, commercial interests relating to the value of the CONTAMINATED LAND, or to its the ownership or occupation, are disregarded (*section 78T(11)*). This means that information cannot be excluded from the REGISTER solely on the basis that its inclusion might provide information to a prospective buyer of the land, thereby affecting the sale or the sale price.

17.12 In addition, the SECRETARY OF STATE can give directions to ENFORCING AUTHORITIES requiring the inclusion of specified information or descriptions of information, notwithstanding any commercial confidentiality, where he considers that the inclusion of that information would be in the public interest (*section 78T(7)*). No such directions have yet been given.

17.13 If the ENFORCING AUTHORITY considers that any information which it would normally include on its REGISTER could be commercially confidential, it must notify the person concerned in writing. The authority then needs to give that person a reasonable opportunity to make representations requesting the exclusion of the information and explaining why the information is commercially confidential (*section 78T(2)*).

17.14 The ENFORCING AUTHORITY then needs to determine, taking into account any representations received, whether the information is, or is not, commercially confidential.

17.15 If the ENFORCING AUTHORITY determines that the information is commercially confidential, that information is excluded from the REGISTER. However, the authority must include on its REGISTER a statement indicating the existence of excluded information of the relevant kind (*section 78R(7)*). This means, for example, that if details of a REMEDIATION NOTICE are excluded, the statement records that the particulars of such a notice have been excluded.

17.16 If the ENFORCING AUTHORITY determines that the information is not commercially confidential, it notifies the person concerned. That person then has twenty-one days in which he can appeal to the SECRETARY OF STATE (*section 78T(3)*). While any appeal is pending, the information is not included on the REGISTER. If the SECRETARY OF STATE determines that the information is commercially confidential, then the information is excluded with a statement about the exclusion being entered on the REGISTER. If the SECRETARY OF STATE determines that the information is not commercially confidential, or if the appeal is withdrawn, the ENFORCING AUTHORITY includes it on its REGISTER seven days afterwards.

17.17 If no appeal is made within twenty-one days of the date on which the ENFORCING AUTHORITY notified the person concerned of its determination, the ENFORCING AUTHORITY enters the information on its REGISTER.

17.18 Where any information is excluded from a REGISTER on the grounds of commercial confidentiality, that exclusion will generally lapse after four years with the information being treated as no longer being commercially confidential (*section 78T(8)*). This means that where information has been excluded, the ENFORCING AUTHORITY will need to put arrangements in place to ensure that information is included on the REGISTER once the four year period has passed.

17.19 However, the person who furnished the information can apply to the ENFORCING AUTHORITY for information to remain excluded. The authority then determines whether the information is still commercially confidential, and acts accordingly. The same arrangements apply for any appeal against this determination as apply in the case of an original determination (*section 78T(9)*).

18 Procedures relating to special sites

Introduction

18.1 Regulations 2 and 3 of the Contaminated Land (England) Regulations2000, together with Schedule 1 of those Regulations, prescribe various descriptions of CONTAMINATED LAND which are required to be designated as SPECIAL SITES. An explanation of these descriptions is set out in Annex 4 to this circular.

18.2 The actual designation of any individual site is made by the LOCAL AUTHORITY or, in any case where there is a dispute between the LOCAL AUTHORITY and the ENVIRONMENT AGENCY, by the SECRETARY OF STATE, on the basis that the land meets one or more of these descriptions.

18.3 The effect of any such designation is that the ENVIRONMENT AGENCY takes over from the LOCAL AUTHORITY as the ENFORCING AUTHORITY for that site. In carrying out its role as an ENFORCING AUTHORITY, the ENVIRONMENT AGENCY is subject to the same requirements under the primary and secondary legislation and statutory guidance as would be a LOCAL AUTHORITY.

18.4 From the point of view of the OWNER or occupier of the land, or an APPROPRIATE PERSON, the main procedural difference resulting from a designation will be that any appeal against a REMEDIATION NOTICE will be to the Secretary of State and not to the magistrates' court.

The identification of special sites

Identification by the local authority

18.5 Whenever the LOCAL AUTHORITY has identified any CONTAMINATED LAND, it will need to consider whether that land meets one or more of the descriptions prescribed in the Regulations, and should therefore be designated as a SPECIAL SITE (*section 78C(1)*). It will also need to keep this question under review as further information becomes available.

18.6 If the LOCAL AUTHORITY considers, at any time, that some particular CONTAMINATED LAND might be required to be designated as a SPECIAL SITE, it needs to request the advice of the ENVIRONMENT AGENCY (*section 78C(3)*). If the LOCAL AUTHORITY does not consider that the land might be required to be designated, it does not need to consult the ENVIRONMENT AGENCY.

18.7 The LOCAL AUTHORITY then needs to decide, having regard to any such advice received, whether or not the land is required to be designated (*section 78C(3)*). If it decides that it is, the authority must give notice in writing to:

(a) the ENVIRONMENT AGENCY;
(b) the OWNER of the land;
(c) any person who appears to be the occupier of all or part of the land; and

(d) each person who appears to be an APPROPRIATE PERSON (*sections 78C(1)(b) & 78C(2)*).

18.8 The ENVIRONMENT AGENCY then needs to consider whether it agrees with the LOCAL AUTHORITY'S decision that the land should be designated.

18.9 If it does not agree, it must notify the LOCAL AUTHORITY within twenty-one days of the LOCAL AUTHORITY'S notification, giving a statement of its reasons for disagreeing (*section 78D(1)(b)*). It also needs to copy the notification and statement to the SECRETARY OF STATE (*section 78D(2)*). The LOCAL AUTHORITY must then refer its decision to the SECRETARY OF STATE (*section 78D(1)*).

18.10 If the ENVIRONMENT AGENCY agrees with the LOCAL AUTHORITY'S decision, or if it fails to notify its disagreement within the twenty-one days allowed, the CONTAMINATED LAND in question will be designated as a SPECIAL SITE (see paragraphs 18.20 to 18.22 below).

Identification by the Environment Agency

18.11 The ENVIRONMENT AGENCY also needs to consider whether any CONTAMINATED LAND should be designated as a SPECIAL SITE. If at any time it considers that any such land should be designated, it needs to notify in writing the LOCAL AUTHORITY in whose area that land is situated (*section 78C(4)*).

18.12 The ENVIRONMENT AGENCY may take this view on the basis of information received from the LOCAL AUTHORITY or information it obtains itself, for example under its other pollution control functions. However, the basis on which it reaches such a decision must be whether or not it considers that the land meets one or more of the descriptions prescribed in the Regulations. The ENVIRONMENT AGENCY is not entitled to apply any different tests to those which the LOCAL AUTHORITY would apply.

18.13 The LOCAL AUTHORITY must then decide whether or not it agrees with the ENVIRONMENT AGENCY that the CONTAMINATED LAND should be designated a SPECIAL SITE. Once it has reached a decision, it must notify in writing the persons identified in paragraph 18.7 above of its decision (*section 78C(5)*).

18.14 If the LOCAL AUTHORITY agrees with the ENVIRONMENT AGENCY, the land is designated a SPECIAL SITE (see paragraphs 18.20 to 18.22 below).

18.15 If the LOCAL AUTHORITY disagrees with the ENVIRONMENT AGENCY, the Agency has an opportunity to reaffirm its view that the land should be designated. If it wishes to do this, it must notify the LOCAL AUTHORITY, in writing, within twenty-one days of receiving from the LOCAL AUTHORITY notification of its decision. The Agency must provide a statement of the reasons why it considers the land should be designated (*section 78D(1)(b)*) and send this information to the SECRETARY OF STATE (*section 78D(2)*). The LOCAL AUTHORITY must then refer its decision to the SECRETARY OF STATE (*section 78D(1)*).

Referral of decisions to the Secretary of State

18.16 If the LOCAL AUTHORITY receives any notification from the ENVIRONMENT AGENCY that the Agency disagrees with a decision it has made concerning the designation or non-designation of any CONTAMINATED LAND as a SPECIAL SITE, the LOCAL AUTHORITY must refer that decision to the SECRETARY OF STATE.

18.17 In doing so, the LOCAL AUTHORITY must send the SECRETARY OF STATE a statement setting out the reasons why it reached its decision (*section 78D(1)*). It must also notify in writing the persons identified in paragraph 18.7 above of the fact that it has referred its decision to the SECRETARY OF STATE (*section 78D(3)*).

18.18 The SECRETARY OF STATE then decides whether he considers that all, or part, of the CONTAMINATED LAND in question meets one or more of the descriptions prescribed in the Regulations as being required to be designated a SPECIAL SITE. If he decides that some land should be designated, then it is so designated (*section 78D(4)(a)*).

18.19 The SECRETARY OF STATE is under a duty to notify in writing the LOCAL AUTHORITY and the persons identified in paragraph 18.7 above of his decision (*section 78D(4)(b)*).

The actual designation as a special site

18.20 In any case where the LOCAL AUTHORITY'S decision that land should be designated a SPECIAL SITE has not been referred to the SECRETARY OF STATE, the notification it gives of that decision takes effect as the designation on the following basis:

(a) if the ENVIRONMENT AGENCY notifies the LOCAL AUTHORITY that it agrees with its decision, the designation takes effect on the day after that notification; or

(b) if no such notification is given, the designation takes effect on the day after a period of twenty-one days has elapsed since the LOCAL AUTHORITY notified the ENVIRONMENT AGENCY of its original decision (*section 78C(6)*).

18.21 Where a designation takes effect in this way, the LOCAL AUTHORITY must notify in writing the same categories of person as it notified of its original decision (*section 78C(6)*). It must also enter the relevant particulars on its REGISTER (*section 78R(1)(e)*; see paragraphs 17.1 to 17.19 above).

18.22 In any case where a decision has been referred to the SECRETARY OF STATE, and he decides that some CONTAMINATED LAND should be designated a SPECIAL SITE, the notice he gives of this decision to the LOCAL AUTHORITY and the persons identified in paragraph 18.7 above serves as the actual designation. The designation takes effect on the day after he gives the notification (*sections 78D(5) & (6)*). The LOCAL AUTHORITY and the ENVIRONMENT AGENCY must enter the relevant particulars of the SECRETARY OF STATE'S notification onto their respective REGISTERS (see paragraphs 17.1 to 17.19 above).

Remediation of special sites

18.23 In general, the procedures relating to the REMEDIATION of a SPECIAL SITE are the same as for any other CONTAMINATED LAND, with the exception that the ENVIRONMENT AGENCY is the ENFORCING AUTHORITY, rather than the LOCAL AUTHORITY. In particular, the ENVIRONMENT AGENCY is required to have regard to the statutory guidance on remediation *(Chapter C)* and the recovery of costs *(Chapter E)*, and to act in accordance with the statutory guidance on EXCLUSIONS from, and APPORTIONMENT of, liability *(Chapter D)*.

18.24 In some cases the designation of a SPECIAL SITE may be made after a REMEDIATION NOTICE has been served or after the LOCAL AUTHORITY has started carrying out REMEDIATION itself.

18.25 If a REMEDIATION NOTICE has already been served, the ENVIRONMENT AGENCY needs to decide whether or not to adopt the existing REMEDIATION NOTICE (*section 78Q(1)*). For example, it may consider that:

(a) the REMEDIATION ACTIONS specified in the existing notice are still appropriate;

(b) those REMEDIATION ACTIONS should not be carried out; or

(c) additional, or alternative, REMEDIATION ACTIONS should be carried out.

18.26 If the ENVIRONMENT AGENCY decides to adopt the REMEDIATION NOTICE, it must notify in writing the LOCAL AUTHORITY which originally served the notice, and the person or persons on whom the notice was served (*section 78Q(1)(a)*). The notice then has effect as if it had been given by the Agency (*section 78Q(1)(b)*). It is also good practice to send a copy of such a notification to anyone else to whom a copy of the original REMEDIATION NOTICE was sent (*regulation 5*).

18.27 The adoption of a REMEDIATION NOTICE by the ENVIRONMENT AGENCY means that the Agency has the power to enforce it, bringing a prosecution and carrying out the REMEDIATION itself if the notice is not complied with.

18.28 If the ENVIRONMENT AGENCY does not adopt a REMEDIATION NOTICE, that notice ceases to have effect, and the person on whom it was served is no longer obliged to comply with its requirements. But the ENVIRONMENT AGENCY then needs to decide whether it is required to serve a further REMEDIATION NOTICE. In doing so, it must consult in the same manner as would a LOCAL AUTHORITY for any CONTAMINATED LAND which is not a SPECIAL SITE. Except where urgency is involved, the ENVIRONMENT AGENCY is prevented from serving any REMEDIATION NOTICE until three months have elapsed since the LOCAL AUTHORITY, or the Secretary of State, gave notification that the land was designated a SPECIAL SITE (*sections 78H(3)(b) & (c)*).

18.29 In any case where the ENVIRONMENT AGENCY does not adopt a REMEDIATION NOTICE, it is good practice for the Agency to notify the LOCAL AUTHORITY which originally served the notice, any person on whom the notice was served and anyone else to whom a copy of the notice was sent.

18.30 If the LOCAL AUTHORITY has begun to carry out any REMEDIATION itself before the land is designated a SPECIAL SITE, the LOCAL AUTHORITY needs to decide whether to continue carrying out that REMEDIATION (*section 78Q(2)(a)*). Whatever it decides, it is entitled to recover the reasonable costs it incurs, or has already incurred, in carrying out the REMEDIATION, even though it is no longer the ENFORCING AUTHORITY (*section 78Q(2)(b)*).

18.31 As an ENFORCING AUTHORITY, the ENVIRONMENT AGENCY is under a duty to maintain a REGISTER (*section 78R(1)*), with an entry for each SPECIAL SITE. Each time it enters any particulars onto its REGISTER, the ENVIRONMENT AGENCY must send a copy of those particulars to the LOCAL AUTHORITY in whose area the land is situated (*section 78R(4)*; see paragraphs 17.4 to 17.6 above). The LOCAL AUTHORITY then must enter those particulars onto its own REGISTER (section 78R(6)).

Termination of a designation

18.32 The ENVIRONMENT AGENCY can inspect the SPECIAL SITE from time to time, in order to keep its condition under review (*section 78Q(3)*). In particular, the ENVIRONMENT AGENCY needs to consider whether the land still meets one or more of the descriptions of land prescribed in the Regulations.

18.33 If it decides that the land no longer meets one or more of those descriptions, it must also decide whether it wishes to terminate that land's designation as a SPECIAL SITE. It is not obliged to terminate the designation as soon as the land ceases to meet any of the descriptions of land prescribed in the Regulations (*section 78Q(4)*). It may choose, for example, to wait until REMEDIATION has been completed on the land.

18.34 If the ENVIRONMENT AGENCY decides to terminate any designation, it must notify in writing the SECRETARY OF STATE and the LOCAL AUTHORITY in whose area the land is situated. The termination takes effect from whatever date is specified by the ENVIRONMENT AGENCY (*section 78Q(4)*). Both the ENVIRONMENT AGENCY and the LOCAL AUTHORITY then need to enter particulars of this notification onto their respective REGISTERS (*section 78R(1)(g)*). It is also good practice to notify everyone else who was notified of the original designation of the land as a SPECIAL SITE (see paragraph 18.7 above).

Annex 3 Chapter A – Statutory guidance on the definition of contaminated land

Part 1 Scope of the Chapter

A.1 The statutory guidance in this Chapter is issued under section 78A(2), (5) and (6) of Part IIA of the Environmental Protection Act 1990 and provides guidance on applying the definition of contaminated land.

A.2 "Contaminated land" is defined at section 78A(2) as:

"any land which appears to the local authority in whose area it is situated to be in such a condition, by reason of substances in, on or under the land, that–

(a) significant harm is being caused or there is a significant possibility of such harm being caused; or
(b) pollution of controlled waters is being, or is likely to be caused; . . ."

A.3 Section 78A(5) further provides that:
"the questions—
(a) what harm is to be regarded as "significant"
(b) whether the possibility of significant harm being caused is "significant"
(c) whether pollution of controlled waters is being, or is likely to be caused,

"shall be determined in accordance with guidance issued . . . by the Secretary of State".

A.4 In determining these questions the local authority is therefore required to act in accordance with the guidance contained in this Chapter.

A.5 As well as defining contaminated land, section 78A(2) further provides that:

" . . . in determining whether any land appears to be such land, a local authority shall . . . act in accordance with guidance issued by the Secretary of State . . . with respect to the manner in which that determination is to be made"

A.6 Guidance on the manner in which that determination is to be made is set out in Part 3 of the statutory guidance in Chapter B.

Part 2 Definitions of terms and general material

A.7 Unless otherwise stated, any word, term or phrase given a specific meaning in Part IIA of the Environmental Protection Act 1990 has the same meaning for the purposes of the guidance in this Chapter.

A.8 Any reference to "Part IIA" means "Part IIA of the Environmental Protection Act 1990". Any reference to a "section" in primary legislation means a section of the Environmental Protection Act 1990, unless it is specifically stated otherwise.

Risk Assessment

A.9 The definition of contaminated land is based upon the principles of risk assessment. For the purposes of this guidance, "risk" is defined as the combination of:

(a) the probability, or frequency, of occurrence of a defined hazard (for example, exposure to a property of a substance with the potential to cause harm); and
(b) the magnitude (including the seriousness) of the consequences.

A.10 The guidance below follows established approaches to risk assessment, including the concept of contaminant-pathway-receptor. (In the technical literature, this is sometimes referred to as source-pathway-target.)

A.11 There are two steps in applying the definition of contaminated land. The first step is for the local authority to satisfy itself that a "contaminant", a "pathway" (or pathways), and a "receptor" have been identified with respect to that land. These three concepts are defined for the purposes of this Chapter in paragraphs A.12 to A.14 below.

A.12 A contaminant is a substance which is in, on or under the land and which has the potential to cause harm or to cause pollution of controlled waters.

A.13 A receptor is either:

(a) a living organism, a group of living organisms, an ecological system or a piece of property which

(i) is in a category listed in Table A (see below) as a type of receptor, and

(ii) is being, or could be, harmed, by a contaminant; or

(b) controlled waters which are being, or could be, polluted by a contaminant.

A.14 A pathway is one or more routes or means by, or through, which a receptor:

(a) is being exposed to, or affected by, a contaminant, or

(b) could be so exposed or affected.

A.15 It is possible for a pathway to be identified for this purpose on the basis of a reasonable assessment of the general scientific knowledge about the nature of a particular contaminant and of the circumstances of the land in question. Direct observation of the pathway is not necessary.

A.16 The identification of each of these three elements is linked to the identification of the others. A pathway can only be identified if it is capable of exposing an identified receptor to an identified contaminant. That particular contaminant should likewise be capable of harming or, in the case of controlled waters, be capable of polluting that particular receptor.

A.17 In this Chapter, a "pollutant linkage" means the relationship between a contaminant, a pathway and a receptor, and a "pollutant" means the contaminant in a pollutant linkage. Unless all three elements of a pollutant linkage are identified in respect of a piece of land, that land should not be identified as contaminated land. There may be more than one pollutant linkage on any given piece of land.

A.18 For the purposes of determining whether a pollutant linkage exists (and for describing any such linkage), the local authority may treat two or more substances as being a single substance, in any case where:

(a) the substances are compounds of the same element, or have similar molecular structures; and

(b) it is the presence of that element, or the particular type of molecular structures, that determines the effect that the substances may have on the receptor which forms part of the pollutant linkage.

A.19 The second step in applying the definition of contaminated land is for the local authority to satisfy itself that both:

(a) such a pollutant linkage exists in respect of a piece of land; and

(b) that pollutant linkage:

(i) is resulting in significant harm being caused to the receptor in the pollutant linkage,

(ii) presents a significant possibility of significant harm being caused to that receptor,

(iii) is resulting in the pollution of the controlled waters which constitute the receptor, or
(iv) is likely to result in such pollution.

A.20 In this Chapter, a "significant pollutant linkage" means a pollutant linkage which forms the basis for a determination that a piece of land is contaminated land. A "significant pollutant" is a pollutant in a "significant pollutant linkage".

A.21 The guidance in Part 3 below relates to questions about significant harm and the significant possibility of such harm being caused. The guidance in Part 4 below relates to the pollution of controlled waters.

Part 3 Significant harm and the significant possibility of significant harm

A.22 Section 78A(4) defines "harm" as meaning "harm to the health of living organisms or other interference with the ecological systems of which they form part and, in the case of man, includes harm to his property". Section 78A(5) provides that what harm is to be regarded as "significant" and whether the possibility of significant harm being caused is significant shall be determined in accordance with this guidance.

What Harm is to be regarded as "significant"

A.23 The local authority should regard as significant only harm which is both:
(a) to a receptor of a type listed in Table A, and
(b) within the description of harm specified for that type of receptor in that Table.

A.24 The local authority should not regard harm to receptors of any type other than those mentioned in Table A as being significant harm for the purposes of Part IIA. For example, harm to ecological systems outside the descriptions in the second entry in the table should be disregarded. Similarly, the authority should not regard any other description of harm to receptors of the types mentioned in Table A as being significant harm.

A.25 The authority should disregard any receptors which are not likely to be present, given the "current use" of the land or other land which might be affected.

A.26 For the purposes of this guidance, the "current use" means any use which is currently being made, or is likely to be made, of the land and which is consistent with any existing planning permission (or is otherwise lawful under town and country planning legislation). This definition is subject to the following qualifications:
(a) the current use should be taken to include any temporary use, permitted under town and country planning legislation, to which the land is, or is likely to be, put from time to time;
(b) the current use includes future uses or developments which do not require a new, or amended, grant of planning permission (but see also paragraph A.34 below);
(c) the current use should, nevertheless, be taken to include any likely informal recreational use of the land, whether authorised by the owners or occupiers or not, (for example, children playing on the land); however, in assessing the likelihood of any such informal use, the local authority should give due attention to measures taken to prevent or restrict access to the land; and
(d) in the case of agricultural land, however, the current agricultural use should not be taken to extend beyond the growing or rearing of the crops or animals which are habitually grown or reared on the land.

Table A—Categories of Significant Harm

Type of Receptor	**Description of harm to that type of receptor that is to be regarded as significant harm**
1 Human beings	Death, disease, serious injury, genetic mutation, birth defects or the impairment of reproductive functions. For these purposes, disease is to be taken to mean an unhealthy condition of the body or a part of it and can include, for example, cancer, liver dysfunction or extensive skin ailments. Mental dysfunction is included only insofar as it is attributable to the effects of a pollutant on the body of the person concerned. In this Chapter, this description of significant harm is referred to as a "human health effect".
2 Any ecological system, or living organism forming part of such a system, within a location which is: • an area notified as an area of special scientific interest under section 28 of the Wildlife and Countryside Act 1981; • any land declared a national nature reserve under section 35 of that Act; • any area designated as a marine nature reserve under section 36 of that Act; • an area of special protection for birds, established under section 3 of that Act; • any European Site within the meaning of regulation 10 of the Conservation (Natural Habitats etc) Regulations 1994 (*i.e.* Special Areas of Conservation and Special Protection Areas); • any candidate Special Areas of Conservation or potential Special Protection Areas given equivalent protection; • any habitat or site afforded policy protection under paragraph 13 of Planning Policy Guidance Note 9 (PPG9) on nature conservation (*i.e.* candidate Special Areas of Conservation, potential Special Protection Areas and listed Ramsar sites); or • any nature reserve established under section 21 of the National Parks and Access to the Countryside Act 1949.	For <u>any</u> protected location: • harm which results in an irreversible adverse change, or in some other substantial adverse change, in the functioning of the ecological system within any substantial part of that location; or • harm which affects any species of special interest within that location and which endangers the long-term maintenance of the population of that species at that location. In addition, in the case of a protected location which is a European Site (or a candidate Special Area of Conservation or a potential Special Protection Area), harm which is incompatible with the favourable conservation status of natural habitats at that location or species typically found there. In determining what constitutes such harm, the local authority should have regard to the advice of English Nature and to the requirements of the Conservation (Natural Habitats etc.) Regulations 1994. In this Chapter, this description of significant harm is referred to as an "ecological system effect".

Table A—Categories of Significant Harm—*continued*

Type of Receptor	Description of harm to that type of receptor that is to be regarded as significant harm
3 Property in the form of: • crops, including timber; • produce grown domestically, or on allotments, for consumption; • livestock; • other owned or domesticated animals; • wild animals which are the subject of shooting or fishing rights.	For crops, a substantial diminution in yield or other substantial loss in their value resulting from death, disease or other physical damage. For domestic pets, death, serious disease or serious physical damage. For other property in this category, a substantial loss in its value resulting from death, disease or other serious physical damage. The local authority should regard a substantial loss in value as occurring only when a substantial proportion of the animals or crops are dead or otherwise no longer fit for their intended purpose. Food should be regarded as being no longer fit for purpose when it fails to comply with the provisions of the Food Safety Act 1990. Where a diminution in yield or loss in value is caused by a pollutant linkage, a 20% diminution or loss should be regarded as a benchmark for what constitutes a substantial diminution or loss. In this Chapter, this description of significant harm is referred to as an "animal or crop effect".
4 Property in the form of buildings. For this purpose, "building" means any structure or erection, and any part of a building including any part below ground level, but does not include plant or machinery comprised in a building.	Structural failure, substantial damage or substantial interference with any right of occupation. For this purpose, the local authority should regard substantial damage or substantial interference as occurring when any part of the building ceases to be capable of being used for the purpose for which it is or was intended. Additionally, in the case of a scheduled Ancient Monument, substantial damage should be regarded as occurring when the damage significantly impairs the historic, architectural, traditional, artistic or archaeological interest by reason of which the monument was scheduled. In this Chapter, this description of significant harm is referred to as a "building effect".

Whether the possibility of significant harm being caused in significant

A.27 As stated in paragraph A.9 above, the guidance on determining whether a particular possibility is significant is based on the principles of risk assessment, and in particular on considerations of the magnitude or consequences of the different types of significant harm caused. The term "possibility of significant harm being caused" should be taken as referring to a measure of the probability, or frequency, of the occurrence of circumstances which would lead to significant harm being caused.

A.28 The local authority should take into account the following factors in deciding whether the possibility of significant harm being caused is significant:

(a) the nature and degree of harm;
(b) the susceptibility of the receptors to which the harm might be caused; and
(c) the timescale within which the harm might occur.

A.29 In considering the timescale, the authority should take into account any evidence that the current use of the land (as defined in paragraphs A.25 and A.26 above) will cease in the foreseeable future.

A.30 The local authority should regard as a significant possibility any possibility of significant harm which meets the conditions set out in Table B for the description of significant harm under consideration.

A.31 In Table B, references to "relevant information" mean information which is:

(a) scientifically based;
(b) authoritative;
(c) relevant to the assessment of risks arising from the presence of contaminants in soil; and
(d) appropriate to the determination of whether any land is contaminated land for the purposes of Part IIA, in that the use of the information is consistent with providing a level of protection of risk in line with the qualitative criteria set out in Tables A and B.

A.32 In general, when considering significant harm to non-human receptors, the local authority should apply the tests set out in the relevant entries in Table B to determine whether there is a significant possibility of that harm being caused. However, the local authority may also determine that there is a significant possibility of significant harm with respect to a non-human receptor in any case where the conditions in the third, fourth and fifth entries in Table B are not met, but where:

(a) there is a reasonable possibility of significant human being caused; and
(b) that harm would result from either:

(i) a single incident such as a fire or explosion, or
(ii) a short-term (that is, less than 24 hour) exposure of the receptor to the pollutant.

Table B—Significant Possibility of Significant Harm	
Descriptions Of Significant Harm (As Defined in Table A)	**Conditions For There Being A Significant Possibility Of Significant Harm**
1 Human health effects arising from • the intake of a contaminant, or • other direct bodily contact with a contaminant.	If the amount of the pollutant in the pollutant linkage in question: • which a human receptor in that linkage might take in, or • to which such a human might otherwise be exposed, as a result of the pathway in that linkage, would represent an unacceptable intake or direct bodily contact, assessed on the basis of relevant information on the toxicological properties of that pollutant. Such an assessment should take into account. • the likely total intake of, or exposure to, the substance or substances which form the pollutant, from all sources including that from the pollutant linkage in question; • the relative contribution of the pollutant linkage in question to the likely aggregate intake of, or exposure to, the relevant substance or substances, and • the duration of intake or exposure resulting from the pollutant linkage in question. The question of whether an intake or exposure is unacceptable is independent of the number of people who might experience or be affected by that intake or exposure. Toxicological properties should be taken to include carcinogenic, mutagenic, teratogenic, pathogenic, endocrine-disrupting and other similar properties.
2 All other human health effects (particularly by way of explosion or fire).	If the probability, or frequency, of occurrence of significant harm of that description is unacceptable, assessed on the basis of relevant information concerning: • that type of pollutant linkage, or • that type of significant harm arising from other causes. In making such an assessment, the local authority should take into account the levels of risk which have been judged unacceptable in other similar contexts and should give particular weight to cases where the pollutant linkage might cause significant harm which: • would be irreversible or incapable of being treated; • would affect a substantial number of people; • would result from a single incident such as a fire or an explosion; or • would be likely to result from a short-term (that is, less than 24-hour) exposure to the pollutant.

Table B—Significant Possibility of Significant Harm—*continued*	
Descriptions Of Significant Harm (As Defined in Table A)	**Conditions For There Being A Significant Possibility Of Significant Harm**
3. All ecological system affects.	If either: ● significant harm of that description is more likely than not to result from the pollutant linkage in question; or ● there is a reasonable possibility of significant harm of that description being caused, and if that harm were to occur, it would result in such a degree of damage to features of special interest at the location in question that they would be beyond any practicable possibility of restoration. Any assessment made for these purposes should take into account relevant information for that type of pollutant linkage, particularly in relation to the ecotoxicological effects of the pollutant.
4 All animal and crop effects.	If significant harm of that description is more likely than not to result from the pollutant linkage in question, taking into account relevant information for that type of pollutant linkage, particularly in relation to the ecotoxicological effects of the pollutant.
5. All building effects.	If significant harm of that description is more likely than not to result from the pollutant linkage in question during the expected economic life of the building (or, in the case of a scheduled Ancient Monument, the forseeable future), taking into account relevant information for that type of pollutant linkage.

A.33 The possibility of significant harm being caused as a result of any change of use of any land to one which is not a current use of that land (as defined in paragraph A.26 above) should not be regarded as a significant possibility for the purposes of this Chapter.

A.34 When considering the possibility of significant harm being caused in relation to any future use or development which falls within the description of a "current use" as a result of paragraph A.26(b) above, the local authority should assume that if the future use is introduced, or the development carried out, this will be done in accordance with any existing planning permission for that use or development. In particular, the local authority should assume:

(a) that any remediation which is the subject of a condition attached to that planning permission, or is the subject of any planning obligation, will be carried out in accordance with that permission or obligation; and

(b) where a planning permission has been given subject to conditions which require steps to be taken to prevent problems which might be caused by contamination, and those steps are to be approved by the local planning authority, that the local planning authority will ensure that those steps include adequate remediation.

Part 4 The pollution of controlled waters

A.35 Section 78A(9) defines the pollution of controlled waters as:

"the entry into controlled waters of any poisonous, noxious or poluting matter or any solid waste matter".

A.36 Before determining that pollution of controlled waters is being, or is likely to be, caused, the local authority should be satisfied that a substance is continuing to enter controlled waters or is likely to enter controlled waters. For this purpose, the local authority should regard something as being "likely" when they judge it more likely than not to occur.

A.37 Land should not be designated as contaminated land where:

(a) a substance is already present in controlled waters;
(b) entry into controlled waters of that substance from land has ceased; and
(c) it is not likely that further entry will take place.

A.38 Substances should be regarded as having entered controlled waters where:

(a) they are dissolved or suspended in those waters; or
(b) if they are immiscible with water, they have direct contact with those waters on or beneath the surface of the water.

A.39 The term "continuing to enter" should be taken to mean any entry additional to any which has already occurred.

CHAPTER B – STATUTORY GUIDANCE ON THE IDENTIFICATION OF CONTAMINATED LAND

Part 1 Scope of the Chapter

B.1 The statutory guidance in this Chapter is issued under sections 78A(2) and 78B(2) of Part IIA of the Environmental Protection Act 1990, and provides guidance on the inspection of its area by a local authority and the manner in which an authority is to determine whether any land appears to it to be contaminated land.

B.2 Section 78B(1) provides that:

"Every local authority shall cause its area to be inspected from time to time for the purpose.

(a) of identifying contaminated land; and
(b) of enabling the authority to decide whether any such land is land which is required to be designated as a special site."

B.3 Section 78B(2) further provides that:

"In performing [these] functions . . . a local authority shall act in accordance with any guidance issued for the purpose by the Secretary of State."

B.4 Section 78A(2) also provides that:

"'Contaminated land' is any land which appears to the local authority in whose area it is situated to be in such a condition, by reason of substances in, on or under the land, that—

(a) significant harm is being caused or there is a significant possibility of such harm being caused, or
(b) pollution of controlled waters is being, or is likely to be, caused;

and, in determining whether any land appears to be such land, a local authority shall, . . . act in accordance with guidance issued by the Secretary of State . . . with respect to the manner in which that determination is to be made."

B.5 The local authority is therefore required to act in accordance with the statutory guidance contained in this Chapter.

B.6 The questions of what harm is to be regarded as significant, whether the possibility of significant harm being caused is significant, and whether pollution of controlled waters is being or is likely to be caused are to be determined in accordance with guidance contained in Chapter A.

Part 2 Definitions of terms

B.7 Unless otherwise stated, any word, term or phrase given a specific meaning in Part IIA of the Environmental Protection Act 1990, or in the guidance at Chapter A, has the same meaning for the purposes of the guidance in this Chapter.

B.8 Any reference to "Part IIA" means "Part IIA of the Environmental Protection Act 1990". Any reference to a "section" in primary legislation means a section of the Environmental Protection Act 1990, unless it is specifically stated otherwise.

Part 3 The local authority's inspection duty

Strategic Approach to Inspection

B.9 In carrying out its inspection duty under section 78B(1), the local authority should take a strategic approach to the identification of land which merits detailed individual inspection. This approach should:

(a) be rational, ordered and efficient;
(b) be proportionate to the seriousness of any actual or potential risk;
(c) seek to ensure that the most pressing and serious problems are located first;
(d) ensure that resources are concentrated on investigating in areas where the authority is most likely to identify contaminated land; and
(e) ensure that the local authority efficiently identifies requirements for the detailed inspection of particular areas of land.

B.10 In developing this strategic approach the local authority should reflect local circumstances. In particular it should consider:

(a) any available evidence that significant harm or pollution of controlled waters is actually being caused;
(b) the extent to which any receptor (which is either of a type listed in Table A in Chapter A or is controlled waters) is likely to be found in any of the different parts of the authority's area;
(c) the extent to which any of those receptors is likely to be exposed to a contaminant (as defined in Chapter A), for example as a result of

the use of the land or of the geological and hydrogeological features of the area;

(d) the extent to which information on land contamination is already available;

(e) the history, scale and nature of industrial or other activities which may have contaminated the land in different parts of its area;

(f) the nature and timing of past redevelopment in different parts of its area;

(g) the extent to which remedial action has already been taken by the authority or others to deal with land-contamination problems or is likely to be taken as part of an impending redevelopment; and

(h) the extent to which other regulatory authorities are likely to be considering the possibility of harm being caused to particular receptors or the likelihood of any pollution of controlled waters being caused in particular parts of the local authority's area.

B.11 In developing its strategic approach, the local authority should consult the Environment Agency and other appropriate public authorities, such as the county council (where one exists), statutory regeneration bodies, English Nature, English Heritage and the Ministry of Agriculture, Fisheries and Food.

B.12 The local authority should set out its approach as a written strategy, which it should formally adopt and publish. This strategy should be published within 15 months of the issue of this guidance. As soon as its strategy is published, the local authority should send a copy to the Environment Agency.

B.13 The local authority should keep its strategy under periodic review.

B.14 The local authority should not await the publication of its strategy before commencing more detailed work investigating particular areas of land, where this appears necessary.

Contents of the Strategy

B.15 Strategies are likely to vary both between local authorities and between different parts of an authority's area, reflecting the different problems associated with land contamination in different areas. The local authority should include in its strategy:

(a) a description of the particular characteristics of its area and how that influences its approach;

(b) the authority's particular aims, objectives and priorities;

(c) appropriate timescales for the inspection of different parts of its area; and

(d) arrangements and procedures for:

(i) considering land for which it may itself have responsibilities by virtue of its current or former ownership or occupation,

(ii) obtaining and evaluating information on actual harm, or pollution of controlled waters,

(iii) identifying receptors, and assessing the possibility or likelihood that they are being, or could be, exposed to or affected by a contaminant,

(iv) obtaining and evaluating existing information on the possible presence of contaminants and their effects,

(v) liaison with, and responding to information from, other statutory bodies, including, in particular, the Environment Agency, English Nature and the Ministry of Agriculture, Fisheries and Food (see paragraphs B.16 and B.17 below),

(vi) liaison with, and responding to information from, the owners or occupiers of land, and other relevant interested parties,
(vii) responding to information or complaints from members of the public, businesses and voluntary organisations,
(viii) planning and reviewing a programme for inspecting particular areas of land,
(ix) carrying out the detailed inspection of particular areas of land,
(x) reviewing and updating assumptions and information previously used to assess the need for detailed inspection of different areas, and managing new information, and
(xi) managing information obtained and held in the course of carrying out its inspection duties.

Information from other statutory bodies

B.16 Other regulatory authorities may be able to provide information relevant to the identification of land as contaminated land, as a result of their various complementary functions. The local authority should seek to make specific arrangements with such other bodies to avoid unnecessary duplication in investigation.

B.17 For example, the Environment Agency has general responsibilities for the protection of the water environment. It monitors the quality of controlled waters and in doing so may discover land which would appropriately be identified as contaminated land by reason of pollution of controlled waters which is being, or is likely to be, caused.

Inspecting particular areas of land

B.18 Applying the strategy will result in the identification of particular areas of land where it is possible that a pollutant linkage exists. Subject to the guidance in paragraphs B.22 to B.25 and B.27 to B.30 below, the local authority should carry out a detailed inspection of any such area to obtain sufficient information for the authority:

(a) to determine, in accordance with the guidance on the manner of determination in Part 4 below, whether that land appears to be contaminated land; and
(b) to decide whether any such land falls within the definition of a special site prescribed in regulations 2 and 3 of the Contaminated Land (England) Regulations 2000, and is therefore required to be designated as a special site.

B.19 To be sufficient for the first of these purposes the information should include, in particular, evidence of the actual presence of a pollutant.

B.20 Detailed inspection may include any or all of the following:

(a) the collation and assessment of documentary information, or other information from other bodies;
(b) a visit to the particular area for the purposes of visual inspection and, in some cases, limited sampling (for example of surface deposits); or
(c) intrusive investigation of the land (for example by exploratory excavations).

B.21 Section 108 of the Environment Act 1995 gives the local authority the power to authorise a person to exercise specific powers of entry. For the purposes of this Chapter, any detailed inspection of land carried out through use of this power by the local authority is referred to as an "inspection using statutory powers of entry".

B.22 Before the local authority carries out an inspection using statutory powers of entry, it should be satisfied, on the basis of any information already obtained:

(a) in all cases, that there is a reasonable possibility that a pollutant linkage (as defined in Chapter A) exists on the land; this implies that not only must the authority be satisfied that there is a reasonable possibility of the presence of a contaminant, a receptor and a pathway, but also that these would together create a pollutant linkage; and

(b) further, in cases involving an intrusive investigation,

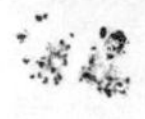

(i) that it is likely that the contaminant is actually present, and

(ii) given the current use of the land as defined at paragraph A.26, that the receptor is actually present or is likely to be present.

B.23 The local authority should not carry out any inspection using statutory powers of entry which takes the form of intrusive investigation if:

(a) it has already been provided with detailed information on the condition of the land, whether by the Environment Agency or some other person such as the owner of the land, which provides an appropriate basis upon which the local authority can determine whether the land is contaminated land in accordance with the requirements of the guidance in this Chapter; or

(b) a person offers to provide such information within a reasonable and specified time, and then provides such information within that time.

B.24 The local authority should carry out any intrusive investigation in accordance with appropriate technical procedures for such investigations. It should also ensure that it takes all reasonable precautions to avoid harm, water pollution or damage to natural resources or features of historical or archaeological interest which might be caused as a result of its investigation. Before carrying out any intrusive investigation on any area notified as an area of special scientific interest (SSSI), the local authority should consult English Nature on any action which, if carried out by the owner or occupier, would require the consent of English Nature under section 28 of the Wildlife and Countryside Act 1981.

B.25 If at any stage, the local authority considers, on the basis of information obtained from a detailed inspection, that there is no longer a reasonable possibility that a particular pollutant linkage exists on the land, the authority should not carry out any further detailed inspection for that pollutant linkage.

Land which may be a special site

B.26 If land has been determined to be contaminated land and it also falls within one or more of the "special sites" descriptions prescribed in regulations made under Part IIA, it is required to be designated as a special site. The Environment Agency then becomes the enforcing authority for that land. It is therefore helpful for the Environment Agency to have a formal role at the inspection stage for any such land.

B.27 Before authorising or carrying out on any land an inspection using statutory powers of entry, the local authority should consider whether, if that land were found to be contaminated land, it would meet any of the descriptions of land prescribed in the Regulations as requiring to be designated a special site.

B.28 If the local authority already has information that this would be the case, the authority should always seek to make arrangements with the Environment Agency for that Agency to carry out the inspection of the land on behalf of the local authority. This might occur, for example, where the prescribed description of land in the Regulations relates to its current or former use, such as land on which a process designated for central control under the Integrated Pollution Control regime has been carried out, or land which is occupied by the Ministry of Defence.

B.29 If the local authority considers that there is a reasonable possibility that a particular pollutant linkage is present, and the presence of a linkage of that kind would require the designation of the land as a special site (were that linkage found to be a significant pollutant linkage), the authority should seek to make arrangements with the Environment Agency for the Agency to carry out the inspection of the land. An example of this kind of pollutant linkage would be the pollution of waters in the circumstances described in regulation 3(b) of the Contaminated Land (England) Regulations 2000.

B.30 Where the Environment Agency is to carry out an inspection on behalf of the local authority, the authority should, where necessary, authorise a person nominated by the Agency to exercise the powers of entry conferred by section 108 of the Environment Act 1995. Before the local authority gives such an authorisation, the Environment Agency should satisfy the local authority that the conditions for the use of the statutory powers of entry set out in paragraphs B.22 to B.25 above are met.

Part 4 Determining whether land appears to be contaminated land

B.31 The local authority has the sole responsibility for determining whether any land appears to be contaminated land. It cannot delegate this responsibility (except in accordance with section 101 of the Local Government Act 1972), although in discharging it the local authority can choose to rely on information or advice provided by another body such as the Environment Agency, or by a consultant appointed for that purpose. This applies even where the Agency has carried out the inspection of land on behalf of the local authority (see paragraphs B.26 to B.30 above).

Physical extent of land

B.32 A determination that land is contaminated land is necessarily made in respect of a specific area of land. In deciding what that area should be, the primary consideration is the extent of the land which is contaminated land. However, there may be situations in which the local authority may consider that separate designations of parts of a larger area of contaminated land may simplify the administration of the consequential actions. In such circumstances, the local authority should do so, taking into account:

(a) the location of significant pollutants in, on or under the land;
(b) the nature of the remediation which might be required; and
(c) the likely identity of those who may be the appropriate persons to bear responsibility for the remediation (where this is reasonably clear at this stage).

B.33 If necessary, the local authority should initially review a wider area, the history of which suggests that contamination problems are likely. It can subsequently refine this down to the precise areas which meet the statutory tests for identification as contaminated land, and use these as the basis for its determination.

B.34 In practice, the land to be covered by a single determination is likely to be the smallest area which is covered by a single remediation action which cannot sensibly be broken down into smaller actions. Subject to this, the land is likely to be the smaller of:

(a) the plots which are separately recorded in the Land Register or are in separate ownership or occupation; and
(b) the area of land in which the presence of significant pollutants has been established.

B.35 The determination should identify the area of contaminated land clearly, including reference to a map or plan at an appropriate scale.

B.36 The local authority should also be prepared to review the decision on the physical extent of the land to be identified in the light of further information.

Making the determination

B.37 In determining whether any land appears to the local authority to be contaminated land, the authority is required to act in accordance with the guidance on the definition of contaminated land set out in Chapter A. Guidance on the manner in which the local authority should determine whether land appears to it to be contaminated land, by reason of substances in, on or under the land, is set out in paragraphs B.39 to B.51 below.

B.38 There are four possible grounds for the determination, (corresponding to the parts of the definition of contaminated land in section 78A(2)) namely that:

(a) significant harm is being caused (see paragraph B.44 below);
(b) there is a significant possibility of significant harm being caused (see paragraphs B.45 to B.49 below);
(c) pollution of controlled waters is being caused (see paragraph B.50 below); or
(d) pollution of controlled waters is likely to be caused (see paragraph B.51 below).

B.39 In making any determination, the local authority should take all relevant and available evidence into account and carry out an appropriate scientific and technical assessment of that evidence.

B.40 The local authority should identify a particular pollutant linkage or linkages (as defined in Chapter A) as the basis for the determination. All three elements of any pollutant linkage (pollutant, pathway and receptor) should be identified. A linkage which forms a basis for the determination that land is contaminated land is then a "significant pollutant linkage"; and any pollutant which forms part of it is a "significant pollutant".

B.41 The local authority should consider whether:

(a) there is evidence that additive or synergistic effects between potential pollutants, whether between the same substance on different areas of land or between different substances, may result in a significant pollutant linkage;
(b) a combination of several different potential pathways linking one or more potential pollutants to a particular receptor, or to a particular class of receptors, may result in a significant pollutant linkage; and
(c) there is more than one significant pollutant linkage on any land; if there are, each should be considered separately, since different people may be responsible for the remediation.

Consistency with other statutory bodies

B.42 In making a determination which relates to an "ecological system effect" as defined in Table A of Chapter A, the local authority should adopt an approach consistent with that adopted by English Nature. To this end, the local authority should consult that authority and have regard to its comments in making its determination.

B.43 In making a determination which relates to pollution of controlled waters the local authority should adopt an approach consistent with that adopted by the Environment Agency in applying relevant statutory provisions. To this end, where the local authority is considering whether pollution of controlled waters is being or is likely to be caused, it should consult the Environment Agency and have regard to its comments before determining whether pollution of controlled waters is being or is likely to be caused.

Determining that "significant harm is being caused"

B.44 The local authority should determine that land is contaminated land on the basis that significant harm is being caused where:

(a) it has carried out an appropriate scientific and technical assessment of all the relevant and available evidence; and
(b) on the basis of that assessment, it is satisfied on the balance of probabilities that significant harm is being caused.

Determining that "there is a significant possibility of significant harm being caused"

B.45 The local authority should determine that land is contaminated land on the basis that there is a significant possibility of significant harm being caused (as defined in Chapter A), where:

(a) it has carried out a scientific and technical assessment of the risks arising from the pollutant linkage, according to relevant, appropriate, authoritative and scientifically based guidance on such risk assessments;
(b) that assessment shows that there is a significant possibility of significant harm being caused; and
(c) there are no suitable and sufficient risk management arrangements in place to prevent such harm.

B.46 In following any such guidance on risk assessment, the local authority should be satisfied that it is relevant to the circumstances of the pollutant linkage and land in question, and that any appropriate allowances have been made for particular circumstances.

B.47 To simplify such an assessment of risks, the local authority may use authoritative and scientifically based guideline values for concentrations of the potential pollutants in, on or under the land in pollutant linkages of the type concerned. If it does so, the local authority should be satisfied that:

(a) an adequate scientific and technical assessment of the information on the potential pollutant, using the appropriate, authoritative and scientifically based guideline values, shows that there is a significant possibility of significant harm; and
(b) there are no suitable and sufficient risk management arrangements in place to prevent such harm.

B.48 In using any guideline values, the local authority should be satisfied that:

(a) the guideline values are relevant to the judgement of whether the effects of the pollutant linkage in question constitute a significant possibility of significant harm;
(b) the assumptions underlying the derivation of any numerical values in the guideline values (for example, assumptions regarding soil conditions, the behaviour of potential pollutants, the existence of pathways, the land-use patterns, and the availability of receptors) are relevant to the circumstances of the pollutant linkage in question;
(c) any other conditions relevant to the use of the guideline values have been observed (for example, the number of samples taken or the methods of preparation and analysis of those samples); and
(d) appropriate adjustments have been made to allow for the differences between the circumstances of the land in question and any assumptions or other factors relating to the guideline values.

B.49 The local authority should be prepared to reconsider any determination based on such use of guideline values if it is demonstrated to the authority's satisfaction that under some other more appropriate method of assessing the risks the local authority would not have determined that the land appeared to be contaminated land.

Determining that "pollution of controlled waters is being caused"

B.50 The local authority should determine that land is contaminated land on the basis that pollution of controlled waters is being caused where:

(a) it has carried out an appropriate scientific and technical assessment of all the relevant and available evidence, having regard to any advice provided by the Environment Agency; and
(b) on the basis of that assessment, it is satisfied on the balance of probabilities that both of the following circumstances apply:

(i) a potential pollutant is present in, on or under the land in question, which constitutes poisonous, noxious or polluting matter, or which is solid waste matter, and
(ii) that potential pollutant is entering controlled waters by the pathway identified in the pollutant linkage.

Determining that "pollution of controlled waters is likely to be caused"

B.51 The local authority should determine that land is contaminated land on the basis that pollution of controlled waters is likely to be caused where:

(a) it has carried out an appropriate scientific and technical assessment of all the relevant and available evidence, having regard to any advice provided by the Environment Agency; and
(b) on the basis of that assessment it is satisfied that, on the balance of probabilities, all of the following circumstances apply:

(i) a potential pollutant is present in, on or under the land in question, which constitutes poisonous, noxious or polluting matter, or which is solid waste matter,
(ii) the potential pollutant in question is in such a condition that it is capable of entering controlled waters,
(iii) taking into account the geology and other circumstances of the land in question, there is a pathway (as defined in Chapter A) by which the potential pollutant can enter identified controlled waters,

(iv) the potential pollutant in question is more likely than not to enter these controlled waters and, when it enters the controlled waters, will be in a form that is poisonous, noxious or polluting, or solid waste matter, and

(v) there are no suitable and sufficient risk management arrangements relevant to the pollution linkage in place to prevent such pollution.

Record of the determination that land is contaminated land

B.52 The local authority should prepare a written record of any determination that particular land is contaminated land. The record should include (by means of a reference to other documents if necessary):

(a) a description of the particular significant pollutant linkage, identifying all three components of pollutant, pathway and receptor;
(b) a summary of the evidence upon which the determination is based;
(c) a summary of the relevant assessment of this evidence; and
(d) a summary of the way in which the authority considers that the requirements of the guidance in this Part and in Chapter A of the guidance have been satisfied.

Chapter C – Statutory guidance on the remediation of contaminated land

Part 1 Scope of the Chapter

C.1 The statutory guidance in this Chapter is issued under section 78E(5) of Part IIA of the Environmental Protection Act 1990, and provides guidance on the remediation which may be required for any contaminated land.

C.2 Section 78E provides:

"(4) The only things by way of remediation which the enforcing authority may do, or require to be done, under or by virtue of [Part IIA of the Environmental Protection Act 1990] are things which it considers reasonable, having regard to–

(a) the cost which is likely to be involved; and
(b) the seriousness of the harm, or pollution of controlled waters, in question.

(5) In determining for any purpose of this Part–

(a) what is to be done (whether by an appropriate person, the enforcing authority, or any other person) by way of remediation in any particular case,
(b) the standard to which any land is, or waters are, to be remediated pursuant to [a remediation] notice, or
(c) what is, or is not, to be regarded as reasonable for the purposes of subsection (4) above,

"the enforcing authority shall have regard to any guidance issued for the purpose by the Secretary of State".

C.3 The enforcing authority is therefore required to have regard to this guidance when it is:

(a) determining what remediation action it should specify in a remediation notice as being required to be carried out (section 78E(1));

(b) satisfying itself that appropriate remediation is being, or will be, carried out without the service of a notice (section 78H(5)(b)); or
(c) deciding what remediation action it should carry out itself (section 78N).

C.4 The guidance in this Chapter does not attempt to set out detailed technical procedures or working methods. For information on these matters, the enforcing authority may wish to consult relevant technical documents prepared under the contaminated land research programmes of DETR and the Environment Agency, and by other professional and technical organisations.

Part 2 Definitions of Terms

C.5 Unless otherwise stated, any word, term or phrase given a specific meaning in Part IIA of the Environmental Protection Act 1990, or in the statutory guidance in Chapters A or B, has the same meaning for the purposes of the guidance in this Chapter.

C.6 "Remediation" is defined in section 78A(7) as meaning:

"(a) the doing of anything for the purpose of assessing the condition of–

(i) the contaminated land in question;
(ii) any controlled waters affected by that land; or
(iii) any land adjoining or adjacent to that land;

(b) the doing of any works, the carrying out of any operations or the taking of any steps in relation to any such land or waters for the purpose–

(i) of preventing or minimising, or remedying or mitigating the effects of, any significant harm, or any pollution of controlled waters, by reason of which the contaminated land is such land; or
(ii) of restoring the land or waters to their former state; or

(c) the making of subsequent inspections from time to time for the purpose of keeping under review the condition of the land or waters."

C.7 The definition of remediation given in section 78A extends more widely than the common usage of the term, which more normally relates only to the actions defined as "remedial treatment actions" below.

C.8 For the purposes of the guidance in this Chapter, the following definitions apply:

(a) a "remediation action" is any individual thing which is being, or is to be, done by way of remediation;
(b) a "remediation package" is the full set or sequence of remediation actions, within a remediation scheme, which are referable to a particular significant pollutant linkage;
(c) a "remediation scheme" is the complete set or sequence of remediation actions (referable to one or more significant pollutant linkages) to be carried out with respect to the relevant land or waters;
(d) "relevant land or waters" means the contaminated land in question, any controlled waters affected by that land and any land adjoining or adjacent to the contaminated land on which remediation might be required as a consequence of the contaminated land being such land;

(e) an "assessment action" means a remediation action falling within the definition of remediation in section 78A(7)(a) (see paragraph C.6 above);

(f) a "remedial treatment action" means a remediation action falling within the definition in section 78A(7)(b) (see paragraph C.6 above); and

(g) a "monitoring action" means a remediation action falling within the definition in section 78A(7)(c) (see paragraph C.6 above).

C.9 Any reference to "Part IIA" means "Part IIA of the Environmental Protection Act 1990". Any reference to a "section" in primary legislation means a section of the Environmental Protection Act 1990, unless it is specifically stated otherwise.

Part 3 Securing remediation

C.10 When the enforcing authority is serving a remediation notice, it will need to specify in that notice any remediation action which is needed in order to achieve remediation of the relevant land or waters to the standard described in Part 4 of this Chapter and which is reasonable for the purposes of section 78E(4) (see Part 5 of this Chapter). Part 6 of this Chapter provides further guidance relevant to determining the necessary standard of remediation. Part 7 provides guidance on the circumstances in which different types of remediation action may, or may not, be required.

C.11 The enforcing authority should be satisfied that appropriate remediation is being, or will be, carried out without the service of a remediation notice if that remediation would remediate the relevant land or waters to an equivalent, or better, standard than would be achieved by the remediation action or actions that the authority could, at that time, otherwise specify in a remediation notice.

Phased remediation

C.12 The overall process of remediation on any land or waters may require a phased approach, with different remediation actions being carried out in sequence. For example, the local authority may have obtained sufficient information about the relevant land or waters to enable it to identify the land as falling within the definition of contaminated land, but that information may not be sufficient information for the enforcing authority to be able to specify any particular remedial treatment action as being appropriate. Further assessment actions may be needed in any case of this kind as part of the remediation scheme. In other cases, successive phases of remedial treatment actions may be needed.

C.13 The phasing of remediation is likely to follow a progression from assessment actions, through remedial treatment actions and onto monitoring actions. However, this will not always be the case, and the phasing may omit some stages or revisit others. For example, in some circumstances it may be possible for a remedial treatment action to be carried out without any previous assessment action (because sufficient information is already available). But, conversely, in some instances additional assessment action may be found to be necessary only in the light of information derived during the course of a first phase of a required assessment action or the carrying out of required remedial treatment actions.

C.14 Where it is necessary for the remediation scheme as a whole to be phased, a single remediation notice may not be able to include all of the remediation actions which could eventually be needed. In these circumstances, the enforcing authority should specify in the notice the remediation action or actions which, on the basis of the information available at that time, it considers to be appropriate, taking into account in particular the guidance in Part 7 of this Chapter. In due course, the authority may need to serve further remediation notices which include remediation actions for further phases of the scheme.

C.15 However, before serving any further remediation notice, the enforcing authority must be satisfied that the contaminated land which was originally identified still appears to it to meet the definition in section 78A(2). If, for example, the information obtained as a result of an assessment action reveals that there is not, in fact, a significant possibility of significant harm being caused, nor is there a likelihood of any pollution of controlled waters being caused, then no further assessment, remedial treatment or monitoring action can be required under section 78E(1).

Part 4 The standard to which land or waters should be remediated

C.16 The statutory guidance in this Part is issued under section 78E(5)(b) and provides guidance on the standard to which land or waters should be remediated.

The Standard of Remediation

C.17 The Government's intention is that any remediation required under this regime should result in land being "suitable for use". The aim of any remediation should be to ensure that the circumstances of the land are such that, in its current use (as defined in paragraph A.26 of Chapter A) it is no longer contaminated land (as defined in section 78A(2)), and that the effects of any significant harm or pollution of controlled waters which has occurred are remedied. However, it is always open to the appropriate person to carry out remediation on a broader basis than this, if he considers it in his interests to do so, for example if he wishes to prepare the land for redevelopment.

C.18 The standard to which the relevant land or waters as a whole should be remediated should be established by considering separately each significant pollutant linkage identified on the land in question. For each such linkage, the standard of remediation should be that which would be achieved by the use of a remediation package which forms the best practicable techniques of remediation for:

(a) ensuring that the linkage is no longer a significant pollutant linkage, by doing any one or more of the following:

(i) removing or treating the pollutant;
(ii) breaking or removing the pathway; or
(iii) protecting or removing the receptor; and

(b) remedying the effect of any significant harm or pollution of controlled waters which is resulting, or has already resulted from, the significant pollutant linkage.

C.19 In deciding what represents the best practicable technique for any particular remediation, the enforcing authority should look for the method of achieving the desired results which, in the light of the nature and volume of the significant pollutant concerned and the timescale within which remediation is required:

(a) is reasonable, taking account of the guidance in Part 5; and
(b) represents the best combination of the following qualities:
    (i) practicability, both in general and in the particular circumstances of the relevant land or waters;
    (ii) effectiveness in achieving the aims set out in paragraph C.18 above; and
    (iii) durability in maintaining that effectiveness over the timescale within which the significant harm or pollution of controlled waters may occur.

C.20 Further guidance on how the factors set out in sub-paragraph (b) above should be considered is set out in Part 6. The determination of what, in any particular case, represents the best practicable technique of remediation may require a balance to be struck between these factors.

C.21 When considering what would be the best practicable techniques for remediation in any particular case, the enforcing authority should work on the basis of authoritative scientific and technical advice. The authority should consider what comparable techniques have recently been carried out successfully on other land, and also any technological advances and changes in scientific knowledge and understanding.

C.22 Where there is established good practice for the remediation of a particular type of significant pollutant linkage, the authority should assume that this represents the best practicable technique for remediation for a linkage of that type, provided that:

(a) it is satisfied that the use of that means of remediation is appropriate, given the circumstances of the relevant land or waters; and
(b) the remediation actions involved would be reasonable having regard to the cost which is likely to be involved and the seriousness of the harm or pollution of controlled waters in question.

C.23 In some instances, the best practicable techniques of remediation with respect to any significant pollutant linkage may not fully achieve the aim in subparagraph C.18(a), that is to say that if the remediation were to be carried out the pollutant linkage in question would remain a significant pollutant linkage. Where this applies, the standard of remediation with respect to that significant pollutant linkage should be that which, by the use of the best practicable techniques:

(a) comes as close as practicable to achieving the aim in subparagraph C.18(a);
(b) achieves the aim in subparagraph C.18(b); and
(c) puts arrangements in place to remedy the effect of any significant harm or pollution of controlled waters which may be caused in the future as a consequence of the continued existence of the pollutant linkage.

C.24 In addition, the best practicable techniques for remediation with respect to a significant pollutant linkage may, in some circumstances, not fully remedy the effect of past or future significant harm or pollution of controlled waters. Where this is the case the standard of remediation should be that which, by the use of the best practicable techniques, mitigates as far as practicable the significant harm or pollution of controlled waters which has been caused as a consequence of the existence of that linkage, or may be caused in the future as a consequence of its continued existence.

C.25 For any remediation action, package or scheme to represent the best practicable techniques, it should be implemented in accordance with best practice, including any precautions necessary to prevent damage to the environment and any other appropriate quality assurance procedures.

Multiple pollutant linkages

C.26 Where more than one significant pollutant linkage has been identified on the land, it may be possible to achieve the necessary overall standard of remediation for the relevant land or waters as a whole by considering what remediation actions would form part of the appropriate remediation package for each linkage (ie, representing the best practicable techniques of remediation for that linkage) if it were the only such linkage, and then carrying out all of these remediation actions.

C.27 However, the enforcing authority should also consider whether there is an alternative remediation scheme which would, by dealing with the linkages together, be cheaper or otherwise more practicable to implement. If such a scheme can be identified which achieves an equivalent standard of remediation with respect to all of the significant pollutant linkages to which it is referable, the authority should prefer that alternative scheme.

Volunteered remediation

C.28 In some cases, the person carrying out remediation may wish to adopt an alternative remediation scheme to that which could be required in a remediation notice. This might occur, in particular, if the person concerned wished also to prepare the land for redevelopment. The enforcing authority should consider such a remediation scheme as appropriate remediation provided the scheme would achieve at least the same standard of remediation with respect to each of the significant pollutant linkages identified on the land as would be achieved by the remediation scheme which the authority would otherwise specify in a remediation notice.

Part 5 The reasonableness of remediation

C.29 The statutory guidance in this Part is issued under section 78E(5)(c) and provides guidance on the determination by the enforcing authority of what remediation is, or is not, to be regarded as reasonable having regard to the cost which is likely to be involved and the seriousness of the harm or of the pollution of controlled waters to which it relates.

C.30 The enforcing authority should regard a remediation action as being reasonable for the purpose of section 78E(4) if an assessment of the costs likely to be involved and of the resulting benefits shows that those benefits justify incurring those costs. Such an assessment should include the preparation of an estimate of the costs likely to be involved and of a statement of the benefits likely to result. This latter statement need not necessarily attempt to ascribe a financial value to these benefits.

C.31 For these purposes, the enforcing authority should regard the benefits resulting from a remediation action as being the contribution that the action makes, either on its own or in conjunction with other remediation actions, to:

(a) reducing the seriousness of any harm or pollution of controlled waters which might otherwise be caused; or

(b) mitigating the seriousness of any effects of any significant harm or pollution of controlled waters.

C.32 In assessing the reasonableness of any remediation, the enforcing authority should make due allowance for the fact that the timing of expenditure and the realisation of benefits is relevant to the balance of costs and benefits. In particular, the assessment should recognise that:

(a) expenditure which is delayed to a future date will have a lesser impact on the person defraying it than would an equivalent cash sum to be spent immediately;
(b) there may be a gain from achieving benefits earlier but this may also involve extra expenditure; the authority should consider whether the gain justifies the extra costs. This applies, in particular, where natural processes, managed or otherwise, would over time bring about remediation; and
(c) there may be evidence that the same benefits will be achievable in the foreseeable future at a significantly lower cost, for example, through the development of new techniques or as part of a wider scheme of development or redevelopment.

C.33 The identity or financial standing of any person who may be required to pay for any remediation action are not relevant factors in the determination of whether the costs of that action are, or are not, reasonable for the purposes of section 78E(4). (These factors may however be relevant in deciding whether or not the enforcing authority can impose the cost of remediation on that person, either through the service of a remediation notice or through the recovery of costs incurred by the authority; see section 78P and the guidance in Chapter E.)

The cost of remediation

C.34 When considering the costs likely to be involved in carrying out any remediation action, the enforcing authority should take into account:

(a) all the initial costs (including tax payable) of carrying out the remediation action, including feasibility studies, design, specification and management, as well as works and operations, and making good afterwards;
(b) any on-going costs of managing and maintaining the remediation action; and
(c) any relevant disruption costs.

C.35 For these purposes, "relevant disruption costs" mean depreciation in the value of land or other interests, or other loss or damage, which is likely to result from the carrying out of the remediation action in question. The enforcing authority should assess these costs as their estimate of the amount of compensation which would be payable if the owner of the land or other interest had granted rights under section 78G(2) to permit the action to be carried out and had claimed compensation under section 78G(5) and regulation 6 of the Contaminated Land England Regulations 2000 (whether or not such a claim could actually be made).

C.36 Each of the types of cost set out in paragraph C.34 above should be included even where they would not result in payments to others by the person carrying out the remediation. For example, a company may choose to use its own staff or equipment to carry out the remediation, or the person carrying out the remediation may already own the land in question and would therefore not be entitled to receive compensation under section 78G(5). The evaluation of the cost involved in remediation should not be affected by the identity of the person carrying it out, or internal resources available to that person.

C.37 The enforcing authority should furthermore regard it as a necessary condition of an action being reasonable that:

(a) where two or more significant pollutant linkages have been identified on the land in question, and the remediation action forms part of a wider remediation scheme which is dealing with two or more of those linkages, there is no alternative scheme which would achieve the same purposes for a lower overall cost; and

(b) subject to subparagraph (a) above, where the remediation action forms part of a remediation package dealing with any particular significant pollutant linkage, there is no alternative package which would achieve the same standard of remediation at a lower overall cost.

C.38 In addition, for any remediation action to be reasonable there should be no alternative remediation action which would achieve the same purpose, as part of any wider remediation package or scheme, to the same standard for a lower cost (bearing in mind that the purpose of any remediation action may relate to more than one significant pollutant linkage).

The seriousness of harm or of pollution of controlled waters

C.39 When evaluating the seriousness of any significant harm, for the purposes of assessing the reasonableness of any remediation, the enforcing authority should consider:

(a) whether the significant harm is already being caused;
(b) the degree of the possibility of the significant harm being caused;
(c) the nature of the significant harm with respect, in particular, to:

(i) the nature and importance of the receptor,
(ii) the extent and type of any effects on that receptor of the significant harm,
(iii) the number of receptors which might be affected, and
(iv) whether the effects would be irreversible; and

(d) the context in which the effects might occur, in particular:

(i) whether the receptor has already been damaged by other means and, if so, whether further effects resulting from the harm would materially affect its condition, and
(ii) the relative risk associated with the harm in the context of wider environmental risks.

C.40 Where the significant harm is an "ecological system effect" as defined in Chapter A, the enforcing authority should take into account any advice received from English Nature.

C.41 In evaluating for this purpose the seriousness of any pollution of controlled waters, the enforcing authority should consider:

(a) whether the pollution of controlled waters is already being caused;
(b) the likelihood of the pollution of controlled waters being caused;
(c) the nature of the pollution of controlled waters involved with respect, in particular, to:

(i) the nature and importance of the controlled waters which might be affected,
(ii) the extent of the effects of the actual or likely pollution on those controlled waters, and

(iii) whether such effects would be irreversible; and

(d) the context in which the effects might occur, in particular:

(i) whether the waters have already been polluted by other means and, if so, whether further effects resulting from the water pollution would materially affect their condition, and

(ii) the relative risk associated with the water pollution in the context of wider environmental risks.

C.42 Where the enforcing authority is the local authority, it should take into account any advice received from the Environment Agency when it is considering the seriousness of any pollution of controlled waters.

C.43 In some instances, it may be possible to express the benefits of addressing the harm or pollution of controlled waters in direct financial terms. For example, removing a risk of explosion which renders a building unsafe for occupation could be considered to create a benefit equivalent to the cost of acquiring a replacement building. Various Government departments have produced technical advice, which the enforcing authority may find useful, on the consideration of non-market impacts of environmental matters.

Part 6 The practicability, effectiveness and durability of remediation

C.44 The statutory guidance in this Part is issued under section 78E(5)(b) and is relevant to the guidance given in Part 4 on the standard to which land and waters should be remediated.

General considerations

C.45 In some instances, there may be little firm information on which to assess particular remediation actions, packages or schemes. For example, a particular technology or technique may not have been subject previously to field-scale pilot testing in circumstances comparable to those to be found on the contaminated land in question. Where this is the case, the enforcing authority should consider the effectiveness and durability which it appears likely that any such action would achieve, and the practicability of its use, on the basis of information which it does have at that time (for example information derived from laboratory or other "treatability" testing).

C.46 If the person who will be carrying out the remediation proposes the use of an innovative approach to remediation, the enforcing authority should be prepared to agree to that approach being used (subject to that person obtaining any other necessary permits or authorisations), notwithstanding the fact that there is little available information on the basis of which the authority can assess its likely effectiveness. If the approach to remediation proves to be ineffective, further remediation actions may be required, for which the person proposing the innovative approach will be liable.

C.47 However, the enforcing authority should not, under the terms of a remediation notice, require any innovative remediation action to be carried out for the purposes of establishing its effectiveness in general, unless either the person carrying out the remediation agrees or there is clear evidence that it is likely that the action would be effective on the relevant land or waters and it would meet all other requirements of the statutory guidance in this Chapter.

The practicability of remediation

C.48 The enforcing authority should consider any remediation as being practicable to the extent that it can be carried out in the circumstances of the relevant land or waters. This applies both to the remediation scheme as a whole and the individual remediation actions of which it is comprised.

C.49 In assessing the practicability of any remediation, the enforcing authority should consider, in particular, the following factors:

(a) technical constraints, for example whether:

(i) any technologies or other physical resources required (for example power or materials) are commercially available, or could reasonably be made available, on the necessary scale, and

(ii) the separate remediation actions required could be carried out given the other remediation actions to be carried out, and without preventing those other actions from being carried out;

(b) site constraints, for example whether:

(i) the location of and access to the relevant land or waters, and the presence of buildings or other structures in, on or under the land, would permit the relevant remediation actions to be carried out in practice, and

(ii) the remediation could be carried out, given the physical or other condition of the relevant land or waters, for example the presence of substances, whether these are part of other pollutant linkages or are not pollutants;

(c) time constraints, for example whether it would be possible to carry out the remediation within the necessary time period given the time needed by the person carrying out the remediation to:

(i) obtain any necessary regulatory permits and constraints, and

(ii) design and implement the various remediation actions; and

(d) regulatory constraints, for example whether:

(i) the remediation can be carried out within the requirements of statutory controls relating to health and safety (including engineering safety) and pollution control,

(ii) any necessary regulatory permits or consents would reasonably be expected to be forthcoming,

(iii) any conditions attached to such permits or consents would affect the practicability or cost of the remediation, and

(iv) adverse environmental impacts may arise from carrying out the remediation (see paragraphs C.51 to C.57 below).

C.50 The responsibility for obtaining any regulatory permits or consents necessary for the remediation to be carried out rests with the person who will actually be carrying out the remediation, and not with the enforcing authority. However, the authority may in some circumstances have particular duties to contribute to health and safety in the remediation work, under the Construction (Design and Management) Regulations 1994 (S.I. 1994/3140).

Adverse environmental impacts

C.51 Although the objective of any remediation is to improve the environment, the process of carrying out remediation may, in some circumstances, create adverse environmental impacts. The possibility of such impacts may affect the determination of what remediation package represents the best practicable techniques for remediation.

C.52 Specific pollution control permits or authorisations may be needed for some kinds of remediation processes, for example:

(a) authorisations under Part I of the Environmental Protection Act 1990 (Integrated Pollution Control and Local Authority Air Pollution Control);
(b) site or mobile plant licences under Part II of the Environmental Protection Act 1990 (waste management licensing); or
(c) abstraction licences under Part II, or discharge consents under Part III, of the Water Resources Act 1991.

C.53 Permits or authorisations of these kinds may include conditions controlling the manner in which the remediation is to be carried out, intended to prevent or minimise adverse environmental impacts. Where this is the case, the enforcing authority should assume that these conditions provide a suitable level of protection for the environment.

C.54 Where this is not the case, the enforcing authority should consider whether the particular remediation package can be carried out without damaging the environment, and in particular:

(a) without risk to water, air, soil and plants and animals;
(b) without causing a nuisance through noise or odours;
(c) without adversely affecting the countryside or placesof special interest; and
(d) without adversely affecting a building of special architectural or historic interest (that is, a building listed under town and country planning legislation or a building in a designated Conservation Area) or a site of archaeological interest (as defined in article 1(2) of the Town and Country Planning (General Permitted Development) Order 1995).

C.55 If the enforcing authority considers that there is some risk that the remediation might damage the environment, it should consider whether:

(a) the risk is sufficiently great to mean that the balance of advantage, in terms of improving and protecting the environment, would lie with adopting an alternative approach to remediation, even though such an alternative may not fully achieve the objectives for remediation set out at paragraph C.18 above; or
(b) the risk can be sufficiently reduced by including, as part of the description of what is to be done by way of remediation, particular precautions designed to prevent the occurrence of such damage to the environment (for example, precautions analogous to the conditions attached to a waste management licence).

C.56 In addition, the enforcing authority should consider whether it is likely that the process of remediation might lead to a direct or indirect discharge into groundwater of a substance in either List I or List II of the Schedule to the Groundwater Regulations 1998 (S.I. 1998/1006). (For these purposes, the terms direct discharge, indirect discharge and groundwater have the meanings given to them in the 1998 Regulations.)

C.57 If the enforcing authority considers that such a discharge is likely, it should (where that authority is not the Environment Agency) consult the Environment Agency, and have regard to its advice on whether an alternative remediation package should be adopted or precaution required as to the way that remediation is carried out.

The effectiveness of remediation

C.58 The enforcing authority should consider any remediation as being effective to the extent to which the remediation scheme as a whole, and its component remediation packages, would achieve the aims set out in paragraph C.18 above in relation to each of the significant pollutant linkages identified on the relevant land or waters. The enforcing authority should consider also the extent to which each remediation action, or group of actions required for the same particular purpose, would achieve the purpose for which it was required.

C.59 Within this context, the enforcing authority should consider also the time which would pass before the remediation would become effective. In particular, the authority should establish whether the remediation would become effective sufficiently soon to match the particular degree of urgency resulting from the nature of the significant pollutant linkage in question. However, the authority may also need to balance the speed in reaching a given level of effectiveness against higher degrees of effectiveness which may be achievable, but after a longer period of time, by the use of other remediation methods.

C.60 If any remedial treatment action representing the best practicable techniques will not fully achieve the standard set out in paragraph C.18 above, the enforcing authority should consider whether additional monitoring actions should be required.

The durability of remediation

C.61 The enforcing authority should consider a remediation scheme as being sufficiently durable to the extent that the scheme as a whole would continue to be effective with respect to the aims in paragraph C.18 above during the time over which the significant pollutant linkage would otherwise continue to exist or recur. Where other action (such as redevelopment) is likely to resolve or control the problem within that time, a shorter period may be appropriate. The durability of an individual remediation action is a measure of the extent to which it will continue to be effective in meeting the purpose for which it is to be required taking into account normal maintenance and repair.

C.62 Where a remediation scheme cannot reasonably and practicably continue to be effective during the whole of the expected duration of the problem, the enforcing authority should require the remediation to continue to be effective for as long as can reasonably and practicably be achieved. In these circumstances, additional monitoring actions may be required.

C.63 Where a remediation method requires on-going management and maintenance in order to continue to be effective (for example, the maintenance of gas venting or alarm systems), these on-going requirements should be specified in any remediation notice as well as any monitoring actions necessary to keep the effectiveness of the remediation under review.

Part 7 What is to be done by way of remediation

C.64 The statutory guidance in this Part is issued under section 78E(5)(a) and provides guidance on the determination by the enforcing authority of what is to be done by way of remediation—in particular, on the circumstances in which any action within the three categories of remediation action (that is, assessment, remedial treatment and monitoring actions) should be required.

Assessment action

C.65 The enforcing authority should require an assessment action to be carried out where this is necessary for the purpose of obtaining information on the condition of the relevant land or waters which is needed:

(a) to characterise in detail a significant pollutant linkage (or more than one such linkage) identified on the relevant land or waters for the purpose of enabling the authority to establish what would need to be achieved by any remedial treatment action;
(b) to enable the establishment of the technical specifications or design of any particular remedial treatment action which the authority reasonably considers it might subsequently require to be carried out; or
(c) where, after remedial treatment actions have been carried out, the land will still be in such a condition that it would still fall to be identified as contaminated land, to evaluate the condition of the relevant land or waters, or the incidence of any significant harm or pollution of controlled waters, for the purpose of supporting future decisions on whether further remediation might then be required (this applies where the remediation action concerned would not otherwise constitute a monitoring action).

C.66 The enforcing authority should not require any assessment action to be carried out unless that action is needed to achieve one or more of the purposes set out in paragraph C.65 above, and it represents a reasonable means of doing so. In particular, no assessment action should be required for the purposes of determining whether or not the land in question is contaminated land. For the purposes of this guidance, assessment actions relate solely to land which has already been formally identified as contaminated land, or to other land or waters which might be affected by it. The statutory guidance in Chapters A and B sets out the requirements for the inspection of land and the manner in which a local authority should determine that land appears to it to be contaminated land.

Remedial treatment action

C.67 The enforcing authority should require a remedial treatment action to be carried out where it is necessary to achieve the standard of remediation described in Part 4, but for no other purpose. Any such remedial treatment action should include appropriate verification measures. When considering what remedial treatment action may be necessary, the enforcing authority should consider also what complementary assessment or monitoring actions might be needed to evaluate the manner in which the remedial treatment action is implemented or its effectiveness or durability once implemented.

Monitoring action

C.68 The enforcing authority should require a monitoring action to be carried out where it is for the purpose of providing information on any changes which might occur in the condition of a pollutant, pathway or receptor, where:

(a) the pollutant, pathway or receptor in question was identified previously as part of a significant pollutant linkage; and
(b) the authority will need to consider whether any further remedial treatment action will be required as a consequence of any change that may occur.

C.69 Monitoring action should not be required to achieve any other purpose, such as general monitoring to enable the enforcing authority to identify any new significant pollutant linkages which might become present in the future. This latter activity forms part of the local authority's duty, under section 78B(1), to cause its area to be inspected from time to time for the purpose of identifying any contaminated land.

What remediation should not be required

C.70 The enforcing authority should not require any remediation to be carried out for the purpose of achieving any aims other than those set out in paragraphs C.18 to C.24 above, or purposes other than those identified in this Part of this Chapter. In particular, it should not require any remediation to be carried out for the purposes of:

(a) dealing with matters which do not in themselves form part of a significant pollutant linkage, such as substances present in quantities or concentrations at which there is neither a significant possibility of significant harm being caused nor a likelihood of any pollution of controlled waters being caused; or

(b) making the land suitable for any uses other than its current use, as defined in paragraphs A.25 and A.26 in Chapter A.

C.71 It is, however, always open to the owner of the land, or any other person who might be liable for remediation, to carry out on a voluntary basis remediation to meet these wider objectives.

CHAPTER D – STATUTORY GUIDANCE ON EXCLUSION FROM, AND APPORTIONMENT OF, LIABILITY FOR REMEDIATION

Part 1 Scope of the Chapter

D.1 The statutory guidance in this Chapter is issued under sections 78F(6) and 78F(7) of the Environmental Protection Act 1990. It provides guidance on circumstances where two or more persons are liable to bear the responsibility for any particular thing by way of remediation. It deals with the questions of who should be excluded from liability, and how the cost of each remediation action should be apportioned between those who remain liable after any such exclusion.

D.2 Section 78F provides that:

"(6) Where two or more persons would, apart from this subsection, be appropriate persons in relation to any particular thing which is to be done by way of remediation, the enforcing authority shall determine in accordance with guidance issued for the purpose by the Secretary of State whether any, and if so which, of them is to be treated as not being an appropriate person in relation to that thing.

(7) Where two or more persons are appropriate persons in relation to any particular thing which is to be done by way of remediation, they shall be liable to bear the cost of doing that thing in proportions determined by the enforcing authority in accordance with guidance issued for the purpose by the Secretary of State".

D.3 The enforcing authority is therefore required to act in accordance with the guidance in this Chapter. Introductory summaries are included to various parts and sections of the guidance: these do not necessarily give the full detail of the guidance; the section concerned should be consulted.

Part 2 Definitions of terms

D.4 Unless otherwise stated, any word, term or phrase given a specific meaning in Part IIA of the Environmental Protection Act 1990, or in the statutory guidance in Chapters A or B, has the same meaning for the purpose of the guidance in this Chapter.

D.5 In addition, for the purposes of this Chapter, the following definitions apply:

(a) a person who is an appropriate person by virtue of section 78F(2) (that is, because he has caused or knowingly permitted a pollutant to be in, on or under the land) is described as a "Class A person";
(b) a person who is an appropriate person by virtue of section 78F(4) or (5) (that is, because he is the owner or occupier of the land in circumstances where no Class A person can be found with respect to a particular remediation action) is described as a "Class B person";
(c) collectively, the persons who are appropriate persons with respect to any particular significant pollutant linkage are described as the "liability group" for that linkage; a liability group consisting of one or more Class A persons is described as a "Class A liability group", and a liability group consisting of one or more Class B persons is described as a "Class B liability group";
(d) any determination by the enforcing authority under section 78F(6) (that is, a person is to be treated as not being an appropriate person) is described as an "exclusion";
(e) any determination by the enforcing authority under section 78F(7) (dividing the costs of carrying out any remediation action between two or more appropriate persons) is described as an "apportionment"; the process of apportionment between liability groups is described as "attribution";
(f) a "remediation action" is any individual thing which is being, or is to be, done by way of remediation;
(g) a "remediation package" is all the remediation actions, within a remediation scheme, which are referable to a particular significant pollutant linkage; and
(h) a "remediation scheme" is the complete set or sequence of remediation actions (referable to one or more significant pollutant linkages) to be carried out with respect to the relevant land or waters.

D.6 Any reference to "Part IIA" means "Part IIA of the Environmental Protection Act 1990". Any reference to a "section" in primary legislation means a section of the Environmental Protection Act 1990, unless it is specifically stated otherwise.

Part 3 The procedure for determining liabilities

D.7 For most sites, the process of determining liabilities will consist simply of identifying either a single person (either an individual or a corporation such as a limited company) who has caused or knowingly permitted the presence of a single significant pollutant, or the owner of the site. The history of other sites may be more complex. A succession of different occupiers or of different industries, or a variety of substances may all have contributed to the problems which have made the land "contaminated land" as defined for the purposes of Part IIA. Numerous separate remediation actions may be required, which may not correlate neatly with those who are to bear responsibility for the costs. The degree of responsibility for the state of the land may vary widely. Determining liability for the costs of each remediation action can be correspondingly complex.

D.8 The statutory guidance in this Part sets out the procedure which the enforcing authority should follow for determining which appropriate persons should bear what responsibility for each remediation action. It refers forward to the other Parts of this Chapter, and describes how they should be applied. Not all stages will be relevant to all cases, particularly where there is only a single significant pollutant linkage, or where a liability group has only one member.

First stage – Identifying potential appropriate persons and liability groups

D.9 As part of the process of determining that the land is "contaminated land" (see Chapters A and B), the enforcing authority will have identified at least one significant pollutant linkage (pollutant, pathway and receptor), resulting from the presence of at least one significant pollutant.

Where there is a single significant pollutant linkage

D.10 The enforcing authority should identify all of the persons who would be appropriate persons to pay for any remediation action which is referable to the pollutant which forms part of the significant pollutant linkage. These persons constitute the "liability group" for that significant pollutant linkage. (In this guidance the term "liability group" is used even where there is only a single appropriate person who is a "member" of the liability group.)

D.11 To achieve this, the enforcing authority should make reasonable enquiries to find all those who have caused or knowingly permitted the pollutant in question to be in, on or under the land. Any such persons constitute a "Class A liability group" for the significant pollutant linkage.

D.12 If no such Class A persons can be found for any significant pollutant, the enforcing authority should consider whether the significant pollutant linkage of which it forms part relates solely to the pollution of controlled waters, rather than to any significant harm. If this is the case, there will be no liability group for that significant pollutant linkage, and it should be treated as an "orphan linkage" (see paragraphs D.103 to D.109 below).

D.13 In any other case where no Class A persons can be found for a significant pollutant, the enforcing authority should identify all of the current owners or occupiers of the contaminated land in question. These persons then constitute a "Class B liability group" for the significant pollutant linkage.

D.14 If the enforcing authority cannot find any Class A persons or any Class B persons in respect of a significant pollutant linkage, there will be no liability group for that linkage and it should be treated as an orphan linkage (see paragraphs D.103 to D.109 below).

Where there are two or more significant pollutant linkages

D.15 The enforcing authority should consider each significant pollutant linkage in turn, carrying out the steps set out in paragraphs D.10 to D.14 above, in order to identify the liability group (if one exists) for each of the linkages.

In all cases

D.16 Having identified one or more liability groups, the enforcing authority should consider whether any of the members of those groups are exempted from liability under the provisions in Part IIA. This could apply where:

(a) a person who would otherwise be a Class A person is exempted from liability arising with respect to water pollution from an abandoned mine (see section 78J(3));

(b) a Class B person is exempted from liability arising from the escape of a pollutant from one piece of land to other land (see section 78K); or

(c) a person is exempted from liability by virtue of his being a person "acting in a relevant capacity" (such as acting as an insolvency practitioner), as defined in section 78X(4).

D.17 If all of the members of any liability group benefit from one or more of these exemptions, the enforcing authority should treat the significant pollutant linkage in question as an orphan linkage (see paragraphs D.103 to D.109 below).

D.18 Persons may be members of more than one liability group. This might apply, for example, if they caused or knowingly permitted the presence of more than one significant pollutant.

D.19 Where the membership of all of the liability groups is the same, there may be opportunities for the enforcing authority to abbreviate the remaining stages of this procedure. However, the tests for exclusion and apportionment may produce different results for different significant pollutant linkages, and so the enforcing authority should exercise caution before trying to simplify the procedure in any case.

Second stage – Characterising remediation actions

D.20 Each remediation action will be carried out to achieve a particular purpose with respect to one or more defined significant pollutant linkages. Where there is a single significant pollutant linkage on the land in question, all the remediation actions will be referable to that linkage, and there is no need to consider how the different actions relate to different linkages. This stage and the third stage of the procedure therefore do not need to be carried out in where there is only a single significant pollutant linkage.

D.21 However, where there are two or more significant pollutant linkages on the land in question, the enforcing authority should establish whether each remediation action is:

(a) referable solely to the significant pollutant in a single significant pollutant linkage (a "single-linkage action"); or

(b) referable to the significant pollutant in more than one significant pollutant linkage (a "shared action").

D.22 Where a remediation action is a shared action, there are two possible relationships between it and the significant pollutant linkages to which it is referable. The enforcing authority should establish whether the shared action is:

(a) a "common action"—that is, an action which addresses together all of the significant pollutant linkages to which it is referable, and which would have been part of the remediation package for each of those linkages if each of them had been addressed separately; or

(b) a "collective action"—that is, an action which addresses together all of the significant pollutant linkages to which it is referable, but which would not have been part of the remediation package for every one of those linkages if each of them had been addressed separately, because:

(i) the action would not have been appropriate in that form for one or more of the linkages (since some different solution would have been more appropriate),

(ii) the action would not have been needed to the same extent for one or more of the linkages (since a less far-reaching version of that type of action would have sufficed), or

(iii) the action represents a more economic way of addressing the linkages together which would not be possible if they were addressed separately.

D.23 A collective action replaces actions that would have been appropriate for the individual significant pollutant linkages if they had been addressed separately, as it achieves the purposes which those other actions would have achieved.

Third stage – Attributing responsibility between liability groups

D.24 This stage of the procedure does not apply in the simpler cases. Where there is only a single significant pollutant linkage, the liability group for that linkage bears the full cost of carrying out any remediation action. (Where the linkage is an orphan linkage, the enforcing authority has the power to carry out the remediation action itself, at its own cost.)

D.25 Similarly, for any single-linkage action, the liability group for the significant pollutant linkage in question bears the full cost of carrying out that action

D.26 However, the enforcing authority should apply the guidance in Part 9 with respect to each shared action, in order to attribute to each of the different liability groups their share of responsibility for that action.

D.27 After the guidance in Part 9 has been applied to all shared actions, it may be the case that a Class B liability group which has been identified does not have to bear the costs for any remediation actions. Where this is the case, the enforcing authority does not need to apply any of the rest of the guidance in this Chapter to that liability group.

Fourth stage – Excluding members of a liability group

D.28 The enforcing authority should now consider, for each liability group which has two or more members, whether any of those members should be excluded from liability:

(a) for each Class A liability group with two or more members, the enforcing authority should apply the guidance on exclusion in Part 5; and

(b) for each Class B liability group with two or more members, the enforcing authority should apply the guidance on exclusion in Part 7.

Fifth stage – Apportioning liability between members of a liability group

D.29 The enforcing authority should now determine how any costs attributed to each liability group should be apportioned between the members of that group who remain after any exclusions have been made.

D.30 For any liability group which has only a single remaining member, that person bears all of the costs falling to that liability group, that is both the cost of any single-linkage action referable to the significant pollutant linkage in question, and the share of the cost of any shared action attributed to the group as a result of the attribution process set out in Part 9.

D.31 For any liability group which has two or more remaining members, the enforcing authority should apply the relevant guidance on apportionment between those members. Each of the remaining members of the group will then bear the proportion determined under that guidance of the total costs falling to the group, that is both the cost of any single-linkage action referable to the significant pollutant linkage in question, and the share of the cost of any shared action attributed to the group as a result of the attribution process set out in Part 9. The relevant apportionment guidance is:

(a) for any Class A liability group, the guidance set out in Part 6; and
(b) for any Class B liability group, the guidance set out in Part 8.

Part 4 General considerations relating to the exclusion, apportionment and attribution procedures

D.32 This Part sets out general guidance about the application of the exclusion, apportionment and attribution procedures set out in the rest of this Chapter. It is accordingly issued under both section 78F(6) and section 78F(7).

D.33 The enforcing authority should ensure that any person who might benefit from an exclusion, apportionment or attribution is aware of the guidance in this Chapter, so that they may make appropriate representations to the enforcing authority.

D.34 The enforcing authority should apply the tests for exclusion (in Parts 5 and 7) with respect to the members of each liability group. If a person, who would otherwise be an appropriate person to bear responsibility for a particular remediation action, has been excluded from the liability groups for all of the significant pollutant linkages to which that action is referable, he should be treated as not being an appropriate person in relation to that remediation action.

Financial circumstances

D.35 The financial circumstances of those concerned should have no bearing on the application of the procedures for exclusion, apportionment and attribution in this Chapter, except where the circumstances in paragraph D.85 below apply (the financial circumstances of those concerned are taken into account in the separate consideration under section 78P(2) on hardship and cost recovery). In particular, it should be irrelevant in the context of decisions on exclusion and apportionment:

(a) whether those concerned would benefit from any limitation on the recovery of costs under the provisions on hardship and cost recovery in section 78P(2); or
(b) whether those concerned would benefit from any insurance or other means of transferring their responsibilities to another person.

Information and decisions

D.36 The enforcing authority should make reasonable endeavours to consult those who may be affected by any exclusion, apportionment or attribution. In all cases, however, it should seek to obtain only such information as it is reasonable to seek, having regard to:

(a) how the information might be obtained;
(b) the cost of obtaining the information for all parties involved; and
(c) the potential significance of the information for any decision.

D.37 The statutory guidance in this Chapter should be applied in the light of the circumstances as they appear to the enforcing authority on the basis of the evidence available to it at that time. The enforcing authority's judgements should be made on the basis of the balance of probabilities. The enforcing authority should take into account the information that it has acquired in the light of the guidance in the previous paragraph, but the burden of providing the authority with any further information needed to establish an exclusion or to influence an apportionment or attribution should rest on any person seeking such a benefit. The enforcing authority should consider any relevant information which has been provided by those potentially liable under these provisions. Where any such person provides such information, any other person who may be affected by an exclusion, apportionment or attribution based on that information should be given a reasonable opportunity to comment on that information before the determination is made.

Agreements on liabilities

D.38 In any case where:

(a) two or more persons are appropriate persons and thus responsible for all or part of the costs of a remediation action;
(b) they agree, or have agreed, the basis on which they wish to divide that responsibility; and
(c) a copy of the agreement is provided to the enforcing authority and none of the parties to the agreement informs the authority that it challenges the application of the agreement;

the enforcing authority should generally make such determinations on exclusion, apportionment and attribution as are needed to give effect to this agreement, and should not apply the remainder of this guidance for exclusion, apportionment or attribution between the parties to the agreement. However, the enforcing authority should apply the guidance to determine any exclusions, apportionments or attributions between any or all of those parties and any other appropriate persons who are not parties to the agreement.

D.39 However, where giving effect to such an agreement would increase the share of the costs theoretically to be borne by a person who would benefit from a limitation on recovery of remediation costs under the provision on hardship in section 78P(2)(a) or under the guidance on cost recovery issued under section 78P(2)(b), the enforcing authority should disregard the agreement.

Part 5 Exclusion of members of a Class A liability group

D.40 The guidance in this Part is issued under section 78F(6) and, with respect to effects of the exclusion tests on apportionment (see paragraph D.43 below in particular), under section 78F(7). It sets out the tests for determining whether to exclude from liability a person who would otherwise be a Class A person (that is, a person who has been identified as responsible for remediation costs by reason of his having "caused or knowingly permitted" the presence of a significant pollutant). The tests are intended to establish whether, in relation to other members of the liability group, it is fair that he should bear any part of that responsibility.

D.41 The exclusion tests in this Part are subject to the following overriding guidance:

(a) the exclusions that the enforcing authority should make are solely in respect of the significant pollutant linkage giving rise to the liability of the liability group in question; an exclusion in respect of one significant pollutant linkage has no necessary implication in respect to any other such linkage, and a person who has been excluded with respect to one linkage may still be liable to meet all or part of the cost of carrying out a remediation action by reason of his membership of another liability group;

(b) the tests should be applied in the sequence in which they are set out; and

(c) if the result of applying a test would be to exclude all of the members of the liability group who remain after any exclusions resulting from previous tests, that further test should not be applied, and consequently the related exclusions should not be made.

D.42 The effect of any exclusion made under Test 1, or Tests 4 to 6 below should be to remove completely any liability that would otherwise have fallen on the person benefiting from the exclusion. Where the enforcing authority makes any exclusion under one of these tests, it should therefore apply any subsequent exclusion tests, and make any apportionment within the liability group, in the same way as it would have done if the excluded person had never been a member of the liability group.

D.43 The effect of any exclusion made under Test 2 ("Payments Made for Remediation") or Test 3 ("Sold with Information"), on the other hand, is intended to be that the person who received the payment or bought the land, as the case may be, (the "payee or buyer") should bear the liability of the person excluded (the "payer or seller") in addition to any liability which he is to bear in respect of his own actions or omissions. To achieve this, the enforcing authority should:

(a) complete the application of the other exclusion tests and then apportion liability between the members of the liability group, as if the payer or seller were not excluded as a result of Test 2 or Test 3; and

(b) then apportion any liability of the payer or seller, calculated on this hypothetical basis, to the payee or buyer, in addition to the liability (if any) that the payee or buyer has in respect of his own actions or omissions; this should be done even if the payee or buyer would otherwise have been excluded from the liability group by one of the other exclusion tests.

Related companies

D.44 Before applying any of the exclusion tests, the enforcing authority should establish whether two or more of the members of the liability group are "related companies".

D.45 Where the question to be considered in any exclusion test concerns the relationship between, or the relative positions of, two or more related companies, the enforcing authority should not apply the test so as to exclude any of the related companies. For example, in Test 3 ("Sold with Information"), if the "seller" and the "buyer" are related companies, the "seller" would not be excluded by virtue of that Test.

D.46 For these purposes, "related companies" are those which are, or were at the "relevant date", members of a group of companies consisting of a "holding company" and its "subsidiaries". The "relevant date" is that on which the enforcing authority first served on anyone a notice under section 78B(3) identifying the land as contaminated land, and the terms "holding company" and "subsidiaries" have the same meaning as in section 736 of the Companies Act 1985.

The exclusion tests for Class A persons

Test 1 – "Excluded activities"

D.47 The purpose of this test is to exclude those who have been identified as having caused or knowingly permitted the land to be contaminated land solely by reason of having carried out certain activities. The activities are ones which, in the Government's view, carry such limited responsibility, if any, that exclusion would be justified even where the activity is held to amount to "causing or knowingly permitting" under Part IIA. It does not imply that the carrying out of such activities necessarily amounts to "causing or knowingly permitting".

D.48 In applying this test with respect to any appropriate person, the enforcing authority should consider whether the person in question is a member of a liability group solely by reason of one or more of the following activities (not including any associated activity outside these descriptions):

(a) providing (or withholding) financial assistance to another person (whether or not that other person is a member of the liability group), in the form of any one or more of the following:

(i) making a grant,
(ii) making a loan or providing any other form of credit, including instalment credit, leasing arrangements and mortgages,
(iii) guaranteeing the performance of a person's obligations,
(iv) indemnifying a person in respect of any loss, liability or damage,
(v) investing in the undertaking of a body corporate by acquiring share capital or loan capital of that body without thereby acquiring such control as a "holding company" has over a "subsidiary" as defined in section 736 of the Companies Act 1985, or
(vi) providing a person with any other financial benefit (including the remission in whole or in part of any financial liability or obligation);

(b) underwriting an insurance policy under which another person was insured in respect of any occurrence, condition or omission by reason of which that other person has been held to have caused or knowingly permitted the significant pollutant to be in, on or under the land in question; for the purposes of this sub-paragraph:

(i) underwriting an insurance policy is to be taken to include imposing any conditions on the person insured, for example relating to the manner in which he carries out the insured activity, and
(ii) it is irrelevant whether or not the insured person can now be found;

(c) as a provider of financial assistance or as an underwriter, carrying out any action for the purpose of deciding whether or not to provide such financial assistance or underwrite such an insurance policy as is mentioned above; this sub-paragraph does not apply to the carrying out of any intrusive investigation in respect of the land in question for the purpose of making that decision where:

(i) the carrying out of that investigation is itself a cause of the existence, nature or continuance of the significant pollutant linkage in question, and
(ii) the person who applied for the financial assistance or insurance is not a member of the liability group;

(d) consigning, as waste, to another person the substance which is now a significant pollutant, under a contract under which that other person knowingly took over responsibility for its proper disposal or other management on a site not under the control of the person seeking to be excluded from liability; (for the purpose of this sub-paragraph, it is irrelevant whether or not the person to whom the waste was consigned can now be found);

(e) creating at any time a tenancy over the land in question in favour of another person who has subsequently caused or knowingly permitted the presence of the significant pollutant linkage in question (whether or not the tenant can now be found);

(f) as owner of the land in question, licensing at any time its occupation by another person who has subsequently caused or knowingly permitted the presence of the significant pollutant in question (whether or not the licensee can now be found); this test does not apply in a case where the person granting the licence operated the land as a site for the disposal or storage of waste at the time of the grant of the licence;

(g) issuing any statutory permission, licence or consent required for any action or omission by reason of which some other person appears to the enforcing authority to have caused or knowingly permitted the presence of the significant pollutant in question (whether or not that other person can now be found); this test does not apply in the case of statutory undertakers granting permission for their contractors to carry out works;

(h) taking, or not taking, any statutory enforcement action:

  (i) with respect to the land, or
  (ii) against some other person who appears to the enforcing authority to have caused or knowingly permitted the presence of the significant pollutant in question, whether or not that other person can now be found;

(i) providing legal, financial, engineering, scientific or technical advice to (or design, contract management or works management services for) another person (the "client"), whether or not that other person can now be found:

  (i) in relation to an action or omission (or a series of actions and/or omissions) by reason of which the client has been held to have caused or knowingly permitted the presence of the significant pollutant,
  (ii) for the purpose of assessing the condition of the land, for example whether it might be contaminated, or
  (iii) for the purpose of establishing what might be done to the land by way of remediation;

(j) as a person providing advice or services as described in sub-paragraph (i) above carrying out any intrusive investigation in respect of the land in question, except where:

  (i) the investigation is itself a cause of the existence, nature or continuance of the significant pollutant linkage in question, and
  (ii) the client is not a member of the liability group; or

(k) performing any contract by providing a service (whether the contract is a contract of service (employment), or a contract for services) or by supplying goods, where the contract is made with another person who is also a member of the liability group in question; for the purposes of this sub-paragraph and paragraph D.49 below, the person providing the service or supplying the goods is referred to as the "contractor" and the other party as the "employer"; this sub-paragraph applies to subcontracts where either the ultimate employer or an intermediate contractor is a member of the liability group; this sub-paragraph does not apply where:

  (i) the activity under the contract is of a kind referred to in a previous sub-paragraph of this paragraph,

(ii) the action or omission by the contractor by virtue of which he has been identified as an appropriate person was not in accordance with the terms of the contract, or

(iii) the circumstances in paragraph D.49 below apply.

D.49 The circumstances referred to in paragraph D.48(k)(iii) are:

(a) the employer is a body corporate;

(b) the contractor was a director, manager, secretary or other similar officer of the body corporate, or a person purporting to act in any such capacity, at the time when the contract was performed; and

(c) the action or omissions by virtue of which the employer has been identified as an appropriate person were carried out or made with the consent or connivance of the contractor, or were attributable to any neglect on his part.

D.50 If any of the circumstances in paragraph D.48 above apply, the enforcing authority should exclude the person in question.

Test 2 – "Payments made for remediation"

D.51 The purpose of this test is to exclude from liability those who have already, in effect, met their responsibilities by making certain kinds of payment to some other member of the liability group, which would have been sufficient to pay for adequate remediation.

D.52 In applying this test, the enforcing authority should consider whether all the following circumstances exist:

(a) one of the members of the liability group has made a payment to another member of that liability group for the purpose of carrying out particular remediation on the land in question; only payments of the kinds set out in paragraph D.53 below are to be taken into account;

(b) that payment would have been sufficient at the date when it was made to pay for the remediation in question;

(c) if the remediation for which the payment was intended had been carried out effectively, the land in question would not now be in such a condition that it has been identified as contaminated land by reason of the significant pollutant linkage in question; and

(d) the remediation in question was not carried out or was not carried out effectively.

D.53 Payments of the following kinds alone should be taken into account:

(a) a payment made voluntarily, or to meet a contractual obligation, in response to a claim for the cost of the particular remediation;

(b) a payment made in the course of a civil legal action, or arbitration, mediation or dispute resolution procedure, covering the cost of the particular remediation, whether paid as part of an out-of-court settlement, or paid under the terms of a court order; or

(c) a payment as part of a contract (including a group of interlinked contracts) for the transfer of ownership of the land in question which is either specifically provided for in the contract to meet the cost of carrying out the particular remediation or which consists of a reduction in the contract price explicitly stated in the contract to be for that purpose.

D.54 For the purposes of this test, payments include consideration of any form.

D.55 However, no payment should be taken into account where the person making the payment retained any control after the date of the payment over the condition of the land in question (that is, over whether or not the substances by reason of which the land is regarded as contaminated land were permitted to be in, on or under the land). For this purpose, neither of the following should be regarded as retaining control over the condition of the land:

(a) holding contractual rights to ensure the proper carrying out of the remediation for which the payment was made; nor
(b) holding an interest or right of any of the following kinds:

(i) easements for the benefit of other land, where the contaminated land in question is the servient tenement, and statutory rights of an equivalent nature,
(ii) rights of statutory undertakers to carry out works or install equipment,
(iii) reversions upon expiry or termination of a long lease, or
(iv) the benefit of restrictive covenants or equivalent statutory agreements.

D.56 If all of the circumstances set out in paragraph D.52 above apply, the enforcing authority should exclude the person who made the payment in respect of the remediation action in question. (See paragraph D.43 above for guidance on how this exclusion should be made.)

Test 3 – "Sold with information"

D.57 The purpose of this test is to exclude from liability those who, although they have caused or knowingly permitted the presence of a significant pollutant in, on or under some land, have disposed of that land in circumstances where it is reasonable that another member of the liability group, who has acquired the land from them, should bear the liability for remediation of the land.

D.58 In applying this test, the enforcing authority should consider whether all the following circumstances exist:

(a) one of the members of the liability group (the "seller") has sold the land in question to a person who is also a member of the liability group (the "buyer");
(b) the sale took place at arms' length (that is, on terms which could be expected in a sale on the open market between a willing seller and a willing buyer);
(c) before the sale became binding, the buyer had information that would reasonably allow that particular person to be aware of the presence on the land of the pollutant identified in the significant pollutant linkage in question, and the broad measure of that presence; andthe seller did nothing material to misrepresent the implications of that presence; and
(d) after the date of the sale, the seller did not retain any interest in the land in question or any rights to occupy or use that land.

D.59 In determining whether these circumstances exist:

(a) a sale of land should be regarded as being either the transfer of the freehold or the grant or assignment of a long lease; for this purpose, a "long lease" means a lease (or sub-lease) granted for a period of more than 21 years under which the lessee satisfies the definition of "owner" set out in section 78A(9);
(b) the question of whether persons are members of a liability group should be decided on the circumstances as they exist at the time of the determination (and not as they might have been at the time of the sale of the land);
(c) where there is a group of transactions or a wider agreement (such as the sale of a company or business) including a sale of land, that

sale of land should be taken to have been at arms' length where the person seeking to be excluded can show that the net effect of the group of transactions or the agreement as a whole was a sale at arms' length;

(d) in transactions since the beginning of 1990 where the buyer is a large commercial organisation or public body, permission from the seller for the buyer to carry out his own investigations of the condition of the land should normally be taken as sufficient indication that the buyer had the information referred to in paragraph D.58(c) above; and

(e) for the purposes of paragraph D.58(d) above, the following rights should be disregarded in deciding whether the seller has retained an interest in the contaminated land in question or rights to occupy or use it:

(i) easements for the benefit of other land, where the contaminated land in question is the servient tenement, and statutory rights of an equivalent nature,
(ii) rights of statutory undertakers to carry out works or install equipment,
(iii) reversions upon expiry or termination of a long lease, and
(iv) the benefit of restrictive covenants or equivalent statutory agreements.

D.60 If all of the circumstances in paragraph D.58 above apply, the enforcing authority should exclude the seller. (See paragraph D.43 above for guidance on how this exclusion should be made.)

D.61 This test does not imply that the receipt by the buyer of the information referred to in paragraph D.58(c) above necessarily means that the buyer has "caused or knowingly permitted" the presence of the significant pollutant in, on or under the land.

Test 4 – "Changes to substances"

D.62 The purpose of this test is to exclude from liability those who are members of a liability group solely because they caused or knowingly permitted the presence in, on or under the land of a substance which has only led to the creation of a significant pollutant linkage because of its interaction with another substance which was later introduced to the land by another person.

D.63 In applying this test, the enforcing authority should consider whether all the following circumstances exist:

(a) the substance forming part of the significant pollutant linkage in question is present, or has become a significant pollutant, only as the result of a chemical reaction, biological process or other change (the "intervening change") involving:

(i) both a substance (the "earlier substance") which would not have formed part of the significant pollutant linkage if the intervening change had not occurred, and
(ii) one or more other substances (the "later substances");

(b) the intervening change would not have occurred in the absence of the later substances;

(c) a person (the "first person") is a member of the liability group because he caused or knowingly permitted the presence in, on or under the land of the earlier substance, but he did not cause or knowingly permit the presence of any of the later substances;

(d) one or more other persons are members of the liability group because they caused or knowingly permitted the later substances to be in, on or under the land;

(e) before the date when the later substances started to be introduced in, on or under the land, the first person:

(i) could not reasonably have foreseen that the later substances would be introduced onto the land,
(ii) could not reasonably have foreseen that, if they were, the intervening change would be likely to happen, or
(iii) took what, at that date, were reasonable precautions to prevent the introduction of the later substances or the occurrence of the intervening change, even though those precautions have, in the event, proved to be inadequate; and

(f) after that date, the first person did not:

(i) cause or knowingly permit any more of the earlier substance to be in, on or under the land in question,
(ii) do anything which has contributed to the conditions that brought about the intervening change, or
(iii) fail to do something which he could reasonably have been expected to do to prevent the intervening change happening.

D.64 If all of the circumstances in paragraph D.63 above apply, the enforcing authority should exclude the first person (or persons, if more than one member of the liability group meets this description).

Test 5 – "Escaped substances"

D.65 The purpose of this test is to exclude from liability those who would otherwise be liable for the remediation of contaminated land which has become contaminated as a result of the escape of substances from other land, where it can be shown that another member of the liability group was actually responsible for that escape.

D.66 In applying this test, the enforcing authority should consider whether all the following circumstances exist:

(a) a significant pollutant is present in, on or under the contaminated land in question wholly or partly as a result of its escape from other land;
(b) a member of the liability group for the significant pollutant linkage of which that pollutant forms part:

(i) caused or knowingly permitted the pollutant to be present in, on or under that other land (that is, he is a member, of that liability group by reason of section 78K(1)), and
(ii) is a member of that liability group solely for that reason; and

(c) one or more other members of that liability group caused or knowingly permitted the significant pollutant to escape from that other land and its escape would not have happened but for their actions or omissions.

D.67 If all of the circumstances in paragraph D.66 above apply, the enforcing authority should exclude any person meeting the description in paragraph D.66(b) above.

Test 6 – "Introduction of pathways or receptors"

D.68 The purpose of this test is to exclude from liability those who would otherwise be liable solely because of the subsequent introduction by others of the relevant pathways or receptors (as defined in Chapter A) in the significant pollutant linkage.

D.69 In applying this test, the enforcing authority should consider whether all the following circumstances exist:

(a) one or more members of the liability group have carried out a relevant action, and/or made a relevant omission ("the later actions"), either

(i) as part of the series of actions and/or omissions which amount to their having caused or knowingly permitted the presence of the pollutant in a significant pollutant linkage, or

(ii) in addition to that series of actions and/or omissions;

(b) the effect of the later actions has been to introduce the pathway or the receptor which form part of the significant pollutant linkage in question;

(c) if those later actions had not been carried out or made, the significant pollutant linkage would either not have existed, or would not have been a significant pollutant linkage, because of the absence of a pathway or of a receptor; and

(d) a person is a member of the liability group in question solely by reason of his carrying out other actions or making other omissions ("the earlier actions") which were completed before any of the later actions were carried out or made.

D.70 For the purpose of this test:

(a) a "relevant action" means:

(i) the carrying out at any time of building, engineering, mining or other operations in, on, over or under the land in question, and/or

(ii) the making of any material change in the use of the land in question for which a specific application for planning permission was required to be made (as opposed to permission being granted, or deemed to be granted, by general legislation or by virtue of a development order, the adoption of a simplified planning zone or the designation of an enterprise zone) at the time when the change in use was made; and

(b) a "relevant omission" means:

(i) in the course of a relevant action, failing to take a step which would have ensured that a significant pollutant linkage was not brought into existence as a result of that action, and/or

(ii) unreasonably failing to maintain or operate a system installed for the purpose of reducing or managing the risk associated with the presence on the land in question of the significant pollutant in the significant pollutant linkage in question.

D.71 This test applies only with respect to developments on, or changes in the use of, the contaminated land itself; it does not apply where the relevant acts or omissions take place on other land, even if they have the effect of introducing pathways or receptors.

D.72 If all of the circumstances in paragraph D.69 above apply, the enforcing authority should exclude any person meeting the description at paragraph D.69(d) above.

Part 6 Apportionment between members of any single Class A liability group

D.73 The statutory guidance in this Part is issued under section 78F(7) and sets out the principles on which liability should be apportioned within each Class A liability group as it stands after any members have been excluded from liability with respect to the relevant significant pollutant linkage as a result of the application of the exclusion tests in Part 5.

D.74 The history and circumstances of different areas of contaminated land, and the nature of the responsibility of each of the members of any Class A liability group for a significant pollutant linkage, are likely to vary greatly. It is therefore not possible to prescribe detailed rules for the apportionment of liability between those members which would be fair and appropriate in all cases.

General principles

D.75 In apportioning costs between the members of a Class A liability group who remain after any exclusions have been made, the enforcing authority should follow the general principle that liability should be apportioned to reflect the relative responsibility of each of those members for creating or continuing the risk now being caused by the significant pollutant linkage in question. (For these purposes, "risk" has the same meaning as that given in Chapter A.) In applying this principle, the enforcing authority should follow, where appropriate, the specific approaches set out in paragraphs D.77 to D.86 below.

D.76 If appropriate information is not available to enable the enforcing authority to make such an assessment of relative responsibility (and, following the guidance at paragraph D.36 above, such information cannot reasonably be obtained) the authority should apportion liability in equal shares among the remaining members of the liability group for any significant pollutant linkage, subject to the specific guidance in paragraph D.85 below.

Specific approaches

Partial applicability of an exclusion test

D.77 If, for any member of the liability group, the circumstances set out in any of the exclusion tests in Part 5 above apply to some extent, but not sufficiently to mean that the an exclusion should be made, the enforcing authority should assess that person's degree of responsibility as being reduced to the extent which is appropriate in the light of all the circumstances and the purpose of the test in question. For example, in considering Test 2, a payment may have been made which was sufficient to pay for only half of the necessary remediation at that time—the authority could therefore reduce the payer's responsibility by half.

The entry of a substance vs. its continued presence

D.78 In assessing the relative responsibility of a person who has caused or knowingly permitted the entry of a significant pollutant into, onto or under land (the "first person") and another person who has knowingly permitted the continued presence of that same pollutant in, on or under that land (the "second person"), the enforcing authority should consider the extent to which the second person had the means and a reasonable opportunity to deal with the presence of the pollutant in question or to reduce the seriousness of the implications of that presence. The authority should then assess the relative responsibilities on the following basis:

(a) if the second person had the necessary means and opportunity, he should bear the same responsibility as the first person;

(b) if the second person did not have the means and opportunity, his responsibility relative to that of the first person should be substantially reduced; and

(c) if the second person had some, but insufficient, means or opportunity, his responsibility relative to that of the first person should be reduced to an appropriate extent.

Persons who have caused or knowingly permitted the entry of a significant pollutant

D.79 Where the enforcing authority is determining the relative responsibilities of members of the liability group who have caused or knowingly permited the entry of the significant pollutant into, onto or under the land, it should follow the approach set out in paragraphs D.80 to D.83 below.

D.80 If the nature of the remediation action points clearly to different members of the liability group being responsible for particular circumstances at which the action is aimed, the enforcing authority should apportion responsibility in accordance with that indication. In particular, where different persons were in control of different areas of the land in question, and there is no interrelationship between those areas, the enforcing authority should regard the persons in control of the different areas as being separately responsible for the events which make necessary the remediation actions or parts of actions referable to those areas of land.

D.81 If the circumstances in paragraph D.80 above do not apply, but the quantity of the significant pollutant present is a major influence on the cost of remediation, the enforcing authority should regard the relative amounts of that pollutant which are referable to the different persons as an appropriate basis for apportioning responsibility.

D.82 If it is deciding the relative quantities of pollutant which are referable to different persons, the enforcing authority should consider first whether there is direct evidence of the relative quantities referable to each person. If there is such evidence, it should be used. In the absence of direct evidence, the enforcing authority should see whether an appropriate surrogate measure is available. Such surrogate measures can include:

(a) the relative periods during which the different persons carried out broadly equivalent operations on the land;
(b) the relative scale of such operations carried out on the land by the different persons (a measure of such scale may be the quantities of a product that were produced);
(c) the relative areas of land on which different persons carried out their operations; and
(d) combinations of the foregoing measures.

D.83 In cases where the circumstances in neither paragraph D.80 nor D.81 above apply, the enforcing authority should consider the nature of the activities carried out by the appropriate persons concerned from which the significant pollutant arose. Where these activities were broadly equivalent, the enforcing authority should apportion responsibility in proportion to the periods of time over which the different persons were in control of those activities. It would be appropriate to adjust this apportionment to reflect circumstances where the persons concerned carried out activities which were not broadly equivalent, for example where they were on a different scale.

Persons who have knowingly permitted the continued presence of a pollutant

D.84 Where the enforcing authority is determining the relative responsibilities of members of the liability group who have knowingly permitted the continued presence, over a period of time, of a significant pollutant in, on or under land, it should apportion that responsibility in proportion to:

(a) the length of time during which each person controlled the land;

(b) the area of land which each person controlled;
(c) the extent to which each person had the means and a reasonable opportunity to deal with the presence of the pollutant in question or to reduce the seriousness of the implications of that presence; or
(d) a combination of the foregoing factors.

Companies and officers

D.85 If, following the application of the exclusion tests (and in particular the specific guidance at paragraphs D.48(k)(iii) and D.49 above) both a company and one or more of its relevant officers remain as members of the liability group, the enforcing authority should apportion liability on the following bases:

(a) the enforcing authority should treat the company and its relevant officers as a single unit for the purposes of:

(i) applying the general principle in paragraph D.75 above (ie it should consider the responsibilities of the company and its relevant officers as a whole, in comparison with the responsibilities of other members of the liability group), and
(ii) making any apportionment required by paragraph D.76 above; and

(b) having determined the share of liability falling to the company and its relevant officers together, the enforcing authority should apportion responsibility between the company and its relevant officers on a basis which takes into account the degree of personal responsibility of those officers, and the relative levels of resources which may be available to them and to the company to meet the liability.

D.86 For the purposes of paragraph D.85 above, the "relevant officers" of a company are any director, manager, secretary or other similar officer of the company, or any other person purporting to act in any such capacity.

Part 7 Exclusion of members of a Class B liability group

D.87 The guidance in this Part is issued under section 78F(6) and sets out the test which should be applied in determining whether to exclude from liability a person who would otherwise be a Class B person (that is, a person liable to meet remediation costs solely by reason of ownership or occupation of the land in question). The purpose of the test is to exclude from liability those who do not have an interest in the capital value of the land in question.

D.88 The test applies where two or more persons have been identified as Class B persons for a significant pollutant linkage.

D.89 In such circumstances, the enforcing authority should exclude any Class B person who either:

(a) occupies the land under a licence, or other agreement, of a kind which has no marketable value or which he is not legally able to assign or transfer to another person (for these purposes the actual marketable value, or the fact that a particular licence or agreement may not actually attract a buyer in the market, are irrelevant); or
(b) is liable to pay a rent which is equivalent to the rack rent for such of the land in question as he occupies and holds no beneficial interest in that land other than any tenancy to which such rent relates; where the rent is subject to periodic review, the rent should be

considered to be equivalent to the rack rent if, at the latest review, it was set at the full market rent at that date.

D.90 However, the test should not be applied, and consequently no exclusion should be made, if it would result in the exclusion of all of the members of the liability group.

Part 8 Apportionment between the members of a single Class B liability group

D.91 The statutory guidance in this Part is issued under section 78F(7) and sets out the principles on which liability should be apportioned within each Class B liability group as it stands after any members have been excluded from liability with respect to the relevant significant pollutant linkage as a result of the application of the exclusion test in Part 7.

D.92 Where the whole or part of a remediation action for which a Class B liability group is responsible clearly relates to a particular area within the land to which the significant pollutant linkage as a whole relates, liability for the whole, or the relevant part, of that action should be apportioned amongst those members of the liability group who own or occupy that particular area of land.

D.93 Where those circumstances do not apply, the enforcing authority should apportion liability for the remediation actions necessary for the significant pollutant linkage in question amongst all of the members of the liability group.

D.94 Where the enforcing authority is apportioning liability amongst some or all of the members of a Class B liability group, it should do so in proportion to the capital values of the interests in the land in question, which include those of any buildings or structures on the land:

(a) where different members of the liability group own or occupy different areas of land, each such member should bear responsibility in the proportion that the capital value of his area of land bears to the aggregate of the capital values of all the areas of land; and

(b) where different members of the liability group have an interest in the same area of land, each such member should bear responsibility in the proportion which the capital value of his interest bears to the aggregate of the capital values of all those interests; and

(c) where both the ownership or occupation of different areas of land and the holding of different interests come into the question, the overall liability should first be apportioned between the different areas of land and then between the interests within each of those areas of land, in each case in accordance with the last two sub-paragraphs.

D.95 The capital value used for these purposes should be that estimated by the enforcing authority, on the basis of the available information, disregarding the existence of any contamination. The value should be estimated in relation to the date immediately before the enforcing authority first served a notice under section 78B(3) in relation to that land. Where the land in question is reasonably uniform in nature and amenity and is divided among a number of owner-occupiers, it can be an acceptable approximation of this basis of apportionment to make the apportionment on the basis of the area occupied by each.

D.96 Where part of the land in question is land for which no owner or occupier can be found, the enforcing authority should deduct the share of costs attributable to that land on the basis of the respective capital values of that land and the other land in question before making a determination of liability.

D.97 If appropriate information is not available to enable the enforcing authority to make an assessment of relative capital values (and, following the guidance at paragraph D.36 above, such information cannot reasonably be obtained), the enforcing authority should apportion liability in equal shares among all the members of the liability group.

Part 9 Attribution of responsibility between liability groups

D.98 The statutory guidance in this Part is issued under section 78F(7) and applies where one remediation action is referable to two or more significant pollutant linkages (that is, it is a "shared action"). This can occur either where both linkages require the same action (that is, it is a "common action") or where a particular action is part of the best combined remediation scheme for two or more linkages (that is, it is a "collective action"). This Part provides statutory guidance on the attribution of responsibility for the costs of any shared action between the liability groups for the linkages to which it is referable.

Attributing responsibility for the cost of shared actions between liability groups

D.99 The enforcing authority should attribute responsibility for the costs of any common action among the liability groups for the significant pollutant linkages to which it is referable on the following basis:

(a) if there is a single Class A liability group, then the full cost of carrying out the common action should be attributed to that group, and no cost should be attributed to any Class B liability group);
(b) if there are two or more Class A liability groups, then an equal share of the cost of carrying out the common action should be attributed to each of those groups, and no cost should be attributed to any Class B liability group); and
(c) if there is no Class A liability group and there are two or more Class B liability groups, then the enforcing authority should treat those liability groups as if they formed a single liability group, attributing the cost of carrying out the common action to that combined group, and applying the guidance on exclusion and apportionment set out in Parts 7 and 8 of this Chapter as between all of the members of that combined group.

D.100 The enforcing authority should attribute responsibility for the cost of any collective action among the liability groups for the significant pollutant linkages to which it is referable on the same basis as for the costs of a common action, except that where the costs fall to be divided among several Class A liability groups, instead of being divided equally, they should be attributed on the following basis:

(a) having estimated the costs of the collective action, the enforcing authority should also estimate the hypothetical cost for each of the liability groups of carrying out the actions which are subsumed by the collective action and which would be necessary if the significant pollutant linkage for which that liability group is responsible were to be addressed separately; these estimates are the "hypothetical estimates" of each of the liability groups;

(b) the enforcing authority should then attribute responsibility for the cost of the collective action between the liability groups in the proportions which the hypothetical estimates of each liability group bear to the aggregate of the hypothetical estimates of all the groups.

Confirming the attribution of responsibility

D.101 If any appropriate person demonstrates, before the service of a remediation notice, to the satisfaction of the enforcing authority that the result of an attribution made on the basis set out in paragraphs D.99 and D.100 above would have the effect of the liability group of which he is a member having to bear a liability which is so disproportionate (taking into account the overall relative responsibilities of the persons or groups concerned for the condition of the land) as to make the attribution of responsibility between all the liability groups concerned unjust when considered as a whole, the enforcing authority should reconsider the attribution. In doing so, the enforcing authority should consult the other appropriate persons concerned.

D.102 If the enforcing authority then agrees that the original attribution would be unjust it should adjust the attribution between the liability groups so that it is just and fair in the light of all the circumstances. An adjustment under this paragraph should be necessary only in very exceptional cases.

Orphan linkages

D.103 As explained above at paragraphs D.12, D.14 and D.17 above, an orphan linkage may arise where:

(a) the significant pollutant linkage relates solely to the pollution of controlled waters (and not to significant harm) and no Class A person can be found;
(b) no Class A or Class B persons can be found; or
(c) those who would otherwise be liable are exempted by one of the relevant statutory provisions (ie sections 78J(3), 78K or 78X(3)).

D.104 In any case where only one significant pollutant linkage has been identified, and that is an orphan linkage, the enforcing authority should itself bear the cost of any remediation which is carried out.

D.105 In more complicated cases, there may be two or more significant pollutant linkages, of which some are orphan linkages. Where this applies, the enforcing authority will need to consider each remediation action separately.

D.106 For any remediation action which is referable to an orphan linkage, and is not referable to any other linkage for which there is a liability group, the enforcing authority should itself bear the cost of carrying out that action.

D.107 For any shared action which is referable to an orphan linkage and also to a single significant pollutant linkage for which there is a Class A liability group, the enforcing authority should attribute all of the cost of carrying out that action to that Class A liability group.

D.108 For any shared action which is referable to an orphan linkage and also to two or more significant pollutant linkages for which there are Class A liability groups, the enforcing authority should attribute the costs of carrying out that action between those liability groups in the same way as it would do if the orphan linkage did not exist.

D.109 For any shared action which is referable to an orphan linkage and also to a significant pollutant linkage for which there is a Class B liability group (and not to any significant pollutant linkage for which there is a Class A liability group) the enforcing authority should adopt the following approach:

(a) where the remediation action is a common action the enforcing authority should attribute all of the cost of carrying out that action to the Class B liability group; and

(b) where the remediation action is a collective action, the enforcing authority should estimate the hypothetical cost of the action which would be needed to remediate separately the effects of the linkage for which that group is liable. The enforcing authority should then attribute the costs of carrying out the collective action between itself and the Class B liability group so that the expected liability of that group does not exceed that hypothetical cost.

Chapter E – Statutory guidance on the recovery of the costs of remediation

Part 1 Scope of the Chapter

E.1 The statutory guidance in this Chapter is issued under section 78P(2) of the Environmental Protection Act 1990. It provides guidance on the extent to which the enforcing authority should seek to recover the costs of remediation which it has carried out and which it is entitled to recover.

E.2 Section 78P provides that:

"(1) Where, by virtue of section 78N(3)(a), (c), (e) or (f) . . . the enforcing authority does any particular thing by way of remediation, it shall be entitled, subject to sections 78J(7) and 78K(6) . . . to recover the reasonable cost incurred in doing it from the appropriate person or, if there are two or more appropriate persons in relation to the thing in question, from those persons in proportions determined pursuant to section 78F(7) . . .

(2) In deciding whether to recover the cost, and, if so, how much of the cost, which it is entitled to recover under subsection (1) above, the enforcing authority shall have regard—

(a) to any hardship which the recovery may cause to the person from whom the cost is recoverable; and

(b) to any guidance issued by the Secretary of State for the purposes of this subsection."

E.3 The guidance in this Chapter is also crucial in deciding when the enforcing authority is prevented from serving a remediation notice. Under section 78H(5), the enforcing authority may not serve a remediation notice if the authority has the power to carry out remediation itself, by virtue of section 78N. Under that latter section, the authority asks the hypothetical question of whether it would seek to recover all of the reasonable costs it would incur if it carried out the remediation itself. The authority then has the power to carry out that remediation itself if it concludes that, having regard to hardship and the guidance in this chapter, it would either not seek to recover its costs, or seek to recover only a part of its costs.

E.4 Section 78H(5) provides that:

"(5) The enforcing authority shall not serve a remediation notice on a person if and so long as . . .

(d) the authority is satisfied that the powers conferred on it by section 78N below to do what is appropriate by way of remediation are exercisable . . ."

E.5 Section 78N(3) provides that the enforcing authority has the power to carry out remediation:

"(e) where the enforcing authority considers that, were it to do some particular thing by way of remediation, it would decide, by virtue of subsection (2) of section 78P . . . or any guidance issued under that subsection, –

"(i) not to seek to recover under subsection (1) of that section any of the reasonable cost incurred by it in doing that thing; or

"(ii) to seek so to recover only a portion of that cost;. . ."

E.6 The enforcing authority is required to have regard to the statutory guidance in this Chapter.

Part 2 Definition of terms

E.7 Unless otherwise stated, any word, term or phrase given a specific meaning in Part IIA of the Environmental Protection Act 1990, or in the statutory guidance in Chapters A, B, C, or D has the same meaning for the purpose of the guidance in this Chapter.

E.8 In addition, for the purposes of the statutory guidance in this Chapter, the term "cost recovery decision" is used to describe any decision by the enforcing authority, for the purposes either of section 78P or of sections 78H and 78N, whether:

(a) to recover from the appropriate person all of the reasonable costs incurred by the authority in carrying out remediation; or

(b) not to recover those costs or to recover only part of those costs (described below as "waiving or reducing its cost recovery").

E.9 Any reference to "Part IIA" means "Part IIA of the Environmental Protection Act 1990". Any reference to a "section" in primary legislation means a section of the Environmental Protection Act 1990, unless it is specifically stated otherwise.

Part 3 Cost recovery decisions

Cost recovery decisions in general

E.10 The statutory guidance in this Part sets out considerations to which the enforcing authority should have regard when making any cost recovery decision. In view of the wide variation in situations which are likely to arise, including the history and ownership of land, and liability for its remediation, the statutory guidance in this Chapter sets out principles and approaches, rather than detailed rules. The enforcing authority will need to have regard to the circumstances of each individual case.

E.11 In making any cost recovery decision, the enforcing authority should have regard to the following general principles:

(a) the authority should aim for an overall result which is as fair and equitable as possible to all who may have to meet the costs of remediation, including national and local taxpayers; and

(b) the "polluter pays" principle, by virtue of which the costs of remediating pollution are to be borne by the polluter; the authority should therefore consider the degree and nature of responsibility of the appropriate person for the creation, or continued existence, of the circumstances which lead to the land in question being identified as contaminated land.

E.12 In general, this will mean that the enforcing authority should seek to recover in full its reasonable costs. However, the authority should waive or reduce the recovery of costs to the extent that the authority considers this appropriate and reasonable, either:

(a) to avoid any hardship which the recovery may cause to the appropriate person; or

(b) to reflect one or more of the specific considerations set out in the statutory guidance in Parts 4, 5 and 6 below.

E.13 When deciding how much of its costs it should recover in any case, the enforcing authority should consider whether it could recover more of its costs by deferring recovery and securing them by a charge on the land in question under section 78P. Such deferral may lead to payment from the appropriate person either in instalments (see section 78P(12)) or when the land is next sold.

Information for making decisions

E.14 In general, the enforcing authority should expect anyone who is seeking a waiver or reduction in the recovery of remediation costs to present any information needed to support his request.

E.15 In making any cost recovery decision, the authority should always consider any relevant information provided by the appropriate person. The authority should also seek to obtain such information as is reasonable, having regard to:

(a) how the information might be obtained;

(b) the cost, for all the parties involved, of obtaining the information; and

(c) the potential significance of the information for any decision.

E.16 The enforcing authority should, in all cases, inform the appropriate person of any cost recovery decisions taken, explaining the reasons for those decisions.

Cost recovery policies

E.17 In order to promote transparency, fairness and consistency, an enforcing authority which is a local authority may wish to prepare, adopt and make available as appropriate a policy statement about the general approach it intends to follow in making cost recovery decisions. This would outline circumstances in which it would waive or reduce cost recovery (and thereby, by inference, not serve a remediation notice because it has the powers to carry out the remediation itself), having had regard to hardship and the statutory guidance in this Chapter.

E.18 Where the Environment Agency is making a cost recovery decision with respect to a special site falling within the area of a local authority which has adopted such a policy statement, the Agency should take account of that statement.

Part 4 Considerations applying both to Class A & Class B persons

E.19 The statutory guidance in this Part sets out considerations to which the enforcing authority should have regard when making any cost recovery decisions, irrespective of whether the appropriate person is a Class A person of a Class B person (as defined in Chapter D). They apply in addition to the general issue of the "hardship" which the cost recovery may cause to the appropriate person.

Commercial enterprises

E.20 Subject to the specific guidance elsewhere in this Chapter, the enforcing authority should adopt the same approach to all types of commercial or industrial enterprises which are identified as appropriate persons. This applies whether the appropriate person is a public corporation, a limited company (whether public or private), a partnership (whether limited or not) or an individual operating as a sole trader.

Threat of business closure or insolvency

E.21 In the case of a small or medium-sized enterprise which is the appropriate person, or which is run by the appropriate person, the enforcing authority should consider:

(a) whether recovery of the full cost attributable to that person would mean that the enterprise is likely to become insolvent and thus cease to exist; and

(b) if so, the cost to the local economy of such a closure.

E.22 Where the cost of closure appears to be greater than the costs of remediation which the enforcing authority would have to bear themselves, the authority should consider waiving or reducing its costs recovery to the extent needed to avoid making the enterprise insolvent.

E.23 However, the authority should not waive or reduce its costs recovery where:

(a) it is clear that an enterprise has deliberately arranged matters so as to avoid responsibility for the costs of remediation;

(b) it appears that the enterprise would be likely to become insolvent whether or not recovery of the full cost takes place; or

(c) it appears that the enterprise could be kept in, or returned to, business even it does become insolvent under its current ownership.

E.24 For these purposes, a "small or medium-sized enterprise" is as defined in the European Commission's Community Guidelines on State Aid for Small and Medium-Sized Enterprises, published in the Official Journal of the European Communities (the reference number for the present version of the guidelines is OJ C213 1996 item 4). This can be summarised as an independent enterprise with fewer than 250 employees, and either an annual turnover not exceeding 40 million, or an annual balance sheet total not exceeding 27 million.

E.25 Where the enforcing authority is a local authority, it may wish to take account in any such cost recovery decisions of any policies it may have for assisting enterprise or promoting economic development (for example, for granting financial or other assistance under section 33 of the Local Government and Housing Act 1989, including any strategy which it has published under section 35 of that Act concerning the use of such powers).

E.26 Where the Environment Agency is the enforcing authority, it should seek to be consistent with the local authority in whose area the contaminated land in question is situated. The Environment Agency should therefore consult the local authority, and should take that authority's views into consideration in making its own cost recovery decision.

Trusts

E.27 Where the appropriate persons include persons acting as trustees, the enforcing authority should assume that such trustees will exercise all the powers which they have, or may reasonably obtain, to make funds available from the trust, or from borrowing that can be made on behalf of the trust, for the purpose of paying for remediation. The authority should, nevertheless, consider waiving or reducing its costs recovery to the extent that the costs of remediation to be recovered from the trustees would otherwise exceed the amount that can be made available from the trust to cover those costs.

E.28 However, as exceptions to the approach set out in the preceding paragraph, the authority should not waive or reduce its costs recovery:

(a) where it is clear that the trust was formed for the purpose of avoiding paying the costs of remediation; or
(b) to the extent that trustees have personally benefited, or will personally benefit, from the trust.

Charities

E.29 Since charities are intended to operate for the benefit of the community, the enforcing authority should consider the extent to which any recovery of costs from a charity would jeopardise that charity's ability to continue to provide a benefit or amenity which is in the public interest. Where this is the case, the authority should consider waiving or reducing its costs recovery to the extent needed to avoid such a consequence. This approach applies equally to charitable trusts and to charitable companies.

Social housing landlords

E.30 The enforcing authority should consider waiving or reducing its costs recovery if:

(a) the appropriate person is a body eligible for registration as a social housing landlord under section 2 of the Housing Act 1996 (for example, a housing association);
(b) its liability relates to land used for social housing; and
(c) full recovery would lead to financial difficulties for the appropriate person, such that the provision or upkeep of the social housing would be jeopardised.

E.31 The extent of the waiver or reduction should be sufficient to avoid any such financial difficulties.

Part 5 Specific considerations applying to Class A persons

E.32 The statutory guidance in this Part sets out specific considerations to which the enforcing authority should have regard in cost recovery decisions where the appropriate person is a Class A person, as defined in Chapter D (that is, a person who has caused or knowingly permitted the significant pollutant to be in, on or under the contaminated land).

E.33 In applying the approach in this Part, the enforcing authority should be less willing to waive or reduce its costs recovery where it was in the course of carrying on a business that the Class A person caused or knowingly permitted the presence of the significant pollutants, than where he was not carrying on a business. This is because in the former case he is likely to have earned profits from the activity which created or permitted the presence of those pollutants.

Where other potentially appropriate persons have not been found

E.34 In some cases where a Class A person has been found, it may be possible to identify another person who caused or knowingly permitted the presence of the significant pollutant in question, but who cannot now be found for the purposes of treating him as an appropriate person. For example, this might apply where a company has been dissolved.

E.35 The authority should consider waiving or reducing its costs recovery from a Class A person if that person demonstrates to the satisfaction of the enforcing authority that:

(a) another identified person, who cannot now be found, also caused or knowingly permitted the significant pollutant to be in, on or under the land; and

(b) if that other person could be found, the Class A person seeking the waiver or reduction of the authority's costs recovery would either:

(i) be excluded from liability by virtue of one or more of the exclusion tests set out in Part 5 of Chapter D, or

(ii) the proportion of the cost of remediation which the appropriate person has to bear would have been significantly less, by virtue of the guidance on apportionment set out in Part 6 of Chapter D.

E.36 Where an appropriate person is making a case for the authority's costs recovery to be waived or reduced by virtue of paragraph E.35 above, the enforcing authority should expect that person to provide evidence that a particular person, who cannot now be found, caused or knowingly permitted the significant pollutant to be in, on or under the land. The enforcing authority should not regard it as sufficient for the appropriate person concerned merely to state that such a person must have existed.

Part 6 Specific considerations applying to Class B persons

E.37 The statutory guidance in this Part sets out specific considerations relating to cost recovery decisions where the appropriate person is a Class B person, as defined in Chapter D (that is, a person who is liable by virtue or their ownership or occupation of the contaminated land, but who has not caused or knowingly permitted the significant pollutant to be in, on or under the land).

Costs in relation to land values

E.38 In some cases, the costs of remediation may exceed the value of the land in its current use (as defined in Chapter A) after the required remediation has been carried out.

E.39 The enforcing authority should consider waiving or reducing its costs recovery from a Class B person if that person demonstrates to the satisfaction of the authority that the costs of remediation are likely to exceed the value of the land. In this context, the "value" should be taken to be the value that the remediated land would have on the open market, at the time the cost recovery decision is made, disregarding any possible blight arising from the contamination

E.40 In general, the extent of the waiver or reduction in costs recovery should be sufficient to ensure that the costs of remediation borne by the Class B person do not exceed the value of the land. However, the enforcing authority should seek to recover more of its costs to the extent that the remediation would result in an increase in the value of any other land from which the Class B person would benefit.

Precautions taken before acquiring a freehold or a leasehold interest

E.41 In some cases, the appropriate person may have been reckless as to the possibility that land he has acquired may be contaminated, or he may have decided to take a risk that the land was not contaminated. On the other hand, he may have taken precautions to ensure that he did not acquire land which is contaminated.

E.42 The authority should consider reducing its costs recovery where a Class B person who is the owner of the land demonstrates to the satisfaction of the authority that:

(a) he took such steps prior to acquiring the freehold, or accepting the grant of assignment of a leasehold, as would have been reasonable at that time to establish the presence of any pollutants;
(b) when he acquired the land, or accepted the grant of assignment of the leasehold, he was nonetheless unaware of the presence of the significant pollutant now identified and could not reasonably have been expected to have been aware of their presence; and
(c) it would be fair and reasonable, taking into account the interests of national and local taxpayers, that he should not bear the whole cost of remediation.

E.43 The enforcing authority should bear in mind that the safeguards which might reasonably be expected to be taken will be different in different types of transaction (for example, acquisition of recreational land as compared with commercial land transactions) and as between buyers of different types (for example, private individuals as compared with major commercial undertakings).

Owner-occupiers of dwellings

E.44 Where a Class B person owns and occupies a dwelling on the contaminated land in question, the enforcing authority should consider waiving or reducing its costs recovery where that person satisfies the authority that, at the time the person purchased the dwelling, he did not know, and could not reasonably have been expected to have known, that the land was adversely affected by presence of a pollutant.

E.45 Any such waiver or reduction should be to the extent needed to ensure that the Class B person in question bears no more of the cost of remediation than it appears reasonable to impose, having regard to his income, capital and outgoings. Where the appropriate person has inherited the dwelling or received it as a gift, the approach in paragraph E.44 above should be applied with respect to the time at which he received the property.

E.46 Where the contaminated land in question extends beyond the dwelling and its curtilage, and is owned or occupied by the same appropriate person, the approach in paragraph E.44 above should be applied only to the dwelling and its curtilage.

The housing renewal grant analogy

E.47 In judging the extent of a waiver or reduction in costs recovery from an owner-occupier of a dwelling, an enforcing authority which is a local authority may wish to apply an approach analogous to that used for applications for housing renovation grant (HRG). These grants are assessed on a means-tested basis, as presently set out in the Housing Renewal Grants Regulations 1996 (SI 1996/2890, as amended). The HRG test determines how much a person should contribute towards the cost of necessary renovation work for which they are responsible, taking into account income, capital and outgoings, including allowances for those with particular special needs.

E.48 The HRG approach can be applied as if the appropriate person were applying for HRG and the authority had decided that the case was appropriate for grant assessment. Using this analogy, the authority would conclude that costs recovery should be waived or reduced to the extent that the appropriate person contributes no more than if the work were house renovations for which HRG was being sought. For this purpose, any upper limits for grants payable under HRG should be ignored.

E.49 Where the Environment Agency is the enforcing authority, it should seek to be consistent with the local authority in whose area the contaminated land in question is situated. The Environment Agency should therefore consult the local authority, and should take that authority's views into consideration in making its own cost recovery decision.

ANNEX 4

Guide to the Contaminated Land (England) Regulations 2000

Introduction

1 This annex provides additional material to help with the understanding of the Contaminated Land (England) Regulations 2000 (SI 2000/227), which are referred to in it as "the Regulations".

2 Cross-references to the other parts of this circular help to show how the Regulations relate to the rest of the new contaminated-land regime.

3 The Regulations should always be consulted for the precise legal requirements and meanings. What follows is merely an informal guide.

4 The Regulations set out detailed provisions on parts of the regime which Part IIA of the Environmental Protection Act 1990 leaves to be specified in secondary legislation. In addition to the necessary general provisions, the Regulations deal with five main subjects:

(a) special sites (see paragraphs 7 to 15 below);
(b) remediation notices (see paragraphs 16 to 20 below);
(c) compensation (see in paragraphs 21 to 38 below);
(d) appeals (see paragraphs 39 to 78 below); and
(e) public registers (see in paragraphs 79 to 100 below).

General provisions

5 Regulation 1 contains the usual provisions on citation and references. Any reference to a numbered "section" in this guide refers to that section in Part IIA of the Environmental Protection Act 1990.

6 Since the primary legislation applies to the whole of Great Britain, regulation 1 specifically provides that these regulations apply only to England. The Scottish Executive and the National Assembly for Wales are responsible for any provision that will be made for Scotland or Wales.

Special sites

7 Section 78C(8) provides that land is to be a special site if it is land of a description prescribed in regulations. Regulations 2 and 3, with Schedule 1, provide the necessary descriptions. The procedures related to special sites are described in section 18 of Annex 2 to this circular.

8 There are three main groups of cases where a description of land is prescribed for this purpose. The individual descriptions of land to be

designated are contained in paragraphs (a) to (j) of regulation 2(1). If land is contaminated land ***and*** it falls within one of the descriptions, it must be designated as a special site. Otherwise, it cannot be so designated. The descriptions of land do not imply that land of that type is more likely to constitute contaminated land. They identify cases where, ***if*** the land is contaminated land, the Environment Agency is best placed to be the enforcing authority.

Water-pollution cases

9 Regulations 2(1)(a) and 3 ensure that the Environment Agency becomes the enforcing authority in three types of case where the contaminated land is affecting controlled waters and their quality, and where the Environment Agency will also have other concerns under other legislation. These cases are set out in regulation 3, and are broadly as follows:

(a) *Wholesomeness of drinking water:* Regulation 3(a) covers cases where contaminated land affects controlled waters used, or intended to be used, for the supply of drinking water. To meet the description, the waters must be affected by the land in such a way that a treatment process or a change in treatment process is needed in order for such water to satisfy wholesomeness requirements. The standards of wholesomeness are currently set out in the Water Supply (Water Quality) Regulations 1989 (SI 1989/1147 as amended by SI 1989/1384, SI 1991/1837 and SI 1999/1524), and the Private Water Supplies Regulations 1991 (SI 1991/2790). An intention to use water for the supply of drinking water would be demonstrated by the existence of a water abstraction licence for that purpose, or an application for such a licence.

(b) *Surface-water classification criteria:* Regulation 3(b) covers cases where controlled waters are being affected so that those waters do not meet or are not likely to meet relevant surface water criteria. These are currently set out in four sets of Surface Waters (Dangerous Substances) (Classification) Regulations: SI 1989/2286, SI 1992/337, SI 1997/2560 and SI 1998/389.

(c) *Major aquifers:* Regulation 3(c) covers cases where particularly difficult pollutants are affecting major aquifers. The Environment Agency will already be concerned both with pollutants of this type and with managing water resources. The list of pollutants is set out in paragraph 1 of Schedule 1. It corresponds to List I of the Groundwater Directive (80/86/EEC). The major aquifers are described in paragraph 2 of Schedule 1 by reference to the underground strata in which they are contained. The British Geological Survey publishes maps which show the location and boundaries of such strata.

10 For the purposes of regulation 3(c), the fact that contaminated land may be located over one of the listed underground strata does not by itself make the land a special site. The land must be contaminated land on the basis that it is causing, or is likely to cause, pollution of controlled waters; the pollution must be by reason of one or more substances from schedule 1; and the waters being or likely to be polluted must be contained within the strata.

Industrial cases

11 The subsequent items in regulation 2(1) ensure that the Environment Agency is the enforcing agency in respect of contaminated land which is,

or has been, used as a site for industrial activities that either pose special remediation problems or are subject to regulation under other national systems, either by the Environment Agency itself, or by some other national agency. The designation of such sites as special sites is intended to deploy the necessary expertise and to help co-ordination between the various regulatory systems. The descriptions are in respect of:

(a) *Waste acid tar lagoons (regulation 2(1)(b)):* Regulation 2(2) defines what falls into this description. The retention basins (or lagoons) concerned typically involve cases where waste acid tar arose from the use of concentrated sulphuric acid in the production of lubricating oils and greases or the reclamation of base lubricants from mineral oil residues. The description is not intended to include cases where the tars resulted from coal product manufacture, or where these tars were placed in pits or wells.

(b) *Oil refining (regulation 2(1)(c)(i)):* The problems resulting from this are again considered more appropriate for the expertise of the Environment Agency. As for waste acid tar lagoons, activities related to coal are not included.

(c) *Explosives (regulation 2(1)(c)(ii)):* The relatively few sites in this category pose specific problems, which are more appropriately handled by the Environment Agency.

(d) *IPC (Integrated Pollution Control) sites (regulation 2(1)(d)):* Sites which are regulated under Part I of the 1990 Act and which have become contaminated will generally be regulated under those powers. But there may be situations where Part IIA powers will be needed. This item ensures that the Environment Agency will be the enforcing authority under Part IIA where it is already the regulatory authority under Part I. The description therefore refers to a "prescribed process designated for central control". In England, this means a Part A process. This description covers:-

    (i) land on which past activities were authorised under "central control" but which have ceased;
    (ii) land where the activities are continuing but the contamination arises from a non-"central control" process on the land; and
    (iii) land where the contamination arises from an authorised "central control" process but a remediation notice could nevertheless be served. (Section 78YB(1) precludes the service of a remediation notice in cases where it appears to the authority that the powers in section 27 of the 1990 Act may be exercised.)

    This description does not cover land where the Part I authorisation is obtained in order to carry out remediation required under Part IIA. It also does not cover land which has been contaminated by an activity which ceased before the application of "central controls", but would have been subject to those controls if it had continued after they came into force. (Legislation to implement the Integrated Pollution Prevention and Control Directive (96/61/EC) may lead to an additional special site description being prescribed in the future concerning activities authorised under the proposed Pollution Prevention and Control regime.)

(e) *Nuclear sites (regulation 2(1)(e)):* Regulation 2(4) defines what is to be treated as a nuclear site for this purpose. The designation of a nuclear site as contaminated land under these regulations will have effect only in relation to non-radioactive contamination. Any harm, or pollution of controlled waters, attributable to radioactivity will be dealt with under a separate regime to be introduced by regulations to be made under section 78YC. Consultation is under way on the form that this separate regime should take.

Defence cases

12 Regulation 2(1)(f), (h) and (i) ensures that the Environment Agency deals with most cases where contaminated land involves the Ministry of Defence (MOD) estate. Broadly speaking, the descriptions include any contaminated land at current military, naval and airforce bases and other properties, including those of visiting forces; the Atomic Weapons Establishment; and certain lands at Greenwich Hospital (section 30 of the Armed Forces Act 1996). However, off-base housing or off-base NAAFI premises are not included, and nor is property which has been disposed of to civil ownership or occupation. Training areas and ranges that MOD does not own or occupy but may use occasionally do not fall within the descriptions. Regulation 2(1)(g) describes land formerly used for the manufacture, production or disposal of chemical and biological weapons and related materials, regardless of current ownership. In all these cases, the Environment Agency is best placed to ensure uniformity across the country and liaison with the Ministry of Defence and the armed forces.

Other aspects of special sites

13 *Adjoining/adjacent land (regulation 2(1)(j)):* Where the conditions on a special site lead to adjacent or adjoining land also being contaminated land by reason of the presence of substances which appear to have escaped from the special site, that adjacent or adjoining land is also to be a special site. This does not apply where the special site is one of the water-pollution cases described in regulations 2(1)(a) and 3. With this exception, the Environment Agency will be the enforcing authority for the adjoining land as well as for the special site that has caused the problem. This approach is intended to avoid regulatory control being split.

14 *Waste management sites:* Land used for waste management activity, such as landfill, is not as such designated as a special site. This is because Part II of the 1990 Act already contains wide powers for the Environment Agency to ensure that problems are tackled. However, such land may fall within one or more of the special site descriptions, for example if pollution of controlled waters is being caused. The interface between Part IIA controls and waste management controls is described at Annex 1, paragraphs 55 to 58.

15 *Role of the Environment Agency:* It remains the task of the local authority to decide, in the first instance, whether land within the description of a special site is contaminated land or not. The work of the Environment Agency as enforcing authority only starts once that determination is made. However, the statutory guidance on the identification of contaminated land says that, in making that determination, local authorities should consider whether, if land were designated, it would be a special site. If that is the case, the local authority should always seek to make arrangements with the Environment Agency to carry out any inspections of the land that may be needed, on behalf of the local authority (see Annex 3, paragraphs B.26 to B.30).

Remediation notices

16 Section 78E(1) requires a remediation notice to specify what each person who is an appropriate person to bear responsibility for remediation is to

do by way of remediation and the timescale for that remediation. Where several people are appropriate persons, section 78E(3) requires the remediation notice to state the proportion which each of them is to bear of the costs of that remediation (see Chapter D of Annex 2). Section 78E(6) then provides that regulations may lay down other requirements on the form and content of remediation notices and the associated procedure.

17 Regulation 4 sets out the **additional** requirements about the content of a remediation notice. The overall intention is to make the notice informative and self-contained. There should be a clear indication of what is to be done; by whom; where; by when; in relation to what problem; the basis for the authority's actions; who else is involved; the rights of appeal; that a notice is suspended if there is an appeal; and other key information.

Copying remediation notices to others

18 As well as serving the remediation notice on the appropriate person or persons, regulation 5 requires the enforcing authority, at the same time as it serves a remediation notice on the appropriate person(s), to send a copy of the notice to:

(a) anyone whom the authority considers to be the owner or occupier of any of the relevant land or waters, and whom they have therefore consulted under section 78G(3)(a) about rights that may need to be granted to enable the work to be done;

(b) anyone whom the authority considers will be required to grant rights over the land or waters to enable the work to be done, and whom they have therefore consulted under section 78G(3)(b) about such rights;

(c) anyone whom the authority considers to be the owner or occupier of any of the land to which the notice relates and whom they have therefore consulted under section 78H(1)about the remediation to be required; and

(d) the Environment Agency, where the local authority is the enforcing authority, or the local authority, where the Environment Agency is the enforcing authority.

19 It will be good practice for the authority to indicate to the recipient in which capacity they are being sent a copy of the notice. Where a remediation notice is served without consultation because of imminent danger of serious harm (see sections 78G(4) and 78H(4)), the copies should be sent to those who would have been consulted if there had not been an emergency.

Model notices

20 Although the Regulations prescribe the content of remediation notices, they do not prescribe the form of the remediation notice. However, the Department and the Environment Agency aim to draw up a model form which all enforcing authorities can use, in the interests of consistency and minimising preparatory work.

Compensation for rights of entry etc

21 Under section 78G(2), any person (the "grantor") whose consent is required before any thing required by a remediation notice may be done

must grant (or join in granting) the necessary rights in relation to land or waters. For example, an appropriate person may be required to carry out remediation actions upon land which he does not own, perhaps because it has been sold since he caused or knowingly permitted its contamination. Another example may be where access to adjoining land owned or occupied by the grantor's land is needed to carry out the necessary works.

22 The rights that the grantor must grant (or join in granting) are not some special statutory right, but a licence or similar permission of the kind which any person would need to enter on land which they do not own or occupy and carry out works on it.

23 Regulation 6 and Schedule 2 set out a code for compensation payable to those who are required to grant such rights and who thereby suffer detriment. The provisions are closely modelled on those which apply for compensation payable in relation to works required in connection with waste management licences.

Applications for compensation

24 Under paragraph 2 of schedule 2, applications must be made by grantors within:

(a) twelve months of the date of the grant of any rights;
(b) twelve months of the final determination or abandonment of an appeal, or
(c) six months of the first exercise of the rights,

whichever is the latest.

25 Paragraph 3 requires applications to be made in writing and delivered at or sent by pre-paid post to the last known address of the appropriate person to whom the rights were granted. They must include a copy of the grant of rights and any plans attached to it; a description of the exact nature of any interest in the land concerned; and a statement of the amount being claimed, distinguishing between each of the descriptions of loss or damage in the Regulations and showing how each amount has been calculated.

26 Paragraph 4 of the Schedule sets out the various descriptions of loss or damage for which compensation may be claimed. Distinctions are drawn between the grantor's land out of which the rights are granted, any other land of the grantor which might be affected, and other forms of loss. They can be summarised broadly as:

(a) *depreciation:* depreciation in the value of

(i) any relevant interest (that is, the interest in land as a result of holding which the grantor is able to make the grant) which results from the ***grant*** of the rights; or
(ii) any other interest in land, which results from the exercise of the rights;

(b) *disturbance:* loss or damage sustained in relation to the grantor's relevant interest, equivalent to the compensation for "disturbance" under compulsory purchase legislation; this might arise where for example there was damage to the land itself or things on it as a consequence of the exercise of the rights, or a loss of income or a loss of profits resulted from the grant of the right or its exercise;
(c) *injurious affection:* damage to or injurious affection of the grantor's interest in any other land (that is, land not subject to the grant of rights); this again is analogous to the compensation for "injurious affection" under compulsory-purchase legislation; this might arise

where the works on the contaminated land had some permanent adverse effect on adjoining land; and

(d) *abortive work:* loss in respect of carried out by, or on behalf of, the grantor which is rendered abortive as a result of the grant or the work done under it; this might arise where, for example, access to a newly erected building on the land was no longer possible after the grant of the rights, so that the building could no longer be used (paragraph 5(4) of Schedule 2 ensures that this can include expenditure on drawing up plans etc).

Professional fees

27 Compensation can also be claimed for any reasonable expenses incurred in getting valuations or carrying out legal work in order to make or pursue the application itself (paragraph 5(6) of Schedule 2).

Rules for assessing compensation

28 Paragraph 5 of Schedule 2 ensures that the basic rules in section 5 of the Land Compensation Act 1961 apply to these cases. In particular, this section indicates what is meant by "value" when assessing depreciation.

29 To guard against the possibility of unnecessary things being done on land in order to claim or inflate compensation, paragraph 5(3) requires the value of such things to be ignored in assessing compensation.

Position of mortgagees

30 There may be cases where mortgagees join in with mortgagors in the grant of rights, or grant such rights themselves. This might be because they are a mortgagee in possession, or they may have reserved the right to join in the grant of any rights. In these cases, mortgagees fall within section 78G(5) and are able to obtain compensation in their own right.

31 The effect of paragraph 6(1) of Schedule 2 is that in all cases where there is a mortgage, the compensation is paid to the mortgagee (to the first one, if there are several mortgagees), but that it is then applied as if it were the proceeds of sale. This ensures that the mortgagor, or any other mortgagee, will get any appropriate share. Paragraph 5(5) prevents two payments of compensation (ie one each to mortgagee and owner) for the same interest in land.

Disputes

32 Disputes about compensation may be referred, by either party, to the Lands Tribunal (paragraph 6(3)). The Tribunal's procedure rules (SI 1996/1022) enable the Tribunal, with the consent of the parties, to determine a case on the basis of written representations, without the need for an oral hearing (rule 27). Rule 28 provides for a simplified procedure aimed at enabling certain cases to be dealt with speedily and at minimum expense to the parties. In such cases, the hearing takes place before a single Member of the Tribunal acting as arbitrator. Parties may in straightforward cases, and with the Tribunal's permission, be represented at hearings by a non-lawyer, such as a professional valuer.

Payment

33 Payments are to be made on the date or dates agreed by the parties (paragraph 6(2)) or as soon as practicable after the determination in cases where there is a dispute.

Interest

34 Interest may be payable on compensation, for example where applications take a long time to resolve. The Planning and Compensation Act 1991 makes provision for the calculation of interest on compensation. It is proposed to apply this to compensation applications made under these Regulations by an Order amending of Schedule 18 of the 1991 Act. This will also provide the date from which interest is to be payable for the various types of compensation.

Other cases

35 Compensation under Part IIA is not available for any loss resulting from remediation work other than in relation to the heads of compensation specified in the Regulations. Nor is it available in cases where there is no remediation notice—for example where remediation is carried out voluntarily, without a remediation notice being served. In such cases, there is no requirement for the grant of rights: any rights that are needed must be acquired by negotiation in the usual way.

36 Where a local authority exercises powers of entry under section 108 of the Environment Act 1995 in connection with its contaminated land functions, the relevant compensation provisions are those at Schedule 18 of the 1995 Act.

Role of the enforcing authority

37 Arrangements for compensation under Part IIA are a matter for the grantor and the appropriate persons concerned, and the enforcing authority is not involved. However, it is required to consult those who may have to grant rights and to send them a copy of the remediation notice (see paragraph 18(b) above).

38 In addition, it is good practice for authorities to let those who they have consulted because they may be required to grant rights to the appropriate person(s) know the final outcome of the determination of any appeal against the remediation notice, so that they are alerted to the need to be ready to apply for compensation.

Appeals against remediation notices

39 Remediation notices must include information on the right to appeal against them (see paragraph 17 above). This section of this guide shows how the provisions in Part IIA fit together with the provisions in regulations 7 to 14 and the normal practice of the Department of the Environment, Transport and the Regions in handling appeals.

Matters affecting appeals generally

Time-limit for appeals

40 Any appeal must be made within twenty-one days of receiving the remediation notice (*section 78L(1)*). There is no provision for extending this time-limit.

The grounds for appeal

41 Any appeal against a remediation notice must be made on one or more of the grounds set out in regulation 7(1). In broad terms, the grounds concern the following matters:

(a) whether the land is contaminated land as defined; this ground may arise either because of failure to act in accordance with the statutory guidance in Chapters A and B of Annex 3 or because the identification is otherwise unreasonable;
(b) what is required to be done by way of remediation; this ground may arise either because of failure to have regard to the statutory guidance in Chapter C of Annex 3 or because the requirements are otherwise unreasonable;
(c) whether an appellant is an appropriate person to bear responsibility for a remediation action; section 78F is relevant;
(d) whether someone else is also an appropriate person for a remediation action; section 78F is relevant; under this ground, the appellant must claim either to have found someone else who has caused or knowingly permitted the pollution or that someone else is also an owner or occupier of all or part of the land;
(e) whether the appellant should have been excluded from responsibility for a remediation action; this ground may arise because of failure to act in accordance with the statutory guidance in Chapter D of Annex 3;
(f) the proportion of cost to be borne by the appellant; this ground may arise either because of failure to act in accordance with the statutory guidance in Chapter D of Annex 3 or because the determination of the appellant's share is otherwise unreasonable;
(g) whether the notice complies with restrictions in the Act on the serving of notices; section 78H(1) and (3) is relevant;
(h) whether the case is one of imminent danger of serious harm from the contaminated land; section 78H(4) is relevant;
(i) whether remediation is taking, or will take, place without a remediation notice; section 78H(5) of the Act is relevant;
(j) whether remediation requirements breach restrictions on liability for pollution of controlled waters; section 78J is relevant;
(k) whether remediation requirements breach restrictions on liability relating to escaping substances; section 78K is relevant;
(l) whether the authority has itself agreed to carry out the remediation at the cost of the person served with the remediation notice; section 78N(3)(b) of the Act is relevant;
(m) whether the authority should have decided that the recipient of the remediation notice would benefit from waiver or reduction of cost recovery on grounds of hardship or in line with the statutory guidance in Chapter E of Annex 3, that it therefore had power itself to carry out the remediation and that it was thus precluded from serving a remediation notice; sections 78N(3)(e) and 78P(1) and (2) are relevant;
(n) whether the authority's powers to remediate were exercisable because this was a case where hardship or the statutory guidance in Chapter E of Annex 3 should lead to a waiver or reduction in cost recovery; this ground may arise either because of failure to have regard to hardship or the statutory guidance in Chapter E or because the decision was otherwise unreasonable; sections 78N(3)(e) and 78P(1) and (2) are relevant;
(o) whether regard was had to site-specific guidance from the Environment Agency; section 78V(1) is relevant;
(p) whether enough time was allowed for remediation; the statutory guidance in Chapter C of Annex 3 to this Circular may be relevant;
(q) whether the notice would make an insolvency practitioner, an official receiver or other receiver or manager personally liable in breach of the limits on such liability; section 78X(3)(a) and (4) is relevant;

(r) whether certain powers under the Integrated Pollution Control system (Part I of the Environmental Protection Act 1990) or under the waste management licensing system (Part II of that Act) were available to the authority; section 78YB(1) and (3) are relevant; the powers concerned are those in section 27 (Part I) and section 59 (Part II); and

(s) whether there is some informality, defect or error concerning the notice, not covered above; in an appeal on this ground, the appellate authority must dismiss the appeal if it is satisfied that the informality, defect or error was not a material one.

Suspension of remediation notice upon appeal

42 Once an appeal has been duly made, the remediation notice concerned is suspended (regulation 14). It remains suspended either until the appeal is finally determined or is withdrawn (abandoned) by the appellant. "Duly made" for this purpose means that an appeal must be made within the time limit, and in accordance with the Regulations.

Appeals relating to land which is not a special site

43 If the remediation notice was served by a local authority, appeals are to a magistrates' court (see section 78L(1)(a)). (However, if the land has subsequently been designated a special site and the notice has been adopted by the Environment Agency, any appeal would be to the Secretary of State.)

44 Regulation 8 sets out the requirements for making such an appeal. It provides that the appeal must be by way of a complaint for an order. At the same time as submitting the complaint to the justices' clerk, the appellant must:

(a) file (that is, deposit with the justices' clerk) a notice of appeal (which is different from the complaint) and serve a copy of this notice on:

(i) the enforcing authority;

(ii) any other appropriate person named in the remediation notice;

(iii) any person who is named in the appeal as an appropriate person; this relates to appeal ground (d);

(iv) any person named in the remediation notice as the owner or occupier of the land.

(b) file a copy of the remediation notice, and (because they will not have had it previously) serve a copy on any person named in the appeal as an appropriate person who was not named as such in that remediation notice;

(c) file a statement of the names and addresses of the above persons (except for the enforcing authority). These will normally be found in the remediation notice, except for details of any additional person named in the appeal as an appropriate person. This ensures that the court has a list of all those to whom notice may need to be given at a later stage.

45 The notice of appeal must state the appellant's name and address, and the grounds of the appeal.

46 The justices' clerk or the court may give directions for the handling of the case including the timetable, documents and evidence (for example, it will be helpful to arrange for exchanging evidence). This function may be delegated to other court staff in accordance with relevant magistrates' court rules (see regulation 8(5)).

47 Any of the persons involved in paragraph 44(a) above will be given notice of the hearing and have an opportunity to be heard, including the appellant and the enforcing authority.

48 In accordance with the usual practices for determining who hears cases in magistrates' courts, it is expected that most appeals of this kind will be heard by a stipendiary magistrate.

49 An appellant who wishes to abandon (withdraw) an appeal to the court may request the court's permission to do so (but see paragraph 77 below, if there is a proposed modification to the remediation notice).

50 After the appeal has been determined or abandoned, the court has power to award costs in accordance with section 64 of the Magistrates Courts Act 1980. This provides that the court has the power to award such costs as are just and reasonable.

51 The appellant and the local authority have a right to appeal against the decision of the magistrates' court on the appeal. Under regulation 13, the appeal is made to the High Court. The procedure on such a further appeal is governed by the Rules of the Supreme Court on statutory appeals to the High Court. If any other appropriate person named in the remediation notice or the notice of appeal exercised their right to appear at the hearing before the magistrates' court, they will also have a similar right of appeal.

Special sites appeals

52 If land is a special site, and the remediation notice was served or adopted by the Environment Agency, appeals are to the Secretary of State for the Environment, Transport and the Regions.

53 The appellant in a special site case must appeal by submitting a "notice of appeal" to the Secretary of State. No particular form is prescribed for such a notice of appeal but, in accordance with regulation 9, it must state:

(a) the appellant's name and address;
(b) the grounds of appeal; and
(c) whether the appellant wishes the appeal to be in the form of a hearing, or alternatively have the appeal decided on the basis of written representations.

54 The appellant must at the same time serve a copy of the notice of appeal on

(a) the Environment Agency;
(b) any other appropriate person named in the remediation notice;
(c) any person who is named in the appeal as an appropriate person; this relates to appeal ground (d);
(d) any person named in the remediation notice as the owner or occupier of the land.

55 The appellant must also send to the Secretary of State

(a) a list of the names and addresses of the above persons (except for the Agency); these will normally be found in the remediation notice, except for details of any additional person named in the appeal as an appropriate person; and
(b) a copy of the remediation notice.

56 The appellant must also (because they will not have had it previously) serve a copy of the remediation notice on any person named in the appeal as an appropriate person, or as an owner or occupier, who was not named as such in that remediation notice.

57 Appeals to the Secretary of State should be submitted to the Planning Inspectorate. Their current address and telephone number are as follows

The Planning Inspectorate,
Room 1413,
Tollgate House,
Houlton Street,
BRISTOL BS2 9DJ
Tel: 0117 987 8812

Initial procedure on an appeal to the Secretary of State

58 Within 14 days of receiving a copy of the notice of appeal, the Environment Agency will notify all others whom the appellant was required to send a copy of the appeal. This notification will ensure that they know there is an appeal, and will make them aware that:

(a) written representations to the Secretary of State may be made within 21 days from the receipt of the Environment Agency's notice;
(b) such representations will be copied to the appellant and the Agency; and
(c) those who make representations will be informed about any public hearing.

59 All written representations made to the Secretary of State at any time throughout the appeal should be dated with the date on which they are submitted.

Delegation to inspectors

60 Most cases will be decided by Inspectors appointed on the Secretary of State's behalf, under the provisions of section 78L(6) which allow for appeal decisions to be delegated to them. References to the Secretary of State in the procedures set out below may be taken to include the inspector, except where the context indicates otherwise.

61 Some cases may, however, be recovered for decision by the Secretary of State. In these "recovered" cases, the Secretary of State will determine the appeal on the basis of a written report from the inspector. In accordance with regulation 10(4), this report must contain conclusions and recommendations, or reasons for not making recommendations.

62 Each special-site appeal will be looked at individually to decide whether it should be "recovered". The categories most likely to be recovered are as follows

(a) cases involving special sites of major importance or having more than local significance;
(b) cases giving rise to significant local controversy;
(c) cases which raise significant legal difficulties; and
(d) cases which raise major, novel issues and which could therefore set a precedent.

63 Other special site appeal cases may on occasion merit being "recovered" for decision by the Secretary of State.

Deciding an appeal to the Secretary of State

64 A hearing will be arranged if either of the parties asks for that to be done. Otherwise, the appeal will be decided on the basis of written representations, unless the Secretary of State decides, in accordance with regulation 10, that it is desirable to hold a hearing or a public local inquiry.

Written representations

65 If the appeal is being decided by written representations, the procedure will normally be as follows:–

Step 1
The Secretary of State will invite the Agency to respond to the grounds of appeal; to provide any other information that it relies on to support its decision to serve the remediation notice within 28 days; and to send the appellant, and any other appropriate person on whom the notice was served, a copy of its response at the same time as it is submitted to the Secretary of State.
Step 2
The appellant, and any other appropriate person on whom the notice was served, will then be given an opportunity to comment on the representations from the Agency. These should be made within 14 days of the date of submission of the Agency's representations and must be copied to the Agency and any appropriate person on whom the notice was served, at the same time. The Secretary of State will also send to the appellant, any other appropriate person on whom the notice was served and the Agency copies of the representations received under regulation 10 (other than the copy of the Agency response mentioned in step 1 above, which will already have been copied to the appellant and any other appropriate person on whom the notice was served). The Secretary of State will seek their comments, which should also be given within 14 days.
Step 3
Arrangements will be made for an Inspector to visit the appeal site. As far as possible, a mutually convenient time will be arranged. The Agency, the appellant and any other person sent a copy of the notice of appeal under regulation 9(2) will be invited to attend. No representations about the appeal can be made during the visit but must be made in writing under the procedures for making representations and within the appropriate time limits.

66 This procedure is intended to allow the determination of appeals as expeditiously as possible. However, the Secretary of State may in certain exceptional cases set time limits which differ from those above, or may extend a time limit either before or after it has expired. The Secretary of State may also request exchanges of information in addition to those mentioned above.

Hearings

67 Where an appeal is to be decided after a hearing, in accordance with regulation 10(1)—(3), the appellant, the Agency and those required to be sent a copy of the notice of appeal under regulation 9(2) will be invited to make representations at the hearing. Other persons may be heard at the discretion of the Inspector. The Agency will inform other persons of the date of the hearing where they have previously expressed an interest in the case.

68 A pre-hearing timetable will be provided for the submission of written statements. Failure to provide this information, within the specified timescales, could lead to hearings being adjourned resulting in unnecessary delays. The conduct of the hearing will be for the Inspector to determine, and will generally follow the Code of Practice for Hearings given at Annex 2 of DOE Circular 15/96. It may sometimes be necessary to hold a pre-hearing meeting to discuss the nature of the evidence to be given, who is likely to participate and the programme to be adopted.

69 The presumption is that hearings will be held in public. However, a hearing, or any part of it, may be held in private if the Inspector hearing the appeal decides that there are particular and special grounds for doing so, such as reasons of commercial confidentiality, or national security.

Public inquiries

70 The holding of a public local inquiry under regulation 10(1)(b) is expected to be more appropriate for particularly complex or locally controversial cases. A pre-inquiry timetable will be provided for the submission of statements and proofs of evidence. It is important that this is adhered to. Inquiry proceedings are more formal in nature than the majority of hearings. Inquiries will be conducted in accordance with the spirit of the Town and Country Planning (Inquiries Procedures) Rules 1992. The rules require details of the inquiry to be posted locally. As in the case of hearings, it may sometimes be necessary to hold a pre-inquiry meeting.

Abandonment of appeals

71 An appellant who wishes to abandon (withdraw) a special site appeal must notify the Secretary of State in writing, who will in turn notify all those who have received notice of the appeal in accordance with regulation 9(2) and (5). The appeal is deemed to be abandoned on the day the Secretary of State receives the notice of the abandonment. Abandonment may be refused by the Secretary of State under regulation 9(4) if the appellant has been notified of a modification to the remediation notice under regulation 12 (see paragraph 77 below).

Notification of appeal decision

72 Regulation 11 requires that the appellant must be notified in writing of the decision on the appeal, and sent a copy of any report made to the Secretary of State by an inspector. The decision letter, and the report if any, must be copied by the Secretary of State to the Agency and to anyone who was entitled to receive a copy of the notice of the appeal.

73 Details of decision letters on special-site cases will be placed on the register. As long as they do not contain confidential information or trade secrets, copies will also be available for a small charge from:

The Decision Library
The Planning Inspectorate,
Room 1508,
Tollgate House,
Houlton Street,
BRISTOL BS2 9DJ
Tel: 0117 987 8759

Further information can also be obtained from the same source.

Award of costs

74 Costs may be awarded where there is a hearing or a public local inquiry. Awards of costs will follow existing general guidance in Department of the Environment Circular 8/93, which governs planning appeals and similar cases. This means that each party will bear their own costs unless there has been unreasonable behaviour leading to unnecessary expense, as described in that Circular. In cases decided by written representations, the parties must meet their own expenses.

Appeals or complaints against the decision

75 There is no statutory right of appeal against a decision made on appeal by the Secretary of State. Once a decision letter has been issued, the

decision is final, and the Secretary of State and the inspector can no longer consider any representations or make any further comments on the merits or otherwise of the case. A party to the appeal may be able to seek judicial review of the decision in the High Court. If they consider that there has been maladministration in reaching the decision, they may also ask an MP to take up the matter with the Parliamentary Commissioner for Administration (the Ombudsman), though the Ombudsman cannot re-open the appeal.

76 If anyone has a complaint about the handling of an appeal by the Planning Inspectorate, they should write to the Complaints Officer at the address shown in paragraph 57 above.

Modification of remediation notices

77 Section 78L(2)(b) enables an appellate authority to modify the remediation notice which is the subject of the appeal. If it proposes to do so in a way which is less favourable to the appellant, or any other appropriate person on whom the notice was served but who may not have appealed, then regulation 12 applies. The appellate authority must notify those persons of the proposed modification, and also notify any other persons who were required to be sent a copy of the notice of appeal under regulations 8(2) or 9(2) (see paragraphs 44 and 54 to 56 above). Any of those persons have a right to make representations. The appellant or any other appropriate person on whom the remediation notice was served has a right to be heard, and if this right to be heard is exercised, the enforcing authority (but no other person) also has the right to be heard. The appellate authority may refuse to permit an appeal to be withdrawn if it has given notice of a proposed modification (regulations 8(4)(c) and 9(4)).

Additional remediation notices to reflect an appeal decision

78 A decision by the appellate authority to quash or modify a remediation notice on appeal may also have implications for a person who has not been served with a remediation notice. This might arise where, in particular, an appeal succeeds on the grounds that there is another person who should be held liable instead of or as well as the appellant. In such cases the enforcing authority will need to consider serving a further remediation notice(s) which take(s) into account the appellate authority's decision. Such additional notices would need to fulfil all the relevant requirements of the Act, regulations, and the statutory guidance, in the usual way. They would attract the normal rights of appeal.

Public registers

79 Section 78R requires each enforcing authority to keep a public register. The public register is intended to act as a full and permanent record, open for public inspection, of all regulatory action taken by the enforcing authority in respect of the remediation of contaminated land, and will include information about the condition of land.

80 As records of regulatory activity, registers are broadly similar in purpose to, and part of the suite of, registers kept in relation to other environmental protection controls, including those kept under Part I and Part II of the Act (IPC etc, and waste regulation); and planning registers kept under the Town and Country Planning Acts, which may also contain valuable information relevant to the condition of land in particular locations.

81 The Agency register is to be kept at the Agency office for the area in question, and the local authority register is kept at the authority's principal office (regulation 15(3)).

Content of the registers

82 Section 78R(1) specifies what material is to be entered on the register. It leaves the details of that material to be prescribed in regulations. These details are set out in Schedule 3.

83 It is good practice to ensure that the register is so organised that all the entries relating to a particular site can be readily consulted in connection with each other.

84 Schedule 3 requires registers to include *full particulars* of certain matters, rather than *copies of* the various forms of notice and other documents listed. However, there is no legal objection to authorities placing a copy of the various documents on the register. Any document not placed on the register may, in any case, be accessible under the Environmental Information Regulations 1992 (SI 1992/3240, as amended).

Information to be placed on the register

Information about remediation

85 For a ***remediation notice***, the effect of regulation 15 and Schedule 3 is that the following information must be placed on the register:

Site Information

(a) the location and extent of the contaminated land sufficient to enable it to be identified; this requirement would ideally be met by showing its address and the estimated area in hectares, together with a plan to a suitable scale and a National Grid reference;

(b) the significant harm or pollution of controlled waters by reason of which the land is contaminated land;

(c) the substances by reason of which the land is contaminated land and, if any of the substances have escaped from other land, the location of that other land;

(d) the current use of the land in question;

Remediation information

(e) the name and address of the person on whom the notice is served;

(f) what each appropriate person is to do by way of remediation, and the periods within which they are required to do each of the things;

86 In cases where site investigation reports obtained by or provided to the authority, which relate to the condition of land or any remediation action, are likely to be publicly accessible under the Environmental Information Regulations, it would also be good practice to include a reference to such information. The entry could include:

(a) a description of the information,

(b) the date on which it was prepared,

(c) the person by whom and for whom it was prepared, and

(d) where it is available to be inspected or copied.

87 It is also good practice for the remediation particulars referred to in paragraph 85(f) above to include an indication of whether the action required was "assessment action", "remedial treatment action" or "monitoring action" (see the definitions of these terms in paragraph C.8 of Chapter C of the statutory guidance, reflecting section 78A(7)).

88 For ***remediation declarations, remediation statements*** and ***notifications of claimed remediation*** (that is notifications for the purposes of section 78R(1)(h) or (j)), the requirement is to enter full particulars of the instrument in question, together with the site information described at paragraphs 85(a)-(d) above. This means that the registers should show, in addition to the date of the instrument and the site information, at least:

(a) ***for remediation declarations*** (see paragraphs 4 and 5 of Schedule 3): the reason why the authority was precluded from specifying a particular remediation action (where, therefore, in the case of pollution of controlled waters, the authority considered that remediation of pollution was precluded on the basis that it would be unreasonable, having regard to the nature of that pollution, the register will show why the authority considered that the contamination was not significant);

(b) ***for remediation statements*** (see paragraphs 6 and 7 of Schedule 3): the remediation action that has been, is being or will be taken, the timescale for that action and the details of the person who is taking it;

(c) ***for notifications of claimed remediation*** (see regulation 15(2) and paragraph 11 of Schedule 3): the remediation action that is claimed to have been taken, the timescale of that action and the details of the person who claims to have taken it.

89 In respect of notifications of claimed remediation, it is open to the person giving the notification to include additional material. In particular, it will be in the interests of both regulators and those giving the notifications to include, in addition, an indication of what the work carried out was intended to achieve; a description of any appropriate quality assurance procedure adopted relating to what has been claimed to be done; and a description of any verification measures carried out for the purpose of assessing the effectiveness of the remediation in relation to the particular significant harm or pollution of controlled waters to which it was referable.

90 Section 78R(3) makes clear that an entry in the register relating to notifications of claimed remediation in no way represents any endorsement or confirmation by the authority maintaining the register that remediation measures have been carried out nor, therefore, that land is no longer contaminated land. It would be good practice to ensure that this disclaimer is clearly associated with all entries of this kind.

91 ***Other environmental controls:*** The register is required, by paragraphs 14 and 15 of Schedule 3, to include information in cases of the two situations where a site may be formally identified as contaminated land but is dealt with under other environmental controls, instead of under Part IIA (see section 78YB(1) and (3)). These other powers are section 27 in Part I of the Environmental Protection Act 1990 (Integrated Pollution Control) and section 59 in Part II of that Act (waste management licensing). In both cases, the register is required to include, in addition to the site information described in paragraphs 85(a)-(d) above particulars of any steps about which the enforcing authority knows that have been taken under those other powers.

92 The register is also required, by paragraph 16 of Schedule 3, to include information about any cases where particular remediation actions cannot be specified in a remediation notice because they would have the effect of interfering with a discharge into controlled waters for which consent has been given under Chapter II of Part III of the Water Resources Act 1991 (see section 78YB(4)). In addition to the site information described in paragraphs 85(a)-(d) above, the register is required to give particulars of the discharge consent.

Other information

Special sites

93 Where the land is a special site, the register should include the information required in respect of any other site. In addition, under paragraph 10 of Schedule 3, the register is required to include:

(a) the notice designating it as such (given by a local authority under section 78C(1)(b) or 78C(5)(a), or by the Secretary of State under section 78D(4)(b));
(b) an identification of the description of land under which it is a special site (see regulations 2 or 3 and Schedule 1)
(c) any notice given by the appropriate Agency of its decision to adopt a remediation notice;
(d) any notice given by or to the enforcing authority under section 78Q(4) terminating the designation.

Agency site-specific guidance

94 Under paragraph 13 of Schedule 3, the register is required to include the date of any site-specific guidance issued by the Environment Agency under section 78V(1). Where such site-specific guidance exists, information in it may be required to be available to the public under the Environmental Information Regulations. Where this is likely, it would be good practice to include a reference to where it is available to be inspected or copied.

Appeals against a remediation notice

95 Where a person on whom a remediation notice has been served appeals against that notice, the register is required, under paragraphs 2 and 3 of Schedule 3, to include full particulars of:

(a) any appeal against a remediation notice, including the date and the name and address of the appellant; and
(b) the decision on such an appeal.

96 If there is an appeal to the High Court against the judgement of a magistrates' court on an appeal against a remediation notice, the requirement to include the decision on the appeal extends to including on the register the decision of the High Court on that further appeal. It would also be good practice to include on the register any judgement of the High Court, or subsequent appeal judgement, on an application for judicial review of the determination of the Secretary of State on an appeal against a remediation notice.

Appeals against a charging notice

97 Where the owner or occupier of any land appeals to the county court under section 78P(8) against a notice charging costs to be recovered by the enforcing authority on his land, the register is required to contain full particulars of:

(a) any appeal against a charging notice; including the date and the name and address of the appellant; and
(b) the decision on such an appeal.

Convictions

98 Under paragraph 12 of Schedule 3, the register is required to include full particulars of any conviction under section 78M (failure to comply with a

remediation notice), including the name of the offender, the date of conviction, the penalty imposed, and the name of the Court.

99 Authorities should have regard to the provisions of the Rehabilitation of Offenders Act 1974, under which convictions of individuals can become spent. The Department understands that it would not be unlawful under that Act to retain details of a spent conviction on the register, but nonetheless retention would seem contrary to its spirit. The Department recommends therefore that authorities should regularly review their registers with the aim of identifying and removing spent convictions, although it may be desirable to continue to record that an offence has taken place. In the case of convictions of a body corporate, the 1974 Act does not apply, but it would seem equitable for the same approach to be applied as for the spent convictions of individuals.

Confidentiality

100 Sections 78S and 78T set out restrictions on information to be placed on the register because of considerations of national security or commercial confidentiality. The effect of these provisions is explained in Annex 2, paragraphs 17.8 to 17.19.

ANNEX 5

Guide to the Environment Act 1995 (Commencement No.16 and Saving Provision) (England) Order 2000

Commencement of Part IIA Environmental Protection Act 1990

1 The Environment Act 1995 (Commencement No.1) Order 1995 (S.I. 1995/1983) brought into force section 57 of the Environment Act 1995 ("the 1995 Act"), in so far as was necessary to enable the Secretary of State to consult on and issue statutory guidance and make regulations.

2 The main effect of the Environment Act 1995 (Commencement No.16 and Saving Provision) (England) Order 2000 (SI 2000/340(C.8)) is to bring the remainder of section 57 of the 1995 Act into force in England on 1 April 2000. This, in turn, brings the Part IIA regime into force.

Repeals and other amendments to the 1990 Act

3 The Order also brings into force the following amendments to the 1990 Act:

(a) amendments to the definition of a statutory nuisance in section 79, so as to exclude any matter which consists of, or is caused by, land in a contaminated state;

(b) the repeal of the following sections (neither of which ever came into force):

(i) section 61, which would have created specific duties for waste regulation authorities as respects closed landfills, and

(ii) section 143, which would have required local authorities to compile registers of land which may be contaminated; and

(c) an amendment to section 161, relating to the use of the affirmative resolution procedure for any order under the new section 78M(4) (which deals with changes to the maximum level of fines for non-compliance with remediation notices).

Saving provision relating to statutory nuisance

4 Article 3 of the Order makes a saving provision with respect to the dis-application of the statutory nuisance regime from land contamination problems. This has the effect of ensuring that any regulatory action which had commenced before 1 April 2000 can continue.

Annex 6—Glossary of Terms

The statutory guidance (and other parts of this Circular) uses a number of terms which are defined in Part IIA of the 1990, other Acts or in the guidance itself. The meanings of the most important of these terms are set out below, along with a reference to the section in the Act or the paragraph in which the relevant term is defined.
Terms which are defined in statutes (mostly in section 78A of the 1990 Act) are shown with underlining.

Animal or crop effect: significant harm of a type listed in box 3 of Table A of Chapter A.

Apportionment: any determination by the enforcing authority under section 78F(7) (that is, a division of the costs of carrying out any remediation action between two or more appropriate persons). *Paragraph D.5(e)*

Appropriate person: defined in section 78A(9) as:

> "any person who is an appropriate person, determined in accordance with section 78F . . ., to bear responsibility for any thing which is to be done by way of remediation in any particular case."

Assessment action: a remediation action falling within the definition of remediation in section 78A(7)(a), that is the doing of anything for the purpose of assessing the condition of the contaminated land in question, or any controlled waters affected by that land or any land adjoining or adjacent to that land. *Paragraph C.8(e)*

Attribution: the process of apportionment between liability groups. *Paragraph D.5(e)*

Building: any structure or erection, and any part of a building including any part below ground, but not including plant or machinery comprised in a building. *Table A*

Building effect: significant harm of a type listed in box 4 of Table A of Chapter A.

Caused or knowingly permitted: test for establishing responsibility for remediation, under section 78F(2); see paragraphs 9.8 to 9.14 of Annex 2 for a discussion of the interpretation of this term.

Changes to Substances: an exclusion test for Class A persons set out in Part 5 of Chapter D. *Paragraphs D.62 to D.64.*

Charging notice: a notice placing a legal charge on land served under section 78P(3)(b) by an enforcing authority to enable the authority to recover from the appropriate person any reasonable cost incurred by the authority in carrying out remediation.

Class A liability group: a liability group consisting of one or more Class A persons. *Paragraph D.5(c)*

Class A person: a person who is an appropriate person by virtue of section 78F(2) (that is, because he has caused or knowingly permitted a pollutant to be in, on or under the land). *Paragraph D.5(a)*

Class B liability group: a liability group consisting of one or more Class B persons. *Paragraph D.5(c)*

Class B person: a person who is an appropriate person by virtue of section 78F(4) or (5) (that is, because he is the owner or occupier of the land in circumstances where no Class A person can be found with respect to a particular remediation action). *Paragraph D.5(b)*

Collective action: a remediation action which addresses together all of the significant pollution linkages to which it is referable, but which would not have been part of the remediation package for every one of those linkages if each of them had been addressed separately. *Paragraph D.22(b)*

Common action: a remediation action which addresses together all of the significant pollution linkages to which it is referable, and which would have been part of the remediation package for each of those linkages if each of them had been addressed separately. *Paragraph D.22(a)*

Contaminant: a substance which is in, on or under the land and which has the potential to cause harm or to cause pollution of controlled waters. *Paragraph A.12*

Contaminated land: defined in section 78A(2) as

"any land which appears to the local authority in whose area it is situated to be in such a condition, by reason of substances in, on or under the land, that —

(a) significant harm is being caused or there is a significant possibility of such harm being caused, or;
(b) pollution of controlled waters is being, or is likely to be, caused."

Contaminated Land (England) Regulations 2000: regulations (SI 2000/227) made under Part IIA — described in *Annex 4*.

Controlled waters: defined in section 78A(9) by reference to Part III (section 104) of the Water Resources Act 1991; this embraces territorial and coastal waters, inland fresh waters, and ground waters.

Cost recovery decision: any decision by the enforcing authority whether:

(a) to recover from the appropriate person all the reasonable costs incurred by the authority in carrying out remediation, or
(b) not to recover those costs or to recover only part of those costs. *Paragraph E.8*

Current use: any use which is currently being made, or is likely to be made, of the land and which is consistent with any existing planning permission (or is otherwise lawful under town and country planning legislation). This definition is subject to the following qualifications:

(a) the current use should be taken to include any temporary use, permitted under town and country planning legislation, to which the land is, or is likely to be, put from time to time;
(b) the current use includes future uses or developments which do not require a new, or amended, grant of planning permission;
(c) the current use should, nevertheless, be taken to include any likely informal recreational use of the land, whether authorised by the owners or occupiers or not, (for example, children playing on the land); however, in assessing the likelihood of any such informal use, the local authority should give due attention to measures taken to prevent or restrict access to the land; and
(d) in the case of agricultural land, however, the current agricultural use should not be taken to extend beyond the growing or rearing of the crops or animals which are habitually grown or reared on the land. *Paragraph A.26.*

Ecological system effect: significant harm of a type listed in box 2 of Table A of Chapter A.

Enforcing authority: defined in section 78A(9) as:

(a) in relation to a special site, the Environment Agency;
(b) in relation to contaminated land other than a special site, the local authority in whose area the land is situated.

Escaped Substances: an exclusion test for Class A persons set out in Part 5 of Chapter D. *Paragraphs D.65 to D.67*

Excluded Activities: an exclusion test for Class A persons set out in Part 5 of Chapter D. *Paragraphs D.47 to D.50*

Exclusion: any determination by the enforcing authority under section 78F(6) (that is, that a person is to be treated as not being an appropriate person). *Paragraph D.5(d)*

Favourable conservation status: defined in Article 1 of Council Directive 92/43/EEC on the conservation of natural habitats and of wild fauna and flora.

Hardship: a factor underlying any cost recovery decision made by an enforcing authority under section 78P(2). See paragraphs 10.8 to 10.10 of Annex 2 for a discussion of the interpretation of this term.

Harm: defined in section 78A(4) as:

> "harm to the health of living organisms or other interference with the ecological systems of which they form part and, in the case of man, includes harm to his property."

Human health effect: significant harm of a type listed in box 1 of Table A of Chapter A.

Industrial, trade or business premises: defined in section 78M(6), for the purpose of determining the penalty for failure to comply with a remediation notice, as:

> "premises used for any industrial, trade or business purposes or premises not so used on which matter is burnt in connection with any industrial, trade or business process, and premises are used for industrial purposes where they are used for the purposes of any treatment or process as well as where they are used for the purpose of manufacturing."

Inspection using statutory powers of entry: any detailed inspection of land carried out through use of powers of entry given to an enforcing authority by section 108 of the Environment Act 1995. *Paragraph B.21*

Introduction of Pathways or Receptors: an exclusion test for Class A persons set out in Part 5 of Chapter D. *Paragraphs D.68 to D.72.*

Intrusive investigation: an investigation of land (for example by exploratory excavations) which involves actions going beyond simple visual inspection of the land, limited sampling or assessment of documentary information. *Paragraph B.20(c)*

Liability group: the persons who are appropriate persons with respect to a particular significant pollutant linkage. *Paragraph D.5(c)*

Local authority: defined in section 78A(9) as meaning any unitary authority, district council, the Common Council of the City of London, the Sub-Treasurer of the Inner Temple and the Under-Treasurer of the Middle Temple.

Monitoring action: a remediation action falling within the definition in section 78A(7)(c), that is "making of subsequent inspections from time to time for the purpose of keeping under review the condition of the land or waters". *Paragraph C.8(g)*

Orphan linkage: a significant pollutant linkage for which no appropriate person can be found, or where those who would otherwise be liable are exempted by one of the relevant statutory provisions. *Paragraphs D.12*, *D.14* and *D.17*

Owner: defined in section 78A(9) as:

> "a person (other than a mortgagee not in possession) who, whether in his own right or as trustee for any other person, is entitled to receive the rack rent of the land, or where the land is not let at a rack rent, would be so entitled if it were so let."

Part IIA: Part IIA of the Environmental Protection Act 1990.

Pathway: one or more routes or means by, or through, which a receptor:

(a) is being exposed to, or affected by, a contaminant, or
(b) could be so exposed or affected. *Paragraph A.14*

Payments Made for Remediation: an exclusion test for Class A persons set out in Part 5 of Chapter D. *Paragraphs D.51 to D.56*

Person acting in a relevant capacity: defined in section 78X(4), for the purposes of limiting personal liability, as any of the following:

"(a) a person acting as an insolvency practitioner, within the meaning of section 388 of the Insolvency Act 1986 (including that section as it applies in relation to an insolvent partnership by virtue of any order made under section 421 of that Act;
(b) the official receiver acting in a capacity in which he would be regarded as acting as an insolvency practitioner within the meaning of section 388 of the Insolvency Act 1986 if subsection (5) of that section were disregarded;
(c) the official receiver acting as a receiver or manager;
(d) a person acting as a special manager under section 177 or 370 of the Insolvency Act 1986; . . .
(f) a person acting as a receiver or receiver and manager under or by virtue of any enactment, or by virtue of his appointment as such by an order of a court or by any other instrument."

Pollutant: a contaminant which forms part of a pollutant linkage. *Paragraph A.17*

Pollutant linkage: the relationship between a contaminant, a pathway and a receptor. *Paragraph A.17*

Pollution of controlled waters: defined in section 78A(9) as:

"the entry into controlled waters of any poisonous, noxious or polluting matter or any solid waste matter."

Possibility of significant harm: a measure of the probability, or frequency, of the occurrence of circumstances which would lead to significant harm being caused. *Paragraph A.27*

Receptor: either:

(a) a living organism, a group of living organisms, an ecological system or a piece of property which:

 (i) is in a category listed in Table A in Chapter A as a type of receptor, and
 (ii) is being, or could be, harmed, by a contaminant; or

(b) controlled waters which are being, or could be, polluted by a contaminant. *Paragraph A.13*

Register: the public register maintained by the enforcing authority under section 78R of particulars relating to contaminated land.

Related companies: are those which are, or were at the "relevant date", members of a group of companies consisting of a "holding company" and its "subsidiaries". The "relevant date" is that on which the enforcing authority first served on anyone a notice under section 78B(3) identifying the land as contaminated land, and the terms "holding company" and "subsidiaries" have the same meaning as in section 736 of the Companies Act 1985. *Paragraph D.46.*

Relevant information: information relating to the assessment of whether there is a significant possibility of significant harm being caused, which is:

(a) scientifically-based;
(b) authoritative;
(c) relevant to the assessment of risks arising from the presence of contaminants in soil; and
(d) appropriate to the determination of whether any land is contaminated land for the purposes of Part IIA, in that the use of the information is consistent with providing a level of protection of risk in line with the qualitative criteria set out in Tables A and B of Chapter A. *Paragraph A.31*

Relevant land or waters: the contaminated land in question, any controlled waters affected by that land and any land adjoining or adjacent to the contaminated land on which remediation might be required as a consequence of the contaminated land being such land. *Paragraph C.8(d)*

Remedial treatment action: a remediation action falling within the definition in section 78A (7)(b), that is the doing of any works, the carrying out of any operations or the taking of any steps in relation to any such land or waters for the purpose:

(a) of preventing or minimising, or remedying or mitigating the effects of any significant harm, or any pollution of controlled waters, by reason of which the contaminated land is such land, or
(b) of restoring the land or waters to their former state. *Paragraph C.8(f)*

Remediation: defined in section 78A(7) as

"(a) the doing of anything for the purpose of assessing the condition of –

(i) the contaminated land in question;
(ii) any controlled waters affected by that land; or
(iii) any land adjoining or adjacent to that land;

(b) the doing of any works, the carrying out of any operations or the taking of any steps in relation to any such land or waters for the purpose –

(i) of preventing or minimising, or remedying or mitigating the effects of any significant harm, or any pollution of controlled waters, by reason of which the contaminated land is such land; or
(ii) of restoring the land or waters to their former state; or

(c) the making of subsequent inspections from time to time for the purpose of keeping under review the condition of the land or waters."

Remediation action: any individual thing which is being, or is to be, done by way of remediation. *Paragraph C.8(a)*

Remediation declaration: defined in section 78H(6). It is a document prepared and published by the enforcing authority recording remediation actions which it would have specified in a remediation notice, but which it is precluded from specifying by virtue of sections 78E(4) or (5), the reasons why it would have specified those actions and the grounds on which it is satisfied that it is precluded from specifying them in a notice.

Remediation notice: defined in section 78E(1) as a notice specifying what an appropriate person is to do by way of remediation and the periods within which he is required to do each of the things so specified.

Remediation package: the full set or sequence of remediation actions, within a remediation scheme, which are referable to a particular significant pollutant linkage. *Paragraph C.8(b)*

Remediation scheme: the complete set or sequence of remediation actions (referable to one or more significant pollutant linkages) to be carried out with respect to the relevant land or waters. *Paragraph C.8(c)*

Remediation statement: defined in section 78H(7). It is a statement prepared and published by the responsible person detailing the remediation actions which are being, have been, or are expected to be, done as well as the periods within which these things are being done.

Risk: the combination of:

(a) the probability, or frequency, of occurrence of a defined hazard (for example, exposure to a property of a substance with the potential to cause harm); and
(b) the magnitude (including the seriousness) of the consequences. *Paragraph A.9*

Shared action: a remediation action which is referable to the significant pollutant in more than one significant pollutant linkage. *Paragraph D.21(b)*

Single-linkage action: a remediation action which is referable solely to the significant pollutant in a single significant pollutant linkage. *Paragraph D.21(a)*

Significant harm: defined in section 78A(5). It means any harm which is determined to be significant in accordance with the statutory guidance in Chapter A (that is, it meets one of the descriptions of types of harm in the second column of Table A of that Chapter).

Significant pollutant: a pollutant which forms part of a significant pollutant linkage. *Paragraph A.20*

Significant pollutant linkage: a pollutant linkage which forms the basis for a determination that a piece of land is contaminated land. *Paragraph A.20*

Significant possibility of significant harm: a possibility of significant harm being caused which, by virtue of section 78A(5), is determined to be significant in accordance with the statutory guidance in Chapter A.

Sold with Information: an exclusion test for Class A persons set out in Part 5 of Chapter D. *Paragraph D.57 to D.61*

Special site: defined by section 78A(3) as:

"any contaminated land —

(a) which has been designated as such a site by virtue of section 78C(7) or 78D(6) . . .; and
(b) whose designation as such has not been terminated by the appropriate Agency under section 78Q(4) . . .".

The effect of the designation of any contaminated land as a special site is that the Environment Agency, rather than the local authority, becomes the enforcing authority for the land.

Substance: defined in section 78A(9) as:

"any natural or artificial substance, whether in solid or liquid form or in the form of a gas or vapour."

INDEX